METHODS & NEW FRONTIERS IN NEUROSCIENCE

Series Editors
Sidney A. Simon, Ph.D.
Miguel A.L. Nicolelis, M.D., Ph.D.

Published Titles

Apoptosis in Neurobiology
Yusuf A. Hannun, M.D., Professor of Biomedical Research and Chairman/Department of Biochemistry and Molecular Biology, Medical University of South Carolina
Rose-Mary Boustany, M.D., tenured Associate Professor of Pediatrics and Neurobiology, Duke University Medical Center

Methods for Neural Ensemble Recordings
Miguel A.L. Nicolelis, M.D., Ph.D., Professor of Neurobiology and Biomedical Engineering, Duke University Medical Center

Methods of Behavioral Analysis in Neuroscience
Jerry J. Buccafusco, Ph.D., Alzheimer's Research Center, Professor of Pharmacology and Toxicology, Professor of Psychiatry and Health Behavior, Medical College of Georgia

Neural Prostheses for Restoration of Sensory and Motor Function
John K. Chapin, Ph.D., Professor of Physiology and Pharmacology, State University of New York Health Science Center
Karen A. Moxon, Ph.D., Assistant Professor/School of Biomedical Engineering, Science, and Health Systems, Drexel University

Computational Neuroscience: Realistic Modeling for Experimentalists
Eric DeSchutter, M.D., Ph.D., Professor/Department of Medicine, University of Antwerp

Methods in Pain Research
Lawrence Kruger, Ph.D., Professor of Neurobiology (Emeritus), UCLA School of Medicine and Brain Research Institute

Motor Neurobiology of the Spinal Cord
Timothy C. Cope, Ph.D., Professor of Physiology, Emory University School of Medicine

Nicotinic Receptors in the Nervous System
Edward D. Levin, Ph.D., Associate Professor/Department of Psychiatry and Pharmacology and Molecular Cancer Biology and Department of Psychiatry and Behavioral Sciences, Duke University School of Medicine

Methods in Genomic Neuroscience
Helmin R. Chin, Ph.D., Genetics Research Branch, NIMH, NIH
Steven O. Moldin, Ph.D, Genetics Research Branch, NIMH, NIH

Methods in Chemosensory Research
Sidney A. Simon, Ph.D., Professor of Neurobiology, Biomedical Engineering, and Anesthesiology, Duke University
Miguel A.L. Nicolelis, M.D., Ph.D., Professor of Neurobiology and Biomedical Engineering, Duke University

The Somatosensory System: Deciphering the Brain's Own Body Image
Randall J. Nelson, Ph.D., Professor of Anatomy and Neurobiology,
  University of Tennessee Health Sciences Center

New Concepts in Cerebral Ischemia
Rick C. S. Lin, Ph.D., Professor of Anatomy, University of Mississippi Medical Center

DNA Arrays: Technologies and Experimental Strategies
Elena Grigorenko, Ph.D., Technology Development Group, Millennium Pharmaceuticals

Methods for Alcohol-Related Neuroscience Research
Yuan Liu, Ph.D., National Institute of Neurological Disorders and Stroke, National Institutes of Health
David M. Lovinger, Ph.D., Laboratory of Integrative Neuroscience, NIAAA

In Vivo Optical Imaging of Brain Function
Ron Frostig, Ph.D., Associate Professor/Department of Psychobiology,
  University of California, Irvine

Primate Audition: Behavior and Neurobiology
  Asif A. Ghazanfar, Ph.D., Primate Cognitive Neuroscience Lab, Harvard University

Methods in Drug Abuse Research: Cellular and Circuit Level Analyses
  Dr. Barry D. Waterhouse, Ph.D., MCP-Hahnemann University

FUNCTIONAL AND NEURAL MECHANISMS OF INTERVAL TIMING

Edited by Warren H. Meck

CRC PRESS

Boca Raton   London   New York   Washington, D.C.

Library of Congress Cataloging-in-Publication Data

Functional and neural mechanisms of interval timing / edited by Warren H. Meck.
    p. cm.
Includes bibliographical references and index.
ISBN 0-8493-1109-8
1. Time perception. I. Meck, Warren H.

QP445 .F865 2003
153.7′53—dc21   2002035040

This book contains information obtained from authentic and highly regarded sources. Reprinted material is quoted with permission, and sources are indicated. A wide variety of references are listed. Reasonable efforts have been made to publish reliable data and information, but the authors and the publisher cannot assume responsibility for the validity of all materials or for the consequences of their use.

Neither this book nor any part may be reproduced or transmitted in any form or by any means, electronic or mechanical, including photocopying, microfilming, and recording, or by any information storage or retrieval system, without prior permission in writing from the publisher.

All rights reserved. Authorization to photocopy items for internal or personal use, or the personal or internal use of specific clients, may be granted by CRC Press LLC, provided that $1.50 per page photocopied is paid directly to Copyright Clearance Center, 222 Rosewood Drive, Danvers, MA 01923 USA. The fee code for users of the Transactional Reporting Service is ISBN 0-8493-1109-8/02/$0.00+$1.50. The fee is subject to change without notice. For organizations that have been granted a photocopy license by the CCC, a separate system of payment has been arranged.

The consent of CRC Press LLC does not extend to copying for general distribution, for promotion, for creating new works, or for resale. Specific permission must be obtained in writing from CRC Press LLC for such copying.

Direct all inquiries to CRC Press LLC, 2000 N.W. Corporate Blvd., Boca Raton, Florida 33431.

Trademark Notice: Product or corporate names may be trademarks or registered trademarks, and are used only for identification and explanation, without intent to infringe.

Visit the CRC Press Web site at www.crcpress.com

© 2003 by CRC Press LLC

No claim to original U.S. Government works
International Standard Book Number 0-8493-1109-8
Library of Congress Card Number 2002035040
Printed in the United States of America 1 2 3 4 5 6 7 8 9 0
Printed on acid-free paper

April 23, 2003

Dear Customer:

Thank you for your purchase of *Functional and Neural Mechanisms of Interval Timing* (Cat. #1109) edited by Warren Meck.

On this page and the reverse are corrections to three figures in the first printing of this book. We regret any inconvenience this may have caused you. Please let us know if we can be of any assistance regarding this title or any other titles CRC Press publishes.

Thank you,

CRC Press LLC

Page 66—Below is the corrected Figure 3.2

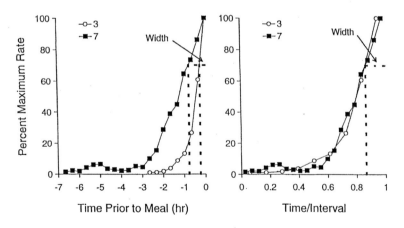

#1109/0-8493-1109-8

Below is corrected Figure 3.3.

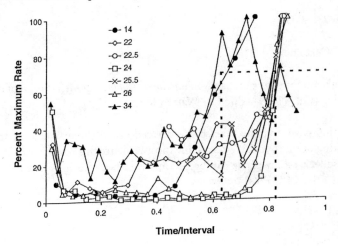

Below is corrected Figure 3.6.

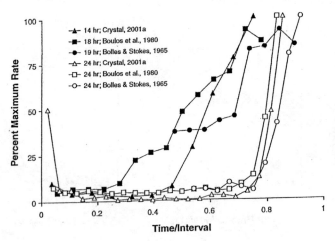

Dedication

This book is dedicated to:

Janice E. Meck and Hubert A. Meck
With love and gratitude for giving me the time of my life.

Russell M. Church
With appreciation for training me how to conduct research on time.

Christina L. Williams
With recognition of her time and persistence.

Foreword

For more than a century, time has been an object of study in experimental psychology. In his *Experimental Psychology*, Titchener (1905) wrote, "A student who knows his *time sense* ... has a good idea of what experimental psychology has been and of what it has come to be." At the dawn of the 21st century, I believe that Titchener's judgment about the status of timing and time perception in psychology is still appropriate. As was the case a century ago, knowledge of the current research on timing gives a sense of what cognition, cognitive psychology, and cognitive neuroscience have come to be and will become.

The beginning of the modern era in timing and time perception was signaled by John Gibbon and Lorraine Allan's (1984) seminal volume in the *Annals of the New York Academy of Sciences*. Since then, the field has exploded. As is the case in other domains of cognitive and behavioral psychology, a very large corpus of descriptive data has accumulated and numerous experimental methods have been developed. For a number of years, prominent authors have claimed that cognitive psychology is in need of theoretical unification. Others have also called for going further than describing *what* are the critical phenomena by focusing on understanding *how* these phenomena are produced. The book is certainly the most impressive attempt at achieving the goals of both unifying the field of timing and time perception and understanding the timing process.

Functional and Neural Mechanisms of Interval Timing acknowledges the need to provide a way of bringing together different research areas that have historically evolved independently. Research on the analysis of animal timing behavior and work on human timing and time perception have to be systematically connected. A common understanding of the traditions of timing research with different time scales has to be provided. It has been generally accepted that timing of circadian periods operated under mechanisms different from those controlling interval timing in the range of seconds and minutes or in the range of milliseconds (e.g., Hinton and Meck, 1997). Again, evidence that auditory stimuli are timed differently from visual stimuli (e.g., Grondin and Rousseau, 1991; Penney et al., 2000) has to be accounted for. On top of addressing these issues, the book gives a striking demonstration that timing is pervasive across species, developmental stages, and forms and complexity of behaviors. One cannot help but be convinced that the study of timing is at the heart of the enterprise of understanding behavior and cognition.

Furthermore, the book has a unique way of promoting unification of the field through the adoption of a specific theoretical point of view rooted in the family of pacemaker-counter models of timing and based on the principle of scalar timing. The scalar expectancy theory has been, for the last 20 years, the most influential theory in that family. It provides a common theoretical basis for most of the chapters in this book. The reader is shown that timing in very different forms of behavior in animals and humans follows the principle of scalar timing.

Russell Church (1984) proposed that a description of *how* time was processed had to rely on at least three basic levels of analysis: psychological, formal, and biological. That approach has been generally adopted, but to my knowledge, the present book represents by far the best example of considering the three-prong approach. In this manner, *Functional and Neural Mechanisms of Interval Timing* makes a concerted effort to give timing research a driving thrust using all the existing power of cognitive science. Setting most chapters in the theoretical context of pacemaker-counter models has enabled a common way of addressing the psychological level. The concept of an internal clock linked with processes like attention, memory, and decision making form the current conceptual basis in the field, and that is skillfully reflected in the book (e.g., Fortin et al., 1993; Rousseau and Rousseau, 1996). Furthermore, the book gives the reader the current status of the development of formal models of timing. Probabilistic models, connectionist models, and oscillator models are described both in simple and advanced ways, giving an up-to-date comprehension of timing models.

The biological level of *how* time is processed has been heavily influenced by the recent development of various methods in the field of cognitive neuroscience. Techniques for recording electrical activity or blood flow of the brain, brain lesions and pharmacological interventions; study of patients with neurological problems; and even genetic research have been applied to the study of cognition. The book provides the reader with chapters covering the most exhaustive set of these methods applied to timing and time perception. It includes one of the first attempts at considering the genetic analysis of interval timing. The reader is also presented with the most advanced description of the brain structures supporting timing. The way the biological level is treated in the book is an invaluable contribution to the progress of the field.

The volume, edited by Warren Meck, presents cutting-edge scientific work in a way that promotes a concerted view of timing and time perception. The chapters systematically introduce and explain in detail the recent progress that has been made in identifying the functional and neural mechanisms of interval timing. Never before has such a large array of phenomena and methods been put together with such a coherent analysis. It is a remarkable achievement for both unity and synthesis in the domains of behavioral and cognitive neuroscience. The book is certainly a major step in giving the field of timing and time perception a theoretical focus of great scientific power.

REFERENCES

Church, R.M., Properties of the internal clock, in *Timing and Time Perception*, Gibbon, J. and Allan, L., Eds., The New York Academy of Sciences, New York, 1984, pp. 566–582.

Fortin, C., Rousseau, R., Bourque, P., and Kirouac, E., Time estimation and concurrent nontemporal processing: specific interference from short-term-memory demands, *Percept. Psychophys.*, 53, 536–548, 1993.

Gibbon, J. and Allan, L., Eds., *Timing and Time Perception*, The New York Academy of Sciences, New York, 1984.

Grondin, S. and Rousseau, R., Judging the relative duration of multimodal short empty time intervals, *Percept. Psychophys.*, 49, 245–256, 1991.

Hinton, S.H. and Meck, W.H., The "internal clocks" of circadian and interval timing, *Endeavour*, 21, 82–87, 1997.

Penney, T.B., Gibbon, J., and Meck, W.H., Differential effects of auditory and visual signals on clock speed and temporal memory, *J. Exp. Psychol. Hum. Percept. Performance*, 26, 1770–1787, 2000.

Rousseau, L. and Rousseau, R., Stop-reaction time and the internal clock, *Percept. Psychophys.*, 58, 434–448, 1996.

Titchener, E.B., *Experimental Psychology*, Vol. II, MacMillan Co., London, 1905.

Robert Rousseau
Laval University
Sainte-Foy, Quebec
August 20, 2002

Contributors

Melissa Bateson, D.Phil.
School of Biology
University of Newcastle
Newcastle upon Tyne, United Kingdom

Elizabeth M. Brannon, Ph.D.
Center for Cognitive Neuroscience
Duke University
Durham, North Carolina

Catalin V. Buhusi, Ph.D.
Psychological and Brain Sciences
Duke University
Durham, North Carolina

Münire Özlem Çevik, Ph.D.
Department of Psychology
Abant Izzet Baysal University
Bolu, Turkey

Russell M. Church, Ph.D.
Department of Psychology
Brown University
Providence, Rhode Island

Jonathon D. Crystal, Ph.D.
Department of Psychology
University of Georgia
Athens, Georgia

Jörn Diedrichsen, B.S.
Department of Psychology
University of California, Berkeley
Berkeley, California

Sylvie Droit-Volet, Ph.D.
Department of Psychology
Blaise Pascal University
Clermont-Ferrand, France

Claudette Fortin, Ph.D.
School of Psychology
Laval University
Sainte-Foy, Quebec, Canada

Thomas T. Hills, Ph.D.
Department of Biology
University of Utah
Salt Lake City, Utah

Sean C. Hinton, Ph.D.
Department of Neurology
Medical College of Wisconsin
Milwaukee, Wisconsin

John W. Hopson, Ph.D.
Psychological and Brain Sciences
Duke University
Durham, North Carolina

Richard B. Ivry, Ph.D.
Department of Psychology
University of California, Berkeley
Berkeley, California

Penelope A. Lewis, D.Phil.
Department of Physiology
Oxford University
Oxford, United Kingdom

Cindy Lustig, Ph.D.
Department of Psychology
Washington University
Saint Louis, Missouri

Françoise Macar, Ph.D.
Center for Cognitive Neuroscience
CNRS — 31 Chemin Joseph-Aiguier
Marseille, France

Christopher J. MacDonald, B.S.
Psychological and Brain Sciences
Duke University
Durham, North Carolina

Chara Malapani, M.D., Ph.D.
Division of Biopsychology
New York State Psychiatric Institute
New York, New York

Matthew S. Matell, Ph.D.
Department of Neurology
University of Michigan
Ann Arbor, Michigan

J. Devin McAuley, Ph.D.
Department of Psychology
Bowling Green State University
Bowling Green, Ohio

Warren H. Meck, Ph.D.
Psychological and Brain Sciences
Duke University
Durham, North Carolina

R. Chris Miall, Ph.D.
Department of Physiology
Oxford University
Oxford, United Kingdom

Miguel A.L. Nicolelis, M.D., Ph.D.
Department of Neurobiology
Duke University Medical Center
Durham, North Carolina

Keiichi Onoda, B.S.
School of Biosphere Sciences
Hiroshima University
Higashi-Hiroshima, Hiroshima, Japan

Kevin C.H. Pang, Ph.D.
Department of Psychology
Bowling Green State University
Bowling Green, Ohio

Trevor B. Penney, Ph.D.
Department of Psychology
Chinese University of Hong Kong
Shatin, New Territories, Hong Kong

Jeff Pressing, Ph.D.
Department of Psychology
University of Melbourne
Melbourne, Australia

Viviane Pouthas, Ph.D.
LENA-CNRS, UPR 640
Salpetriere Hospital
Paris, France

Brian C. Rakitin, Ph.D.
Department of Neurology
Columbia University
New York, New York

Jamie D. Roitman, Ph.D.
Department of Neurobiology
Duke University
Durham, North Carolina

Robert Rousseau, Ph.D.
School of Psychology
Laval University
Sainte-Foy, Quebec, Canada

Shogo Sakata, Ph.D.
Department of Behavioral Sciences
Hiroshima University
Higashi-Hiroshima, Hiroshima, Japan

Contents

Foreword vii
Robert Rousseau

Introduction: The Persistence of Time xvii
Warren H. Meck

SECTION I Functional Mechanisms

Chapter 1
A Concise Introduction to Scalar Timing Theory 3
Russell M. Church

Chapter 2
General Learning Models: Timing without a Clock.. 23
John W. Hopson

Chapter 3
Nonlinearities in Sensitivity to Time: Implications for Oscillator-Based
Representations of Interval and Circadian Clocks... 61
Jonathon D. Crystal

Chapter 4
Toward a Unified Theory of Animal Event Timing... 77
Thomas T. Hills

Chapter 5
Interval Timing and Optimal Foraging 113
Melissa Bateson

Chapter 6
Nonverbal Representations of Time and Number in Animals and Human
Infants 143
Elizabeth M. Brannon and Jamie D. Roitman

Chapter 7
Temporal Experience and Timing in Children 183
Sylvie Droit-Volet

Chapter 8
Modality Differences in Interval Timing: Attention, Clock Speed, and
Memory 209
Trevor B. Penney

Chapter 9
Attentional Time-Sharing in Interval Timing 235
Claudette Fortin

Chapter 10
Grandfather's Clock: Attention and Interval Timing in Older Adults 261
Cindy Lustig

SECTION II Neural Mechanisms

Chapter 11
Neurogenetics of Interval Timing 297
Münire Özlem Çevik

Chapter 12
Dopaminergic Mechanisms of Interval Timing and Attention 317
Catalin V. Buhusi

Chapter 13
Electrophysiological Correlates of Interval Timing 339
Shogo Sakata and Keiichi Onoda

Chapter 14
Importance of Frontal Motor Cortex in Divided Attention and Simultaneous
Temporal Processing 351
Kevin C.H. Pang and J. Devin McAuley

Chapter 15
Integration of Behavior and Timing: Anatomically Separate Systems or
Distributed Processing? 371
Matthew S. Matell, Warren H. Meck, and Miguel A.L. Nicolelis

Chapter 16
Time Flies and May Also Sing: Cortico-Striatal Mechanisms of Interval Timing and Birdsong 393
Christopher J. MacDonald and Warren H. Meck

Chapter 17
Neuroimaging Approaches to the Study of Interval Timing 419
Sean C. Hinton

Chapter 18
Electrophysiological Evidence for Specific Processing of Temporal Information in Humans 439
Viviane Pouthas

Chapter 19
Cerebellar and Basal Ganglia Contributions to Interval Timing 457
Jörn Diedrichsen, Richard B. Ivry, and Jeff Pressing

Chapter 20
Interval Timing in the Dopamine-Depleted Basal Ganglia: From Empirical Data to Timing Theory 485
Chara Malapani and Brian C. Rakitin

Chapter 21
Overview: An Image of Human Neural Timing 515
Penelope A. Lewis and R. Chris Miall

Afterword
Timing in the New Millenium: Where Are We Now? 533
Françoise Macar

Index 541

Introduction: The Persistence of Time

Humans and other animals engage in a startlingly diverse array of behaviors that depend critically on the time of day or the ability to time short intervals. Timing intervals on the scale of many hours to around a day are mediated by the circadian timing system, while in the range of seconds to minutes a different system, known as interval timing, is used. Nonlinearities in the sensitivity to time as well as a dependence on the normal functioning of the suprachiasmatic region of the hypothalamus have been observed for both systems, suggesting a greater degree of commonality between circadian and interval timing than is often appreciated (see Cohen et al., 1997; Crystal, 1999, 2001, this volume; Crystal et al., 1997; Killeen, 2002). Fortunately, recent advances have illuminated some of the functional and neural mechanisms underlying the "internal clocks" of these two different timing systems, including the identification of molecular circuitry associated with circadian and interval timing (see Allada et al., 2001; Blakeslee, 1998; Cevik, this volume; Hills, this volume; Hinton and Meck, 1997b; King et al., 2001; Meck, 2001; Morell, 1996; Travis, 1996; Wright, 2002; as well as the British Broadcasting Company documentary "The Body Clock: What Makes Us Tick?" 1999).

The term *interval timing* is used to describe the temporal discrimination processes involved in the estimation and reproduction of relatively short durations in the seconds-to-minutes range that form the fabric of our everyday existence and unite our mental representations of action sequences and rhythmical structures (e.g., Fraisse, 1963; Gallistel, 1990; Gibbon and Allan, 1984; Krampe et al., 2002; Macar et al., 1992; McAuley and Jones, 2002; Pressing, 1999; Rousseau and Rousseau, 1996). A classic example of interval timing comes from the fixed-interval (FI) procedure in which a subject is reinforced for the first response it makes after a programmed interval has elapsed since the previous reinforcement (Skinner, 1936). Subjects (e.g., primates, rodents, birds, and fish) trained on this procedure typically show what is known as the fixed-interval scallop. This pattern of behavior involves pausing after the delivery of reinforcement and starting to respond after a fixed proportion of the interval has elapsed despite the absence of any external time cues. Interval timing of this type has been identified in the majority of vertebrate animals in which it has been tested for (Richelle and Lejeune, 1980, 1984) and has been shown to be exquisitely sensitive to the effect of neurotoxicants and psychoactive drugs (e.g., Meck, 1996; Paule et al., 1999). The FI procedure gave rise to a discrete trial variant known as the peak-interval (PI) procedure (Catania, 1970; Roberts, 1981), which is now widely used in animal (e.g., Buhusi and Meck, 2000, 2002; Hinton and Meck, 1997a; Liu et al., 2002; Matell and Meck, 1999; Meck, 2000a, 2000b; Meck and Church, 1984; Olton et al., 1988; Ohyama et al., 2000; Pang et al., 2001; Penney et al., 1996) and human (e.g., Malapani et al., 1998b, 2002; Rakitin et al., 1998) studies of interval timing. In this procedure a stimulus such as a tone or light is turned on to signal the beginning of the interval, and in a proportion of trials the subject is reinforced for the first response it makes after the criterion time. In the remainder of the trials, known as probe trials, no reinforcement is given and the stimulus remains on for two or three times the criterion time. When the mean response rate in many probe trials is calculated, an approximately Gaussian peak of responses is seen centered on the criterion. The time at which this timing function is at its maximum, also known as the peak time, gives an estimate of how accurately the subject is timing; precision is indicated by the spread of the timing function. These quantitative measures make the PI procedure an attractive tool for the study of interval timing. In addition, the temporal relations defined by the start and stop times for responding on individual trials have been used to identify sources of variability contributed by clock, memory, and response thresholds (e.g., Abner et al., 2001; Cheng and Westwood, 1993; Cheng et al., 1993; Church et al., 1994; Rakitin et al., 1998).

An example of interval timing data collected from an adult participant (ALB) diagnosed with attentional-deficit disorder (ADD) using the PI procedure is shown in Figure 1. The top panel illustrates percent maximum response rate for ALB in an unmedicated state (NicPre) plotted as a function of 7-s and 17-s criteria trained using methods reported by Levin et al, 1996, 1998; Rakitin et al., 1998. In this procedure, a blue square presented on a computer monitor is transformed to magenta at the appropriate criterion time during fixed-time training trials. Thereafter, participants are requested to reproduce the temporal criterion for a sequence of test trials for which a distribution of their responses is plotted on a relative time scale immediately following the trial during the inter-trial interval (ITI). This ITI feedback is displayed on the computer monitor and provides the participant with information concerning the relative accuracy and precision of their temporally-controlled responding on the preceding trial. ITI feedback can be randomly presented following a fixed proportion of trials (in this case 25% and 100%). As can be seen in the top panel of the figure, when the participant is provided with ITI feedback on 100% of the trials the PI functions are centered at the correct times showing excellent accuracy of the reproduced intervals. In contrast, when ITI feedback is provided on only 25% of the trials a proportional rightward shift is observed in the timing of the 7-s and 17-s intervals, reflecting a discrepancy in the accuracy of temporal reproductions that is not observed in normal participants. This rightward shift is accompanied by a broadening of the PI functions indicating a decrease in temporal precision with lower levels of feedback. Both of these findings are consistent with a slowing of the internal clock as a function of the probability of feedback and may be the result of an attentional deficit (e.g., flickering mode switch) as described by Lustig (this volume), Meck and Benson, 2002, and Penney (this volume). Interestingly, the bottom panel indicates that when the participant is given a stimulant drug (transdermal nicotine skin patch) that increases dopamine levels in the brain during the NicPost condition the effects of 25% ITI feedback are enhanced and produce levels of temporal accuracy

Introduction: The Persistence of Time

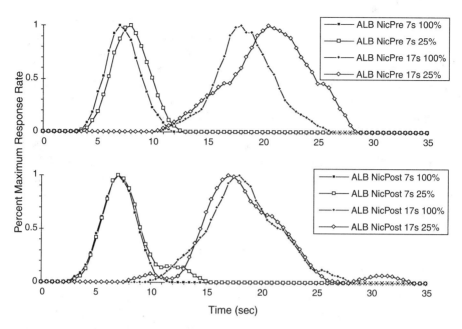

FIGURE 1 Peak-interval procedure: Percent maximum response rate plotted as a function of signal duration for a single adult participant (ALB) diagnosed with attention deficit disorder (ADD) trained at two criterion times (7s and 17s) under two conditions of intertrial interval (ITI) feedback (25% and 100%). The top panel shows performance at these two criterion times under the two ITI feedback conditions in an unmedicated state (NicPre) and the bottom panel shows performance in a medicated state (NicPost). See Levin et al. (1996, 1998) for additional procedural details.

and precision that are equivalent to the 100% ITI feedback condition in both the medicated and unmedicated states. These results suggest an equivalence of the ITI feedback effects and the types of pharmacological stimulation provided to ADD patients by drugs such as nicotine (see Levin et al., 1996, 1998). These findings also support the proposal that attentional deficits can lead to the underestimation of signal durations in a manner that is consistent with a slowing of an internal clock that is sensitive to dopaminergic manipulations whether they are produced by behavioral (ITI feedback) or pharmacological (nicotine patch) means.

Behavioral data derived from timing tasks such as the PI procedure have influenced the development of a number of different psychological theories of interval timing (for a review, see Grondin, 2001; Matell and Meck, 2000). Of these theories, scalar timing theory, or scalar expectancy theory (SET), stands out because not only does it explain much behavioral data, but it has also been useful in interpreting and guiding anatomical and pharmacological work in the attempt to identify the neuropsychological mechanisms responsible for these behaviors (e.g., Allan, 1998; Gibbon, 1977, 1991; Gibbon et al., 1984, 1997; Gibbon and Malapani, 2002; Killeen, 2002; Wearden, 1999). SET provides both a formal quantitative and an information-processing account of interval timing that postulates three distinct stages: a clock, a memory, and a decision stage. The clock stage is hypothesized to consist of a pacemaker that emits pulses which are gated to an accumulator by a switch. When reinforcement occurs the current count in the accumulator is transferred to reference memory. As training with a particular interval progresses, a distribution of values in the reference memory is formed. Finally, if the participant needs to estimate or produce the learned interval, this is done in the decision stage of the system by making a ratio comparison between the current value in the accumulator and a random sample drawn from reference memory (see Church, this volume). At its heart, SET is a model in which Poisson, constant, and scalar sources of variability compete for control over temporal discrimination. In most cases it is the scalar source of variability that comes to dominate behavior, but the other sources also play a critical role in the clock, memory, and decision stages of interval timing. The interested reader should refer to Gibbon (1992), Killeen (2002), and Rousseau et al. (1984) for cogent discussions of the ubiquity of a Poisson process underlying time estimation.

THE IMPORTANCE OF INTERVAL TIMING IN ADAPTATION AND LEARNING

It is becoming increasingly evident that interval timing is crucial for many basic forms of adaptation and learning (e.g., Staddon and Higa, 1996, 1999). One of the clearest cases comes from the field of optimal foraging, which studies the extent to which animals' foraging decisions are the direct product of natural selection (for a review, see Krebs and Kacelnik, 1984). In most cases, to make decisions that maximize fitness, an animal needs to measure its rate of food intake in one or more environments, and measuring rate requires measurement of time. Recent work has shown that European starlings are deftly sensitive to their rate of food intake and appear to record the interval of time between each prey they capture and consume (see Bateson, this volume; Bateson and Kacelnik, 1997, 1998). Even nematodes such as *Caenorhabditis elegans* engage in complex foraging behaviors that are temporally sensitive — suggesting that interval timing is a very basic process that can be fruitfully explored at the molecular level (e.g., Brockie et al., 2001).

Song learning in passerine birds (songbirds) has also been a central focus of ethological research, and the study of the neurobiology of song learning, song production, and song perception has stimulated many seminal contributions to our understanding of brain–behavior relationships and how learning influences these patterns. The study of the temporal hierarchical control of singing in birds is clearly an area ripe for the application of principles learned from the psychophysical analysis of timing and time perception (see Hahnloser et al., 2002; Hills, this volume; Fee et al., 2002; MacDonald and Meck, this volume; Yu and Margoliash, 1996).

Associative learning or conditioning is another widespread and elemental form of animal behavior for which computational accounts of interval timing are becoming enormously influential (see Arcediano and Miller, 2002; Bugmann, 1998; Gallistel, 1990; Gallistel and Gibbon, 2000, 2001; Hopson, 1999, this volume; Migliore et al., 2001). For example, in classical conditioning it is well established that the efficiency of learning about a conditioned stimulus (CS) is affected by the time interval between the CS and the unconditioned stimulus (US): in general terms it is found that the shorter the CS-US interval, the faster and better the conditioning that occurs (Balsam, 1984). In fact, it is not the absolute duration of the CS-US interval that is important, as was initially thought, but the ratio of the CS-US interval to the interval between successive US (e.g., Balsam, 1984; Gibbon et al., 1977; Gibbon and Balsam, 1981; Jenkins, 1984; Meck, 1985). The observation that temporal factors of this sort dramatically impact the acquisition of conditioned responding suggests that learning about time may be a necessary condition for associative learning (e.g., Balsam et al., 2002; Roberts and Holder, 1984).

THE IMPORTANCE OF INTERVAL TIMING IN MEMORY AND ATTENTION

Human learning and memory is also highly sensitive to temporal factors, and oscillator-based models have been proposed for the coding of serial order in memory (e.g., Brown and Chater, 2001; Brown et al., 2000; McCormack and Hoerl, 2001). In addition, deficits in learning, memory, set shifting, and interval timing have been observed in a variety of patient populations with damage to the basal ganglia, including Parkinson's disease and Huntington's disease patients, as well as other cortical and subcortical brain structures affected by Alzheimer's disease, injury, and stroke (see Cronin-Golomb et al., 1994; Diedrichsen et al., this volume; Gabrieli et al., 1996; Harrington and Haaland, 1999; Malapani and Rakitin, this volume; Nichelli et al., 1993; Stebbins et al., 1999). Temporal bisection data obtained from age-matched controls ($n = 6$), cerebellar lesion patients ($n = 3$), Alzheimer's disease patients ($n = 4$), and Parkinson's disease patients ($n = 4$) tested off of their medication are presented in Figure 2 (for procedural details, see Penney, this volume; Penney et al., 2000). These temporal bisection functions show the typical modality difference in which sounds are judged longer than lights of equal physical durations for all groups of participants. The major difference among the groups is in the quality of the temporal discriminations. Patients with cerebellar damage are little affected on this task, whereas Alzheimer's disease patients show a similar decline in sensitivity for both auditory and visual stimuli that may reflect deficits in attention and memory. Interestingly, Parkinson's disease patients tested off of their medication are the most disrupted in their sensitivity to signal duration and show a differential effect as a function of signal modality, with an exaggerated bias for calling auditory stimuli long and visual stimuli short. In either case, Parkinson's disease patients are severely compromised in their ability to accurately judge signal duration and resort to a greater reliance on response biases consistent with the involvement of the basal ganglia in timing and time perception (see Malapani and Rakitin, this volume; Malapani et al., 1998b).

As a complement to these neuropsychological studies, specialized techniques have been developed to study interval timing in humans using the brain imaging technologies of functional magnetic resonance imaging and event-related scalp potentials (see Brannon and Roitman, this volume; Hinton, this volume; Hinton et al., 1996; Hinton and Meck, 1997b; Lejeune et al., 1997; Lewis and Miall, this volume; Maquet et al., 1996; Meck et al., 1998; Pouthas, this volume; Pouthas et al., 1999, 2000; Rao et al., 2001).

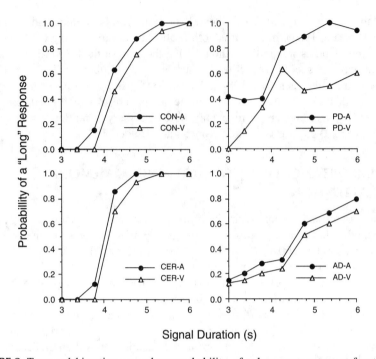

FIGURE 2 Temporal bisection procedure: probability of a long response as a function of signal duration (3.0, 3.37, 3.78, 4.25, 4.77, 5.35, and 6.0 sec) for auditory (A) and visual (V) stimuli presented within the same session. Participants were age-matched controls (n = 6), cerebellar lesion patients (n = 3), Parkinson's disease (PD) patients (n = 4), or Alzheimer's disease (AD) patients (n = 4). See Penney et al. (2000) for additional procedural details.

A number of researchers have presented psychophysical data suggesting that duration judgments depend on the amount of attentional resources allocated to a temporal processor or internal clock (e.g., Buhusi, this volume; Droit-Volet, this volume; Fortin, this volume; Lustig, this volume; Penney, this volume). For example, Fortin and Massé (1999, 2000) and Fortin and Rousseau (1987) have shown that if participants expect the timing of an ongoing signal to be interrupted by a gap or another type of task, they will divide attention between timing the ongoing signal and monitoring for the onset of the interrupting event. This division of attention leads to shorter-than-normal duration judgments as a function of the location of the interruption, suggesting that the internal clock runs at a slower rate when attention is divided in this manner. Because monitoring continues until the interruption is completed, the maximal effect will be observed when an interruption is expected, but none actually occurs — thus requiring the participant to monitor the entire interval. These data provide strong and convincing evidence for the role of attentional time-sharing in interval timing (see also Lejeune, 1998; Zakay and Block, 1996). Some electrophysiological support for this proposal has been provided by showing a relationship between the amplitude of brain wave activity and performance when participants focus their attention on the temporal parameters of a task (e.g., Casini and Macar, 1996a, 1996b, 1999; Pouthas, this volume). Identifying the ways in which attentional processes come under temporal control (and vice versa) is an exciting direction for future research efforts with important neuropsychological applications in both young and aged populations (see Casini and Ivry, 1999; Droit-Volet, this volume; Ferrandez and Pouthas, 2001; Harrington et al., 1998; Lustig, this volume; Lustig and Meck, 2001; Pang and McAuley, this volume; Perbal et al., 2001; Vanneste and Pouthas, 1999). Other nontemporal variables affecting attention and temporal judgments (e.g., stimulus modality and salience) are now being incorporated into SET as well as other theories of interval timing (see Buhusi, this volume; Buhusi and Meck, 2000, 2002; Meck, 1984; Penney, this volume; Penney et al., 1996, 2000).

The ability of organisms to time and coordinate temporal sequences of events and to select particular aspects of their internal and external environments to which they will attend has inspired some investigators to propose ways in which the same frontal-striatal circuits keep time and shift attention using the gating properties of this system (see Meck and Benson, 2002; Pang and McAuley, this volume). The attentional control of the perception and production of short durations appears to be better accounted for by an interval-based, as opposed to a beat-based, timing system, suggesting that the timing system is rapidly reset upon command (e.g., Pashler, 2001). It is also interesting to note that some researchers have argued that a primary function of the internal clock is to allow for the efficient transfer of information from one stage of information processing to another at regularly spaced intervals. This oscillatory pattern of information transfer between processing stages is achieved by the internal clock producing periodic inhibition with each clock pulse, thereby temporarily increasing the signal-to-noise ratio (e.g., Burle and Bonnet, 1999).

THE IMPORTANCE OF INTERVAL TIMING IN COGNITIVE DEVELOPMENT

Interest in comparative cognition, mental processes subserving thought without language, and the development of temporal knowledge have contributed to the recent excitement surrounding studies of interval timing and counting in preverbal infants and young children (see Brannon and Roitman, this volume; Droit-Volet, this volume). One of the central issues is whether children are able to time with the same accuracy and precision as adults or whether there are developmental transitions in their timing abilities (e.g., Droit-Volet and Wearden, 2001; McCormack et al., 1999). Other issues are concerned with the way in which children understand and abstract knowledge of duration from other stimulus attributes such as force and deal with attentional distraction during a timing task (e.g., Droit-Volet, 1998, this volume; Gautier and Droit-Volet, 2002). For example, when 3- and $5^1/_2$-year-old children are asked to carry out a response duration task in which they are instructed to press longer or harder than in a previous session, the results indicate that 3-year-olds rely on a certain amount of force to produce the correct response duration. In contrast, the $5^1/_2$-year-olds are able to dissociate force and duration; i.e., when they are asked to press harder, they do not press longer (Droit-Volet, 1998, this volume). These findings indicate that a marked dissociation between force and duration only emerges between the ages of 3 and 5½ and suggest different conceptualizations of time as a function of experience and maturation.

Analog representations of stimulus magnitudes have also been explored using mode-control models of temporal integration as a guide for understanding nonsymbolic counting and timing processes in animals, nonverbal infants, young children, and adults (see Brannon and Roitman, this volume; Clement and Droit-Volet, 2001; Dehaene et al., 1999; Gallistel and Gelman, 1992; Meck and Church, 1983; Meck et al., 1985; Wynn, 1995, 1998). The mode-control model posits that magnitude estimations of time and number are mediated by the same pacemaker–accumulator system, but operated in different pulse accumulation modes (e.g., a run mode for time and an event mode for number). This unified model of temporal integration has become influential in the debate surrounding the foundations of numerical thinking and the evidence for nonverbal counting ability in a variety of animals, including monkeys and human infants (e.g., Brannon and Roitman, this volume; Brannon and Terrace, 1998; Brannon et al., 2001; Breukelaar and Dalrymple-Alford, 1999; Broadbent et al., 1993; Church and Meck, 1984; Dehaene et al., 1999; Gallistel, 1990; Gallistel and Gelman, 1992; Grondin et al., 1999; Roberts, 1995; Roberts and Mitchell, 1994; Starkey et al., 1990; Whalen et al., 1999; Wynn, 1992, 1995, 1998; Wynn and Chiang, 1998).

In order to illustrate, it is sometimes helpful to count only when a certain condition occurs. For example, if timing pulses are counted when a machine is running (i.e., the ON signal is true), you can calculate the percentage of time the machine is in use just by comparing the total count with the elapsed time. This calculation is simple even though the machine may start and stop many times during the monitoring. In order to do this, the counter must have a switch input, as well as its normal count input. You connect the machine's ON signal to the switch and a source of clock pulses to the normal count input, making the measurements within the counter. The way that this mechanism is applied in the mode-control model is that at the onset of a relevant stimulus, pulses are directed into an accumulator so that they can be integrated over time. This is accomplished by a mode switch that allows pulses to flow into the accumulator in one of three different modes, depending on the nature of the stimulus (see Brannon and Roitman, this volume; Meck, 1997; Meck and Church, 1983; Meck et al., 1985). In this model numerosity is represented by the linear magnitude of an internal variable (e.g., pulse accumulation), and it is this value that is remembered and entered into calculation. The mode-control model is assumed to be a model of counting when the switch is set to the *event* mode. In this condition discreet stimuli are marked by a fixed increment in the accumulator. It is this temporal integration process that represents the numerosity of events or objects and thus constitutes this model's proposed numeron, just as this same temporal integration process represents duration when pulses are gated through the switch in the *run* or *stop* mode.

The mode-control model provides a unified theory of counting and timing by positing the existence of an isomorphism between number and duration. The model incorporates the idea that the nervous system inverts the representational convention whereby numbers are used to represent linear magnitudes. Instead of using number to represent magnitude, it is proposed that the nervous system uses magnitude to represent number (see Carey, 2001; Gallistel and Gelman, 2000; Gelman and Cordes, 2001).

CONNECTIONS BETWEEN INTERVAL TIMING, NEUROPHARMACOLOGY, AND DRUG ABUSE

Drugs that increase the effective level of dopamine in the brain, such as cocaine and methamphetamine, are among the most commonly abused drugs today (Stahl, 2000). The connection between interval timing and drug abuse comes from the fact that dopaminergic drugs cause predictable distortions in timing and time perception. The dopamine agonist, methamphetamine, causes a leftward shift in psychophysical timing functions that is proportional in size to the duration of the interval being timed (e.g., Meck, 1983, 1996). This result is compatible with the hypothesis that increasing the effective level of dopamine in the brain causes an increase in the speed of the pacemaker or oscillatory processes used for timing: if the pacemaker is caused to run faster than when a time interval was first learned, then participants will think that feedback is due earlier than it actually is, and consequently timing functions will be horizontally shifted to the left.

Genetic modifications of the dopamine system can be made in various ways, including the deletion of the gene coding for the dopamine reuptake transporter (DAT), which leads to an increase in the synaptic levels of dopamine. The DAT mediates uptake of dopamine into neurons and is a major target for cocaine and amphetamine (e.g., Carboni et al., 2001). Since its cloning, much information has been obtained regarding the structure and function of the DAT. Binding domains for dopamine and various blocking drugs (e.g., cocaine) are likely formed by interactions with multiple amino acid residues, some of which are separate in the primary structure, but lie close together in the still unknown tertiary structure. The DAT gene is expressed only in the central nervous system within a small subset of neurons (i.e., dopamine-containing neurons) and not in glia cells. DAT expression is more restricted, for instance, than the expression of genes encoding dopamine biosynthetic enzymes (e.g., tyrosine hydroxylase and aromatic amino acid decarboxylase) or dopamine receptors. DAT therefore provides an excellent marker for most dopamine neurons and their projections. In the rodent, DAT mRNA is found in great abundance within midbrain dopamine neurons of the substantia nigra, with somewhat lower expression in the ventral tegmental nuclei and adjacent nuclei. Within the hypothalamus, DAT is expressed within the A13 (zona incerta) and, to a lesser extent, the A14 (periventricular) and A12 (arcuate nucleus) cell groups, but not other tyrosine hydroxylase–positive cell groups. Moderate DAT expression is also seen in the A16 cell group of the olfactory bulb. DAT mRNA is not found in regions devoid of dopamine cell bodies or within dopamine nerve terminals. The major sites of DAT expression correspond well with the brain regions known to be involved in interval timing (Gibbon et al., 1997).

Gene-dosage effects of the DAT can be observed. Wild-type mice (+/+) have a normal (100%) level of the DAT. There is 50% expression of the DAT in heterozygous mice (+/–) and a total lack of the DAT in homozygous mice (–/–). Dopamine stays in the synapse 100 times longer in –/– mice than in +/+ mice. Mice without the DAT are five to six times more active than wild-type mice and have been used as an animal model of cocaine and amphetamine addiction, attention deficit hyperactivity disorders, and schizophrenia (see Cevik, this volume; Gainetdinov and Caron, 2001; Gainetdinov et al., 1999, 2001a, 2001b). Recently, mice deficient in the DAT have been shown to have altered timing behavior consistent with the dopaminergic regulation of temporal integration in the seconds-to-minutes range (see Cevik, this volume; Sasaki et al., 2002).

Other molecular mechanisms underlying the role of dopamine in interval timing may be studied using catechol-o-methyltransferase (COMT) deficient mice. Although there are many proteins involved in the biological actions of dopamine, COMT, because it metabolizes released dopamine primarily in the prefrontal cortex may be a critical marker for schizophrenia and cognitive functions such as interval timing (Egan et al., 2001).

It is reasonable to assume that abuse of methamphetamine by humans causes a similar speeding up of the clock to the increases observed in the laboratory with birds and rodents. Given the above evidence that interval timing is important both in the assessment of rate of reinforcement and in classical conditioning, it has been hypothesized that timing distortions could be important in understanding the mind-altering properties of dopaminergic drugs (see Buhusi, this volume; Cevik, this volume; Paule et al., 1999). Other psychoactive substances (e.g., marijuana and its active ingredient tetrahydrocannabinol) have been shown to alter time perception as a function of their effects on cortical and cerebellar blood flow (e.g., Mathew et al., 1998).

DISCREPANCIES IN THE CONTENT OF TEMPORAL MEMORY

Individual differences in the content of temporal memory can be evaluated by the horizontal placement of psychophysical functions that relate signal duration to the probability of a response (Church and Meck, 1988; Gibbon et al., 1984). Data obtained from the PI procedure have shown that discrepancies in the content of temporal memory produce stable horizontal displacements of timing functions such that they can be centered at times that are either less than or greater than the programmed time of reinforcement (Church, 1989; Meck, 2002a, b). Typically, the average remembered time of reinforcement for a group of mature rats would be very close to the programmed time of reinforcement, with a symmetrical distribution of individual peak times centered around that time. In contrast, as rats age they demonstrate a proportional rightward shift in their timing functions, indicating that their remembered durations reflect a constant percentage overestimate of the programmed time of reinforcement (see Lustig, this volume; Meck, 2002a; Meck et al., 1986).

In rats, the observed differences in this remembered time of reinforcement have been shown to be proportional to the programmed time of reinforcement and can be related to modifications in cholinergic (Ch) function as determined by drug and lesion studies. For example, increasing the effective levels of acetylcholine in the brain by systemic administration of physostigmine has been shown to produce a maintained proportional leftward shift in timing functions, whereas decreasing the effective levels of acetylcholine by systemic administration of atropine has been shown to produce a maintained proportional rightward shift in timing functions, relating the probability of some response to signal duration (Meck, 1983, 1996; Meck and Church, 1987). The behavioral effects of selective brain damage have also indicated differential modifications in the content of temporal memory. Lesions of the fimbria-fornix or the Ch cell bodies in the medial septal area projecting to the hippocampus have been shown to produce a maintained proportional leftward shift in timing functions, whereas lesions of the frontal cortex or the Ch cell bodies in the nucleus basalis magnocellularis projecting to the frontal cortex have been shown to produce a maintained proportional rightward shift in timing functions (Hills, this volume; Meck et al., 1987; Olton et al., 1987, 1988).

This discrepancy in the content of temporal memory has been described in terms of a multiplicative translation constant that is responsible for producing scalar transforms of sensory input taken from an internal clock (e.g., Church, 1989; Gibbon et al., 1984; King et al., 2001; Meck, 1983, 1996). An example of how this might work is as follows: The number of pacemaker pulses integrated by an accumulator during the presentation of a signal can be considered to serve as a clock reading that provides a representation of the perceived duration for the current trial. When this clock reading is transferred to reference memory (presumably following feedback), the transfer may occur with some bias that we have referred to as a memory storage or a memory translation constant. In scalar timing theory (see Gibbon and Church, 1990), this has formally been referred to as the K^* parameter, which is a multiplicative constant in the equations describing how the reference memory process works during encoding and retrieval when animals make a temporal discrimination. If the remembered time of reinforcement reliably differed from the obtained clock reading, then an animal would consistently expect the feedback to occur later than the programmed time if its memory storage constant was greater than 1.0. Alternatively, it would consistently expect feedback to occur before the programmed time if its memory storage constant was less than 1.0. As indicated by Church (1989, p. 64),

> Without a memory storage concept, the behavior of the rats that act as if they expect an event earlier or later than it regularly occurs would seem to violate the general principle of reinforcement in which differential reinforcement of responding at the correct time would eventually lead to time estimates that are accurate in the mean.

Memory storage speed has been used as a proposed mechanism for this translation constant whereby sensory input from the internal clock is transferred to reference memory at some modifiable baud rate that ultimately influences the quantitative aspects of the represented signal duration (e.g., Church, 1989; Gibbon et al., 1984; Hinton and Meck, 1997a; Matell and Meck, 2000; Meck, 1983, 1996). The main idea here is that the content of temporal memory (i.e., the remembered duration of an event) is based on the amount of time required to transfer the number of pulses in the accumulator (i.e., this process functions as an "up counter") to reference memory (i.e., this process functions as a "down counter"). As a consequence, the transfer time (and hence the remembered duration) would be directly related to the number of pulses in the accumulator and the speed of transfer. Presumably, the nervous system can function as if it has a conversion table that relates the amount of time the memory storage network is activated during the transfer of a specific signal duration. If the speed of this memory storage process were to deviate from normal values (i.e., conditions under which the remembered duration of an event is equal to the programmed duration of an event), then it would be possible for proportionally shorter or longer values to be represented in memory. Unlike changes in clock speed, changes in memory storage speed would not be self-correcting even in the case where animals were "surprised" by the mismatch between their clock readings and a value sampled from reference memory. This is because any "updating" of memory in this case would lead to the continued distortion of stored values for as long as the modification in memory storage speed remained in effect, i.e., as long as K^* was $>$ or < 1.0 (e.g., Church and Meck, 1988; Meck, 1983, 1996, 2002a, 2002b, 2002c).

Sodium-dependent high-affinity choline uptake (SDHACU) reflects the activity of cholinergic neurons in brain regions during behavioral activation. Using this measure, Meck (2002a) attempted to establish the quantitative relation between SDHACU in the frontal cortex and hippocampus as a function of the discrepancies in the remembered times of reinforcement. SDHACU levels should be proportional to the magnitude of the discrepancy in temporal memory in those brain areas that monitor the comparison between the current clock reading and a sample of the expected time of reinforcement taken from reference memory. This discrepancy should reflect the participant's degree of surprise at the mismatch between these two variables when feedback is provided. The concepts of surprise and expectancy are major features of contemporary theories of animal learning and memory (e.g., Gibbon, 1977; Gibbon and Balsam, 1981; Kamin, 1969; Rescorla and Wagner, 1972; Schultz et al., 1997; Wagner, 1981).

Theta rhythm activity has also been investigated as a potential index of the content of temporal memory (Meck, 2002b). Atropine-sensitive theta rhythm (4 to 12 Hz) in the frontal cortex and hippocampus appears to be necessary for certain learning and memory processes (e.g., Givens and Olton, 1990). Differences in electrical patterns (e.g., amplitude and modal frequency) can be related to individual differences in the remembered times of reinforcement stored in working and reference memory. It is also possible to impair or facilitate temporal memory by driving hippocampal rhythmical slow-wave activity at frequencies that differ significantly from those of normal patterns of rhythmic excitation (Meck, 2002b). Taken together, the results of a series of experiments suggest that theta rhythm frequency in the frontal cortex serves as an indicator of the *direction of the discrepancy* in the content of temporal memory and that SDHACU amplitude in the frontal cortex serves as an indicator of the *magnitude of the discrepancy* in the content of temporal memory (Meck, 2002a, b).

NEURAL BASIS OF INTERVAL TIMING

A rich tradition of normative psychophysics has identified two ubiquitous properties of interval timing: the scalar property, a strong form of Weber's law, and ratio comparison mechanisms (Gallistel and Gibbon, 2001; Gibbon, 1977). Temperature and reinforcement density effects on the speed of the internal clock have also been studied in order to determine mechanisms of compensation and regulation (see Hills, this volume). Isolating the neural substrate of these properties is a major challenge for neurobiology (Gibbon et al., 1997). On the basis of the accumulation of evidence from drug and lesion studies, a potential mapping between the information-processing elements of SET and structures in the brain has been proposed (see MacDonald and Meck, this volume; Malapani and Rakitin, this volume; Matell and Meck, 2000; Matell et al., this volume; Meck, 1996; Meck and Benson, 2002). Specifically, the output from dopaminergic neurons in the substantia nigra pars compacta is proposed to play a central role in initiating and maintaining the temporal integration process involving cortico-striatal circuits. This hypothesis is supported by the observation that methamphetamine, a stimulant drug that acts by facilitating the synaptic release of dopamine, speeds up the clock, whereas haloperidol, which acts by blocking dopamine receptors, slows down the clock (e.g., Abner et al., 2001; Maricq et al., 1981; Meck, 1983). Dopamine D2 receptors are specifically implicated in the function of the internal clock by a study showing that the *in vitro* binding affinity of different neuroleptic drugs for the D2 receptor predicts the size of the rightward shift in timing functions they produce (Meck, 1986).

Much progress has been made in the last 25 years in terms of our understanding of the psychological processes involved in timing and time perception. Nevertheless, our understanding of the neural basis of interval timing is far from complete. The ability of the brain to process time in the seconds-to-minutes range remains a fascinating problem given that the basic electrophysiological properties of neurons operate on a millisecond time scale. One physiologically realistic model of interval timing integrates a multitude of cortical and thalamic oscillations with a "perceptron" processing system in the basal ganglia to arrive at the detection of intervals much larger than the oscillation periods (see Matell and Meck, 2000; Matell et al., this volume). This model is based on the observation that striatal spiny neurons receive input from 10,000 to 30,000 separate inputs from a wide variety of cortical and thalamic areas. These cortical and thalamic neurons oscillate with a mean periodicity of 10 Hz. The striatal spiny neurons have been hypothesized to be capable of detecting and responding to select patterns of cortical input. The particular pattern of excitatory input is selected by long-term potentiation and long-term depression, which are believed to result from dopaminergic activity from the midbrain (e.g., substantia nigra pars compacta and the ventral tegmental area) following the delivery of feedback (e.g., Houk, 1995). Additionally, these dopamine neurons have been shown to transfer their activation onset to the signals that predict subsequent feedback (e.g., Schultz et al., 1993). These neurobiological properties of the cortico-striatal circuitry can be combined with a "beat frequency" model of timing (Miall, 1989, 1996) that suggests that after resetting a range of oscillatory inputs, a specific time can be encoded by selectively weighting which inputs are currently active at the criterion time. This model's time coding is similar to the idea that one can code the number 15 by asking for the lowest common multiple of 3 and 5, thereby coding large numbers with much smaller numbers. Thus, the model provides a manner to encode a long interval with very short neuronal mechanisms using the concept of coincidence detection, which has been hypothesized as a function of basal ganglia information processing (see Houk, 1995; Matell and Meck, 2000; Matell et al., this volume).

Specifically, upon onset of a meaningful signal (e.g., a cue that predicts important outcomes), dopamine neurons fire in a burst pattern that transiently synchronizes the cortical and thalamic oscillations, as well as hyperpolarizes the striatal membrane, thereby resetting the integrating mechanism. The cortical and thalamic neurons begin to oscillate at their inherent periods, thus eliminating their synchronization and allowing particular patterns of activity to become meaningful. Upon detection of a previously reinforced pattern of input, via the crossing of a coherent activity threshold (set by baseline levels of dopamine input and striatal interneurons), an ensemble of striatal spiny neurons fire, thereby engendering a response that the encoded time has been reached. This striatal activity passes out of the basal ganglia to the thalamus, and from there back to the cortex and striatum, thereby impinging on the current oscillatory inputs, allowing alterations of timing and time perception. Such information flow through cortico-striato-thalamo-cortical loops has been observed in functional magnetic resonance imaging data during psychophysical timing tasks with human participants (see Hinton, this volume; Hinton et al., 1996; Meck et al., 1998). Alternative views concerning the roles of the dorsolateral prefrontal cortex (dlPFC) and striatum in timing have been proposed that rely on decay processes within the dlPFC to form the central clock process (e.g., Lewis, 2002). Such "decay models" do not attribute a significant role in interval timing to the striatum, although they may do so once they are fully specified. It is clear, however, that the types of inhibitory delay circuits described by Constantinidis et al. (2002) are an important feature of temporal processing in the dlPFC and may provide the basis for an internal clock, as suggested by Lewis and Miall (this volume).

Additional studies of central nervous system electrophysiology have suggested an important role for oscillatory neuronal activity in sensory perception, sensorimotor integration, and movement timing. These studies have demonstrated significant structure in basal ganglia neuron spiking activity at relatively long time scales. The modulation of multisecond periodicities in the firing rate of basal ganglia neurons by dopaminergic agonists and their correlation with theta bursts in transcortical and hippocampal electroencephalographs (EEGs) suggest the involvement of these and other brain structures (e.g., frontal cortex, hippocampus, and cerebellum) in the coordination of cognitive processes (e.g., Allers et al., 2002; Ruskin et al., 1999; Sakata and Onoda, this volume).

In addition to the basal ganglia, the cerebellum has been proposed to constitute an important component of a neural circuit including the cerebral cortex that is involved in timing and time perception (see Diedrichsen et al., this volume). Cerebellar damage alters the cerebral metabolism in the prefrontal cortex (e.g., Botez et al., 1991; Junck et al., 1988). In humans, frontal, temporal, occipital, and also subcortical lesions affect temporal processing, which is supported by the observation that during tasks where time is a crucial parameter, the contingent negative variation, a slow variation of the cerebral potentials correlated with the estimation of the time separating two stimuli, has a maximum amplitude over the prefrontal areas (see Macar et al., 1990; Monfort et al., 2000; Pouthas, this volume; Vidal et al., 1992).

On the basis of their analysis of patients with cerebellar lesions of various etiology, Ivry and Keele (1989) and Keele and Ivry (1990) proposed that the cerebellum serves as a biological clock. Patients with cerebellar damage had deficits in a tapping task performed in the absence of an external cue, their interresponse variance times being much higher than those of normal subjects. The deficits of cerebellar patients were not limited to tasks requiring movement, as these patients also had difficulty in tasks requiring the estimation of the duration of an auditory stimulus and of the velocity of a visual stimulus. The latter results are especially convincing in demonstrating a role for the cerebellum in time estimation, as the required responses were emitted in the absence of external cues and in the absence of movement.

Just as in the case of the basal ganglia, however, there are some uncertainties about the precise role of the cerebellum in interval timing. Surgical intervention or diseases of the cerebellum generally result in increased variability in temporal processing, whereas both clock and memory effects are seen for pharmacological interventions, lesions, and diseases of the basal ganglia. Some theorists have argued that cerebellar dysfunction may induce deregulation of tonic thalamic tuning, which disrupts gating of the mnemonic temporal information generated in the basal ganglia through striato-thalamo-cortical loops (see Casini and Ivry, 1999; Diedrichsen et al., this volume; Ivry and Richardson, 2002). Other researchers have claimed somewhat different roles for the cerebellum and basal ganglia in timing and time perception (see Gibbon et al., 1997; Malapani et al., 1998a; Malapani and Rakitin, this volume; Miall and Reckless, 2001). Furthermore, some studies have shown that damage to the cerebellum caused by developmental stunting (Ferguson et al., 2001) or by lesions to the cerebellar vermis and hemispheres (Breukelaar and Dalrymple-Alford, 1999) has little or no effect on time estimation within the range of seconds, but may play a greater role in millisecond timing and counting processes where constant variability is a prominent source of error (e.g., Clarke et al., 1996; Ivry, 1996).

To the extent that there is disagreement in the field as to the nature of the contribution of the basal ganglia and cerebellum to interval timing, it should be kept in mind that different investigators may be optimizing their behavioral tasks to detect the timing contributions of specific brain structures. Consequently, we will conclude this discussion of the neural basis of interval timing by describing the results of a provocative experiment reported by Woodruff-Pak and Papka (1996). In this study Huntington's disease (HD) patients were evaluated for eye blink classical conditioning (EBCC) — a millisecond timing task that shows a high degree of sensitivity to damage to the cerebellum (e.g., Bao et al., 2002; Perrett, 1998). Because HD causes severe atrophy of the basal ganglia and thinning of the cortex, but no disruption or damage to the brain circuits engaged in EBCC, Woodruff-Pak and Papka (1996) predicted that HD patients would perform like normal age-matched controls in a 400-msec delay EBCC paradigm. The findings indicated that there were no differences in the production of conditioned responses between the patients and the controls, but the timing of the conditioned responses was abnormal in HD. Comparison of HD patients to patients with other neurodegenerative diseases (e.g., probable Alzheimer's disease, Down's syndrome) and patients with cerebellar lesions demonstrated significantly better EBCC performance in HD patients, indicating a normal ability to acquire the conditioned response, but an impaired ability to time the conditioned response. Taken together, these data provide support for the striatum having a role in optimizing the timing of the conditioned response in the EBCC paradigm — a task specifically designed to assess the integrity of the cerebellum.

As the interested reader will soon see, the following chapters synthesize the latest information on both human and animal timing behavior as related to both technical and theoretical approaches. Chapters written by the foremost experts in the field provide the necessary background to understanding the psychophysics and neurobiology of interval timing. Such a synthesis sets the stage for an interdisciplinary dialogue among investigators on either side of the behavior–biology divide and leads us in new directions with advances made in molecular biology and neurogenomics. This type of integrated approach has proven invaluable in studies of perceptual systems employed across a wide range of species and will likely do the same for the study of timing and time perception. It is the expectation of all the contributors of this book that understanding temporal integration by the brain will be among the premier topics to unite systems, cellular, computational, and cognitive neuroscience over the next decade.

REFERENCES

Abner, R.T., Edwards, T., Douglas, A., and Brunner, D., Pharmacology of temporal cognition in two mouse strains, *Int. J. Comp. Psychol.*, 14, 189–210, 2001.

Allada, R., Emery, P., Takahashi, J., and Rosbash, M., Stopping time: the genetics of fly and mouse circadian clocks, *Annu. Rev. Neurosci.*, 24, 1091–1119, 2001.

Allan, L.G., The influence of the scalar timing model on human timing research, *Behav. Process.*, 44, 101–117, 1998.

Allers, K.A., Ruskin, D.N., Bergstrom, D.A., Freeman, L.E., Ghazi, L.J., Tierney, P.L., and Walters, J.R., Multisecond periodicities in basal ganglia firing rates correlate with theta bursts in transcortical and hippocampal EEG, *J. Neurophysiol.*, 87, 1118–1122, 2002.

Arcediano, F. and Miller, R.R., Some constraints for models of timing: a temporal coding hypothesis perspective, *Learn. Motiv.*, 33, 105–123, 2002.

Balsam, P., Relative time in trace conditioning, in *Annals of the New York Academy of Sciences: Timing and Time Perception*, Vol. 423, Gibbon, J. and Allan, L., Eds., New York Academy of Sciences, New York, 1984, pp. 211–227.

Balsam, P.D., Drew, M.R., and Yang, C., Timing at the start of associative learning, *Learn. Motiv.*, 33, 141–155, 2002.

Bao, S., Chen, L., Kim, J.J., and Thompson, R.F., Cerebellar cortical inhibition and classical eyeblink conditioning, *Proc. Natl. Acad. Sci. U.S.A.*, 99, 1592–1597, 2002.

Bateson, M. and Kacelnik, A., Starling's preferences for predictable and unpredictable delays to food, *Anim. Behav.*, 53, 1129–1142, 1997.

Bateson, M. and Kacelnik, A., Risk-sensitive foraging: decision making in variable environments, in *Cognitive Ecology: The Evolutionary Ecology of Information Processing and Decision Making*, Dukas, R., Ed., Chicago University Press, Chicago, 1998.

Blakeslee, S., Running late? Researchers blame aging brain, *The New York Times: Science Times*, March 24, 1998, pp. B13, B16.

Botez, M.I., Leveille, J., Lambert, R., and Botez, T., Single photon emission computed tomography (SPECT) in cerebellar disease: cerebello-cerebral diaschisis, *Eur. Neurol.*, 31, 405–412, 1991.

Brannon, E.M. and Terrace, H.S., Ordering of the numerosities 1 to 9 by monkeys, *Science*, 282, 746–749, 1998.

Brannon, E.M., Wusthoff, C.J., Gallistel, C.R., and Gibbon, J., The subjective scaling of number representation, *Psychol. Sci.*, 12, 238–243, 2001.

Breukelaar, J.W.C. and Dalrymple-Alford, J.C., Effects of lesions to the cerebellar vermis and hemispheres on timing and counting in rats, *Behav. Neurosci.*, 113, 78–90, 1999.

Broadbent, H.A., Rakitin, B.C., Church, R.M., and Meck, W.H., Quantitative relationships between timing and counting, in *Numerical Skills in Animals*, Boysen, S. and Capaldi, E.J., Eds., Erlbaum, Hillsdale, NJ, 1993, pp. 171–187.

Brockie, P.J., Mellem, J.E., Hills, T., Madsen, D.M., and Maricq, A.V., The *C. elegans* glutamate receptor subunit NMR-1 is required for slow NMDA-activated currents that regulate reversal frequency during locomotion, *Neuron*, 31, 617–630, 2001.

Brown, G.D.A. and Chater, N., The chronological organization of memory: common psychological foundations for remembering and timing, in *Time and Memory: Issues in Philosophy and Psychology*, Hoerl, C. and McCormack, T., Eds., Oxford University Press, Oxford, 2001, pp. 77–110.

Brown, G.D.A., Preece, T., and Hulme, C., Oscillator-based memory for serial order, *Psychol. Rev.*, 107, 127–181, 2000.

Bugmann, G., Towards a neural model of timing, *Biosystems*, 48, 11–19, 1998.

Buhusi, C.V. and Meck, W.H., Timing for the absence of a stimulus: the gap paradigm reversed, *J. Exp. Psychol. Anim. Behav. Process.*, 26, 305–322, 2000.

Buhusi, C.V. and Meck, W.H., Differential effects of methamphetamine and haloperidol on the control of an internal clock, *Behav. Neurosci.*, 116, 291–297, 2002.

Burle, B. and Bonnet, M., What's an internal clock for? From temporal information processing to temporal processing of information, *Behav. Process.*, 45, 59–72, 1999.

Carboni, E., Spielewoy, C., Vacca, C., Nosten-Bertrand, M., Giros, B., and Di Chiara, G., Cocaine and amphetamine increase extracellular dopamine in the nucleus accumbens of mice lacking the dopamine transporter gene, *J. Neurosci.*, 21, RC14, 2001.

Carey, S., On the very possibility of discontinuities in conceptual development, in *Language, Brain, and Cognitive Development: Essays in Honor of Jacques Mehler*, Dupoux, E., Ed., MIT Press, Cambridge, MA, 2001, pp. 303–324.

Casini, L. and Ivry, R.B., Effects of divided attention on temporal processing in patients with lesions of the cerebellum or frontal lobe, *Neuropsychology*, 13, 10–21, 1999.

Casini, L. and Macar, F., Can the level of prefrontal activity provide an index of performance in humans? *Neurosci. Lett.*, 219, 71–74, 1996a.

Casini, L. and Macar, F., Prefrontal slow way potential in temporal compared to nontemporal tasks, *J. Psychophysiol.*, 10, 252–264, 1996b.

Casini, L. and Macar, F., Multiple approaches to investigate the existence of an internal clock using attentional resources, *Behav. Process.*, 45, 73–85, 1999.

Catania, A.C., Reinforcement schedules and psychophysical judgements: a study of some temporal properties of behavior, in *The Theory of Reinforcement Schedules*, Schoenfeld, W.N., Ed., Appleton-Century-Crofts, New York, 1970, pp. 1–42.

Cheng, K. and Westwood, R., Analysis of single trials in pigeons' timing performance, *J. Exp. Psychol. Anim. Behav. Process.*, 19, 56–67, 1993.

Cheng, K., Westwood, R., and Crystal, J.D., Memory variance in the peak procedure of timing in pigeons, *J. Exp. Psychol. Anim. Behav. Process.*, 19, 68–76, 1993.

Church, R.M., Theories of timing behavior, in *Contemporary Learning Theories: Instrumental Conditioning Theory and the Impact of Biological Constraints on Learning*, Klein, S.B. and Mowrer, R.R., Eds., Erlbaum, Hillsdale, NJ, 1989, pp. 41–71.

Church, R.M. and Meck, W.H., The numerical attribute of stimuli, in *Animal Cognition*, Roitblat, H.L., Bever, T.G., and Terrace, H.S., Eds., Erlbaum, Hillsdale, NJ, 1984, pp. 445–464.

Church, R.M. and Meck, W.H., Biological basis of the remembered time of reinforcement, in *Quantitative Analyses of Behavior: Biological Determinants of Reinforcement*, Vol. 7, Commons, M.L., Church, R.M., Stellar, J.R., and Wagner, A.R., Eds., Erlbaum, Hillsdale, NJ, 1988, pp. 103–119.

Church, R.M., Meck, W.H., and Gibbon, J., Application of scalar timing theory to individual trials, *J. Exp. Psychol. Anim. Behav. Process.*, 20, 135–155, 1994.

Clarke, S., Ivry, R., Grinband, J., Roberts, S., and Shimizu, N., Exploring the domain of cerebellar timing, in *Time, Internal Clocks and Movement*, Pastor, M.A. and Artieda, J., Eds., Elsevier, Amsterdam, 1996, pp. 257–280.

Clement, A. and Droit-Volet, S., Simultaneous or Dissociative Processing of Duration and Number of Sequence of Events? A Developmental Perspective, paper presented at the ESCOP/British Psychological Society Conference, London, UK, September 2002.

Cohen, R.A., Barnes, J., Jenkins, M., and Albers, H.E., Disruption of short-duration timing associated with damage to the suprachiasmatic region of the hypothalamus, *Neurology*, 48, 1533–1539, 1997.

Constantinidis, C., Williams, G.V., and Goldman-Rakic, P.S., A role for inhibition in shaping the temporal flow of information in prefrontal cortex, *Nat. Neurosci.*, 5, 75–80, 2002.

Cronin-Golomb, A., Corkin, S., and Growdon, J.H., Impaired problem solving in Parkinson's disease: impact of a set-shifting deficit, *Neuropsychologia*, 32, 579–593, 1994.

Crystal, J.D., Systematic nonlinearities in the perception of temporal intervals, *J. Exp. Psychol. Anim. Behav. Process.*, 25, 3–17, 1999.

Crystal, J.D., Circadian time perception, *J. Exp. Psychol. Anim. Behav. Process.*, 27, 68–78, 2001.

Crystal, J.D., Church, R.M., and Broadbent, H.A., Systematic nonlinearities in the memory representation of time, *J. Exp. Psychol. Anim. Behav. Process.*, 23, 267–282, 1997.

Dehaene, S., Spelke, E., Pinel, P., Stanescu, R., and Tsivkin, S., Sources of mathematical thinking: behavioral and brain-imaging evidence, *Science*, 284, 970–974, 1999.

Droit-Volet, S., The estimation in young children: an initial force rule governing time production, *J. Exp. Child Psychol.*, 68, 236–249, 1998.

Droit-Volet, S. and Wearden J.H., Temporal bisection in children, *J. Exp. Child Psychol.*, 80, 142–159, 2001.

Egan, M.F., Goldberg, T.E., Kolachana, B.S., Callicott, J.H., Mazzanti, C.M., Straub, R.E., Goldman, D., and Weinberger, D.R., Effect of COMT Val$^{108/158}$ met genotype on frontal lobe function and risk for schizophrenia, *Proceed. Nat. Acad. Sci.*, 98, 6917–6922, 2001.

Fee, M.S., Hahnloser, R.H.R., and Kozhevnikov, A.A., The origin of a representation of time in the brain, *Soc. Neurosci. Abstr.*, 28, 680–683, 2002.

Ferguson, A.A., Cada, A.M., Gray, E.P., and Paule, M.G., No alterations in the performance of two interval timing operant tasks after α-difluoromethylornithine (DFMO)-induced cerebellar stunting, *Behav. Brain Res.*, 126, 135–146, 2001.

Ferrandez, A.M. and Pouthas, V., Does cerebral activity change in middle-aged adults in a visual discrimination task? *Neurobiol. Aging*, 22, 645–657, 2001.

Fortin, C. and Massé, N., Order information in short-term memory and time estimation, *Mem. Cognit.*, 27, 54–62, 1999.

Fortin, C. and Massé, N., Expecting a break in time estimation: attentional time-sharing without concurrent processing, *J. Exp. Psychol. Hum. Percept. Perform.*, 26, 1788–1796, 2000.

Fortin, C. and Rousseau, R., Time estimation as an index of processing demand in memory search, *Percept. Psychophys.*, 42, 377–382, 1987.
Fraisse, P., *The Psychology of Time*, Harper, New York, 1963.
Gabrieli, J.D., Singh, J., Stebbins, G.T., and Goetz, C.G., Reduced working memory span in Parkinson's disease: evidence for the role of a frontostriatal system in working and strategic memory, *Neuropsychology*, 10, 322–332, 1996.
Gainetdinov, R.R. and Caron, M.G., Genetics of childhood disorders: hyperdopaminergic mice as an animal model of ADHD, *J. Am. Acad. Child Adolesc. Psychiatry*, 40, 380–382, 2001.
Gainetdinov, R.R., Mohn, A.R., Bohn, L.M., and Caron, M.G., Glutamatergic modulation of hyperactivity in mice lacking the dopamine transporter, *Proc. Natl. Acad. Sci. U.S.A.*, 98, 11047–11054, 2001a.
Gainetdinov, R.R., Mohn, A.R., and Caron, M.G., Genetic animal models: focus on schizophrenia, *Trends Neurosci.*, 24, 527–533, 2001b.
Gainetdinov, R.R., Wetsel, W.C., Jones, S.R., Levin, E.D., Jaber, M., and Caron, M., Role of serotonin in the paradoxical calming effect of psychostimulants on hyperactivity, *Science*, 282, 397–401, 1999.
Gallistel, C.R., *The Organization of Learning*, MIT Press, Cambridge, MA, 1990.
Gallistel, C.R. and Gelman, R., Preverbal and verbal counting and computation, *Cognition*, 44, 43–74, 1992.
Gallistel, C.R. and Gelman, R., Non-verbal numerical cognition: from reals to integers, *Trends Cognit. Sci.*, 4, 59–65, 2000.
Gallistel, C.R. and Gibbon, J., Time, rate, and conditioning, *Psychol. Rev.*, 107, 289–344, 2000.
Gallistel, C.R. and Gibbon, J., Computational versus associative models of simple conditioning, *Curr. Directions Psychol.*, 10, 146–150, 2001.
Gautier, T. and Droit-Volet, S., Attentional distraction and time perception in children, *Int. J. Psychol.*, 37, 27–34, 2002.
Gelman, R. and Cordes, S., Counting in animals and humans, in *Language, Brain, and Cognitive Development: Essays in Honor of Jacques Mehler*, Dupoux, E., Ed., MIT Press, Cambridge, MA, 2001, pp. 279–301.
Gibbon, J., Scalar expectancy theory and Weber's law in animal timing, *Psychol. Rev.*, 84, 279–325, 1977.
Gibbon, J., Origins of scalar timing, *Learn. Motiv.*, 22, 3–38, 1991.
Gibbon, J., Ubiquity of scalar timing with a Poisson clock, *J. Math. Psychol.*, 36, 283–293, 1992.
Gibbon, J. and Allan, L., *Timing and Time Perception*, New York Academy of Sciences, New York, 1984.
Gibbon, J., Baldock, M.D., Locurto, C., Gold, L., and Terrace, H.S., Trial and intertrial durations in autoshaping, *J. Exp. Psychol. Anim. Behav. Process.*, 3, 264–284, 1977.
Gibbon, J. and Balsam, P.D., The spread of association in time, in *Autoshaping and Conditioning Theory*, Locurto, C., Terrace, H.S., and Gibbon, J., Eds., Academic Press, New York, 1981.
Gibbon, J. and Church, R.M., Representation of time, *Cognition*, 37, 23–54, 1990.
Gibbon, J., Church, R.M., and Meck, W.H., Scalar timing in memory, in *Annals of the New York Academy of Sciences: Timing and Time Perception*, Vol. 423, Gibbon, J. and Allan, L., Eds., New York Academy of Sciences, New York, 1984, pp. 52–77.
Gibbon, J. and Malapani, C., Neural basis of timing and time perception, *Encyclopedia Cognit. Sci.*, in press.
Gibbon, J., Malapani, C., Dale, C.L., and Gallistel, C.R., Toward a neurobiology of temporal cognition: advances and challenges, *Curr. Opin. Neurobiol.*, 7, 170–184, 1997.

Givens, B.S. and Olton, D.S., Cholinergic and GABAergic modulation of medial septal area: effect on working memory, *Behav. Neurosci.*, 104, 849–855, 1990.

Grondin, S., From physical time to the first and second moments of psychological time, *Psychol. Bull.*, 127, 22–44, 2001.

Grondin, S., Meilleur-Wells, G., and Lachance, R., When to start explicit counting in a time-intervals discrimination task: a critical point in the timing process of humans, *J. Exp. Psychol. Hum. Percept. Perform.*, 25, 993–1004, 1999.

Hahnloser, R.H.R., Kozhevnikov, A.A., and Fee, M.S., An ultra-sparse code underlies the generation of neural sequences in a song bird, *Nature,* 419, 65–70, 2002.

Harrington, D.L. and Haaland, K.Y., Neural underpinnings of temporal processing: a review of focal lesion, pharmacological, and functional imaging research, *Rev. Neurosci.*, 10, 91–116, 1999.

Harrington, D.L., Haaland, K.Y., and Knight, R.T., Cortical networks underlying mechanisms of time perception, *J. Neurosci.*, 18, 1085–1095, 1998.

Hinton, S.H. and Meck, W.H., How time flies: functional and neural mechanisms of interval timing, in *Time and Behaviour: Psychological and Neurobiological Analyses*, Bradshaw, C.M. and Szabadi, E., Eds., Elsevier, New York, 1997a, pp. 409–457.

Hinton, S.H. and Meck, W.H., The "internal clocks" of circadian and interval timing, *Endeavour*, 21, 82–87, 1997b.

Hinton, S.C., Meck, W.H., and MacFall, J.R., Peak-interval timing in humans activates frontal-striatal loops, *Neuroimage*, 3, S224, 1996.

Hopson, J.W., Gap timing and the spectral timing model, *Behav. Process.*, 45, 23–31, 1999.

Houk, J.C., Information processing in modular circuits linking basal ganglia and cerebral cortex, in *Models of Information Processing in the Basal Ganglia*, Houk, J.C., Davis, J.L., and Beiser, D.G., Eds., MIT Press, Cambridge, MA, 1995, pp. 3–10.

Ivry, R.B., The representation of temporal information in perception and motor control, *Curr. Opin. Neurobiol.*, 6, 851–857, 1996.

Ivry, R.B. and Keele, S.W., Timing functions of the cerebellum, *J. Cognit. Neurosci.*, 1, 136–153, 1989.

Ivry, R.B. and Richardson, T.C., Temporal control and coordination: the multiple timer mode, *Brain Cognit.*, 48, 117–132, 2002.

Jenkins, H.M., Time and contingency in classical conditioning, in *Annals of the New York Academy of Sciences: Timing and Time Perception*, Vol. 423, Gibbon, J. and Allan, L., Eds., New York Academy of Sciences, New York, 1984, pp. 228–241.

Junck, L., Gilman, S., Rothley, J.R., Betley, A.T., Koeppe, R.A., and Hichwa, R.D., A relationship between metabolism in frontal lobes and cerebellum in normal subjects studies with PET, *J. Cereb. Blood Flow Metab.*, 8, 774–782, 1988.

Kamin, L.J., Predictability, surprise, attention, and conditioning, in *Punishment and Aversive Behavior*, Campbell, B.A. and Church, R.M., Eds., Appleton-Century-Crofts, New York, 1969, pp. 279–296.

Keele, S.W. and Ivry, R.B., Does the cerebellum provide a common computation for diverse tasks? A timing hypothesis, *Ann. N.Y. Acad. Sci.*, 608, 179–211, 1990.

Killeen, P.R., Scalar counters, *Learn. Motiv.*, 33, 63–87, 2002.

King, A., McDonald, R., and Gallistel, C.R., Screening for mice that remember incorrectly, *Int. J. Comp. Psychol.*, 14, 232–257, 2001.

Krampe, R.T., Engbert, R., and Kliegl, R., Representational models and nonlinear dynamics: irreconcilable approaches to human movement timing and coordination or two sides of the same cone? Introduction to the special issue on movement and coordination, *Brain Cognit.*, 48, 1–6, 2002.

Krebs, J.R. and Kacelnik, A., Time horizons of foraging animals, in *Annals of the New York Academy of Sciences: Timing and Time Perception*, Vol. 423, Gibbon, J. and Allan, L., Eds., New York Academy of Sciences, New York, 1984, pp. 278–291.

Lejeune, H., Switching or gating? The attentional challenge in cognitive models of psychological time, *Behav. Process.*, 44, 127–145, 1998.

Lejeune, H., Maquet, P., Bonnet, M., Casini, L., Ferrara, A., Macar, F., Pouthas, V., Timsit-Berthier, M., and Vidal, F., The basic pattern of activation in motor and sensory temporal tasks: positron emission tomography data, *Neurosci. Lett.*, 235, 21–24, 1997.

Levin, E.D., Conners, C.K., Sparrow, E., Hinton, S.C., Erhardt, D., Meck, W.H., Rose, J.E., and March, J., Nicotine effects on adults with attention-deficit/hyperactivity disorder, *Psychopharmacology*, 123, 55–63, 1996.

Levin, E.D., Conners, C.K., Silva, D., Hinton, S.C., Meck, W.H., March, J., and Rose, J.E., Transdermal nicotene effects on attention, *Psychopharmacology*, 140, 135–141, 1998.

Lewis, P.A., Finding the timer, *Trends Cognit. Sci.*, 6, 195–196, 2002.

Liu, J., Head, E., Gharib, A.M., Yuan, W., Ingersoll, R.T., Hagen, T.M., Cotman, C.W., and Ames, B.N., Memory loss in old rats is associated with brain mitochondrial decay and RNA/DNA oxidation: partial reversal by feeding acetyl-L-carnitine and/or R-α-lipoic acid, *Proc. Natl. Acad. Sci. U.S.A.*, 99, 2356–2361, 2002.

Lustig, C. and Meck, W.H., Paying attention to time as one gets older, *Psychol. Sci.*, 12, 478–484, 2001.

Macar, F., Pouthas, V., and Friedman, W.J., *Time, Action, and Cognition: Towards Bridging the Gap*, Kluwer Academic Publishers, Dordrecht, Netherlands, 1992.

Macar, F., Vidal, F., and Bonnet, M., Laplacian derivation of CNV in time programming, in *Psychophysiological Brain Research*, Vol. 1, Brunia, C.H.M., Gaillard, A.W.K., and Kok, A., Eds., Tilburg University Press, Tilburg, Netherlands, 1990, pp. 69–76.

Malapani, C., Deweer, B., and Gibbon, J., Separating storage from retrieval dysfunction of temporal memory in Parkinson's disease, *J. Cognit. Neurosci.*, 14, 311–322, 2002.

Malapani, C., Dubois, B., Rancurel, G., and Gibbon, J., Cerebellar dysfunctions of temporal processing in the seconds range in humans, *Neuroreport*, 9, 3907–3912, 1998a.

Malapani, C., Rakitin, B., Meck, W.H., Deweer, B., Dubois, B., and Gibbon, J., Coupled temporal memories in Parkinson's disease: a dopamine-related dysfunction, *J. Cognit. Neurosci.*, 10, 316–331, 1998b.

Maquet, P., Lejeune, H., Pouthas, V., Bonnet, M., Casini, L., Macar, F., Timsit-Berthier, M., Vidal, F., Ferrara, A., Degueldre, C., Quaglia, L., Delfiore, G., Luxen, A., Woods, R., Mazziotta, J.C., and Comar, D., Brain activation induced by estimation of duration: a PET study, *Neuroimage*, 3, 119–126, 1996.

Maricq, A.V., Roberts, S., and Church, R.M., Methamphetamine and time estimation, *J. Exp. Psychol. Anim. Behav. Process.*, 7, 18–30, 1981.

Matell, M.S. and Meck, W.H., Reinforcement-induced within-trial resetting of an internal clock, *Behav. Process.*, 45, 159–171, 1999.

Matell, M.S. and Meck, W.H., Neuropsychological mechanisms of interval timing behaviour, *Bioessays*, 22, 94–103, 2000.

Mathew, R.J., Wilson, W.H., Turkington, T.G., and Coleman, R.E., Cerebellar activity and disturbed time sense after THC, *Brain Res.*, 797, 183–189, 1998.

McAuley, J.D. and Jones, M.R., Rhythmic expectations in time judgment behavior: implications for entrainment and interval-based models of time perception, *J. Exp. Psychol. Hum. Percept. Perform.*, submitted.

McCormack, T., Brown, G.D.A., Maylor, E.A., Darby, A., and Green, D., Developmental changes in time estimation: comparing childhood and old age, *Dev. Psychol.*, 35, 1143–1155, 1999.

McCormack, T. and Hoerl, C., Eds., *Time and Memory*, Oxford University Press, Oxford, 2001.

Meck, W.H., Selective adjustment of the speed of internal clock and memory storage processes, *J. Exp. Psychol. Anim. Behav. Process.*, 9, 171–201, 1983.

Meck, W.H., Attentional bias between modalities: effect on the internal clock, memory, and decision stages used in animal time discrimination, in *Annals of the New York Academy of Sciences: Timing and Time Perception*, Vol. 423, Gibbon, J. and Allan, L., Eds., New York Academy of Sciences, New York, 1984, pp. 528–541.

Meck, W.H., Postreinforcement signal processing, *J. Exp. Psychol. Anim. Behav. Process.*, 11, 52–70, 1985.

Meck, W.H., Affinity for the dopamine D2 receptor predicts neuroleptic potency in decreasing the speed of an internal clock, *Pharmacol. Biochem. Behav.*, 25, 1185–1189, 1986.

Meck, W.H., Neuropharmacology of timing and time perception, *Cognit. Brain Res.*, 3, 227–242, 1996.

Meck, W.H., Application of a mode-control model of temporal integration to counting and timing behaviour, in *Time and Behaviour: Psychological and Neurobiological Analyses*, Bradshaw, C.M. and Szabadi, E., Eds., Elsevier, New York, 1997, pp. 133–184.

Meck, W.H., Interval timing and genomics: what makes mutant mice tick? *Int. J. Comp. Psychol.*, 14, 211–231, 2001.

Meck, W.H., Choline uptake in the frontal cortex is proportional to the absolute error of a temporal memory translation constant in mature and aged rats, *Learn. Motiv.*, 33, 88–104, 2002a.

Meck, W.H., Distortions in the content of temporal memory: neurobiological correlates, in *Animal Cognition and Sequential Behavior: Behavioral, Biological, and Computational Perspectives*, Fountain, S.B., Bunsey, M.D., Danks, J.H., and McBeath, M.K., Eds., Kluwer Academic Press, Boston, 2002b, pp. 175–200.

Meck, W.H. and Benson, A.M., Dissecting the brain's internal clock: how frontal-striatal circuitry keeps time and shifts attention, *Brain Cognit.*, 48, 195–211, 2002.

Meck, W.H. and Church, R.M., A mode control model of counting and timing processes, *J. Exp. Psychol. Anim. Behav. Process.*, 9, 320–334, 1983.

Meck, W.H. and Church, R.M., Simultaneous temporal processing, *J. Exp. Psychol. Anim. Behav. Process.*, 10, 1–29, 1984.

Meck, W.H. and Church, R.M., Cholinergic modulation of the content of temporal memory, *Behav. Neurosci.*, 101, 457–464, 1987.

Meck, W.H., Church, R.M., and Gibbon, J., Temporal integration in duration and number discrimination, *J. Exp. Psychol. Anim. Behav. Process.*, 11, 591–597, 1985.

Meck, W.H., Church, R.M., and Wenk, G.L., Arginine vasopressin inoculates against age-related increases in sodium-dependent high affinity choline uptake and discrepancies in the content of temporal memory, *Eur. J. Pharmacol.*, 130, 327–331, 1986.

Meck, W.H., Church, R.M., Wenk, G.L., and Olton, D.S., Nucleus basalis magnocellularis and medial septal area lesions differentially impair temporal memory, *J. Neurosci.*, 7, 3505–3511, 1987.

Meck, W.H., Hinton, S.C., and Matell, M.S., Coincidence-detection models of interval timing: evidence from fMRI studies of cortico-striatal circuits, *Neuroimage*, 7, S281, 1998.

Miall, C., The storage of time intervals using oscillating neurons, *Neural Comput.*, 1, 359–371, 1989.

Miall, C., Models of neural timing, in *Time, Internal Clocks and Movement*, Pastor, A. and Artieda, J., Eds., North-Holland/Elsevier Science, Amsterdam, 1996, pp. 69–94.

Miall, R.C. and Reckless, G.Z., The cerebellum and the timing of coordinated eye and hand tracking, *Brain Cognit.*, 48, 212–226, 2002.

Migliore, M., Messineo, L., Cardaci, M., and Ayala, G.F., Quantitative modeling of perception and production of time intervals, *J. Neurophysiol.*, 86, 2754–2760, 2001.

Monfort, V., Pouthas, V., and Ragot, R., Role of frontal cortex in memory for duration: an event-related potential study in humans, *Neurosci. Lett.*, 286, 91–94, 2000.

Morell, V., Setting a biological stopwatch, *Science*, 271, 905–906, 1996.

Nichelli, P., Venneri, A., Molinari, M., Tavani, F., and Grafman, J., Precision and accuracy of subjective time estimation in different memory disorders, *Cognit. Brain Res.*, 1, 87–93, 1993.

Ohyama, T., Horvitz, J.C., Gibbon, J., Malapani, C., and Balsam, P.D., Conditioned and unconditioned behavioral-cognitive effects of a dopamine antagonist in rats, *Behav. Neurosci.*, 114, 1251–1255, 2000.

Olton, D.S., Meck, W.H., and Church, R.M., Separation of hippocampal and amygdaloid involvement in temporal memory dysfunctions, *Brain Res.*, 404, 180–188, 1987.

Olton, D.S., Wenk, G.L., Church, R.M., and Meck, W.H., Attention and the frontal cortex as examined by simultaneous temporal processing, *Neuropsychologia*, 26, 307–318, 1988.

Pang, K.C., Yoder, R.M., and Olton, D.S., Neurons in the lateral agranular frontal cortex have divided attention correlates in a simultaneous temporal processing task, *Neuroscience*, 103, 615–628, 2001.

Pashler, H., Perception and production of brief durations: beat-based versus interval-based timing, *J. Exp. Psychol. Hum. Percept. Perform.*, 27, 485–493, 2001.

Paule, M.G., Meck, W.H., McMillan, D.E., Bateson, M., Popke, E.J., Chelonis, J.J., and Hinton, S.C., The use of timing behaviors in animals and humans to detect drug and/or toxicant effects, *Neurotoxicol. Teratol.*, 21, 491–502, 1999.

Penney, T.B., Gibbon, J., and Meck, W.H., Differential effects of auditory and visual signals on clock speed and temporal memory, *J. Exp. Psychol. Hum. Percept. Perform.*, 26, 1770–1787, 2000.

Penney, T.B., Holder, M.D., and Meck, W.H., Clonidine-induced antagonism of norepinephrine modulates the attentional processes involved in peak-interval timing, *Exp. Clin. Psychopharmacol.*, 4, 82–92, 1996.

Perbal, S., Ehrle, N., Samson, S., Baulac, M., and Pouthas, V., Time estimation in patients with right or left medial-temporal lobe resection, *Neuroreport*, 12, 939–942, 2001.

Perrett, S.P., Temporal discrimination in the cerebellar cortex during conditioned eyelid responses, *Exp. Brain Res.*, 121, 115–124, 1998.

Pouthas, V., Garnero, L., Ferrandez, A.M., and Renault, B., ERPs and PET analysis of time perception: spatial and temporal brain mapping during visual discrimination tasks, *Hum. Brain Mapping*, 10, 49–60, 2000.

Pouthas, V., Maquet, P., Garnero, L., Ferrandez, A.M., and Renault, B., Neural bases of time estimation: a PET and ERP study, *Electroencephalogr. Clin. Neurophysiol. Suppl.*, 50, 598–603, 1999.

Pressing, J., The referential dynamics of cognition and action, *Psychol. Rev.*, 106, 714–747, 1999.

Rakitin, B.C., Gibbon, J., Penney, T.B., Malapani, C., Hinton, S.C., and Meck, W.H., Scalar expectancy theory and peak-interval timing in humans, *J. Exp. Psychol. Anim. Behav. Process.*, 24, 15–33, 1998.

Rao, S.M., Mayer, A.R., and Harrington, D.L., The evolution of brain activation during temporal processing, *Nat. Neurosci.*, 4, 317–323, 2001.

Rescorla, R.A. and Wagner, A.R., A theory of Pavlovian conditioning: variations in the effectiveness of reinforcement and nonreinforcement, in *Classical Conditioning: II. Current Research and Theory*, Black, A.H. and Prokasy, W.F., Eds., Appleton-Century-Crofts, New York, 1972, pp. 64–99.

Richelle, M. and Lejeune, H., *Time in Animal Behavior*, Pergamon Press, New York, 1980.

Richelle, M. and Lejeune, H., Timing competence and timing performance: a cross-species approach, in *Annals of the New York Academy of Sciences: Timing and Time Perception*, Vol. 423, Gibbon, J. and Allan, L., Eds., New York Academy of Sciences, New York, 1984, pp. 254–268.

Roberts, S., Isolation of an internal clock, *J. Exp. Psychol. Anim. Behav. Process.*, 7, 242–268, 1981.

Roberts, S. and Holder, M.D., The function of time discrimination and classical conditioning, in *Annals of the New York Academy of Sciences: Timing and Time Perception*, Vol. 423, Gibbon, J. and Allan, L., Eds., New York Academy of Sciences, New York, 1984, pp. 228–241.

Roberts, W.A., Simultaneous numerical and temporal processing in the pigeon, *Curr. Directions Psychol. Sci.*, 4, 47–51, 1995.

Roberts, W.A. and Mitchell, S., Can a pigeon simultaneously process temporal and numerical information? *J. Exp. Psychol. Anim. Behav. Process.*, 20, 66–78, 1994.

Rousseau, R., Picard, D., and Pitre, E., An adaptive counter model for time estimation, in *Annals of the New York Academy of Sciences: Timing and Time Perception*, Vol. 423, Gibbon, J. and Allan, L., Eds., New York Academy of Sciences, New York, 1984, pp. 639–642.

Rousseau, L. and Rousseau, R., Stop-reaction time and the internal clock, *Percept. Psychophys.*, 58, 434–448, 1996.

Ruskin, D.N, Bergstrom, D.A., Kaneoke, Y., Patel, B.N., Twery, M.J., and Walters, J.R., Multisecond oscillations in firing rate in the basal ganglia: robust modulation by dopamine receptor activation and anesthesia, *J. Neurophysiol.*, 81 2046–2055, 1999.

Sasaki, A., Andrews, J.R., Higashikubo, B.T., and Meck, W.H., Gene-dose dependent effects on interval timing in dopamine-transporter knockout mice, *Soc. Neurosci. Abstr.*, 28, 183.12, 2002.

Schultz, W., Apicella, P., and Ljungberg, T., Responses of monkey dopamine neurons to reward and conditioned stimuli during successive steps of learning a delayed response task, *J. Neurosci.*, 13, 900–913, 1993.

Schultz, W., Dayan, P., and Montague, P.R., A neural substrate of prediction and reward, *Science*, 275, 1593–1599, 1997.

Skinner, B.F., The effect on the amount of conditioning of an interval of time before reinforcement, *J. Gen. Psychol.*, 14, 279–295, 1936.

Staddon, J.E.R. and Higa, J.J., Multiple time scales in simple habituation, *Psychol. Rev.*, 103, 720–733, 1996.

Staddon, J.E.R. and Higa, J.J., Time and memory: towards a pacemaker-free theory of interval timing, *J. Exp. Anal. Behav.*, 71, 215–251, 1999.

Stahl, S.M., *Essential Psychopharmacology: Neuroscientific Basis and Practical Applications*, Cambridge University Press, Cambridge, England, 2000.

Starkey, P., Spelke, E.S., and Gelman, R., Numerical abstraction by human infants, *Cognition*, 36, 97–127, 1990.

Stebbins, G.T., Gabrieli, J.D., Masciari, F., Monti, L., and Goetz, C.G., Delayed recognition memory in Parkinson's disease: a role for working memory? *Neuropsychologia*, 37, 503–510, 1999.

Travis, J., Biological stopwatch found in the brain, *Sci. News*, 149, 101, 1996.

Vanneste, S. and Pouthas, V., Timing in aging: the role of attention, *Exp. Aging Res.*, 25, 49–67, 1999.

Vidal, F., Bonnet, M., and Macar, F., Can duration be a relevant dimension of motor programs? in *Time, Action, and Cognition: Towards Bridging the Gap*, Macar, F., Pouthas, V., and Friedman, W.J., Eds., Kluwer Academic Publishers, Dordrecht, Netherlands, 1992.

Wagner, A.R., SOP: a model of automatic memory processing in animal behavior, in *Information Processing in Animals: Memory Mechanisms*, Spear, N.E. and Miller, R.R., Eds., Erlbaum, Hillsdale, NJ, 1981., pp. 5–47.

Wearden, J.H., "Beyond the fields we know ... ": exploring and developing scalar timing theory, *Behav. Process.*, 45, 3–21, 1999.

Whalen, J., Gallistel, C.R., and Gelman, R., Non-verbal counting in humans: the psychophysics of number representation, *Psychol. Sci.*, 10, 130–137, 1999.

Woodruff-Pak, D.S. and Papka, M., Huntington's disease and eyeblink conditioning: normal learning but abnormal timing, *J. Int. Neuropsychol. Soc.*, 2, 323–334, 1996.

Wright, K., Times of our lives, *Sci. Am.*, 9, 58–65, 2002.

Wynn, K., Addition and subtraction by human infants, *Nature*, 358, 749–750, 1992.

Wynn, K., Infants possess a system of numerical knowledge, *Curr. Directions Psychol. Sci.*, 4, 172–177, 1995.

Wynn, K., Psychological foundations of number: numerical competence in human infants, *Trends Cognit. Sci.*, 2, 296–303, 1998.

Wynn, K. and Chiang, W-C., Limits to infants' knowledge of objects: the case of magical appearance, *Psychol. Sci.*, 9, 448–455, 1998.

Yu, A.C. and Margoliash, D., Temporal hierarchical control of singing in birds, *Science*, 273, 1871–1875, 1996.

Zakay, D. and Block, R.A., The role of attention in time estimation processes, in *Time, Internal Clocks and Movement*, Pastor, M.A. and Artieda, J., Eds., Elsevier, Amsterdam, 1996, pp. 143–164.

Warren H. Meck
Duke University
Durham, North Carolina
January 9, 2003

Section I

Functional Mechanisms

1 A Concise Introduction to Scalar Timing Theory

Russell M. Church

CONTENTS

1.1 Introduction 4
    1.1.1 Interval Timing Procedures 4
    1.1.2 Interval Timing Theories 5
1.2 Three Descriptions of Scalar Timing Theory 5
    1.2.1 The Principles of Scalar Timing Theory 5
    1.2.2 An Information-Processing Metaphor of Scalar Timing Theory 7
    1.2.3 A Formal Model of Scalar Timing Theory 8
        1.2.3.1 Pacemaker 8
        1.2.3.2 Switch and Accumulator 8
        1.2.3.3 Memory Retrieval 9
        1.2.3.4 Threshold 9
        1.2.3.5 Decision Rule 9
        1.2.3.6 Memory Storage 9
        1.2.3.7 Predictions from a Formal Model 9
1.3 An Example of a Timing Experiment 10
    1.3.1 Procedure and Results 10
    1.3.2 Simulation of the Timing Experiment 11
        1.3.2.1 Main Program 12
        1.3.2.2 Initialization of the Fixed-Interval Procedure
             (Initialize_FI) 12
        1.3.2.3 Initialization of the Scalar Timing Model
             (Initialize_SET) 12
        1.3.2.4 The Fixed-Interval Procedure (FI_Procedure) 12
        1.3.2.5 The Scalar Timing Model (Scalar_Timing_Theory) 13
        1.3.2.6 Saving the Data (Record_Data) 13
    1.3.3 Estimation of Parameters 14
1.4 An Evaluation of Scalar Timing Theory 15
    1.4.1 Goodness of Fit 15
    1.4.2 Flexibility 16
    1.4.3 Generality 16

Acknowledgments 17
References 17
Appendix A: Matlab Code 19
Appendix B: Simulated Data 22

1.1 INTRODUCTION

In 1971 John Gibbon wrote, as the first sentence of an article in the *Journal of Mathematical Psychology*, "Scalar timing is proposed as the basic latency mechanism underlying asymptotic free-operant avoidance performance." He described the avoidance latency as a time estimate, and proposed that "time estimates are scale transforms of a single stochastic process." In an influential review, Gibbon (1977) extended this analysis to the behavior of animals in many other timing procedures. The essence of scalar timing is the "scale transform" idea. That is, the pattern of responding in time is the same at all time intervals if time is scaled in relative units (proportion of the interval) rather than absolute units (such as seconds). A process model was developed as a mechanism that would produce quantitative fits of data from animal timing experiments (Gibbon and Church, 1984; Gibbon et al., 1984). With some additional assumptions, scalar timing theory has been extended to account for dynamic effects (acquisition, extinction, and transition effects) as well as mean response rate (Gallistel and Gibbon, 2000; Gallistel and Gibbon, 2002; Gibbon and Balsam, 1981).

This chapter is a tutorial on scalar timing theory. It presents the basic ideas of the theory, the relationship of these ideas to a formal model, and the use of the formal model for making quantitative predictions about data. In this chapter the predictions of the theory are based entirely on computer simulations rather than mathematical derivations. A specific example is given in which scalar timing theory is applied to a single measure of behavior obtained in a single type of procedure. Such simulations can be extended easily to many other procedures and other measures of behavior.

1.1.1 INTERVAL TIMING PROCEDURES

Temporal perception and timed performance procedures consist of a specification of the conditions under which a discrete stimulus (such as a light or noise) will be turned on or off, and a specification of the conditions under which a reinforcer (such as a pellet of food) will be delivered. The procedures within the domain of a timing theory include those with zero, one, or more than one stimulus, and with zero, one, or more than one response contingency for reinforcement. All of these procedures are also in the domain of conditioning theories. They include many procedures from classical conditioning (zero response contingencies), operant conditioning (one response contingency), and choice (more than one response contingency).

Standard interval timing procedures include Pavlovian temporal conditioning in which reinforcement occurs at fixed times (zero stimuli and zero response contingencies), discrete trial fixed-interval (FI) schedules of reinforcement in which food occurs following the first response after a fixed time after stimulus onset (one stimulus and one response contingency), and temporal discrimination in which reinforcement follows one response after one of the stimuli and another response after the other stimulus (two stimuli and two response contingencies). The raw data from experiments in timing and conditioning are the times of the onset and termination of each stimulus, reinforcer, and response. The goal of the theory is to describe the times of the responses based on the specification of the procedure (and some fitted parameters).

1.1.2 INTERVAL TIMING THEORIES

Scalar timing theory is designed to account for the behavior of human beings and other animals in temporal perception and timed performance procedures (Allan, 1998; Gibbon, 1991). Many alternative timing theories have been designed to account for the same facts. These include the behavioral theory of timing (Killeen and Fetterman, 1988), the learning-to-time model (Machado, 1997), the multiple-oscillator model (Church and Broadbent, 1990), the spectral timing theory (Grossberg and Schmajuk, 1989), and the multiple-timescale model (Staddon and Higa, 1999). Although this chapter is restricted to a description of scalar timing theory, the approach that is described can be applied to any timing theory that is completely and precisely described (Church and Kirkpatrick, 2001).

Conditioning theories were also designed to account for the behavior produced by procedures that involve the specification of the times of onset and termination of stimuli and reinforcers, and of responses. The goal of standard conditioning theories (such as Rescorla and Wagner, 1972) was to describe the relative response rates or probabilities averaged over a stimulus, and not the time of responses. But real-time theories of conditioning were developed by Sutton, Barto, and others that were designed to account for the time of responses (Barto and Sutton, 1982; Sutton and Barto, 1981, 1990). Although this chapter will not describe these real-time conditioning theories, the same approach used in this chapter for the analysis of scalar timing theory can be applied to the real-time conditioning theories (Church and Kirkpatrick, 2001). Until recently, theories of timing and conditioning were developed quite separately, but it is no longer clear that separate theories for timing and conditioning are necessary (see Gibbon and Balsam, 1981; Gallistel and Gibbon, 2000, 2002; Hopson, this volume; Kirkpatrick and Church, 1998).

1.2 THREE DESCRIPTIONS OF SCALAR TIMING THEORY

Scalar timing theory can be understood at the level of principles, metaphor, or formal model. The principles are the empirical generalizations, the metaphor is an information-process description of the theory, and the formal equations are the substance of the theory.

1.2.1 THE PRINCIPLES OF SCALAR TIMING THEORY

Some of the principles of scalar timing theory apply to many different procedures. In an article entitled "Origins of Scalar Timing Theory," John Gibbon provided a historical review of the theory (Gibbon, 1991). In the 1950s and 1960s various researchers reported a linear relationship between the mean latency of a response and the interval between stimulus and reinforcement, but the theoretical importance of this empirical regularity was not recognized at the time. This regularity is now referred to as *proportional timing*. Subsequently, a linear relationship between the standard deviation of the latency of a response and the interval between the stimulus and reinforcement was identified. This regularity is now referred to as the *scalar property*. Then it was found that the coefficient of variation (the ratio of the standard deviation to the mean) was approximately constant (Catania, 1970), a result that was logically implied by the linearity of the standard deviation and the mean. This regularity is now referred to as *Weber's law for timing*. The most general rule was superposition — the finding that the response rate as a function of time was approximately the same at all intervals. This result was clearly shown by an examination of the function relating response rate to time since reinforcement in fixed-interval schedules of reinforcement of pigeons of 3, 30, and 300 sec (Dews, 1970). This regularity, which logically implies scalar variance and Weber's law, is now referred to as *timescale invariance*. The principles of proportionality, scalar variance, Weber's law, and timescale invariance apply to a wide range of measures of timing.

For example, consider a two-response temporal discrimination procedure between temporal intervals between the onset and termination of a stimulus of 2 or 8 sec; one response is reinforced following a 2-sec stimulus and a different response is reinforced following an 8-sec stimulus. Unreinforced probe stimuli may be presented for intermediate durations, and the proportion of stimuli to which the rat responds with the "long" response can be the dependent variable (Church and Deluty, 1977). The time at which the rat is equally likely to press the "short" or "long" response (the point of bisection) increases approximately linearly with the geometric mean of the reinforced short and long responses (proportional timing); the standard deviation of this point of bisection increases approximately linearly with stimulus duration (scalar variability). Thus, the ratio of standard deviation to the mean, which is called the coefficient of variation, is approximately constant (Weber's law), and the psychophysical functions at all ranges approximately superimpose when the duration of stimulus is divided by the geometric mean (timescale invariance).

Similar principles apply to a single-response temporal discrimination procedure known as the peak or peak-interval procedure. A peak procedure includes two types of cycles: those with a reinforcer and those without (for an example, see Church et al., 1994). On some proportion of the cycles (normally half), food is delivered immediately after the first response following a fixed interval after stimulus onset (a discriminative fixed-interval schedule); on the other cycles, there is no food and the stimulus terminates after a long presentation of the stimulus (normally four times the duration of the fixed interval). A measure of the temporal discrimination is the time of the maximum response rate, which is called the peak time. The peak time increases linearly with the fixed interval (proportional timing), the standard deviation of the peak time also increases linearly with the fixed interval (scalar property), the coefficient of variation of the peak time is approximately constant (Weber's law), and the response rate as a function of time since stimulus onset is approximately the same at all stimulus intervals when response rate is scaled in terms of proportion of maximum response rate, and time is scaled in proportion of the fixed interval (timescale invariance).

Similar principles also apply to fixed-interval schedules of reinforcement. The time of the median response increases approximately linearly with the time of the reinforced interval (proportional timing); the standard deviation of the median response also increases approximately linearly with stimulus duration (scalar property); the coefficient of variation of the time of the median response is approximately constant (Weber's law); and at a wide range of fixed intervals, the response rate relative to the maximum response rate superposes when the time since stimulus onset is divided by the fixed interval (superposition).

Other principles of scalar timing theory are specific to some class of procedures. For example, the shape of the function relating the probability of a long response to the duration of the stimulus in the temporal bisection procedure is approximately symmetrical on a logarithmic scale of time; but the shape of the function relating the response rate to the time since stimulus onset in the peak procedure is approximately symmetrical on a linear scale of time. The time at which the rat is equally likely to press the short or long response (the point of bisection) is approximately at the geometric mean between the durations of the long and short reinforced stimuli in the two-response bisection procedure. A goal of the information-processing model was to explain such results as proportionality, the scalar rule, constant coefficient of variation, superposition, bisection at the geometric mean, and other results, such as preference for more variable intervals when the mean interval was fixed.

1.2.2 AN INFORMATION-PROCESSING METAPHOR OF SCALAR TIMING THEORY

The model consists of three processes: perception, memory, and decision (as illustrated in Figure 1.1). Perception of time consists of a pacemaker, a switch, and an accumulator. The pacemaker emits pulses according to some simple distribution. If a switch is closed, these pulses enter the accumulator, which can be reset by another switch. When reinforcement occurs, the number of pulses in the accumulator (transformed by a multiplicative constant) is stored in memory. Thus, memory consists of an increasing number of examples of times of reinforcement. The decision process is based on three inputs: the current time (from the accumulator), a remembered time of reinforcement (from a random sample of one element from memory), and a threshold (from a random sample of a distribution of thresholds). A ratio rule is used to determine whether to make a response.

Scalar timing theory is a process model in which there are an experimenter-determined input (the times of onset and termination of the stimuli, and the times of reinforcer delivery) and a subject-determined output (the response). The information-processing metaphor is clearly a cognitive theory with boxes and arrows; the boxes represent functional parts and the arrows represent rules of transformation. It involves multiple intervening variables. The input affects one or more intervening variables; each of the intervening variables affects other intervening variables or the output. The information-processing metaphor helps to organize the intervening variables and their interrelationships, but it is not an essential feature of the theory.

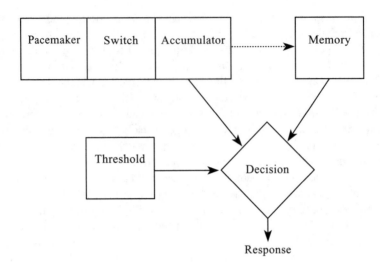

FIGURE 1.1 Information-processing metaphor of scalar timing theory.

1.2.3 A Formal Model of Scalar Timing Theory

The formal model consists of primitive terms (stimulus, reinforcer, and response) and quantitative transformation rules. It is not enough to write, "The pacemaker emits pulses according to some simple distribution." In order to make precise predictions, it is necessary to specify the distribution form and its parameters, such as mean and standard deviation. In scalar timing theory the distribution forms and some of the parameters are assumed, and other parameters are estimated from the data.

1.2.3.1 Pacemaker

Various assumptions have been made about the time series of pulses that are emitted by a pacemaker. These include the assumption of a fixed interpulse interval, a random (exponential) distribution of interpulse intervals, and a fixed interpulse interval during the timing of an interval, but a normal distribution of rates (Gibbon and Church, 1984). Because, at the durations used in most of the experimental research, other sources of scalar variability overwhelm the effects of pacemaker variability, any of these possibilities remains plausible (Gibbon, 1992).

1.2.3.2 Switch and Accumulator

The assumption is that a stimulus onset closes the switch and that pulses flow from the pacemaker to the accumulator when the switch is closed. The informal idea is that the switch starts the clock; i.e., it permits pulses to flow from the pacemaker into the accumulator. There is also the assumption that there is some latency between the occurrence of the stimulus and the closure of the switch; thus, there is some latency between the physical stimulus and the starting of the clock. This is assumed to be a normally distributed variable with some mean and standard deviation, $t_1 = \eta(\mu_{t1}, \sigma_{t1})$. Similarly, at the end of a stimulus the switch opens with some latency,

$t_2 = \eta(\mu_{t2}, \sigma_{t2})$. Thus, the duration of switch closure (and the length of time that pulses are flowing into the accumulator) is $t_2 - t_1$. The value in the accumulator is one of three inputs to a decision. The other two are the memory and the threshold.

1.2.3.3 Memory Retrieval

Memory retrieval consists of a random sample of one example in the memory.

1.2.3.4 Threshold

The threshold distribution is a normal distribution, $b = \eta(\mu_b, \sigma_b)$.

1.2.3.5 Decision Rule

The decision rule is based on a comparison of the perceived time (the accumulator value, a), the remembered time of reinforcement (the sample from memory distribution, m), and a criterion (the sample from the threshold distribution, b). The ratio rule is: if $(|a - m|/m) < b$, then a response will be made; otherwise, no response will be made.

1.2.3.6 Memory Storage

The memory consists of a large number of unorganized examples. Memory storage consists of putting an additional example into the memory. When reinforcement occurs, the value in the accumulator is transformed by multiplication by a memory storage constant, k^*, that has a normal distribution, $\eta(\mu_{k^*}, \sigma_{k^*})$, and put into the memory as another example. This provides a way to separate the perceived time (in the accumulator) from the remembered time, and it has been critical for the analysis of individual differences, drug effects, and lesion effects (see Meck, 1983, 1996, this volume).

1.2.3.7 Predictions from a Formal Model

To make predictions of behavior from scalar timing theory, it is necessary to specify not only the transformation rules precisely, but also the procedure. With both the procedure and the model completely and precisely specified, estimates of the parameters from the data, and the fit of the model to the data, may be based on explicit solutions or simulations.

In an explicit solution, predictions of the model are derived from the assumptions of the models (usually, with the addition of some simplifying assumptions or approximations). Explicit solutions of scalar timing theory have been developed for some procedures. These include the temporal bisection procedure (Gibbon, 1981a, 1981b), temporal generalization (Church and Gibbon, 1982; Gibbon and Church, 1984), time-left (Gibbon and Church, 1981; Gibbon et al., 1984), and the peak-interval procedure (Gibbon et al., 1984). The main advantage of these explicit solutions over simulations is that they provide exact results, and the resulting equations may be used to calculate the consequences of any sets of parameters very

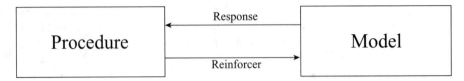

FIGURE 1.2 Relationship between procedure and model. The output of the procedure (stimulus and reinforcement) is the input to the model; the output of the model (responses) is the input to the procedure.

rapidly. There are, however, disadvantages. The explicit solutions have been restricted to specific procedures, so it is difficult to ascertain whether the same model is being used in the explicit solutions for different procedures. They have used some additional simplifying approximations that may not be correct. Although symbolic programs such as Mathematica and Maple provide considerable help, many of the explicit solutions require substantial mathematical knowledge and effort to create, check, or even to understand.

Simulations of the model may also be used to estimate values of the parameters from the data, and the fit of the model to the data. Although the simulations will contain some sampling error, this error can be reduced to any amount by increasing the sample size. No additional simplifying approximations are needed, and the simulations require little mathematical knowledge to create, to avoid error, or to understand. They simply require the ability to follow the logic of following explicit rules to transform the output of the procedure (stimuli and reinforcers) to the output of the model (responses).

With a standard modern computer and a good programming style, the simulations can be done rapidly. Most importantly, the model to be simulated may be kept completely independent of the procedure, as shown in Figure 1.2. The model receives stimuli and reinforcers from the procedure and outputs responses to the procedure. No change in the model is required when a different procedure is used. Thus, one can be sure that the same model is being used in the simulations for different procedures.

1.3 AN EXAMPLE OF A TIMING EXPERIMENT

1.3.1 Procedure and Results

Ten rats were trained on an 82.5-sec fixed-interval schedule of reinforcement for 15 3-h sessions. In this schedule, food was delivered immediately after the first lever response after the fixed interval from the previous delivery of food. The rats were then given five sessions of training on ten different fixed-interval schedules (15, 30, 45, 60, 75, 90, 105, 120, 135, and 150 sec). In each set of five sessions, one rat received each interval. That is, a Latin square design was used. For this example, the analysis will use the last three sessions of each rat from the fixed-interval schedules of 30, 60, and 120 sec.

The mean response rate increased as a function of time since the previous food, as shown in Figure 1.3. In this figure, response rate is shown in absolute units

A Concise Introduction to Scalar Timing Theory

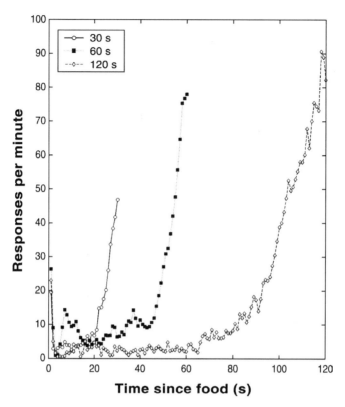

FIGURE 1.3 Response rate as a function of time since food for fixed-interval reinforcement schedules of 30, 60, and 120 sec. (Unpublished data from our laboratory.)

(responses per minute) and time is also expressed in absolute units (seconds). This temporal gradient increased earlier for the 30-sec FI than for the 60-sec interval, and earlier for the 60-sec interval than for the 120-sec interval. The height of the gradient was positively related to the interval duration, and the response rate immediately after food delivery was greater than the response rate a few seconds later.

1.3.2 SIMULATION OF THE TIMING EXPERIMENT

To simulate the behavior of an animal in a particular procedure, both the procedure and the timing model must be specified in a computer program. For some purposes, spreadsheets can be used (Collyer, 1992), but for procedures involving many stimuli, responses, and response contingencies, they may become difficult to write in an orderly manner and may execute slowly. Many different languages, such as C and Pascal, can be used for simulations. MATLAB (MathWorks, Natick, MA) is a particularly good software system to use for simulation of behavior and for comparison of data from theory and model. There are a large number of built-in functions, and other packages of functions are available in toolboxes or can be readily written by the user. With these functions it is easy to write well-organized programs that are easy to read, execute rapidly, and may be accompanied by standard or customized graphs.

The structure of the Matlab program used for the simulation of scalar timing theory has a simple, modular organization and should be clear. This simulation was done with the description of scalar timing theory separate from the description of the fixed-interval procedure. The Matlab code that was used to simulate the procedure and the formal model is in Appendix 1. Comments about this code follow. For large projects in which the speed of operation is critical, a faster speed can easily be achieved with improvements in hardware (such as a faster processor or an increase in random-access memory) or software (such as manipulations of vectors and matrices instead of the use of "for loops," and the compilation of the source code).

1.3.2.1 Main Program

The main program controls five subprograms. First, it initializes the procedure (Initialize_FI) and the model (Initialize_SET). Then it loops for 10,800 virtual seconds (3 h), checking any actions that should be taken by the procedure (FI_procedure) or by the model (Scalar_timing_theory). Finally, it transfers the data to the disk (Record_data).

1.3.2.2 Initialization of the Fixed-Interval Procedure (Initialize_FI)

The initialization of the fixed-interval procedure involves the setting of the interval between the delivery of food and the availability of the next food (30, 60, or 120 sec), assigning initial values to variables, and resetting the uniform and normal random-number generators to a haphazard initial state.

1.3.2.3 Initialization of the Scalar Timing Model (Initialize_SET)

The initialization of the model involves the setting of the clock parameters (mean, standard deviation, and initial speed), the memory parameters (mean and standard deviation of the memory storage constant), and the threshold parameters (mean, standard deviation, and initial threshold). In addition, pointers for memory storage and retrieval are initialized, and one random duration is put into memory.

1.3.2.4 The Fixed-Interval Procedure (FI_Procedure)

At each time unit (here taken as 1-sec intervals) the procedure determines whether food should be delivered. The cycle clock maintains a record of the number of seconds since food was delivered. For the fixed-interval procedure, food is primed if the clock equals or exceeds a fixed interval (set in the initialization of the procedure). If food is primed and a response is made (as determined by the model), an indication that food has been delivered is set, and a record is made of the time that the food was delivered and the event of food delivery (Food_code = 19). If a response is made, a record is made of the time of the response and the response event (Response_code = 1). The data then consist of a matrix, with each row consisting of a time and an event (1 or 19). The number of rows is equal to the number of events (responses plus food deliveries). See Appendix 2.

A Concise Introduction to Scalar Timing Theory

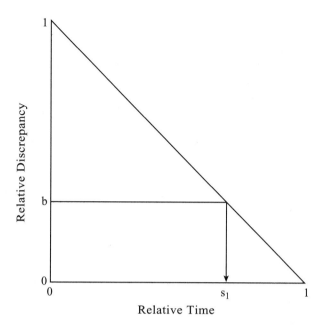

FIGURE 1.4 Basis for a decision in a fixed-interval schedule of reinforcement. Relative discrepancy decreases as a function of relative time. A high rate of responses occur when the relative discrepancy is less than the sample of the threshold (b). This occurs at relative time s_1.

1.3.2.5 The Scalar Timing Model (Scalar_Timing_Theory)

At each time unit (here taken as 1-sec intervals) the model determines whether a left lever response should occur. The value in the accumulator is equal to the clock speed multiplied by the clock time. The decision rule is based on the absolute value of the difference between the value of the sample from memory (m) and the accumulator value (a), divided by the value of the sample from memory (m). If this discrepancy is less than some threshold (b), a response will occur as illustrated in Figure 1.4.

If food is given, the value of the accumulator (multiplied by a random variable k*) will be put into memory. The memory storage constant (k*) is a random sample of one value from a normal distribution with some fixed mean and standard deviation. The speed of the clock is redetermined from a random sample from a normal distribution with some fixed mean and standard deviation; one item in memory is selected at random; and the threshold (b) is redetermined from a random sample from a normal distribution with some fixed mean and standard deviation. The indication that food was delivered is cleared, and the cycle clock is reset. In addition, responses occur at an operant rate of one response per 100 sec.

1.3.2.6 Saving the Data (Record_Data)

The name of the output file is constructed from the name assigned to the procedure, model, and interval; the file is saved on the disk; and the completion of the simulation is indicated on the monitor.

1.3.3 ESTIMATION OF PARAMETERS

The process of estimating the parameters can be described in the following four steps.

First, a simulation was done with the Matlab code in Appendix 1 that led to a simulated data file. The data from the simulation of scalar timing theory, like the data from the experiment, consisted of the times of lever responses and food deliveries of each rat on each session (see Appendix 2). This simulation was based on the specific values for each of the parameters that were set in Initialize_SET. The parameters were the mean and standard deviation of the clock, the mean and standard deviation of the memory storage constant, and the mean and standard deviation of the threshold.

Second, a summary measure was selected. The same data that were shown in Figure 1.3 were replotted in relative units in Figure 1.5. In this figure, response rate is shown as responses per minute divided by the maximum response rate in a given condition; time is shown as time since food was delivered, divided by the fixed interval. The temporal gradients for the three fixed-interval schedules increased in a similar manner. If the relative response rates as a function of the relative time intervals were identical at different intervals, the result would be described in terms of superposition of the functions. Such superposition implies that the mean time of a response is linearly related to the mean interval (i.e., proportionality), the standard deviation of the time of a response is linearly related to the mean interval (i.e., the scalar property), and thus the coefficient of variation (the standard deviation divided by the mean) is a constant. Superposition on a relative scale provides complete information about the temporal gradients on all other intervals based on the gradient on any single interval. The same Matlab program that was used to obtain the relative response rate gradients for the data (shown with the symbols identified in the legend) was used to obtain the relative response rate gradients for the simulation (shown as the solid line).

Third, a goodness-of-fit measure was used to compare the predictions of the theory with the observations from the data. The standard goodness-of-fit measure is ω^2, the percentage of variance accounted for. The sum of squared deviations of the data points from the mean of the data is the total variability; the sum of squared deviations of the data points from the predictions of the model is the unexplained variability. Thus, the explained variability is the difference between the total variability and the unexplained variability. ω^2 is the ratio of the explained variability to the total variability.

Fourth, modifications were made in the values of the parameters to optimize percentage of explained variability. The systematic way of conducting this search is by an exhaustive search of all combinations of parameter values at some resolution, or by a hill-climbing algorithm used for nonlinear fitting problems. The values shown in the program are those that were used for the solid line in Figure 1.5. The mean pacemaker rate was set at five pulses per second, with no variability; the mean memory constant was set at 1.0, with an estimate from the data of memory variability; the mean threshold was estimated from the data, with no threshold variability. Only two parameters were varied for the predictions shown in Figure 1.5: the standard deviation of memory was estimated to be about 0.2, and the mean threshold was estimated to be about 0.1. This accounted for about 96% of the variance.

A Concise Introduction to Scalar Timing Theory

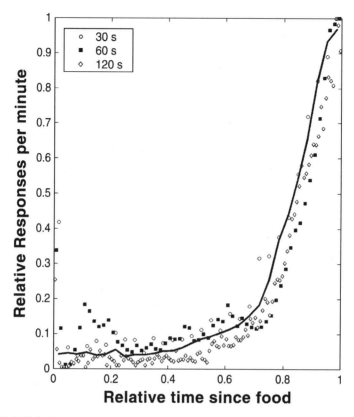

FIGURE 1.5 Relative response rate as a function of relative time since food for fixed-interval schedules of 30, 60, and 120 sec. Relative response rate is the mean number of responses per second, divided by the maximum mean response rate; relative time is the time since stimulus onset divided by the fixed interval. These are the same data shown in Figure 1.3, plotted in relative rather than absolute rate and time. The line is based on the simulation of scalar timing theory.

1.4 AN EVALUATION OF SCALAR TIMING THEORY

1.4.1 Goodness of Fit

A model should provide a good quantitative fit to the data. A failure to explain a high percentage of the variance may be due to variability of the data or a mismatch between the true model and the fitted model. Even if the percentage of variance accounted for is very high, and not due to fitting the noise, the fitted model may not be the true model. This can most readily be determined by examination of the pattern of differences between the data and the model (the residuals). Ideally, the pattern of residuals will be random. If the residuals are systematic, then there is a failure to identify some aspects of the true model. Of course, whether small systematic deviations are important for revealing the underlying mechanism or are due to some disturbance unrelated to the major problem is a matter for scientific judgment.

However, a good quantitative fit of a model to the data is not a sufficient criterion for acceptance of a theory. There is the important concern that a model with a large number of parameters may be fitting random variability in the particular sample of data being examined rather than fitting the true model (Roberts and Pashler, 2000; Zucchini, 2000). This is called overfitting, and various approaches have been used to correct for the number of free parameters. Probably the simplest approach is to estimate the parameters of the model based on some of the data and to apply these estimates to other data. This is called cross-validation.

If the data are reasonably regular, scalar timing theory often accounts for more than 95% of the variance and, in some cases, even more.

1.4.2 Flexibility

Flexibility is not a virtue in a model. Although a theorist may be reluctant to specify a model completely until more data are available, a model that is capable of accounting for many different patterns of behavior is much less useful for predicting outcomes than one that is inflexible. Formally specified theories differ in the number of patterns of behavior that can be predicted, and the ones with the more limited range of predictions are to be preferred. This is a basis for seeking simple, parsimonious theories.

For example, in one case a scalar timing model with two free parameters was judged to be superior to one with five free parameters (Church and Gibbon, 1982).

Quantitative fits are much more satisfactory than qualitative fits that are based on a judgment that the general shape of the observed functions corresponds to the predictions of the theory. A problem with qualitative fits is that they are overly flexible.

With many potential sources of variability in the clock, memory, and decision processes, scalar timing theory has considerable flexibility. Additional flexibility comes from the use of different assumptions regarding sampling from memory and number of thresholds, and the use of different assumptions for different procedures (Brunner et al., 1997). Scalar timing theory has a large number of parameters that can be varied, but typically only a few parameters are varied to account for a large number of observed data points.

1.4.3 Generality

Another basis for evaluating a theory is its generality. This refers both to its ability to account for behavior generated by a large number of different procedures and its ability to account for many different measures of response.

Scalar timing theory has been applied to many different procedures that include both classical conditioning and instrumental training, both perceptual and behavioral, and both appetitive and aversive (Church, 2002). It has been extended from simple procedures to more complex ones involving several stimuli (involving simultaneous temporal processing) and several responses (involving choice). It has also been extended from the analysis of steady-state behavior to the study of dynamics of behavior, including acquisition, extinction, and transition effects (Gallistel and Gibbon, 2000).

Scalar timing theory has been applied to many different dependent variables, including latency, relative response rates as a function of relative time from stimulus onset, and the distribution of the times of transitions from low to high response rates (Church et al., 1994). But it does not account for many other dependent variables. For example, in the fixed-interval schedule of reinforcement, it does not account for the absolute response rates as a function of time since stimulus onset (Figure 1.3) or for the distribution of interresponse times (IRTs). Thus, as it is presently formulated, scalar timing theory could not pass a Turing test (Church, 2001).

ACKNOWLEDGMENTS

Preparation of this chapter was supported by National Institute of Mental Health grant MH44234 to Brown University. The organization of the simulation and many of the ideas in this chapter are based on collaborative research with Kimberly Kirkpatrick, especially Church and Kirkpatrick (2001).

REFERENCES

Allan, L.G., The influence of the scalar timing model on human timing research, *Behav. Process.*, 44, 101–117, 1998.

Barto, A.G. and Sutton, R.S., Simulation of anticipatory responses in classical conditioning by a neuron-like adaptive element, *Behav. Brain Res.*, 4, 221–235, 1982.

Brunner, D., Fairhurst, S., Stolovitzky, G., and Gibbon, J., Mnemonics for variability: remembering food delay, *J. Exp. Psychol. Anim. Behav. Process.*, 23, 68–83, 1997.

Catania, A.C., Reinforcement schedules and psychophysical judgments: a study of some temporal properties of behavior, in *The Theory of Reinforcement Schedules*, Schoenfeld, W.N., Ed., Appleton-Century-Crofts, New York, 1970, pp. 1–42.

Church, R.M., A Turing test for computational and associative theories of learning, *Curr. Directions Psychol. Sci.*, 10, 132–136, 2001.

Church, R.M., A tribute to John Gibbon, *Behav. Process.*, 57, 261–274, 2002.

Church, R.M. and Broadbent, H.A., Alternative representations of time, number, and rate, *Cognition*, 37, 55–81, 1990.

Church, R.M. and Deluty, M.Z., The bisection of temporal intervals, *J. Exp. Psychol. Anim. Behav. Process.*, 3, 216–228, 1977.

Church, R.M. and Gibbon, J., Temporal generalization, *J. Exp. Psychol. Anim. Behav. Process.*, 8, 165–186, 1982.

Church, R.M. and Kirkpatrick, K., Theories of conditioning and timing, in *Handbook of Contemporary Learning Theories*, Mowrer, R.R. and Klein, S.B., Eds., Erlbaum, Mahwah, NJ, 2001, pp. 211–253.

Church, R.M., Meck, W.H., and Gibbon, J., Application of scalar timing theory to individual trials, *J. Exp. Psychol. Anim. Behav. Process.*, 20, 135–155, 1994.

Collyer, C.E., Spreadsheet modeling for research and teaching: programming without programming, *Behav. Res. Methods Instrum. Comput.*, 24, 467–474, 1992.

Dews, P.B., The theory of fixed-interval responding, in *The Theory of Reinforcement Schedules*, Schoenfeld, W.N., Ed., Appleton-Century-Crofts, New York, 1970, pp. 43–61.

Gallistel, C.R. and Gibbon, J., Time, rate and conditioning, *Psychol. Rev.*, 107, 289–344, 2000.

Gallistel, C.R. and Gibbon, J., *The Symbolic Foundations of Conditioned Behavior*, Erlbaum Associates, Mahwah, NJ, 2002.

Gibbon, J., Scalar timing and semi-Markov chains in free-operant avoidance, *J. Math. Psychol.*, 8, 109–138, 1971.

Gibbon, J., Scalar expectancy theory and Weber's law in animal timing, *Psychol. Rev.*, 84, 279–325, 1977.

Gibbon, J., On the form and location of the psychometric function for time, *J. Math. Psychol.*, 24, 58–87, 1981a.

Gibbon, J., Two kinds of ambiguity in the study of psychological time, in *Quantitative Analyses of Behavior*, Vol. 1, Commons, M.L. and Nevin, J., Eds., Ballinger, Cambridge, MA, 1981b, pp. 157–189.

Gibbon, J., Origins of scalar timing, *Learn. Motiv.*, 22, 3–38, 1991.

Gibbon, J., Ubiquity of scalar timing with a Poisson clock, *J. Math. Psychol.*, 36, 283–293, 1992.

Gibbon, J. and Balsam, P., Spreading associations in time, in *Autoshaping and Conditioning Theory*, Locurto, C.M., Terrace, H.S., and Gibbon, J., Eds., Academic Press, New York, 1981, pp. 219–253.

Gibbon, J. and Church, R.M., Time left: linear versus logarithmic subjective time, *J. Exp. Psychol. Anim. Behav. Process.*, 7, 87–108, 1981.

Gibbon, J. and Church, R.M., Sources of variance in an information processing theory of timing, in *Animal Cognition*, Roitblat, H.L., Bever, T.G., and Terrace, H.S., Eds., Erlbaum Associates, Hillsdale, NJ, 1984, pp. 465–488.

Gibbon, J., Church, R.M., and Meck, W.A., Scalar timing in memory, in *Timing and Time Perception*, Gibbon, J. and Allan, L.G., Eds., New York Academy of Sciences, New York, 1984, pp. 52–77.

Grossberg, S. and Schmajuk, N.A., Neural dynamics of adaptive timing and temporal discrimination during associative learning, *Neural Networks*, 2, 79–102, 1989.

Killeen, P.R. and Fetterman, J.G., A behavioral theory of timing, *Psychol. Rev.*, 95, 274–295, 1988.

Kirkpatrick, K. and Church, R.M., Are separate theories of conditioning and timing necessary? *Behav. Process.*, 44, 163–182, 1998.

Machado, A., Learning the temporal dynamics of behavior, *Psychol. Rev.*, 104, 241–265, 1997.

Meck, W.H., Selective adjustment of the speed of internal clock and memory processes, *J. Exp. Psychol. Anim. Behav. Process.*, 9, 171–201, 1983.

Meck, W.H., Neuropharmacology of timing and time perception, *Cognit. Brain Res.*, 3, 227–242, 1996.

Rescorla, R.A. and Wagner, A.R., A theory of Pavlovian conditioning: variations in the effectiveness of reinforcement and nonreinforcement, in *Classical Conditioning II: Current Research and Theory*, Black, A.H. and Prokasy, W.K., Eds., Appleton-Century-Crofts, New York, 1972, pp. 64–99.

Roberts, S. and Pashler, H., How persuasive is a good fit? A comment on theory testing, *Psychol. Rev.*, 107, 358–367, 2000.

Staddon, J.E.R. and Higa, J.J., Time and memory: toward a pacemaker-free theory, *J. Exp. Anal. Behav.*, 71, 215–251, 1999.

Sutton, R.S. and Barto, A.G., Toward a modern theory of adaptive networks: expectation and prediction, *Psychol. Rev.*, 88, 135–170, 1981.

Sutton, R.S. and Barto, A.G., Time-derivative models of Pavlovian reinforcement, in *Learning and Computational Neuroscience: Foundations of Adaptive Networks*, Gabriel, M. and Moore, J., Eds., MIT Press, Cambridge, MA, 1990, pp. 497–537.

Zucchini, W., An introduction to model selection, *J. Math. Psychol.*, 44, 41–61, 2000.

APPENDIX A: MATLAB CODE

```
%Main

Initialize_FI                    %Initialize the fixed-
                                 interval procedure
Initialize_SET                   %Initialize Scalar
                                 expectancy theory
for Session_clock = 1:10800, %Continue until session ends
    FI_procedure                 %Should the procedure deliver
                                 food?
    Scalar_timing_theory         %Should the model make a
                                 response?
end

Record_data                      %Transfer the data to disk

%-----------------------------------------------------------
%Initialize_FI

FI = 30;

clear
Procedure = 'FI';                %Assign procedural name for
                                 saving data file

Seconds = 1; Minutes = 60*Seconds; Hours = 60*Minutes;
Prime_code = 50; Response_code = 1; Food_code = 19;

Prime = 0; Response = 0; Cycle_clock = 0;
Data = 0; Data_pointer = 1; Cycle_count = 0; Food = 0;

randn('state',sum(100*clock));   %Initialize random number
                                 generators
rand('state',sum(100*clock));

%-----------------------------------------------------------
%Initialize_SET

Model = 'SET';

%Clock
Speed_mean = 5;
Speed_sd = 0;
Speed = (Speed_sd * randn) + Speed_mean;
Accumulator = 0;

%Memory
k_mean = 1.0;
k_sd = .2;
```

```
%Decision
Threshold_mean = .1;
Threshold_sd = 0;
Threshold = (Threshold_sd * randn) + Threshold_mean;

Memory_storage_pointer = 0;
Memory_retrieval_pointer = 1;
Memory_sample = ceil(rand*119);

%-----------------------------------------------------------
%FI_procedure

%Input: Response
%Output: Food

if Cycle_clock >= FI*Seconds,
Prime = 1; end                  %Set fixed-interval

if (Prime == 1 & Response == 1)
  Food = 1;
  Cycle_count = Cycle_count + 1;

  Data(Data_pointer,1) =
  round(Session_clock);         %Record time
  Data(Data_pointer,2) =
  Food_code;                    %Record food
  Data_pointer = Data_pointer + 1;
  Prime = 0;                    %Reinitialize prime
end

if (Response),
  Data(Data_pointer,1) =
  round(Session_clock);         %Record time
  Data(Data_pointer,2) =
  Response_code;                %Record response
  Data_pointer = Data_pointer + 1;
  Response = 0;                 %Reinitialize response
end

Cycle_clock = Cycle_clock + 1;

%-----------------------------------------------------------
%Scalar_timing_theory.m

if Food == 1;
  %Memory storage constant to transform perception to
  memory
       Memory_storage_pointer = Memory_storage_pointer + 1;

       k = (k_sd * randn) + k_mean;
       Memory(Memory_storage_pointer) = k * Accumulator;
```

```
   %Get new sample of clock speed
      Speed = (Speed_sd * randn) + Speed_mean;
      Accumulator = Speed * Cycle_clock;

   %Get new sample from memory
      Memory_retrieval_pointer = ceil(rand * Memory_
      storage_pointer);        %Round up to nearest integer
      Memory_sample = Memory(Memory_retrieval_pointer);

   %Get new sample of threshold
   Threshold = (Threshold_sd * randn) + Threshold_mean;

   Food = 0; Cycle_clock = 0;
end

Accumulator = Speed * Cycle_clock;
Discrepancy = abs(Memory_sample - Accumulator) ./ Memory_
sample;
if Discrepancy < Threshold, Response = 1; end

if(rand<.01), Response = 1; end%Mean operant rate of 1
response in 100 s

%----------------------------------------------------------
%Record_data.m

Output_file = [Procedure '_' Model
'_' num2str(FI) '.sim'];       %Name the file
save(Output_file, 'Data', '-ascii',
'-double', '-tabs');           %Save it on disk
['The Simulation of ' Model ' on '
Procedure ' is Done.']         %Report completion
```

APPENDIX B: SIMULATED DATA

Time	Event
10002	19
10002	1
10003	1
10025	1
10029	1
10030	1
10031	1
10032	19
10032	1
10033	1
10046	1
10047	1
10048	1
10049	1
10050	1
10051	1
10052	1
10053	1
10054	1
10055	1
10056	1
10057	1
10058	1
10059	1
10060	1
10061	1
10062	19
10062	1
10063	1
10070	1
10071	1
10081	1
10082	1
10083	1
10084	1
10085	1
10086	1
10087	1
10088	1
10089	1
10090	1
10091	1
10092	19

2 General Learning Models: Timing without a Clock

John W. Hopson

CONTENTS

2.1 Introduction 24
2.2 Background 25
    2.2.1 Clocks 25
    2.2.2 Previous Work 29
        2.2.2.1 Connectionist Models 29
        2.2.2.2 Learning Models 31
2.3 Proposal 32
    2.3.1 Principles 32
        2.3.1.1 An Adapted General Learning Model 32
        2.3.1.2 Minimal Changes from the General 32
        2.3.1.3 Maintain General Learning Performance While Also Allowing Timing 32
        2.3.1.4 Robust Parameters 33
    2.3.2 Architecture 33
    2.3.3 Learning Algorithm 35
    2.3.4 Parameters 37
2.4 Evidence 39
    2.4.1 Basic Timing Phenomena 39
        2.4.1.1 Fixed-Interval Timing 39
        2.4.1.2 Peak-Interval Timing 41
    2.4.2 Learning Phenomena 43
        2.4.2.1 Learning the Time Marker 43
        2.4.2.2 Parsimonious Timing 43
        2.4.2.3 Initial Fixed-Interval Behavior 44
    2.4.3 Complex Timing Phenomena 45
        2.4.3.1 Gap Timing 45
        2.4.3.2 Filled-Gap Timing 49
        2.4.3.3 Scalar Timing 51

		2.4.3.4	Possible Techniques for Scalar Timing	53
		2.4.3.5	Noise and Variation	54
2.5	Discussion			55
	2.5.1	Overall Performance		55
	2.5.2	Difficulties		55
	2.5.3	Future Research		55
	2.5.4	Theoretical Implications		57
References				58

2.1 INTRODUCTION

The senses available to animals tend to be divided into at least two distinct portions. The eye is distinct from the visual cortex, the ear from the auditory cortex. It is possible to study the properties of the eye without reference to the visual areas of the brain. The study of vision as a whole would of course involve both, but the source of visual information is a distinct unit, isolated from the rest of the visual system.

The study of interval timing, how animals learn and remember short periods of time, offers many unique challenges, not least of which is the fact that there is no readily apparent sensory organ for the passage of time. We cannot dissect the "eye" of time to determine its properties and trace its connections to the rest of the mind. Rather, we are forced to infer its existence, properties, and functioning from behavior, and to theorize and test without any certain knowledge of the source of temporal information.

Several types of potential sources of temporal information have been proposed and tested with varying degrees of success. Another common metaphor for these sources is a clock or stopwatch, a mechanism that changes predictably over the course of the interval and can be started, stopped, and reset as necessary. This metaphor is quite natural, since a stopwatch is a mechanical solution to the same problem as our biological sense of time.

The two metaphors of a physical sense and a stopwatch both include the assumption that there is, in addition to the clock, a separate general learning mechanism. The eye itself does not learn to recognize faces; the stopwatch does not learn to turn itself on and off. This learning mechanism has historically received much less attention than the clock mechanism. For example, most timing models presuppose that the animal already knows the signal to start timing and that the problem being presented is exclusively one of timing.

This work is an attempt to cast new light on the field of animal timing by approaching the problem from the opposite direction. If a general learning mechanism is necessary for a complete explanation of animal timing, is it possible that it is sufficient as well? Theoretically, a general learning model should be able to learn patterns of responding in time if it is presented with stimuli that hold different values over the course of the interval. These timing stimuli have previously been produced by a special-purpose clock process, but what if the learning model could find such stimuli within itself? A dynamic model changes over time by definition, raising the possibility that these changes can be used as an internal clock.

General Learning Models: Timing without a Clock

The primary issue addressed by this work is whether a general learning model can fulfill the functions of a clock. If so, it opens the door to some remarkable questions about our understanding of interval timing. Is a separate special-purpose timing model necessary at all? If a general learning mechanism must exist and can be shown to be able to produce interval timing behavior, there would be no need for a separate interval timing clock to evolve. If a general learning model cannot perform any of the functions of a clock, can that failure tell us which aspects of timing are necessarily properties of the clock? If any learning mechanism can time, does that mean all learning mechanisms potentially play a role in interval timing? If successful, this project presents a serious challenge to our current theories of interval timing.

2.2 BACKGROUND

There are a wide variety of models of timing, with different clocks and learning mechanisms, that can serve the current project as criteria for success. By understanding how successful the existing timing models have been in accounting for the experimental data, we can understand how the new model compares to them. Places where they succeed and the new model fails can be taken as evidence that those phenomena might require a special-purpose clock.

These existing models also served as sources of inspiration for the new model. The fact that timing can happen in so many ways can be seen as evidence that it is a more general phenomenon than previously thought. If timing can happen in so many ways, it seems likely that most general learning mechanisms can potentially find a way to keep track of intervals.

2.2.1 Clocks

As a class, interval clocks can be defined as processes that change monotonically as a function of the passage of time and that can act as discriminative stimuli for the animal. In most of the models described below, the clock is entirely internal, receiving no input from the outside world. In fact, information from the outside world is specifically considered to be a contaminant to the pure timing function, in spite of the fact that humans and other animals will use such information at any opportunity.

State-based clocks are perhaps the type of clock closest to standard mechanical clocks. The clock steps through a fixed series of discrete states, with the current active state serving as an indicator of how much time has passed since the clock started. One metaphor for this state-based timing is a counter moving along a chessboard. If the counter always starts from the same place and moves at a consistent rate, the counter's position can be used as a measure of time.

The best-known exemplar of a state-based clock is the behavioral theory of timing (BET) (Killeen and Fetterman, 1988). Explicitly proposed as an alternative to the more cognitive model of scalar expectancy theory (SET) (Gibbon, 1977; Gibbon and Church, 1990), BET hypothesizes that animals keep track of time during an interval by going through a series of discrete behavioral states. For example, a rat might touch its nose, run to the back corner of the cage, stand up, and turn around, and the right amount of time would have elapsed to press the lever and receive a reinforcer. While this example uses large motions, it is perfectly possible that these behavioral states are such that they are not visible to the outside observer (e.g., tense the left calf muscle, tense the right, clamp jaw muscles, etc.). Studies involving observing animals in order to map out these behavioral states have achieved only indifferent success (Lejeune et al., 1998), finding consistent probabilistic patterns but not the consistent deterministic patterns of behavior required for timing under BeT. These experiments are also excruciatingly time-intensive for the experimenters, or more truthfully, for their students and lab assistants.

Two interesting variants of the state-based models are Armando Machado's revised connectionist variant of BET (Machado, 1997) and the diffusion timing models (Higa et al., 1991). The Machado model puts BET on a sounder mathematical footing and extends the model into a wider range of experimental results. It is discussed extensively in the next section.

The diffusion models produced by Staddon and his associates are an extension of sensory generalization models (e.g., Guttman and Kalish, 1956). The animal progresses along a fixed series of states over the course of the interval, just as in other state-based timing models. When the animal is reinforced, the current state gains a quantity of activation that then diffuses along the series of states. This diffused activation causes the model to increase its responding as it nears the reinforcement interval, as seen in animal data. In this model, early or late responding is due to temporal generalization rather than error. These models offer detailed predictions of transient timing phenomena, including learning effects and changes in schedules.

It should be noted that all of these state models include the hidden assumption of a process that bumps the model to the next state at a consistent rate. In a state-based model, the current time is represented by which state the model is in at the moment. For that internal representation to change, the model must move to the next state. This process might be considered similar to the pacemaker process discussed below. In some state-based models, the rate at which the model advances to the next state is a result of the time between reinforcements and is therefore responsible for the scalar property (Killeen and Fetterman, 1988). Experiments to test this assumption have been done, focusing on varying the time between reinforcements without changing the interval to be timed (Bizo and White, 1995). Changes in the interreinforcement interval were found to have no effect on interval timing, and this is generally taken as evidence against state-based timing models. Of course, it may be that the mechanism for adjusting the pace of changing between states is simply more complex than described here and is not fooled by this methodology.

These state-based clocks are perhaps the least likely to be an accurate description of the biological basis of interval timing. These models require a very complex, preexisting special-purpose architecture, and like all previous timing models, they ignore the issue of how the animal learns the relationship between the time marker stimulus and the series of behavioral states.

The connectionist SET model of Church and Broadbent (1990) can be considered a type of complex state-based model, but because it is based on a system of neural oscillators, it will be covered in a later section.

Trace-based clocks are the equivalent of an hourglass, with time measured as a continuous smooth change in a single value. This is most commonly instantiated as a decay process; the time marker causes the value to jump up and then it decays over time, and the model learns to associate a certain level of the stimulus with the reward. A physical metaphor for this process is a leaky bucket. Each time the time marker is presented, the bucket is filled to a preset level and the water begins to leak out. If one notes the level of water in the bucket when the reward is presented, it provides a measure of how much time has passed between the timing signal and the reward. These buckets are sometimes referred to as leaky integrators.

This single-bucket model has been implemented in several versions (e.g., Schmajuk, 1999; and a single-integrator variant of the multiple-timescale model (MTS) presented by Higa and Staddon, 1999b), but a single integrator has several inherent problems. Most importantly, it does not scale properly. At longer durations, the change in the level of the trace becomes extremely small and difficult to measure. This means that a single-trace model would not be able to explain how animals time very long intervals (Gallistel, 1999).

Secondly, there is the issue of the relationship between the strength of the time marker and the strength of the trace. If the animal was trained to start timing when it heard a 30-dB beep, and was later tested with a 60-dB beep, one could reasonably suppose the resulting timing trace would be more intense and thus would take longer to decay to the previously learned level, as shown in Figure 2.1. Experimental evidence suggests that the opposite is true, that increasing the intensity of the signal causes the animals to time faster. In Leonard and Monteau (1971) subjects were presented with a tone, and then a second later, a puff of air was directed at the subjects' eyes. The subjects learned to blink at the correct interval after the tone in

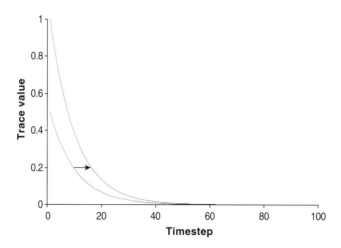

FIGURE 2.1 Activity traces as a function of response strength. If subjects were trained on an interval schedule at one signal strength (the lower curve) and then tested with a more intense signal (the upper curve), a simple trace model of timing would predict that it would take longer for the activity trace to decay to the previously learned level, causing the subjects to respond too late.

order to shield their eyes from the puff of air. If trained with one tone intensity and then tested with a louder tone, subjects blinked sooner than they did normally. This can be taken as evidence that the traces are built up by the timing stimulus over the course of the interval, rather than presented in a single lump and then decaying over the course of the interval. This is a strong point in favor of neural network models of timing, since that is the natural dynamic of neural network models.

These troubles are solved by a more sophisticated type of trace model incorporating multiple traces, such as the spectral timing model (STM) (Grossberg and Schmajuk, 1989) and the multiple-timescale model (Higa and Staddon, 1999b). Each trace decays at its own rate and the model learns to associate a pattern of traces with reinforcement. Because different parts of the model time at different rates, the model never has to adjust its basic functioning to match the timescale of the current task. The portion of the model best suited to the current timescale becomes the primary clock, without any explicit change of scale.

However, the above solution creates a new concern, one more of elegance and parsimony than of functionality. In both STM and MTS, there is an array of integrators with different decay rates. The range of timescale invariance shown by these two models is a function of the range of these decay rates, which are preset by parameter before the model is run. If the model is to be able to learn to time 4-h intervals, it must already have an integrator with an appropriate decay rate ready to go from the beginning. This is usually gotten around by having a large array of these integrators and a correspondingly wide range of decay rates. This is not necessarily a bad thing, but it requires a great deal of redundancy, and in some models (such as the spectral timing model), these rate parameters have to be preset to a very precise set of values, making the model less robust.

At first glance, trace models would seem to automatically solve the assignment-of-credit problem. Whenever any stimulus is presented, the stimulus creates a memory trace that then decays. Only the correct time marker stimulus will consistently decay to the same point in each trial, and therefore it will be the stimulus most strongly associated with reward. However, there is also the issue of *when* the traces decay. Water running out of a leaky bucket is a constant process, limited only by the absence of water in the bucket. It cannot leak out faster or slower or hold. A study by Roberts (1981) showed that animals could pause their timing during the interval and then resume *without any decay* if that was what was required by the task. A trace model could be envisioned with an additional learning process that acts as the equivalent of a thumb over the hole in the bucket, but that thumb would then need to solve the assignment-of-credit problem in order to be able to perform its task. Trace models are particularly suited to neural network modeling and thus can be considered to be the most biologically plausible of the various breeds of timing models.

Pacemaker–accumulator systems lie somewhere in between trace- and state-based timing models. Like a trace-based system, they have an accumulator whose contents are a measure of time. However, a pacemaker adds time to this accumulator in discrete pulses, in a fashion reminiscent of a state-based model. A common metaphor for this type of clock is a leaky faucet dripping into a bucket. If the bucket starts out empty and the faucet drips at a consistent rate, the level of water in the bucket can serve as a measure of time.

The most popular pacemaker–accumulator model is the scalar expectancy theory (Gibbon, 1977; Gibbon et al., 1984; also see Treisman, 1963). In this model, the timing signal closes a switch that allows pulses from the pacemaker to begin building up in the accumulator. When the model is reinforced, the level in the accumulator is recorded. In subsequent trials, the model responds whenever the current level in the accumulator approximates the stored value.

SET has proven to be an enduring and flexible model of interval timing, due to its information-processing approach. By separating the timing mechanism into discrete subunits with clearly defined roles and properties, SET provides a theoretical framework for reducing timing behavior to manageable components (see Church, 1999; this volume; Church et al., 1991; Gibbon, 1992; Gibbon and Church, 1984).

2.2.2 PREVIOUS WORK

2.2.2.1 Connectionist Models

The current project has two sets of direct predecessors. First, there are previous neural network timing models, which offer insights into possible neural timing mechanisms, and the concept that timing and learning can both be accomplished within a single substrate. These models typically use neural nodes linked together in complex and rather specialized ways, a theoretical entanglement that the current project will hopefully avoid.

The spectral timing model of Grossberg and Schmajuk (1989) is certainly the closest model to the current project. It consists of a four-layer neural network in which there is a single input node, a single output node, and two middle layers that produce a predesigned cascade of timing signals.

This cascade is a series of approximately normal curves, each peaking at a different point during the interval. These curves are distributed logarithmically to provide equally accurate timing all along the interval. The model learns to associate the proper curves in the cascade with the reward.

STM's primary strength lies in the use of multiple independent timing traces. This model's design avoids the problems of a single trace described earlier, the inherent scale of a single trace and the relationship between signal strength and timing. The concept of multiple parallel timing traces means the system gains a measure of robustness, because any individual trace could fail or be damaged while the timing system as a whole continues to operate.

The spectral timing model's weaknesses lie in two closely related problems. Most importantly, STM depends on the middle layers having precisely distributed parameter values. Each of the hidden nodes is created with a unique fixed parameter that determines the scale at which the node works. The model is never learning anything about scale; the timing traces exist at their preset scales regardless of the model's experiences and training.

Secondly, in STM learning only occurs in the associational weights between the timing traces and the output node. The model does not learn the association between the input node and the cascade of timing traces; that is also preset. Essentially, the input layer and the cascade serve as a fixed clock, without learning or adjusting to current circumstances. In order for the cascade system to work, there can be only one timing signal that triggers the cascade. In order to time two independent signals, the entire system has to be duplicated.

The connectionist set model (Church and Broadbent, 1990) offers a very different use of neural mechanisms for timing. In this model, the clock is represented as a series of oscillators that alternate back and forth between −1 and 1. These oscillators range from very fast (changing every few milliseconds) to very slow (days, weeks, or even years), allowing the model to work at a wide range of timescales. These oscillators are reset at the onset of each trial, and therefore the total state of all the oscillators serves as a measure of the interval since the trial started.

One of the primary strengths of this model is the fact that oscillators do exist in the animal body. Heartbeats, breathing rhythms, circadian cycles — there are all sorts of rhythms within the body that can be used to help make timing more accurate. Other timing models try to ignore these cycles, considering them contaminants to the pure timing function. Timing experiments do their best to make these cycles less salient. The connectionist SET model embraces these cycles and incorporates them into the timing function. Supporting this idea are experiments that suggest that timing may advance in a sinusoidal fashion, rather than a linear one (Crystal et al., 1997).

While these oscillators are very neural, the learning and memory systems of this model are not. The state of the array of oscillators is stored, retrieved, and acted upon in representational rather than associational manner. This is a product of the original SET's information-processing approach to timing, and the alternative has not been explored. This is not to say that neural learning and memory systems could not be used with an oscillator-based clock. One could easily imagine a three-layer neural network performing all the functions of the Church and Broadbent model. An input layer triggers the oscillator layer, and a series of weights between the oscillators and the output layer learns to associate the appropriate oscillators with reward.

To the best of my knowledge, this modified oscillator model has not been proposed before. This approach was not used in the current project because it requires the oscillators to be special-purpose timing neurons that function according to different rules than the rest of the network. This is both less elegant and a digression from the central concept of this project, that of using the most general model possible.

Another closely related model is Armando Machado's connectionist behavioral theory of timing (CBeT) (Machado, 1999). This model is made up of a series of nodes linked in a one-directional fashion. Activation is introduced into the first node by the timing signal and spreads along the series over the course of the interval. Each node has a degree of association with reward, and the model's output is the sum of each node's activation multiplied by its association. At reinforcement, the nodes' associations are strengthened proportionally to their activation, such that the most active nodes gain the most association with reward. CBeT can time because the activation always diffuses across the simulation at the same rate; thus the same nodes tend to be active at the same points in each trial. The nodes most associated with the reward are those that tend to be active at the reinforcement interval; therefore, the model's output is greatest at that point. The distinctive bell-shaped response curve is a natural product of the diffusion process. A similar process is responsible for the success of the diffusion-based timing models (e.g., Higa and Staddon, 1997).

The CBeT model solves many of the fundamental objections to BeT. Its firm mathematical basis allows it to make quantitative predictions that match the existing animal data very well. Unlike the original BeT model, this model does not need a separate process that bumps it from state to state. The states (nodes) are not exclusive, and activation constantly diffuses across them. This also explains the mixed results of attempts to observe a consistent series of behavioral states, as described above in the analysis of BeT. The fact that we have never been able to observe a consistent monotonic chain of behaviors leading up to reinforcement is not a problem for this version of BeT. The current time is encoded by the relative activation of many behavioral states, not just one. Which particular behavior is exhibited at any given time can be probabilistic without affecting the model's sense of where it is in the interval. This is a step away from the strict behaviorist position that informed the creation of BeT, but not a particularly controversial one.

The primary flaw with CBeT is that there is no good way to solve the assignment-of-credit problem. The system really only works if activation is presented only to the first node and then proceeds uninterrupted across the series. There is no direct connection between the presentation and the reinforcement, so the learning algorithm required to assign credit to the proper timing signal would need to be quite complex. Does every signal presented to the model diffuse across the nodes independently? Does every signal have its own series of nodes? This flaw is closely related to the model's difficulty in timing multiple sources independently. For multiple signals to be timed, there must be either multiple series of states or multiple "flavors" of activation spreading across the same series. Either way, it requires that the timing system be duplicated as a whole. This duplication seems unlikely in the face of research indicating that all signals are timed to some degree. Davis et al. (1989) showed evidence that there was temporal learning on the very first pairing of a light and a shock. For that to be true, the subjects must have started timing the light before they knew that any consequence would be associated with it.

2.2.2.2 Learning Models

There are existing attempts to model timing as a special case of a more general learning system. Arcediano and Miller (2002) attempted to connect classical conditioning theory to interval timing and made a strong case for the idea that timing is a fundamental part of all conditioning, not just the more complex timing procedures. They argue that for any sort of conditioning to occur, the animal must have a sense of the temporal relationships between stimulus and stimulus or stimulus and response. While persuasive, they do not provide a coherent mathematical model for their ideas, and so their ideas are not readily testable in a quantitative fashion. Balsam et al. (2002) support this idea and provide evidence that animals' knowledge about temporal properties exists well before it is demonstrated in their behavior.

Dragoi et al. (in press) propose a theory of timing that is both clockless and mathematical and takes a unique approach to modeling timing. Like the current project, their model is adapted from a general learning model (e.g., Dragoi and Staddon, 1999). Rather than start with a clock or even a particular mechanism such as neural networks, their theory rests on the relationship between two fundamental principles that would be true in a wide variety of learning mechanisms and shows that any model that reflects those principles will be able to demonstrate basic timing phenomena.

The first principle is that responses compete, with the strongest response being the only one emitted by the animal. Second, the competitive strength of a response is a function of its recent reinforcement history. In this model, the common peak-interval timing function is not produced by anticipation of the reinforcement, but by the fact that the operant response becomes more successful in competition over the course of the trial. Unfortunately, this model is somewhat undercut by the fact that it adjusts its parameters as a direct function of the time between reinforcements. This adjustment is necessary for the model to scale, but it is a significant blemish when compared to the elegance of the rest of the model. Significantly, the model does not include any stimulus learning. It does solve one aspect of the assignment-of-credit problem in that it can determine which of its responses is responsible for the rewards it receives. However, it cannot use any time marker but the previous reinforcement, and this prevents it from modeling a number of important interval timing procedures. The model also requires many versions of itself operating in parallel in order to produce animal-like response functions.

2.3 PROPOSAL

2.3.1 PRINCIPLES

The following principles define the goals and design parameters for this project.

2.3.1.1 An Adapted General Learning Model

The starting point for the model should be a well-known general learning model. Instead of first creating a clock and allowing it to dictate the shape of the rest of the model, this project will start with a preexisting model of animal learning.

2.3.1.2 Minimal Changes from the General

If changes are necessary for the model to show timing behavior, they should be as small as possible. The model should remain a learning model used for timing rather than a clock model that learns.

2.3.1.3 Maintain General Learning Performance While Also Allowing Timing

The general learning aspects of the model must be preserved through any changes that are required. If these are compromised, it obviates the entire point of the project.

2.3.1.4 Robust Parameters

The model should be robust, functioning well under a wide range of parameter values. While it is inevitable that some particular set of values will generate the best performance, the model should still produce basic timing behavior within a reasonable range of values.

2.3.2 ARCHITECTURE

For my starting point I choose to use the most common general learning model, a three-layer neural network trained by backpropagation. The choice of using a neural network model was made because of its ubiquity, not because of any particular property of neural networks or of backpropagation. The purpose of this work is to discover if it is possible for timing to occur in a general learning model, and a neural network trained by backpropagation is the single most common exemplar of that category.

As to whether this project could be repeated using other learning mechanisms, there is no *a priori* reason why another learning mechanism such as a reinforcement learning model (Sutton and Barto, 1998) could not also learn to time. Some learning models may do better or worse at timing, but I suspect (but do not assert) that any sufficiently general model could produce the basic results. The key properties necessary to timing are dynamic changes within the model over the course of the interval and the ability to perceive and respond to those changes.

Neural networks are made up of extremely stylized neurons that are referred to as nodes or units. Each node receives inputs, either from other nodes or the outside world, and those inputs contribute to the node's level of activation. The node then passes on some portion of that activation to the nodes it is connected to. The effect of one node upon another is regulated by connection weights. It is these weights that are adjusted as the network learns, eventually reaching a configuration that allows the network to respond to each stimulus with an appropriate output.

Two primary modifications were required to allow the model to time in a fashion similar to that of existing timing models. First, the level of activation in the nodes is carried over from time step to time step. It is common in neural network models for each unit's new state to be purely a function of its inputs, without reference to its current state. For timing to be possible, activation needs to build over time. This is not an uncommon change, but it is a step away from the prototypical artificial neural network. A similar modification was used in the spectral timing model and in many other mainstream neural network models.

This modification is similar to a type of neural network called a recurrent neural network, as shown in Figure 2.2. In these networks, the hidden layer is fed back into itself, its state from the previous time step acting as part of the input layer for the next. This grants the network the ability to respond to its own internal state.

Second, on each time step the typical neural network node is on or off, firing or not firing. Its output is filtered through a sigmoid function of varying steepness. In this model, the neuron's output is directly proportional to its activation, which

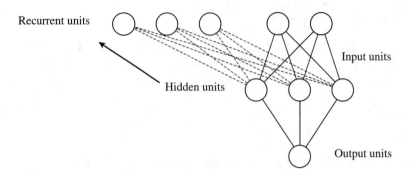

FIGURE 2.2 A recurrent neural network. In a recurrent neural network, the previous activity levels of the hidden nodes are used as inputs. This allows the network to use its own internal state as a stimulus and can serve as a form of short-term memory.

provides the next layer with much more finely grained information. Again, an identical modification was used in the spectral timing model.

The general timing model consists of three layers: an input layer, a hidden layer, and an output layer. Each layer consists of one or more nodes, and each node is connected to every node in the next layer. The layout of the network and sample activation curves for the nodes in each layer over the course of an interval are shown in Figure 2.3. Nodes in the input layer are each directly connected to an outside stimulus, the node's activation being 1 when that stimulus is present and 0 when it is not, as illustrated in Figure 2.4, Equation 1. In a typical simulation, there were two nodes in the input layer, one of which represented the timing signal and was 1 during the interval and 0 otherwise. The other represented background stimuli and was always 1.

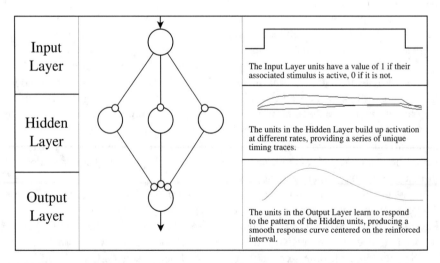

FIGURE 2.3 Diagram of the general timing model. The central panel shows the architecture of a simple version of the general timing model, and each layer of the network is labeled to the left. To the right are sample activation curves for the nodes in each layer, with brief explanations.

General Learning Models: Timing without a Clock

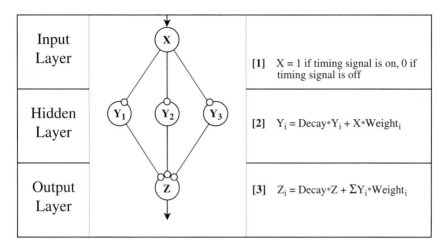

FIGURE 2.4 Activity diagram of the general timing model. This diagram provides a concise summary of the architecture and functioning of the model. Each of the three equations is evaluated at each time step to provide the level of activity for that node for that time step. The decay constant was 0.935 in all simulations.

Nodes in the hidden layer receive activation from the input layer filtered by the weights of the connections between the two layers, as illustrated in Figure 2.4, Equation 2. This causes these nodes to build up activation over the course of the interval, providing a simple trace clock. Because each node starts with a random set of weights, each node builds up activation in a slightly different fashion. The combination of these different activation curves acts as a complex trace clock.

The output layer receives activation from the hidden layer nodes, and the activation level of the output layer is the overall response strength of the model. The weights between the hidden and output layers control what effect the different nodes have on the response strength, as illustrated in Figure 2.4, Equation 3. For example, a hidden layer node that builds up quickly might have a strong negative impact on responding, because responding early in the interval is unlikely to provide food. Similarly, a hidden node that builds up very slowly might be most active after the reinforcement interval has passed and would therefore act to suppress responding.

The level of activation of the output layer can be taken as the general probability of response. It is certainly possible that there is some minimum threshold below which there is no probability of response. Such a threshold was not used in this model because it was not necessary to observe the basic timing phenomena. If a threshold were used, it would produce the break-run-break pattern of responding found in some timing procedures (e.g., Cheng et al., 1993; Church et al., 1994).

2.3.3 Learning Algorithm

The learning algorithm is a slightly modified version of the common backpropagation algorithm. This algorithm works by assessing the amount of error in the model's response and then partitioning the error to determine which weights and nodes need to be adjusted to minimize the error. A more technical review of the

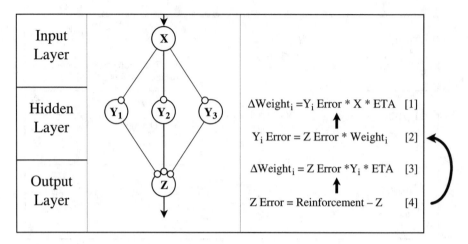

FIGURE 2.5 Learning diagram of the general timing model. The error for the output node is used to adjust the weights between the hidden and output layers. The error is also divided among the hidden layer nodes and used to adjust the weights between the input and hidden layers. ETA is a learning constant.

algorithm and all of its variants can be found in Haykin (1994). The modified backpropagation algorithm takes place in four steps. First, the amount of error of the output node must be calculated as illustrated in Figure 2.5, Equation 4. This error represents the difference between the amount of reinforcement the model received and the amount it expected to receive, which is represented by the current activation of the output node.

This output error is then used to adjust the weights between the hidden and output layers, as illustrated in Figure 2.5, Equation 3. If the model has overpredicted how much reward it would receive, the weights for hidden nodes that stimulated the output node will be decreased and the weights for nodes that suppressed the output node will be increased. If the model has underpredicted reward, the opposite will happen. Over time, these weights become relatively fixed as the model becomes better at predicting reinforcement.

After that, each of the hidden nodes will be apportioned part of the error from the output node, as illustrated in Figure 2.5, Equation 2. These values represent how much each of the hidden nodes contributed to the overall over- or underprediction of the network. They are then used to adjust the weights between the input and hidden layers, as illustrated in Figure 2.5, Equation 1. These weights control how fast activation builds up in the hidden layer. If the hidden layer nodes are underpredicting reward, adjusting these weights upward causes the model's hidden nodes to gain activation faster and thus reach a higher prediction of reward sooner. The difference between this and the standard backpropagation algorithm is that the standard algorithm includes an extra term in the error calculations designed to push the model toward weights of 1 and 0. This was removed because this model is intended to produce a smooth response curve, rather than discrete, all-or-nothing responses.

General Learning Models: Timing without a Clock

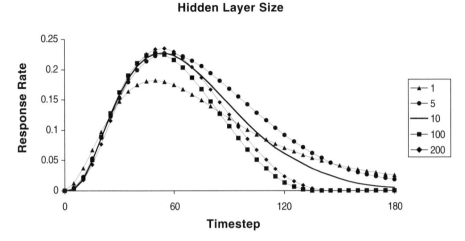

FIGURE 2.6 Hidden layer size. This figure demonstrates the effects of changing the size of the hidden layer. Each line represents the average of five independent peak procedure simulations. The heavy black line represents the standard value for this parameter in all other simulations in this work. In general, increasing the size of the hidden layer improves the discrimination of the timing function.

2.3.4 PARAMETERS

Like any mathematical model, the general timing model includes a number of parameters representing assumptions about the way the model should behave. The parameter values described here were used in all simulations in this work except where specifically stated otherwise.

First, there is the number of nodes in each of the three layers. The model uses two input nodes, ten hidden nodes, and one output node. Of the two input nodes, the first represents the time marker stimulus, going on at the beginning of the interval and off at the end. The second is always on and represents background stimuli in the model's environment. While the input and output layer sizes are determined by the environment and the task, the hidden layer could be any size. The basic peak-interval timing function using various numbers of hidden nodes is illustrated in Figure 2.6.

Second, there are parameters governing the initial values of the weights of the network. The weights between the input and hidden layers began with values randomly distributed between 0.03 and –0.02. The weights between the hidden and output layers began with values randomly distributed between 0.06 and –0.04. These values were then modified by the learning algorithm over the course of the simulation. The initial values were slightly skewed toward the positive in order to create some amount of initial responding that could then be modified by the learning algorithm. How changes in these initial weight ranges can affect interval timing functions is shown is Figures 2.7 and 2.8. Each node in the hidden and output layers also has a decay parameter that controls the rate at which it loses activation. This was set at 0.935 in all simulations described in this work. How altering this value affects the shape of the interval timing functions is shown in Figure 2.9.

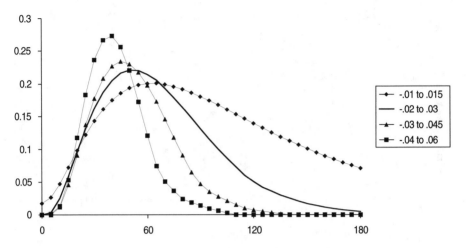

FIGURE 2.7 Initial weights, hidden layer. This figure demonstrates the effects of changing the range of the initial weights between the input and hidden layers. Each line represents the average of five independent peak procedure simulations. The heavy black line represents the standard value for this parameter in all other simulations in this work. In general, increasing the range of initial weights makes the timing function narrower and peak earlier.

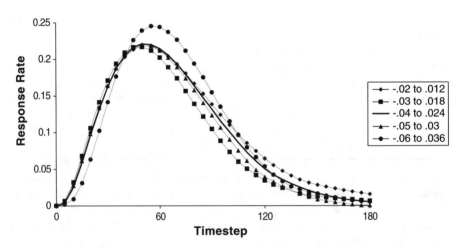

FIGURE 2.8 Initial weights, output layer. This figure demonstrates the effects of changing the range of the initial weights between the hidden and output layers. Each line represents the average of five independent peak procedure simulations. The heavy black line represents the standard value for this parameter in all other simulations in this work. In general, increasing the range of these initial weights does not significantly change the shape of the timing function.

General Learning Models: Timing without a Clock

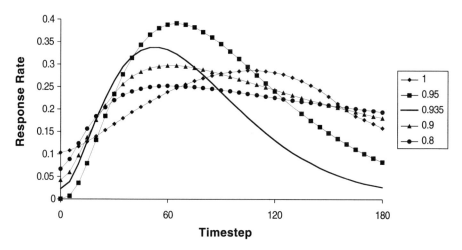

FIGURE 2.9 Decay rates. This figure demonstrates the effects of changing the decay constant for the hidden and output layers. A decay constant of 0.8 means each node loses 20% of its activation each time step; a decay constant of 1 means no loss. Each line represents the average of five independent peak procedure simulations. The heavy black line represents the standard value for this parameter in all other simulations in this work.

Finally, there is a learning parameter that governs how much change can occur in the model's weights each time step. This constant is used in Equations 1 and 3 of Figure 2.5. On nonfood trials the learning constant was 0.01, and on food trials it was 0.1. How various values of this constant affect the interval timing functions is shown in Figures 2.10 and 2.11.

2.4 EVIDENCE

2.4.1 Basic Timing Phenomena

2.4.1.1 Fixed-Interval Timing

The fixed-interval (FI) schedule is perhaps the simplest demonstration of animal timing and was first described in Skinner (1938). The subject is rewarded for the first response that occurs after a fixed period of time has elapsed. Responses before the interval has elapsed have no effect, and once a reinforcement has been given, a new interval begins. This procedure typically produces what is known as the FI scallop, a pattern of responding characterized by an initial slow rate of response that gradually increases over the interval to reach its maximum rate at approximately the end of the interval. This pattern is essentially universal, being found in humans, rats, pigeons, fish (Talton et al., 1999), and many other species. This pattern of response is even seen at very long intervals and has been demonstrated in intervals up to 36 h long (Eckerman, 1999). Over extensive training, the scallop can become increasingly angular, with subjects responding at a very low rate during the first part of the interval and switching to a steady, high rate of response in the latter part of the interval (Schneider, 1969). The performance of the model under an FI schedule is

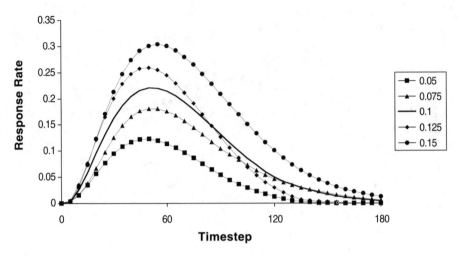

FIGURE 2.10 Learning constant during reinforcement. This figure demonstrates the effects of changing the learning constant that controls the magnitude of the changes in the weights during time steps where the model is being reinforced. Each line represents the average of five independent peak procedure simulations. The heavy black line represents the standard value for this parameter in all other simulations in this work.

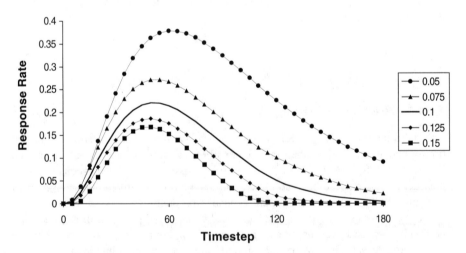

FIGURE 2.11 Learning constant during nonreinforcement. This figure demonstrates the effects of changing the learning constant that controls the magnitude of the changes in the weights during time steps where the model is not being reinforced. Each line represents the average of five independent peak procedure simulations. The heavy black line represents the standard value for this parameter in all other simulations in this work.

General Learning Models: Timing without a Clock

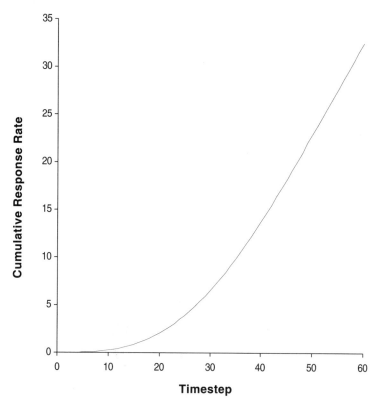

FIGURE 2.12 Fixed-interval simulation, cumulative record. This graph represents the total amount of activity of the output node over the course of the FI. The scallop shape of the function is similar to that found in animals, with a very low level of activity early in the interval and building up to a high, steady rate late in the interval.

shown in Figure 2.12. The scallop is distinct, showing the early pausing typical of FI performance and the steadily increasing rate of response as the interval progresses. The response rate over the course of the interval is plotted in Figure 2.13.

2.4.1.2 Peak-Interval Timing

While the FI schedule is a good simple measure of timing, it only lets us see the subject's behavior leading up to the interval. Once the subject has been reinforced, its internal clock presumably resets, so we cannot see what it would have done after the point where it is normally reinforced. The peak-interval (PI) procedure developed by Catania (1970) and further explored by Roberts (1981) solves this problem as follows. The start of each trial is signaled by the presentation of an explicit timing signal such as a light or a tone, which stays on for the length of the trial. The first response after the interval produces food and the timing signal goes off. There is then a brief intertrial interval (ITI) before the next trial begins. These are referred to as food trials.

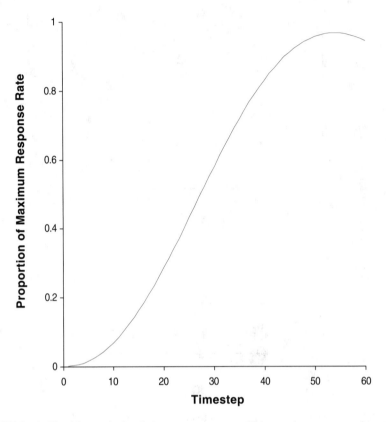

FIGURE 2.13 Fixed-interval simulation, response rate. This graph represents the current amount of activity of the output node over the course of the FI. The shape of the function is similar to that found in animals, with a very low level of activity early in the interval and building up to a high, steady rate late in the interval.

On some portion of the trials, typically 50%, the timing signal stays on for three times the normal interval length and ends without any reinforcement being presented. These are referred to as empty, peak, or probe trials. Because the timing signal stays on and there is no food presented, the animal's clock presumably continues timing and we see how they respond after the time interval has elapsed.

The central finding of the PI procedure is the peak function, a simulation of which is shown in Figure 2.14. As in the FI procedure, responding starts slowly and increases to its maximum rate at approximately the reinforced interval length. However, in the probe trials of the PI procedure we can see that responding then decreases in an almost symmetrical fashion, perhaps just a trifle slower than the rate of increase. The resulting curve is approximately Gaussian with some degree of positive skew.

This clean curve is typically found only after many, many weeks of training and represents averaged data, not individual trials. Typically, on the very first probe trial, the subject's response rate stays high for a long period of time (although it is not clear how general this tendency is). It is only after repeated experiences with

General Learning Models: Timing without a Clock

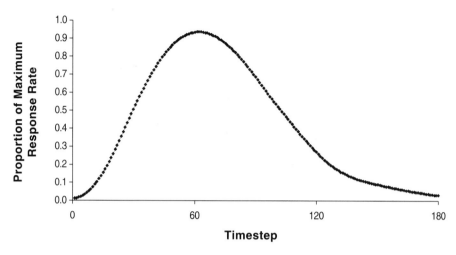

FIGURE 2.14 Peak-interval procedure simulation, response rate. This figure shows the average response rate over the course of a probe trial in the PI procedure. The shape of the function is similar to that found in animals, with an initial slow rate of response building up to a peak at approximately the previously reinforced interval and then falling away.

nonreinforced trials that the portion of the curve after the reinforced interval begins to mirror the first half of the curve. In this model, that is not what is found. On the first probe trial, the peak is already formed. This may indicate a problem with the way the model has been trained for these simulations or with the model itself.

2.4.2 LEARNING PHENOMENA

2.4.2.1 Learning the Time Marker

One of the defining features of this model, compared to other timing models, is that it can determine which of its inputs is the timing signal. This process of deciding which stimuli are related to which outcomes is sometimes referred to as the assignment-of-credit problem (Minsky, 1961). One of the primary issues of artificial intelligence research and all other artificial learning research, it has been essentially ignored by timing researchers. This problem was simulated by presenting the model with ten inputs, one of which was the true timing signal. The others changed between off and on randomly, independent of each other or the reinforcement schedule. As seen in Figure 2.15, the model learned to accurately predict the reinforcement and to ignore the irrelevant stimuli.

2.4.2.2 Parsimonious Timing

One anecdotal finding in timing research is that subjects presented with multiple timing signals will generally use the one closest in time to the reward. This is logical because the absolute size of timing errors will be smaller with the shorter interval between the time marker and the reinforcement. In this simulation, the model was presented with two timing signals. The first started 60 time steps before the reward

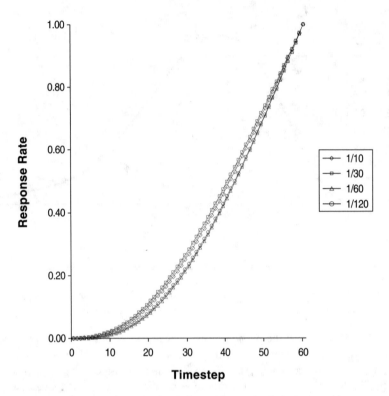

FIGURE 2.15 Assignment-of-credit simulations. These simulations were the same as the fixed-interval simulations with the addition of eight input nodes that turned on and off randomly. Four simulations were run with different probabilities of the random input nodes turning off and on each time step: 1/10, 1/30, 1/60, and 1/120. These data represent the average of the 51st through 55th trials.

was presented; the second started 30 time steps before the reward. Both signals accurately predicted the reward, and both were equally salient. The result can be seen in Figure 2.16. The model begins responding much later than it would if it was trained with only the long signal. This indicates that it does naturally gravitate toward the most parsimonious timing signal.

2.4.2.3 Initial Fixed-Interval Behavior

In general, current timing models excel at describing how animals respond after extensive training. This steady-state behavior is relatively static and predictable and does not require a real-time model. However, there are distinct patterns of responding found in the first few trials under an FI schedule. Before being exposed to the fixed-interval schedule, most animals are trained to respond using a continuous reinforcement (CRF) schedule, in which they are rewarded for every response. Once they are responding at a sufficient rate, the subjects are switched to the FI schedule. On the first trial or two after the switch, subjects typically produce an inverted scallop pattern, initially responding at a very high rate and then tapering off. Over successive

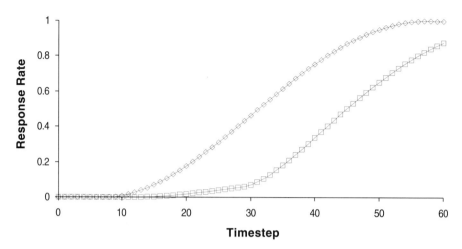

FIGURE 2.16 Parsimonious timing simulation. When presented with a single timing signal that begins 60 time steps before reinforcement, the general timing model responds with a typical fixed-interval response function (diamonds). When trained with a second timing signal that begins 30 sec into the trial, the model relies almost entirely on the second signal (squares).

intervals, responding becomes relatively flat as the animal transitions to the steady-state FI scallop (Ferster and Skinner, 1957).

The performance of the general timing model under these conditions is illustrated in Figure 2.17. The model also begins with an inverted scallop pattern and gradually transitions to the FI scallop, as shown in Figure 2.6. The learning process shown happens over fewer trials than it does in animals, but the progression and dynamics are very similar.

2.4.3 COMPLEX TIMING PHENOMENA

2.4.3.1 Gap Timing

The gap procedure represents one of the most popular paradigms for exploring the connection between learning and timing (see Buhusi, this volume; Fortin this volume). Based on the PI procedure, the gap procedure adds an additional type of probe trial called a gap trial. In a gap trial, the timing stimulus is turned off for a short portion of the interval and then turned back on for the rest of the trial. As in other probe trials, no reinforcement is provided during a gap trial.

The essential result of the gap procedure is that the peak response time is shifted by slightly more than the length of the gap (e.g., Roberts and Church, 1978; Roberts, 1981). This implies that the subject's internal clock is paused or stopped for the duration of the gap and then resumes when the stimulus is turned back on. In longer gaps, the peak is shifted by just less than the length of the gap plus the portion of the interval before the gap, as if the subject's internal clock had reset during the gap and started timing anew once the stimulus was represented.

This stop vs. reset pattern is very attractive to proponents of stopwatch-like models, since those are precisely the functions that a mechanical or digital stopwatch

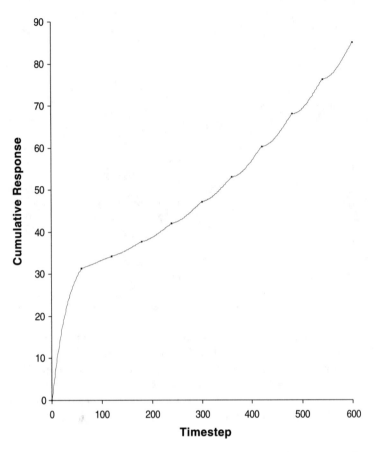

FIGURE 2.17 Initial behavior on a fixed-interval schedule. This figure shows the general timing model's cumulative response function for the first ten intervals on an FI schedule after standard continuous reinforcement training. As in animals, the model begins with an inverted scallop pattern and gradually changes to a normal FI scallop pattern over successive trials. The black circles mark where the model received a reinforcement. The data shown represent the average of five simulations.

needs to function. The fact that the peak shifts are frequently (but not always) more than a stop result and less than a reset result is dismissed by these theorists as the result of a probabilistic system of stopping and resetting the clock. If the clock is automatically stopped during a gap and has a small chance of resetting completely each moment during the gap, one would expect a pattern of data similar to that described above. During short gaps, there would only be on average a few trials where the clock reset completely, so the average peak shift on those trials would be approximately a stop result plus a bit, due to a few resets. On long gaps, most gap trials would result in reset, but there might be a few stops in there to lower the average.

The competing hypothesis is that the subject's memory of the duration before the gap decays during the gap. If the stored duration is reduced by 5% during each second of the gap, only a small portion of the stored time would be lost during a

General Learning Models: Timing without a Clock

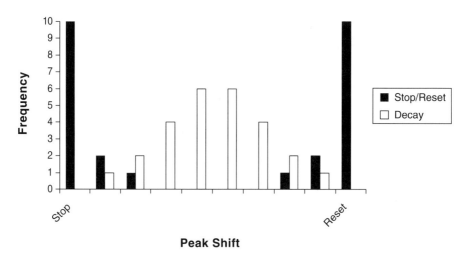

FIGURE 2.18 Peak-interval gap procedure, peak time distributions. This figure demonstrates the differing predictions of the stop–reset and decay hypotheses. If subjects' accumulated subjective time decays smoothly over the course of a gap, one would expect that the amount of peak shift would be normally distributed. If the subjective time is either held or reset, one would expect a bimodal distribution, with some trials showing no loss of subjective time and some showing complete resetting.

short gap, producing something similar to a stop result. In a longer gap, the stored duration would have had time to decay almost completely, producing something like a reset result. The model described here uses a decay system.

These two hypotheses could be tested by looking at the sizes of peak shifts for individual gap trials. If the stop–reset concept was correct, one would expect to see a bimodal distribution of peak shifts, with stop results clustered at one end of the distribution and resets clustered at the other. If the decay hypothesis is correct, the distribution should be relatively normal, as illustrated in Figure 2.18. So far, this analysis has not been done due to the complexity of determining peak times for individual gap trials. Individual trials tend to be noisy, and the methods for determining the peak time of an individual trial are complex and not universally accepted (Church et al., 1994).

A parametric study of the gap procedure done by Cabeza de Vaca et al. (1994) provided three distinct patterns of data. These results, taken together, represent the best opportunity available for comparing the performance of a timing model to animal data in a quantitative fashion. The results of this experiment were very clean and precisely in accord with the predictions of a decay process. In the following simulations, all simulation data represent the peak shift on the very first gap presentation after 100 trials of the PI procedure training.

In the fixed onset series, the subjects were presented with gaps of varying lengths that always started at a fixed point 6 sec into the trial, as shown in Figure 2.19, bottom panel. As described above, the mean peak shifts were almost always somewhere between a stop result and a reset result. The subjects showed a nonlinear pattern of peak shifts, with the peak shifts starting very close to a stop result in the

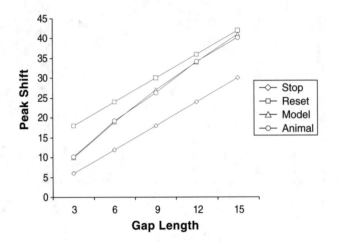

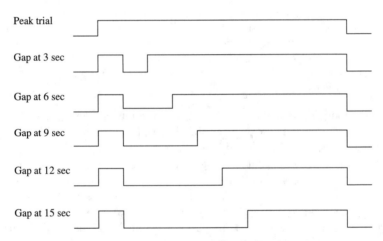

FIGURE 2.19 Peak-interval gap procedure, fixed onset simulations. These data present the amount of peak shift resulting from gaps of different durations but with the same onset time. The lower panel graphically shows the six locations of the gaps within the interval. The animal data presented are taken from Cabeza de Vaca et al. (1994).

short gap trials and asymptotically approaching a reset result in the longer gap trials, as illustrated in Figure 2.19, top panel. The model matches these data extremely well, with a correlation of .9994 ($P < .0001$).

In the fixed offset series, gaps of various lengths were presented that always ended at a fixed point 21 sec into the trial, as shown in the bottom panel of Figure 2.20. The peak shifts produced were again nonlinear, starting close to a stop result and approaching a reset result asymptotically, as shown in Figure 2.20, top panel. The model again matches these data very well, with a correlation of .998 ($P < .0001$).

In the location series, 6-sec gaps were presented at various times within the interval, as shown in Figure 2.21, bottom panel. Subjects produced peak shifts that

General Learning Models: Timing without a Clock

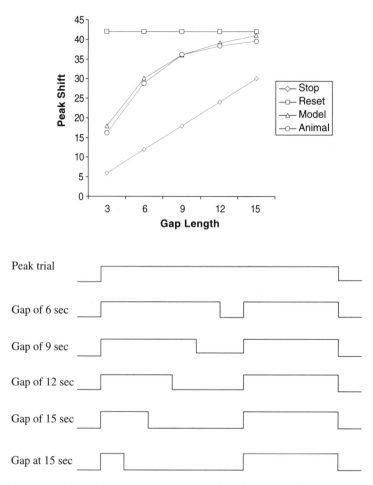

FIGURE 2.20 Peak-interval gap procedure, fixed offset simulations. These data present the amount of peak shift resulting from gaps of different durations but with the same offset time. The lower panel graphically shows the six locations of the gaps within the interval. The animal data presented are taken from Cabeza de Vaca et al. (1994).

increased linearly with the location of the gap, approximately midway between a stop and a reset result, as illustrated in Figure 2.21, top panel. Once again, the model matches the animal data very well, with a correlation of .995 ($P < .0005$).

The excellence of these fits is not unique. Hopson (1999) found that adding a decay mechanism to the spectral timing model allowed it to model these data nearly as well. The success of these fits is most likely due to the fact that the data conform so perfectly to the predictions of a decay process.

2.4.3.2 Filled-Gap Timing

One of the most interesting aspects of the gap procedure is the *lack* of learning that goes on. Gaps occur only during probe trials; therefore, one might expect that the

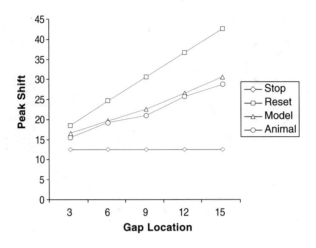

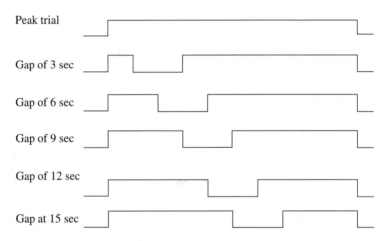

FIGURE 2.21 Peak-interval gap procedure, fixed duration simulations. These data present the amount of peak shift resulting from gaps of the same durations but with different onset times. The lower panel graphically shows the six locations of the gaps within the interval. The animal data presented are taken from Cabeza de Vaca et al. (1994).

animals would learn that a short break in the timing signal predicts that this trial will not produce any reinforcement. If they did learn this, subjects would stop responding entirely after the gap. But even after weeks of training with gap trials, rats and pigeons do not stop responding. As with most experiments involving nonfood trials, overall response rates do go down in gap experiments, but they do not extinguish as they would if subjects were capable of using the gap as a discriminative stimulus.

Roberts (1981) explored what would happen if the timing signal was not removed, but instead a new stimulus was introduced during an equivalent portion of the interval. On a filled-gap trial, the timing signal (a light) is turned on at the beginning of the trial and remains on for three times the reinforced interval length. Ten seconds after the trial begins, a 10-sec tone is presented. The presentation is

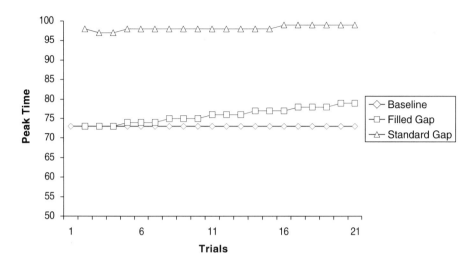

FIGURE 2.22 Filled-gap simulation. The simulation responded differently to a gap filled with a novel stimulus (squares) than to a standard gap (triangles). The baseline data (diamonds) show the normal peak time.

equivalent to a normal gap, but the sound can be learned as a discriminative stimulus. In this experiment, subjects' peak response times were initially the same as in a normal PI trial, but over the course of 2 weeks, their peak times gradually became longer and longer, indicating that the subjects had learned that the sound signaled a nonfood trial. Response rates postgap declined as well, approaching zero by the end of the experiment.

A simulation of this experiment using the general timing model was performed, and the results can be seen in Figure 2.22. The peak times become later and later over successive trials. In the model, this shift takes place within 20 gap trials, unlike the rat experiment, where it occurs over 2 weeks. This can be attributed to the fact that the model simply learns faster than its animal counterpart. One might propose that the model "lives" in a much simplified world, with fewer extraneous stimuli to confuse the issue. Even ignoring the world outside the operant chamber, the model never gets satiated, is never distracted by an odd smell in one corner of the chamber, never misses a signal due to looking the wrong way, or any of ten thousand other variables. The model's learning rate is controlled by a parameter and could be reduced to the point where this learning takes the same number of trials as the animals.

2.4.3.3 Scalar Timing

One of the most important attributes of animal timing is what has been termed the scalar property. Also referred to as timescale invariance, it has been defined as the fact that one should not be able to tell the length of the interval from the shape of the timing function (Gallistel, 1999). In practical terms, this means that the PI functions for a 30-sec schedule and a 300-sec schedule should be of similar proportions. This is not an absolute law of behavior. Studies with very long schedules (over

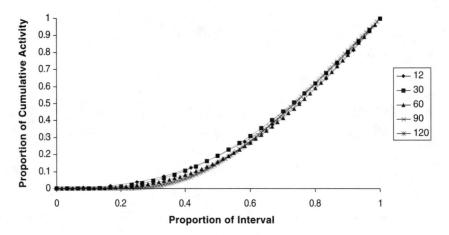

FIGURE 2.23 Scalar property simulation. The model produces an approximately scalar response curve over a range of 12 to 120 time steps. Outside this range, the function becomes distorted.

a day) have shown timing that is more precise than would be predicted by scalar timing. Even at shorter timescales, not all variables are strictly proportional to the reinforcement interval (e.g., Zeiler and Powell, 1993). While scalar timing may not be absolute, it is a useful heuristic in thinking about how animals time. All existing models of interval timing demonstrate this property, and the method for scaling represents one of the primary defining features of any model.

Scaling in the general timing model is accomplished by adjusting the weights between the input layer and the middle layer. These weights govern the rate at which the hidden layer nodes become activated. If they build up quickly, the model can learn to respond to very short intervals. With very small weights they build up slowly, allowing the model to respond at longer intervals.

The model's ability to scale can be seen in Figure 2.23. For reinforcement intervals between 12 and 120 time steps, the model scales fairly well. Response rates are typically low during the first quarter of the interval and then accelerate rapidly. At intervals under 12 time steps, the peak response time is later than the interval. With intervals longer than 120 time steps, the peak response time always occurs before 120 time steps in the interval, as illustrated in Figure 2.24.

At shorter intervals the fact that the simulation must break time into discrete time steps becomes a problem. Because we only have one data point per time step, it makes discerning a clear curve difficult. It also becomes difficult for the model to build up activity in the hidden layer fast enough to produce a full response curve. This is not necessarily a failing of the model as much as an inevitable artifact of computer simulation. It could be corrected by saying that time steps are small enough (say, ten per second) that one would hit reaction time limits before this artifact would matter.

However, this would mean that a 30-sec interval would have 300 time steps, which runs into the upper limit of the model's ability to scale. The learning algorithm seems unable to shift the peak response rate past 120 time steps. This seems to be

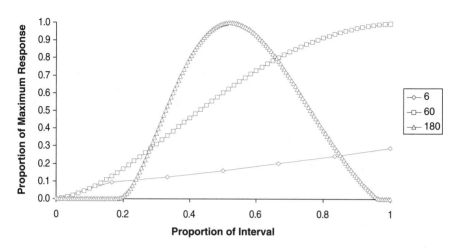

FIGURE 2.24 Scalar property failure. The model produces an approximately scalar response curve over a range of 12 to 120 time steps. Outside this range, the function becomes distorted. This figure shows the distorted peak-interval functions produced by 6 and 180 time step intervals, plus a properly scalar 60 time step interval for comparison.

a product of the fact that the weights between the input and hidden layers would have to be very small in order for activation to build up slowly enough for longer intervals. The learning algorithm seems incapable of adjusting those weights properly at very small values.

2.4.3.4 Possible Techniques for Scalar Timing

The simplest method for creating perfect scaling is to make the time steps of the model proportional to the reinforcement interval. This has been done in several models, most notably in BeT. There is no need to demonstrate that this method would work because it is essentially assuming the scaling happens. By scaling the time steps, we automatically scale the timing function and create the result. This mechanism is ill suited to the general timing model. The length of a time step is not something that can be changed by a neural network while it is running. The time step is analogous to a physical constant, a fixed principle of the universe the model inhabits. The model can change what it does each time step, but not the substrate in which it functions. This is not a constraint of this particular model, but of neural network models in general. There is simply no mechanism within this type of model to adjust the overall rate at which time flows.

A stronger position can also be taken that scaling by adjusting the time steps is theoretically bankrupt. Any mechanism that could adjust the time step length could also be made to adjust the model itself, and adjusting the model is more parsimonious. In terms of the performance of the model, there is no difference between the two. All models already assume that some sort of learning takes place, and scaling should take place within that already existing mechanism for learning. Having scaling time steps take place outside the existing system allows the modeler to make it happen by fiat, rather than by building it into the model.

Another alternative, used by all of the neural timing models mentioned earlier, is to include in the model an array of preset timescales and allow the model to use whichever suits the situation. In the spectral timing model, this is done through having an array of nodes with different parameters, each building up at a slightly slower rate than the last. In the connectionist SET model, each oscillator has its own rate. In the CBeT model, the units are connected in series, so the first few units are only active at short timescales and the last few units are only active at long timescales. This makes learning the correct scale very easy, because the portion of the model that runs at the correct rate will always predict the reinforcement better than the rest of the model.

In the general timing model, the easiest way to do this would be to make some of the hidden nodes act slower than the rest. A node that learned only once every five time steps would essentially live and learn at a timescale five times as long as its neighbors'. For this node, a 300-sec schedule would really be a 60-sec schedule, well within the range for which the model scales. If each node in the hidden layer ran at its own rate, this would provide a wide range of timescales for learning. This solution has not been employed in this model to date because it requires an additional layer of assumptions. These assumptions are reasonable and in accordance with what we know of animal learning, but they would still be adding another layer of complexity to the model. The focus of this chapter is to explore how much of animal timing can be explained with a minimally changed general learning model.

A related solution is that used by artificial neural network researchers attempting to model complex time series. Day and Davenport (1993) incorporated time delays into each node of the network that can be adjusted in a manner similar to that of the network weights. Some of the hidden layer nodes can adjust their delays to match the scale of the reinforcement interval, allowing the model to produce a properly scaled output function. This solution again requires an additional layer of assumptions and is therefore beyond the scope of this project.

The best potential method for creating scaling is modifying the learning algorithm. If the various weights are preset and fixed, the model can produce the correct timing function at a wide variety of timescales. The question then becomes: What about the learning mechanism needs to be changed in order to drive the model toward those solutions? At both extremes, the problem seems to be that the changes to the weights are not proportional to the weights. When the required interval is very small, the weights need to be larger than the learning algorithm seems able to make them. When the interval is very large, the upward adjustment of weights by reinforcement seems to make them too large. To date, attempts to make the changes proportional to their current values have not been successful in allowing the model to scale over a wider range. This does not mean that the solution is wrong, just that implementing it is not trivial. Hopefully, future explorations of this timing model can solve this problem.

2.4.3.5 Noise and Variation

One of the other scalar features of interval timing is that the variation in the shape of the timing function scales as well. This does not occur in this model because the timing function is basically static for any given interval length. However, if the output of the timing function represented the probability of the animal emitting a response, then the variations in the function would scale as well.

2.5 DISCUSSION

2.5.1 OVERALL PERFORMANCE

In terms of its primary goal of testing whether interval timing could occur within a general model of animal learning, this project was quite successful. The timing performance of the model was very good, equivalent to many of our dedicated timing models. By successfully simulating the performance of animals on an FI schedule, a PI schedule, and a gap schedule, the model demonstrates substantial interval timing abilities. In terms of design, the model met all of the criteria. It maintained its learning abilities while timing, and the changes required for timing were small and there is the possibility that they could be made even smaller with further research. The model was robust in its response to different parameter values, especially the number of nodes required for timing.

While the primary goal of this project was to explore whether interval timing could happen at all within a general learning model, the model created to answer that question can now serve with our other existing models of interval timing. It cannot replace them, but it should provide a unique perspective and comparison for them. Its success raises new theoretical questions that can be pursued through other models and methods.

2.5.2 DIFFICULTIES

The principal problem with this general learning model is its failure to scale perfectly. It does scale across a single order of magnitude, but three orders of magnitude are required for the model to approximate the sort of scaling observed in animals. There are two ways to handle this failure to scale. First, it can be taken as a temporary technical setback. It is possible that a general learning model can learn to scale its timing function properly and just has not done so in this case due to problems in the implementation. There are other learning algorithms, and it may simply be a question of hitting upon the right one. This possibility (extant in every computer model) makes it very hard to theorize from the model's inability. As it stands, this model does not scale properly, but that is not to say that a very close cousin could not scale perfectly.

Alternately, this scaling failure could be an inherent limitation of non-clock-based timing systems. If this is so, then it provides us with an interesting theory of the evolutionary pressures that spurred the development of the clock system. If interval timing can be done by any neural system but scaling cannot, then scaling becomes the primary goal of the clock system.

2.5.3 FUTURE RESEARCH

This work is intended as a narrowly focused test of a theoretical position, and so the model created to test the position has not been exhaustively explored. There remain a wide variety of extensions that can be made to the model and other animal experiments that can be simulated. One potential line of research is to bring the model closer to the archetypical neural network model. It should be possible to use the recurrent network techniques to maintain interval timing performance while restoring the all-or-nothing property. The output would no longer be a smooth curve, but averaged response rates could produce something similar. This would not provide the network with any new abilities, but it would help generalize these findings. The more different types of neural networks that can produce these kinds of timing results, the more likely that these results will apply to the quite different neural networks in the animal brain.

The integration of learning and interval timing in this model enables it to predict the results of experiments that involve both learning and timing. We can predict novel experiments in which the clock does not simply gather time linearly. For example, subjects might be trained using a PI procedure. During some of the food trials, a tone is presented at a random point during the trial, signaling that the computer controlling the experiment has halved the amount of time that has elapsed so far in the trial. Food will be presented as normal, but will take longer than it would have on a normal food trial. This procedure requires a mental operation unlike any other existing interval timing procedure, a precise manipulation of the animal's or the model's current stored duration.

Because our interval timing models are strictly timing models, they do not do well when faced with experiments where timing is only one of the cognitive abilities required. With this model, it is possible to begin integrating interval timing with other areas of research such as attention, which has also been explored using neural networks. An additional line of research would be to explore to what degree other general learning models can describe interval timing behavior. Backpropagation is not the most biologically plausible learning mechanism, and artificial neural networks are not necessarily the most accurate models of the mental processes of real animals. These issues do not impact the basic claim of this work — that it is possible for a general learning process to time. A next step could be to come as close as possible to proving that the general learning processes used by real animals can also produce interval timing behavior. Of course, we do not have certain knowledge of the precise learning mechanisms of the mind, but there are models that are more similar to what we do know of those mechanisms. Also, replicating these results with several different types of systems would strengthen the hypothesis that any sufficiently general learning mechanism can time, something suggested but not proven by the success of this project.

An example of alternative systems would be to use neural networks trained via genetic algorithms. In this method, a population of artificial neural networks with identical architecture but different weights would be created, and then this population would be made to evolve to better predict the reinforcement interval. Each individual set of weights would be tested to determine how well it predicted the interval and then assigned a fitness value. Those individuals would then have a better chance of passing their weights on to the next generation, modified by mutation and recombination. As each generation is tested and then bred, the individuals gradually become more adapted to predicting the reinforcer.

This learning mechanism has several advantages compared to the backpropagation algorithm used above, such as the ability to explore multiple types of solutions simultaneously. However, this line of research would not be intended to discover which learning mechanism is most effective, but rather to determine the key aspects necessary for a given learning mechanism to be capable of timing. Similar testing could be done on a wide variety of learning mechanisms.

2.5.4 Theoretical Implications

Perhaps the most powerful idea about interval timing presented by this model is how simple it is. Timing can be accomplished with only three nodes, without any special architecture or processes. This implies that interval timing could potentially be found in any animal with a nervous system, unlike more complex models that require the evolutionary development of special-purpose mechanisms. Interval timing has so far been demonstrated in a variety of mammals and birds, and has recently been shown in fish as well (e.g., Talton et al., 1999). This author is also aware of an unpublished study involving the successful training of crabs under an FI schedule. If this model is correct in predicting that interval timing can happen with very few neurons, there is no reason why insects and other organisms with simpler nervous systems could not produce basic timing behavior.

The true difficulty in exploring this idea will be creating the experimental technology necessary to test such things in insects and other small organisms. Presenting the stimuli in a salient manner, providing rewards of an appropriate size and type, and measuring responses are all nontrivial problems in any new species. As always, it is important to distinguish between the failure of the procedure to accurately measure the animal's capabilities and the absence of those capabilities.

The ability of a generalized neural network to time also suggests that interval timing may be ubiquitous within the brain. If any three neurons hooked end to end can time, then we can reasonably expect that any or all subsystems within the brain can time independently if necessary. Areas of the brain dedicated to visual processing may time visual stimuli; auditory areas, auditory stimuli; and so on (see Penney, this volume).

The best comparison for this ubiquitous timing might be our current paradigms for memory. We no longer think of memories as being kept in a single storehouse, but as associated with the areas involved with the processing, such as memory for faces being stored in the face recognition areas. There are discrete areas of the brain that, if damaged, create global effects on memory, but these are thought of as routing depots rather than storehouses. Similarly, there may be discrete brain areas that are globally involved in timing but are not clocks. A search for a central clock may be futile, but that does not mean that all interval timing functions are perfectly distributed. What we have been looking at as central clocks may in fact be central processing areas for timing related information.

The comparison to memory also raises the idea that there may be qualitatively different types of timing within interval timing. It is commonly understood that interval and circadian timing are distinct, but there may be multiple mechanisms contributing to interval timing in animals. We can distinguish implicit and explicit forms of memory, and there is the fascinating possibility that there may be implicit and explicit varieties of interval timing. Implicit timing would be similar to that shown here, requiring no special architecture, but perhaps limited in scale. Explicit timing would rely on a dedicated clock mechanism that is the focus of the vast majority of research in this field (see Meck, this volume). Each would serve different needs, and it might be possible to damage explicit timing while preserving implicit timing, through precisely targeted lesions or genetic manipulation. This is, of course, pure speculation, but one of the primary purposes of behavioral models is to suggest speculative lines of research that can later be followed up with physiological experiments.

This idea of ubiquitous timing may also predict that the total number of intervals that can be timed simultaneously is very large (see MacDonald and Meck, this volume; Meck and Church, 1984; Pang and McAuley, this volume). The smaller the capacity required to time, the more timing that can be done with a given capacity (for discussions of attentional allocation and timing, see Buhusi, this volume; Fortin, this volume). Our typical research paradigms involve simplifying the experiments as much as possible in order to make analysis of the results clearer. This has meant presenting the subjects with as few intervals as are strictly necessary for the experiment, and to the best of my knowledge, no attempt has been made to explore how many separate intervals animals are capable of tracking simultaneously — although Meck (1987) has shown that three signals of different modalities (auditory, tactile, and visual) can be timed simultaneously. Such experiments would have to be carefully designed to avoid running into other cognitive limitations and therefore incorrectly determining a maximum number of intervals that is too low.

If we carry forth the idea of implicit and explicit timing, it may be that the implicit mechanisms can time an unlimited number of stimuli, while the explicit timing mechanisms can only handle a much smaller number of intervals — perhaps even the classic 7 ± 2. It would make sense that implicit timing abilities, drawing only on properties inherent in every neuron, would be essentially unlimited. But the explicit timing areas, being discrete and therefore of fixed capacity, would only be able to time a smaller number of simultaneous events.

This project has successfully achieved its goals. It has shown that interval timing can occur within a general learning model, requiring only minor, common changes. It has also provided a number of theoretical insights and potential lines of research. Most importantly, it offers a new way to look at one of the most fundamental principles of interval timing theory: the internal clock.

REFERENCES

Arcediano, F. and Miller, R.R., Some constraints for models of timing: a temporal coding hypothesis perspective, *Learn. Motiv.*, 33, 105–123, 2002.

Balsam, P.D., Drew, M.R., and Yang, C., Timing at the start of associative learning, *Learn. Motiv.*, 33, 141–155, 2002.

Bizo, L.A. and White, K.G., Reinforcement context and pacemaker rate in the behavioral theory of timing, *Anim. Learn. Behav.*, 23, 376–382, 1995.

Cabeza de Vaca, S., Brown, B.L., and Hemmes, N.S., Internal clock and memory processes in animal timing, *J. Exp. Psychol. Anim. Behav. Process.*, 20, 184–198, 1994.

Catania, A.C., Reinforcement schedules and psychophysical judgements: a study of some temporal properties of behavior, in *The Theory of Reinforcement Schedules*, Shoenfeld, W.N., Ed., Appelton-Century-Crofts, New York, 1970, pp. 1–42.

Cheng, K., Westwood, R., and Crystal, J., Memory variance in the peak procedure of timing in pigeons, *J. Exp. Psychol. Anim. Behav. Process.*, 19, 68–76, 1993.

Church, R.M., Evaluation of quantitative theories of timing, *J. Exp. Anal. Behav.*, 71, 253–291, 1999.

Church, R.M. and Broadbent, H.A., Alternative representations of time, number, and rate, *Cognition*, 37, 55–81, 1990.

Church, R.M., Meck, W.H., and Gibbon, J., Application of scalar timing theory to individual trials, *J. Exp. Psychol. Anim. Behav. Process.*, 20, 135–155, 1994.

Church, R.M., Miller, K.D., and Meck, W.H., Symmetrical and asymmetrical sources of variance in temporal generalization, *Anim. Learn. Behav.*, 19, 207–214, 1991.

Crystal, J.D., Church, R.M., and Broadbent, H.A., Systematic nonlinearities in the memory representation of time, *J. Exp. Psychol. Anim. Behav. Process.*, 23, 267–282, 1997.

Davis, M., Schlesinger, L.S., and Sorenson, C.A., Temporal specificity of fear conditioning: effects of different conditioned stimulus-unconditioned stimulus intervals on the fear-potentiated startle effect, *J. Exp. Psychol. Anim. Behav. Process.*, 15, 295–310, 1989.

Day, S.P. and Davenport, M.R., Continuous-time temporal backpropagation with adaptive time delays, *IEEE Trans. Neural Networks*, 4, 348–354, 1993.

Dragoi, V. and Staddon, J.E.R., The dynamics of operant conditioning, *Psychol. Rev.*, 106, 20–61, 1999.

Dragoi, V., Staddon, J.E.R., Palmer, R.G., and Buhusi, C.V., Interval timing as an emergent learning property, *Psychol. Rev.*, in press.

Eckerman, D.A., Scheduling reinforcement about once a day, *Behav. Process.*, 45, 101–114, 1999.

Ferster, C.B. and Skinner, B.F., *Schedules of Reinforcement*, Appleton-Century-Crofts, New York, 1957.

Gallistel, C.R., Can a decay process explain the timing of conditioned responses? *J. Exp. Anal. Behav.*, 71, 264–271, 1999.

Gibbon, J., Scalar expectancy theory and Weber's law in animal timing, *Psychol. Rev.*, 84, 270–325, 1977.

Gibbon, J., Ubiquity of scalar timing with a Poisson clock, *J. Math. Psychol.*, 36, 283–293, 1992.

Gibbon, J. and Church, R.M., Sources of variance in an information processing theory of timing, in *Animal Cognition*, Roitblat, H.L., Bever, T.G., and Terrace, H.S., Eds., Erlbaum Associates, Hillsdale, NJ, 1984, pp. 465–488.

Gibbon, J. and Church, R.M., Representation of time, *Cognition*, 37, 23–54, 1990.

Gibbon, J., Church, R.M., and Meck, W.H., Scalar timing in memory, in *Annals of the New York Academy of Sciences: Timing and Time Perception*, Vol. 423, Gibbon, J. and Allan, L., Eds., New York Academy of Sciences, New York, 1984, pp. 52–77.

Grossberg, S. and Schmajuk, N.A., Neural dynamics of adaptive timing and temporal discrimination during associative learning, *Neural Networks*, 2, 79–102, 1989.

Guttman, N. and Kalish, H.I., Discriminability and stimulus generalization, *J. Exp. Psychol.*, 13, 121–128, 1956.

Haykin, S., *Neural Networks*, Macmillan, New York, 1994.

Higa, J.J., Wynne, C.D., and Staddon, J.E.R., Dynamics of time discrimination, *J. Exp. Psychol. Anim. Behav. Process.*, 17, 281–291, 1991.

Higa, J.J. and Staddon, J.E.R., Dynamic models of rapid temporal control in animals, in *Time and Behavior: Psychological and Neurobiological Analysis,* Bradshaw, C.M. and Szabadi, E., Eds., Elsevier Science, Amsterdam, 1997, pp. 1–40.

Hopson, J.W., Gap timing and the spectral timing model, *Behav. Process.,* 45, 23–31, 1999.

Killeen, P.R. and Fetterman, J.G., A behavioral theory of timing, *Psychol. Rev.,* 95, 274–295, 1988.

Lejeune, H., Cornet, S., Ferreira, M.A., and Wearden, J.H., How do Mongolian gerbils (Meriones unguiculatus) pass the time? Adjunctive behavior during temporal differentiation in gerbils, *J. Exp. Psychol. Anim. Behav. Process.,* 24, 325–334, 1998.

Leonard, D.W. and Monteau, J.E., Does CS intensity determine CR amplitude? *Psychonomics Sci.,* 23, 369–371, 1971.

Machado, A., Learning the temporal dynamics of behavior, *Psychol. Rev.,* 104, 241–265, 1997.

Meck, W.H., Vasopressin metabolite neuropeptide facilitates simultaneous temporal processing, *Behav. Brain Res.,* 23, 147–157, 1987.

Meck, W.H. and Church, R.M., Simultaneous temporal processing, *J. Exp. Psychol. Anim. Behav. Process.,* 10, 1–29, 1984.

Minsky, M.L., Steps toward artificial intelligence, *Proc. Inst. Radio Eng.,* 49, 8–30, 1961.

Roberts, S., Isolation of an internal clock, *J. Exp. Psychol. Anim. Behav. Process.,* 7, 242–268, 1981.

Roberts, S. and Church, R.M., Control of an internal clock, *J. Exp. Psychol. Anim. Behav. Process.,* 4, 318–337, 1978.

Schmajuk, N.A., Role of the hippocampus in temporal and spatial navigation: an adaptive neural network, *Behav. Brain Res.,* 39, 205–229, 1990.

Schneider, B.A., A two-state analysis of fixed interval responding in the pigeon, *J. Exp. Anal. Behav.,* 12, 677–687, 1969.

Skinner, B.F., *The Behavior of Organisms,* Appleton-Century-Crofts, New York, 1938.

Staddon, J.E.R. and Higa, J.J., Multiple time scales in simple habituation, *Psychol. Rev.,* 103, 720–733, 1999a.

Staddon, J.E.R. and Higa, J.J., Time and memory: towards a pacemaker-free theory of interval timing, *J. Exp. Anal. Behav.,* 71, 215–251, 1999b.

Sutton, R.S. and Barto, A.G., *Reinforcement Learning: An Introduction,* MIT Press, Cambridge, MA, 1998.

Talton, L.E., Higa, J.J., and Staddon, J.E.R., Interval schedule performance in the goldfish Carassius auratus, *Behavioural Processes,* 45, 193–206, 1999.

Treisman, M., Temporal discrimination and the indifference interval: implications for a model of the "internal clock," *Psychol. Monogr.,* 77, 1–31, 1963.

Zeiler, M.D. and Powell, D.G., Temporal control in fixed-interval schedules, *J. Exp. Anal. Behav.,* 61, 1–9, 1993.

3 Nonlinearities in Sensitivity to Time: Implications for Oscillator-Based Representations of Interval and Circadian Clocks

Jonathon D. Crystal

CONTENTS

3.1 Introduction 61
3.2 Short-Interval Timing 63
3.3 Long-Interval Timing 64
3.4 Circadian Timing 65
3.5 Implications for Theories of Timing 69
    3.5.1 Scalar Timing Theory 70
    3.5.2 Multiple-Oscillator Theory of Timing 72
    3.5.3 Broadcast Theory of Timing 72
    3.5.4 Stochastic Counting Cascades 72
3.6 Conclusions 72
Acknowledgments 73
References 73

3.1 INTRODUCTION

The hallmark of short-interval timing is the scalar property. This property is demonstrated by a constant coefficient of variability (standard deviation divided by the mean of distributions from timing experiments) and is a version of Weber's law. In particular, the distribution of timing data is the same for different target intervals when the data are normalized on x- and y-axes. For the x-axis normalization, time is expressed as a percentage of the different target intervals. For the y-axis normalization, response rate is expressed as a percentage of the maximum rates obtained for each target interval. This empirical feature of timing data will be referred to as the *linear-timing* hypothesis throughout this chapter. This property has been observed in humans (Rakitin et al., 1998; Wearden, 1991; Wearden et al., 1997) and many other animals (Church et al., 1994; Fetterman and Killeen, 1995; Richelle and Lejeune, 1984). For example, in a fixed-interval procedure, food is available for the first response after the fixed interval elapses. Temporal anticipation is documented by the increase in response rate as a function of time. When different fixed intervals are tested, the temporal anticipation functions superimpose when the x- and y-axes are normalized. In contrast, when the data are plotted in absolute rather than relative terms, the data fail to superimpose.

The linear-timing hypothesis has been supported by many experiments using a wide variety of target intervals (e.g., 30, 300, 3000 sec; Dews, 1970), and it has played an important role in the development of research in timing (Allan, 1998; Gibbon, 1991). The linear-timing hypothesis predicts that sensitivity to time is constant (i.e., linear) across a wide range of intervals. A more precise statement of the linear-timing hypothesis is that timing estimates consist of a linear component plus random error. If evaluating the fit of a theoretical function to data reveals nonrandom discrepancies from the average estimate, then the implication is that the function is an unacceptable description of the data. Therefore, it is possible to elaborate the linear-timing hypothesis at two levels of detail. According to the most basic description, the linear-timing hypothesis requires that psychological estimates of time increase as a constant proportion of physical estimates of time. According to a more detailed description, the linear-timing hypothesis predicts that departures from the linear prediction are expected to be random.

A well-established exception to the linear-timing hypothesis is known to occur when relatively short intervals (i.e., in the range of milliseconds) are examined (e.g., Church et al., 1976; Crystal, 1999; Fetterman and Killeen, 1992). In particular, the coefficient of variation of time estimates is relatively large for relatively short intervals. This may be interpreted as a generalized Weber function or an absolute threshold (Church et al., 1976; Crystal, 1999; Fetterman and Killeen, 1992). Another exception to the linear-timing hypothesis occurs when relatively long intervals are examined (e.g., Brunner et al., 1992; Crystal, 2002; Gibbon et al., 1997b; Lejeune and Wearden, 1991; Zeiler, 1991; Zeiler and Powell, 1994). In particular, the coefficient of variation of time estimates is relatively large for relatively long intervals. The objective of this chapter is to review evidence that interval timing is characterized by several temporal regions of improved sensitivity to time. These local maxima in sensitivity to time represent an empirical challenge to the linear-timing hypothesis. The evidence for local maxima in sensitivity to time comes from studies of (a) short-interval timing (i.e., timing of milliseconds, seconds, and minutes), (b) long-interval timing (i.e., timing of hours), and (c) circadian timing (i.e., timing of intervals of approximately 24 h). The chapter concludes with a discussion of implications for theories of timing.

3.2 SHORT-INTERVAL TIMING

A minimum of three target intervals is required to test the basic description of the linear-timing hypothesis. When more than three intervals have been tested, the absolute spacing between conditions generally increased as the magnitude of intervals increased (e.g., Fetterman and Killeen, 1992). Increasing spacing between conditions, as in a geometric series, is useful for evaluating a generalized Weber function because the close spacing between conditions for the shortest intervals permits the documentation of the predicted curve in the data for these short intervals. However, this approach is less appropriate for evaluating *departures* from a theoretical function, particularly if the temporal locations of departures from a linear function are not known *a priori*. In this case, it is necessary to test many closely spaced intervals to detect a local departure from a theoretical function; a systematic, local departure from a theoretical function would not be detected if fewer and more widely spaced interval conditions were examined (e.g., Collyer et al., 1992, 1994; Crystal, 1999, 2001b; Crystal et al., 1997; Kristofferson, 1980, 1984).

Data from temporal discrimination procedures using many closely spaced target intervals are shown in Figure 3.1. The data suggest that multiple temporal ranges are characterized by local maxima in sensitivity to time. The data were obtained from a titration procedure with rat subjects (Crystal, 1999, 2001b). For example, in a discrimination task, a "short" or "long" noise stimulus was presented followed by the insertion of two levers in an operant box. Different responses (left- or right-lever press) were required after short or long stimuli to obtain a food pellet. For a given short-duration condition, the duration of the long signal was adjusted (i.e., titrated) after blocks of discrimination trials to maintain discrimination accuracy at 75% correct, which resulted in a long duration approximately 2 to 2.5 times the short duration. Sensitivity to time was measured using the signal detection theory (Macmillan and Creelman, 1991). Sensitivity to time was approximately constant across short durations of 2 to 34 sec. However, local maxima in sensitivity were observed at approximately 12 and 24 sec (Figure 3.1A and B). When short durations in the millisecond range were examined, local maxima were observed at 0.3 and 1.2 sec (Figure 3.1C).

Local maxima in sensitivity to time violate the linear-timing hypothesis. This phenomenon demonstrates that certain intervals are timed with greater relative precision than other, nearby intervals. Local maxima in sensitivity to time are consistent with an oscillator representation of time in which the location of a local maximum identifies the period of an oscillator. A multiple-oscillator mechanism and other theories are discussed in the section below on implications for theories of timing.

An oscillator interpretation of short-interval timing is supported by recent observations of periodic behavior in the short-interval range. When food is contingent on lever pressing after a random interval, rats search periodically for food (Broadbent, 1994). Thus, behavior was periodic in the absence of periodic stimuli, similar to a free-running rhythm. Furthermore, conditions that promote periodic behavior in the absence of periodic stimuli have recently been identified in the short-interval range (Broadbent, 1994; Kirkpatrick-Steger et al., 1996; Machado and Cevik, 1998).

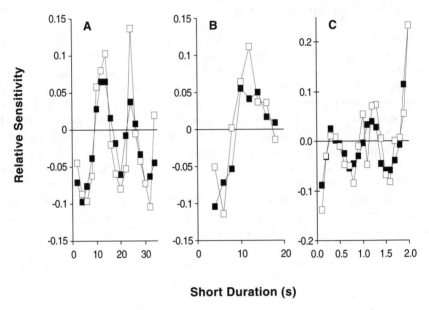

FIGURE 3.1 Sensitivity to time is characterized by local maxima at 12 and 24 sec (A), 12 sec (B), and 0.2 and 1.2 sec (C). Unfilled symbols: average across rats. Filled symbols: a running median was performed on each rat's data, and the smoothed data were averaged across rats to identify the most representative local maxima in sensitivity. (A) Rats discriminated short and long noise durations with the duration adjusted to maintain accuracy at 75% correct. Short durations were tested in ascending order with a step size of 1 sec ($n = 5$) and 2 sec ($n = 5$). Sensitivity was similar across step sizes ($r(15) = .701$, $P < .01$), departed from zero (binomial $z = 5.51$, $P < .001$), and was nonrandom ($r(14)_{lag1} = .710$, $P < .01$). SEM = 0.03. (B) Methods are as described in panel (A), except short durations were tested in random order ($n = 13$) or with each rat receiving a single interval condition ($n = 7$); results from these conditions did not differ. Sensitivity departed from zero (binomial $z = 6.04$, $P < .001$) and was nonrandom ($r(7)_{lag1} = .860$, $P < .01$). SEM = 0.02. (C) Methods are as described in panel (A), except intervals were defined by gaps between 50-msec noise pulses, and short durations were tested in descending order with a step size of 0.1 sec ($n = 6$). Sensitivity departed from zero (binomial $z = 4.84$, $P < .001$) and was nonrandom ($r(18)_{lag1} = .736$, $P < .001$). SEM = 0.04. Sensitivity was measured using d' from the signal detection theory. $d' = z[p(\text{short response} \mid \text{short stimulus})] - z[p(\text{short response} \mid \text{long stimulus})]$. Relative sensitivity is d' − mean d'. (Adapted from Crystal, J.D., *J. Exp. Psychol. Anim. Behav. Process.*, 25, 3–17, 1999; and Crystal, J.D., *Behav. Process.*, 55, 35–49, 2001b).

3.3 LONG-INTERVAL TIMING

The observation that timing data superimpose when plotted in relative time, rather than in absolute time, has played a formative role in the development of theories of interval timing (e.g., Gibbon 1977, 1991; Gibbon and Church, 1984). Most of these data came from short-interval timing (i.e., in the range of seconds and minutes), with relatively less information available about timing of long intervals (i.e., in the range of hours). A central question about long-interval timing is the extent to which the principles of timing that apply to short-interval timing also apply to long-interval timing.

One way to assess the timing of long intervals is to deliver a daily meal at a fixed interval after a salient cue, such as the transition from light to dark in the testing environment. In these experiments, the rats lived in operant boxes continuously for a month at a time and earned all of their food in this context. A standard 12–12 light–dark cycle was in effect (with lights on at 0800 and off at 2000). Meals started 3 or 7 h after light offset for independent groups of rats ($n = 3$ per group). Each 45-mg food pellet was contingent on a photobeam break after a variable interval during 3-h meals. The variable interval was adjusted to maintain food consumption between 15 and 20 g per meal. Anticipation is demonstrated by the fact that the rats started to inspect the food source *before* the start of the meal. Anticipatory responses as a function of time prior to the meal (absolute time, left panel) and as a function of time divided by the intervals 3 and 7 h (relative time, right panel) are plotted in Figure 3.2. Anticipatory responses began earlier (in absolute time) for the 7-h condition than for the 3-h condition (Figure 3.2, left panel). Thus, although the intermeal intervals were 24 h for both groups, the variability of the timing functions was different. The variable that controls this difference was identified by examining response rate as a function of relative time. The data superimposed when plotted in relative time (Figure 3.2, right panel). In particular, the spread of the response distribution (i.e., width) was a constant percentage of the interval between dark onset and the meal. These data suggest that (a) the rats timed with respect to dark onset (i.e., 3- and 7-h intervals), and (b) the rats timed the intervals proportionally (Crystal, 2001a).

The superposition result in Figure 3.2 (right panel) would be a standard finding if the intervals were in the short-interval range (e.g., 30 and 70 sec). The observation that the scalar property applies to intervals in the range of a few hours suggests that similar principles of timing apply to short-interval and long-interval timing. The continuity of principles is noteworthy particularly because the 3- and 7-h intervals were obtained from anticipation of daily (i.e., circadian) meals.

3.4 CIRCADIAN TIMING

Most investigators of interval timing and circadian rhythms regard timing in short-interval and circadian ranges as mediated by separate mechanisms (e.g., Aschoff, 1984; Church, 1984; Gibbon et al., 1997a; Hinton and Meck, 1997), with interactions that may occur at the behavioral level (Gibbon et al., 1984, 1997a; Lustig and Meck, 2001; Meck, 1991; Pizzo and Crystal, in press; Terman et al., 1984) or in the range of hours (Aschoff, 1985a, 1985b; Aschoff and Daan, 1997). However, the observation that the scalar property applies to the anticipation of daily meals, as it does to short-interval timing, raises the prospect that the mechanisms of short-interval and circadian timing may be more related than previously supposed. Moreover, if the location of a local maximum in short-interval sensitivity identifies the period of an oscillator, then a local maximum in sensitivity to approximately 24 h is predicted. Therefore, a series of experiments investigating meal anticipation was undertaken to test the

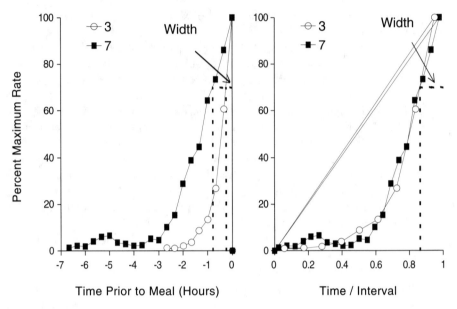

FIGURE 3.2 Anticipatory responses superimpose when plotted as a function of time as a proportion of the interval between light offset and the meal (right panel). Superposition is not observed when anticipation is plotted as a function of time prior to the meal (left panel). Meals started 3 or 7 h after light offset for independent groups of rats ($n = 3$ per group). Variability in timing (i.e., the width of the response rate function indicated by dashed lines) was higher for 7- than for 3-h groups (left panel: $t(4) = 3.75$, $P < .05$). Variability on a proportional timescale (i.e., width/interval) was the same for 3- and 7-h groups (right panel: $t(4) < 1$). This demonstrates that the rats timed with respect to light offset. The same data are plotted in left and right panels as a function of different x-axes. The width of the functions was measured at 70% of the maximum rate. The same conclusions were reached when the width was measured at 25, 50, and 75% of the maximum rate. Left panel: The meal started at 0 h. Right panel: End of the meal corresponds to 1 on the x-axis. (From Crystal, J.D., *J. Exp. Psychol. Anim. Behav. Process.*, 27, 68–78, 2001a. Copyright © 2001 by the American Psychological Association. Reprinted with permission.)

hypothesis that the well-established feeding-entrainable circadian oscillator (for review, see Mistlberger, 1994) is characterized by a local maximum in sensitivity to time (Crystal, 2001a).

Sensitivity to time was examined for intervals in the circadian range by measuring anticipation of restricted feeding in constant darkness. Anticipation of restricted meals was examined for intermeal intervals near the circadian range (22 to 26 h) and outside this range (14 and 34 h). The rats inspected the food trough before meals started (Crystal, 2001a). Response rate increased later into the interval for intermeal intervals near the circadian range than for intervals outside this range, as shown in Figure 3.3. The spread of the response distributions was characterized by two percentages of the interval: low variability for intervals near the circadian range and high variability for intervals outside this range, as shown in Figure 3.4 (Crystal, 2001a). The data plotted in Figure 3.4 document a local maximum in

Nonlinearities in Sensitivity to Time

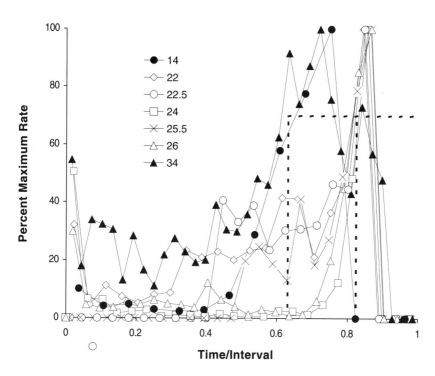

FIGURE 3.3 Response rate increased later into the interval for intermeal intervals near the circadian range (unfilled symbols) relative to intervals outside this range (filled symbols); dashed lines indicate width of response rate functions. Anticipatory responses increase immediately prior to the meal for all intermeal intervals except 34 h. Each 45-mg food pellet was contingent on a photobeam break after a variable interval during 3-h meals. Intermeal intervals were tested in separate groups of rats ($n = 3$ to 5 per group). The end of the meal corresponds to 1 on the x-axis. Testing was conducted in constant darkness. (From Crystal, J.D., *J. Exp. Psychol. Anim. Behav. Process.*, 27, 68–78, 2001a. Copyright © 2001 by the American Psychological Association. Reprinted with permission.)

sensitivity to time near 24 h (Crystal, 2001a). These data are consistent with the hypothesis that a function of an oscillator is improved sensitivity to time.

It is generally accepted that food-anticipatory activity to a daily meal develops only when the interval between successive meals is near 24 h. The limited range of food entrainment has been estimated to be between 22 and 31 h (Aschoff et al., 1983; Boulos et al., 1980; Mistlberger and Marchant, 1995; Stephan, 1981; Stephan et al., 1979a, 1979b; White and Timberlake, 1999) and between 22.17 and 26.67 h (Madrid et al., 1998). It is noteworthy that the animals in the study described above anticipated the 14-h intermeal interval. For additional evidence of timing the 14-h interval, see Figure 3.5, which plots anticipation functions from individual rats (right panel) and activity records (left and center panels). Although it is generally accepted that animals cannot anticipate intermeal intervals outside a limited range near 24 h, this conclusion is based on a relatively limited data set. When these data are reexamined, there is evidence for timing long intermeal intervals that are substantially

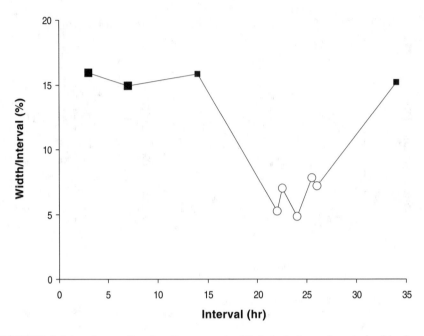

FIGURE 3.4 Intervals near the circadian range (unfilled circles) are characterized by lower variability than intervals outside this range (filled squares). Variability in anticipating a meal was measured as the width of the response distribution prior to the meal at 70% of the maximum rate, expressed as a percentage of the interval. The interval is the time between light offset and meal onset in a 12–12 light–dark cycle (leftmost two squares) or the intermeal interval in constant darkness (all other data). The percentage width was lower in the circadian range than outside this range, $F(1, 20) = 22.65$, $P < .001$. The width/interval did not differ within the circadian, $F(4, 12) = 1$, or noncircadian, $F(3, 8) < 1$, ranges. The same conclusions were reached when the width was measured as 25, 50, and 75% of the maximum rate. (From Crystal, J.D., *J. Exp. Psychol. Anim. Behav. Process.*, 27, 68–78, 2001a. Copyright © 2001 by the American Psychological Association. Reprinted with permission.)

less than 24 h, as shown in Figure 3.6. For example, behaviors that are instrumental in producing food (e.g., approaching the food source or pressing a lever) precede meal availability for intermeal intervals outside this limited range (see Figure 3.5 for 14 h (Crystal, 2001a); see Figure 3.6 for a reanalysis of 18 and 19 h (Bolles and Stokes, 1965; Boulos et al., 1980)). The temporal functions for intervals below the circadian range are less steep and have lower terminal response rates than intervals in the circadian range (see Figure 3.6). These features are characteristic of relatively high variability (i.e., low sensitivity to time), which is consistent with the data illustrated in Figure 3.4 (Crystal, 2001a). In contrast, wheel-running activity does not precede meals at these intervals (Bolles and de Lorge, 1962; Bolles and Stokes, 1965; Mistlberger and Marchant, 1995; Stephan et al., 1979a; White and Timberlake, 1999). Behaviors that are instrumental in producing food may be a more sensitive measure of food anticipation than general activity measures for intervals below the circadian range. Behaviors that serve to produce food may be expected to be par-

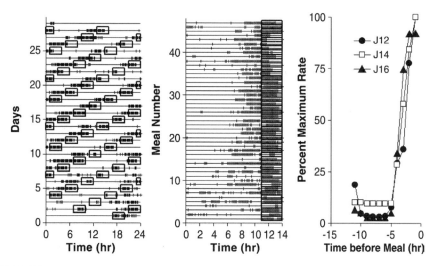

FIGURE 3.5 Activity records for a randomly selected rat from 14-h intermeal intervals showing robust anticipation of 14-h intermeal intervals plotted as a function of days (left panel) and meal intervals (middle panel). Meals (indicated by rectangles) were preceded by a burst of anticipatory responses in constant darkness. Right panel shows mean response rate functions for each individual rat with a 14-h intermeal interval. Data were examined in 10-min bins. If the response rate was greater than one response per minute in the bin, then a vertical deflection was placed on the activity record (left and middle panels). A burst of these anticipatory responses (quantified as at least 4 of 6 bins with responses) occurred in the hour before 70% of the last 10 meals; a lower level of anticipation (16%) was observed 4 h before the meal, $P < .001$. (From Crystal, J.D., *J. Exp. Psychol. Anim. Behav. Process.*, 27, 68–78, 2001a. Copyright © 2001 by the American Psychological Association. Reprinted with permission.)

ticularly relevant in food entrainment from a functional perspective (Timberlake, 1993, 1994).

3.5 IMPLICATIONS FOR THEORIES OF TIMING

Taken together, these results indicate that multiple local maxima in sensitivity to time are observed in the discrimination of time across several orders of magnitude (Figure 3.7; Crystal, 1999, 2001a, 2001b). The existence of a local maximum near a circadian oscillator (Figure 3.7, rightmost peak) and in the short-interval range (Figure 3.7, left side) is consistent with timing based on multiple oscillators (Church and Broadbent, 1990; Crystal, 1999, 2001a; Gallistel, 1990). The location of a local maximum may be used to identify an oscillator's period.

The continuity of timing across several orders of magnitude suggests that principles of short-interval timing may be useful in the analysis of circadian timing, as in the application of scalar variability (see Figures 3.2 and 3.4). Moreover, concepts and methodology from circadian research may be applied to interval timing. For example, documenting free-running behavioral rhythms and phase response curves are standard approaches in circadian research that may be applied to interval timing.

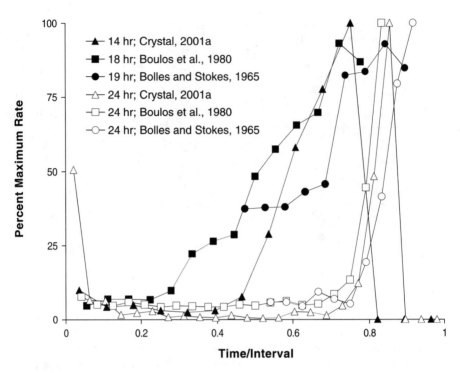

FIGURE 3.6 Rats anticipate intermeal intervals of 14, 18, and 19 h (filled symbols) with less precision (i.e., higher variability) than 24 h (unfilled symbols). Data from Bolles and Stokes (1965) and Boulos et al. (1980), in which meals were earned by pressing a lever, were obtained by enlarging published figures by 200% and measuring each datum at 0.5-mm resolution. (Adapted from Bolles, R.C. and Stokes, L.W., *J. Comp. Physiol. Psychol.*, 60, 290–294, 1965; Boulos, Z. et al., *Behav. Brain Res.*, 1, 39–65, 1980; and Crystal, J.D., *J. Exp. Psychol. Anim. Behav. Process.*, 27, 68–78, 2001a.)

In addition, the identification of additional local maxima in sensitivity to time may be used to identify additional putative oscillators, such as in the temporal region in Figure 3.7 that has not been explored.

The existence of nonlinearities in the sensitivity to time provides constraints for the development of theories of timing. Four theories will be reviewed: scalar timing theory (Gibbon, 1977, 1991), multiple-oscillator theory of timing (Church and Broadbent, 1990), broadcast theory of timing (Rosenbaum, 1998), and stochastic counting cascades (Killeen, 2002; Killeen and Taylor, 2000).

3.5.1 Scalar Timing Theory

Scalar timing theory proposes that a pacemaker sends pulses to an accumulator; the amount accumulated in a currently elapsing interval is compared with a sample from memory of a previously stored reinforced duration, rendering a decision to respond or not to respond. A central feature of scalar timing theory is the prediction that sensitivity to time is constant across a broad range of intervals. The existence of

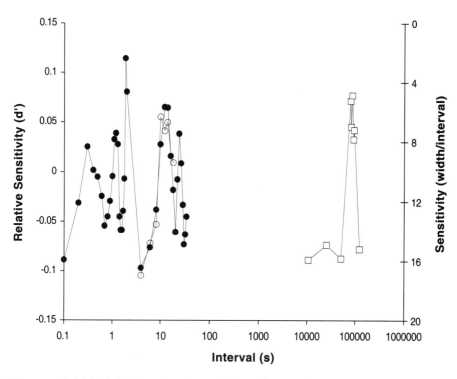

FIGURE 3.7 Multiple local maxima in sensitivity to time are observed in the discrimination of time across several orders of magnitude. The existence of a local maximum near a circadian oscillator (rightmost peak; square symbols) and other local maxima in the short-interval range (left side; filled and unfilled circles) is consistent with the hypothesis that timing is mediated by multiple oscillators. Intervals in the blank region in the center of the figure have not been tested. Left side: Rats discriminated short and long durations, with the long duration adjusted to maintain accuracy at 75% correct. Short durations were tested in sequential order (filled circles; $N = 26$) or independent order (unfilled circles; $N = 20$). Circles represent relative sensitivity using d' from the signal detection theory and are plotted using the y-axis on the left side of the figure. Right side: Rats received food in 3-h meals with fixed intermeal intervals by breaking a photobeam inside the food trough. The rate of photobeam interruption increased before the meal. Squares represent sensitivity, which was measured as the width of the anticipatory function at 70% of the maximum rate prior to the meal, expressed as a percentage of the interval ($N = 29$). The interval is the time between light offset and meal onset in a 12–12 light–dark cycle (leftmost two squares) or the intermeal interval in constant darkness (all other squares). Squares are plotted with respect to the reversed-order y-axis on the right side of the figure. Y-axes use different scales, and the x-axis uses a log scale. (Adapted from Crystal, J.D., *J. Exp. Psychol. Anim. Behav. Process.*, 25, 3–17, 1999; Crystal, J.D., *J. Exp. Psychol. Anim. Behav. Process.*, 27, 68–78, 2001a; and Crystal, J.D., *Behav. Process.*, 55, 35–49, 2001b.)

nonlinearities in sensitivity to time is in conflict with current versions of scalar timing theory.

3.5.2 MULTIPLE-OSCILLATOR THEORY OF TIMING

The multiple-oscillator theory of timing proposes that time is represented by a set of oscillators, each with a unique period (e.g., 100, 200, 400, 800 msec, etc.); rewarded times are stored in an associative matrix and current times are compared with remembered times to render a response decision. The multiple-oscillator theory predicted the existence of nonlinearities in sensitivity to time. The proposal that short-interval timing is mediated by multiple short-interval oscillators provides a basis for the development of a unified theory of timing that can accommodate data ranging from milliseconds to days.

3.5.3 BROADCAST THEORY OF TIMING

The broadcast theory of timing proposes that timing of behavior is based on the time required for neural signals to travel different distances in the nervous system; the variance of delays is proportional to the square of the distance. Therefore, if the delay is subdivided (i.e., time two short intervals instead of one relatively long interval), then the variance of time estimation will be characterized by multiple local peaks. This is consistent with nonlinear sensitivity of time. However, the observation of a local peak in sensitivity at 24 h suggests the involvement of a circadian oscillator rather than the transmission of neural signals across a relatively long distance.

3.5.4 STOCHASTIC COUNTING CASCADES

The model of stochastic counting cascades proposes that counting events (e.g., pulses from a pacemaker) is characterized by failures to set the element in the next stage of a binary counter when a lower element is reset; while counting up, a set failure results in an underestimation of the number of events. Similarly, the resetting of a bit in a binary counter to its zero state may fail to occur; a reset failure results in an overestimation of the number of events. Fallible binary counters can produce nonlinearities in sensitivity to time. Furthermore, the model of stochastic counting cascades may be inserted into other theories of timing as a counting module (Church, 1997; Killeen, 2002; Killeen and Taylor, 2000). Indeed, there is some similarity between the role played by oscillators in the multiple-oscillator theory and Killeen's binary counters; the periods of the oscillators in the multiple-oscillator theory increase by powers of 2, and the behavior of elements in a binary counter is to oscillate with a period corresponding to each element of the counter.

3.6 CONCLUSIONS

The conclusion that emerges from this program of research is that the psychological representation of time is not linearly related to physical time. The observation that the perception of time across many ranges is not linearly related to physical time may provide a basis for the development of a unified theory of timing that encompasses the discrimination of temporal intervals across several orders of magnitude from milliseconds to days.

ACKNOWLEDGMENTS

The preparation of this chapter was partially supported by grants from the National Institute of Mental Health (MH61618) and the National Institute on Drug Abuse (DA13149) to J.D.C. I thank Kenneth W. Maxwell for assistance with reanalyzing published data.

REFERENCES

Allan, L.G., The influence of the scalar timing model on human timing research, *Behav. Process.*, 44, 101–117, 1998.

Aschoff, J., Circadian timing, in *Annals of the New York Academy of Sciences: Timing and Time Perception*, Vol. 423, Gibbon, J. and Allan, L., Eds., New York Academy of Sciences, New York, 1984, pp. 442–468.

Aschoff, J., On the perception of time during prolonged temporal isolation, *Hum. Neurobiol.*, 4, 41–52, 1985a.

Aschoff, J., Time perception and timing of meals during temporal isolation, in *Circadian Clocks and Zeitgebers*, Hiroshige, T. and Honma, K., Eds., Hokkaido University Press, Sapporo, Japan, 1985b, pp. 3–18.

Aschoff, J. and Daan, S., Human time perception in temporal isolation: effects of illumination intensity, *Chronobiol. Int.*, 14, 585–596, 1997.

Aschoff, J., von Goetz, C., and Honma, K.I., Restricted feeding in rats: effects of varying feeding cycles, *Z. Tierpsychol.*, 63, 91–111, 1983.

Bolles, R.C. and de Lorge, J., The rat's adjustment to a-diurnal feeding cycles, *J. Comp. Physiol. Psychol.*, 55, 760–762, 1962.

Bolles, R.C. and Stokes, L.W., Rat's anticipation of diurnal and a-diurnal feeding, *J. Comp. Physiol. Psychol.*, 60, 290–294, 1965.

Boulos, Z., Rosenwasser, A.M., and Terman, M., Feeding schedules and the circadian organization of behavior in the rat, *Behav. Brain Res.*, 1, 39–65, 1980.

Broadbent, H.A., Periodic behavior in a random environment, *J. Exp. Psychol. Anim. Behav. Process.*, 20, 156–175, 1994.

Brunner, D., Kacelnik, A., and Gibbon, J., Optimal foraging and timing processes in the starling, *Sturnus vulgaris*: effect of intercapture interval, *Anim. Behav.*, 44, 597–613, 1992.

Church, R.M., The internal clock: introduction, in *Annals of the New York Academy of Sciences: Timing and Time Perception*, Vol. 423, Gibbon, J. and Allan, L., Eds., New York Academy of Sciences, New York, 1984, p. 469.

Church, R.M., Quantitative models of animal learning and cognition, *J. Exp. Psychol. Anim. Behav. Process.*, 23, 379–389, 1997.

Church, R.M. and Broadbent, H.A., Alternative representations of time, number, and rate, *Cognition*, 37, 55–81, 1990.

Church, R.M., Getty, D.J., and Lerner, N.D., Duration discrimination by rats, *J. Exp. Psychol. Anim. Behav. Process.*, 2, 303–312, 1976.

Church, R.M., Meck, W.H., and Gibbon, J., Application of scalar timing theory to individual trials, *J. Exp. Psychol. Anim. Behav. Process.*, 20, 135–155, 1994.

Collyer, C.E., Broadbent, H.A., and Church, R.M., Categorical time production: evidence for discrete timing in motor control, *Percept. Psychophys.*, 51, 134–144, 1992.

Collyer, C.E., Broadbent, H.A., and Church, R.M., Preferred rates of repetitive tapping and categorical time production, *Percept. Psychophys.*, 55, 443–453, 1994.

Crystal, J.D., Systematic nonlinearities in the perception of temporal intervals, *J. Exp. Psychol. Anim. Behav. Process.*, 25, 3–17, 1999.

Crystal, J.D., Circadian time perception, *J. Exp. Psychol. Anim. Behav. Process.*, 27, 68–78, 2001a.

Crystal, J.D., Nonlinear time perception, *Behav. Process.*, 55, 35–49, 2001b.

Crystal, J.D., Timing inter-reward intervals, *Learn. Motiv.*, 33, 311–326, 2002.

Crystal, J.D., Church, R.M., and Broadbent, H.A., Systematic nonlinearities in the memory representation of time, *J. Exp. Psychol. Anim. Behav. Process.*, 23, 267–282, 1997.

Dews, P.B., The theory of fixed-interval responding, in *The Theory of Reinforcement Schedules*, Schoenfeld, W.N., Ed., Appleton-Century-Crofts, New York, 1970, pp. 43–61.

Fetterman, J.G. and Killeen, P.R., Time discrimination in *Columba livia* and *Homo sapiens*, *J. Exp. Psychol. Anim. Behav. Process.*, 18, 80–94, 1992.

Fetterman, J.G. and Killeen, P.R., Categorical scaling of time: implications for clock-counter models, *J. Exp. Psychol. Anim. Behav. Process.*, 21, 43–63, 1995.

Gallistel, C.R., *The Organization of Learning*, MIT Press, Cambridge, MA, 1990.

Gibbon, J., Scalar expectancy theory and Weber's law in animal timing, *Psychol. Rev.*, 84, 279–325, 1977.

Gibbon, J., Origins of scalar timing, *Learn. Motiv.*, 22, 3–38, 1991.

Gibbon, J. and Church, R.M., Sources of variance in an information processing theory of timing, in *Animal Cognition*, Roitblat, H.L., Bever, T.G., and Terrance, H.S., Eds., Erlbaum, Hillsdale, NJ, 1984, pp. 456–488.

Gibbon, J., Fairhurst, S., and Goldberg, B., Cooperation, conflict and compromise between circadian and interval clocks in pigeons, in *Time and Behavior: Psychological and Neurobehavioural Analyses*, Bradshaw, C.M. and Szabadi, E., Eds., Elsevier, New York, 1997a, pp. 329–384.

Gibbon, J., Malapani, C., Dale, C.L., and Gallistel, C.R., Toward a neurobiology of temporal cognition: advances and challenges, *Curr. Opin. Neurobiol.*, 7, 170–184, 1997b.

Gibbon, J., Morrell, M., and Silver, R., Two kinds of timing in circadian incubation rhythm of ring doves, *Am. J. Physiol. Regul. Integr. Comp. Physiol.*, 247, 1083–1087, 1984.

Hinton, S.H. and Meck, W.H., The "internal clocks" of circadian and interval timing, *Endeavour*, 21, 82–87, 1997.

Killeen, P.R., Scalar counters, *Learn. Motiv.*, 33, 63–87, 2002.

Killeen, P.R. and Taylor, T.J., How the propagation of error through stochastic counters affects time discrimination and other psychophysical judgments, *Psychol. Rev.*, 107, 430–459, 2000.

Kirkpatrick-Steger, K., Miller, S.S., Betti, C.A., and Wasserman, E.A., Cyclic responding by pigeons on the peak timing procedure, *J. Exp. Psychol. Anim. Behav. Process.*, 22, 447–460, 1996.

Kristofferson, A.B., A quantal step function in duration discrimination, *Percept. Psychophys.*, 27, 300–306, 1980.

Kristofferson, A.B., Quantal and deterministic timing in human duration discrimination, *Ann. N.Y. Acad. Sci.*, 423, 3–15, 1984.

Lejeune, H. and Wearden, J.H., The comparative psychology of fixed-interval responding: some quantitative analyses, *Learn. Motiv.*, 22, 84–111, 1991.

Lustig, C. and Meck, W.H., Paying attention to time as one gets older, *Psychol. Sci.*, 12, 478–484, 2001.

Machado, A. and Cevik, M., Acquisition and extinction under periodic reinforcement, *Behav. Process.*, 44, 237–262, 1998.

Macmillan, N.A. and Creelman, C.D., *Detection Theory: A User's Guide*, Cambridge University Press, New York, 1991.

Madrid, J.A., Sanchez-Vazquez, F.J., Lax, P., Matas, P., Cuenca, E.M., and Zamora, S., Feeding behavior and entrainment limits in the circadian system of the rat, *Am. J. Physiol. Regul. Integr. Comp. Physiol.*, 275, 372–383, 1998.

Meck, W.H., Modality-specific circadian rhythmicities influence mechanisms of attention and memory for interval timing, *Learn. Motiv.*, 22, 153–179, 1991.

Mistlberger, R.E., Circadian food-anticipatory activity: formal models and physiological mechanisms, *Neurosci. Biobehav. Rev.*, 18, 171–195, 1994.

Mistlberger, R.E. and Marchant, E.G., Computational and entrainment models of circadian food-anticipatory activity: evidence from non-24-hr feeding schedules, *Behav. Neurosci.*, 109, 790–798, 1995.

Pizzo, M.J. and Crystal, J.D., Representation of time in time-place learning, *Anim. Learn. Behav.*, in press.

Rakitin, B.C., Gibbon, J., Penney, T.B., Malapani, C., Hinton, S.C., and Meck, W.H., Scalar expectancy theory and peak-interval timing in humans, *J. Exp. Psychol. Anim. Behav. Process.*, 24, 15–33, 1998.

Richelle, M. and Lejeune, H., Timing competence and timing performance: a cross-species approach, *Ann. N.Y. Acad. Sci.*, 423, 254–266, 1984.

Rosenbaum, D.A., Broadcast theory of timing, in *Timing of Behavior: Neural, Psychological, and Computational Perspectives*, Rosenbaum, D.A. and Collyer, C.E., Eds., MIT Press, Cambridge, MA, 1998, pp. 215–235.

Stephan, F.K., Limits of entrainment to periodic feeding in rats with suprachiasmatic lesions, *J. Comp. Physiol.*, 143, 401–410, 1981.

Stephan, F.K., Swann, J.M., and Sisk, C.L., Anticipation of 24-hr feeding schedules in rats with lesions of the suprachiasmatic nucleus, *Behav. Neural Biol.*, 25, 346–363, 1979a.

Stephan, F.K., Swann, J.M., and Sisk, C.L., Entrainment of circadian rhythms by feeding schedules in rats with suprachiasmatic lesions, *Behav. Neural Biol.*, 25, 545–554, 1979b.

Terman, M., Gibbon, J., Fairhurst, S., and Waring, A., Daily meal anticipation: interaction of circadian and interval timing, *Ann. N.Y. Acad. Sci.*, 423, 470–487, 1984.

Timberlake, W., Behavior systems and reinforcement: an integrative approach, *J. Exp. Anal. Behav.*, 60, 105–128, 1993.

Timberlake, W., Behavior systems, associationism, and Pavlovian conditioning, *Psychonomic Bull. Rev.*, 1, 405–420, 1994.

Wearden, J.H., Human performance on an analogue of an interval bisection task, *Q. J. Exp. Psychol. Comp. Physiol. Psychol.*, 43, 59–81, 1991.

Wearden, J.H., Rogers, P., and Thomas, R., Temporal bisection in humans with longer stimulus durations, *Q. J. Exp. Psychol. Comp. Physiol. Psychol.*, 50, 79–94, 1997.

White, W. and Timberlake, W., Meal-engendered circadian-ensuing activity in rats, *Physiol. Behav.*, 65, 625–642, 1999.

Zeiler, M.D., Ecological influences on timing, *J. Exp. Psychol. Anim. Behav. Process.*, 17, 13–25, 1991.

Zeiler, M.D. and Powell, D.G., Temporal control in fixed-interval schedules, *J. Exp. Anal. Behav.*, 61, 1–9, 1994.

4 Toward a Unified Theory of Animal Event Timing

Thomas T. Hills

CONTENTS

4.1 Introduction 78
4.2 Lessons from the Psychophysics of Time 79
    4.2.1 Temporal Memory Is Scalar 80
    4.2.2 The Clock Can Discriminate Based on Frequency of Reward 82
    4.2.3 The Clock Can Be Paused 82
    4.2.4 Temperature Affects Clock Speed 83
    4.2.5 The Clock Rate Changes with Reinforcement Rate 84
    4.2.6 Species Comparisons 84
4.3 Circadian and Ultradian Clocks 85
    4.3.1 Circadian Clock Gene mRNA Levels Oscillate with a Circadian Rhythm 85
    4.3.2 Circadian Clocks Are Temperature Compensated 87
    4.3.3 Circadian and Ultradian Rhythms Are Connected, But Not with Event Timing 87
    4.3.4 Circadian Clocks and Event Timers Are Localized to Different Areas of the Nervous System 90
4.4 How Event Timers Might Work 91
4.5 Event Timers in the Natural World 95
    4.5.1 Communication 95
    4.5.2 Navigation 96
    4.5.3 Reproduction 97
    4.5.4 Predator Avoidance 98
    4.5.5 Foraging 99
    4.5.6 Prey Pursuit and Capture 101
    4.5.7 Generalized Learning 102
4.6 Conclusions 103
Acknowledgments 104
References 104

4.1 INTRODUCTION

At the heart of perception lies an animal's ability to recognize change and to make predictions based on the way change has played itself out in the past. Critical to this behavior is the ability to measure time. Without suggesting that it is the basis of predictive science, the perception of time is a primary ingredient in causal understandings.

Animal event timing refers to the process that an animal undergoes in order to recognize an interval of time. In a very anthropomorphic way, one can ask the question this way: How does an animal know the difference between 2 min and 2 h? Or does it know at all? Absolute time may be an artificial consequence of man-made clocks, but animals do behave on temporally defined schedules, and many of them are observed to solve problems in the wild that require a specific estimate of time. Central place foragers must be able to find their way home, and many of them then communicate information about distance to forage sites after they do so (e.g., Kacelnik, 1984; Seeley, 1995). Prey species adjust vigilance schedules to match predator density (e.g., Arenz and Leger, 1997). Prospective mates choose future partners based on the temporal specificity of mating displays (Kyriacou et al., 1992; Michelsen et al., 1985). Foraging animals move on when the rate of resource intake drops too low (Stephens and Krebs, 1986; Charnov, 1976). In all of these cases, animals must perceive the temporal duration of events.

A unified theory of animal event timing requires that we know two things. First, how do nervous systems perceive and learn key aspects of temporal information? What are the molecular and cellular mechanisms required to build perceptual clocks? And second, why have these clocks come to exist? What evolutionary forces have driven their evolution? The answers to these two questions are generally referred to as the proximate and ultimate definitions of a behavior. However, it is a major assumption of this review that, because molecular mechanisms and evolutionary forces feed back on one another, one cannot fully understand one without the other.

To our advantage, psychophysical studies have already established that some animals do measure short intervals of time. This will provide us with a basis for conceptualizing event timing in a definitive way. The purpose of this review will be to formalize what we know about animal event timing by addressing contributions from psychophysics, molecular genetics, neuroscience, and evolutionary ecology, while hopefully combining that information in a way that furthers our understanding of animal event timing.

I begin by reviewing what is known about animal event timing from psychophysics, establishing that animals can measure time and that time measurement appears to obey specific conserved properties over a wide range of species. I then discuss the molecular mechanisms of circadian clocks, which to date are our most well understood molecular clocks. Circadian mechanisms will help us to understand what the event timer is not and also how an event timer might operate at the cellular level. I then suggest possible mechanisms for event timers that agree with the psychophysical and physiological evidence for where these clocks are and how they operate. Finally, I take a tour of ecologically relevant behaviors where we may hope to find event timers at work in the world. This will further our understanding of the evolution of event timers and also allow us to make predictions about what kind of event timers specific animals are likely to have.

4.2 LESSONS FROM THE PSYCHOPHYSICS OF TIME

The Russian physiologist Ivan Pavlov (1849–1936) recognized the learned perception of time in animals when his salivating dogs began to "wait" to salivate after lengthy durations of the conditioned stimuli (story retold by Roberts, 1998). What was at first an unconditioned response to food had become a finely timed prediction about the arrival of food. Since that time, psychophysics has developed a number of techniques for assaying animal event timing. Before covering some of the more established contributions of psychophysics to animal event timing, I present the three most prevalent assays for measuring time perception in animals, as I will refer to them frequently.

One of the first and most easily used assays of animal event timing is the fixed-interval (FI) schedule. FI schedules present a subject with an operandum (e.g., lever) and reward lever presses after a fixed interval of time. There is no deterrent for lever presses prior to the reinforcement, lending subjects to press the lever at will until food is finally procured. Animals that do not temporally regulate behavior based on experience typically show the break-and-run response, which is characterized by a short pause in procuring the reward followed by a steady response rate until the next reward, as illustrated in Figure 4.1a. An alternative to this behavior is the production of scalloped curves in cumulative response records over time, as shown in Figure 4.1b. The scallop is created by the increase in response rate as the reward time approaches, much like the timed increased in salivation observed by Pavlov (1927).

A more informative variation of the FI schedule is the peak-interval (PI) procedure (e.g., Church et al., 1994; Roberts, 1981). In the PI procedure, animals learn that a response will deliver food after a certain interval has passed following the initiation of a signal. The major difference between the FI schedule and the PI procedure is that in the latter, approximately 50% of the time there is no food reward. Instead, the signal stays on for a set period regardless of how the animal responds. It is during these no-reward trials that responses are recorded. Inevitably, the animal increases its response rate until the approximate time of the reward and then decreases its response until the signal is turned off, as shown in Figure 4.2. Movement of the peak function rate left or right is interpreted as a change in the rate of perceived time by the animal.

The temporal bisection procedure requires the subject to discriminate between two signals, long and short in duration (e.g., by pressing the left or right levers, respectively) (Church and Deluty, 1977; Maricq and Church, 1983). This is another psychophysical method for measuring the lengthening or foreshortening of perceived time. The signal duration at which the subject responds long and short with equal probability is called the point of subjective equality (PSE), with typical results illustrated in Figure 4.3.

While the principles that I report here are not established for all animals, they are well established for vertebrates (i.e., birds, mammals, and fish) (see Paule et al., 1999). I will point out exceptional species in "Species Comparisons" (Section 4.2.6).

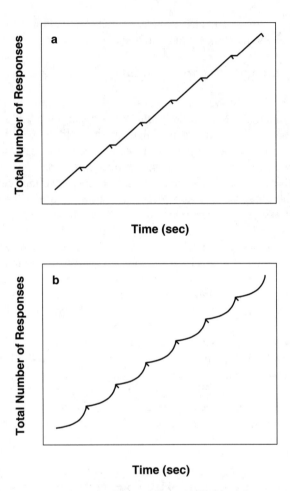

FIGURE 4.1 Typical results from the FI procedure. (a) Break-and-run response. Response rates are constant following a slight delay after the reward (downward slash). (b) Scalloped response. The scalloped curves are the result of increasing response rates near the point of reward. Response rates are not as smooth in real traces, because lever presses lead to discrete jumps in cumulative response number. (Adapted from Ferster, C.B. and Skinner, B.F., *Schedules of Reinforcement*, Appleton-Century-Crofts, New York, 1957.)

The following list of phenomenological characteristics represents a basis for understanding what a mechanistic definition of event timers must ultimately explain. This list will also help us make predictions about how we may expect animals to behave in the wild in response to specific constraints on their perception of temporal events.

4.2.1 Temporal Memory Is Scalar

The ability to discriminate two temporal cues is reduced in a predictable way as the duration of those cues is increased. This property, known as Weber's law, describes

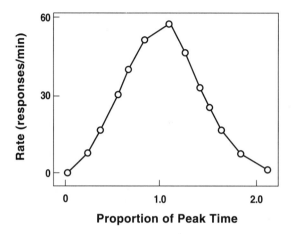

FIGURE 4.2 Typical results from the PI procedure. The animals are trained by rewarding them for the first lever press after a fixed time from the initiation of the signal (e.g., light or tone). Data are recorded on trials without reward. Peak response time is usually near the point of reward. The results here are normalized to the peak time, to show that the relative shape of the curve is conserved for different durations. (Adapted Roberts, S., *J. Exp. Psychol. Anim. Behav. Process.*, 7, 242–268, 1981.)

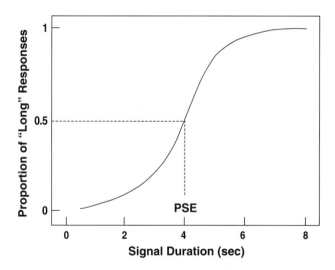

FIGURE 4.3 Typical results from the temporal bisection procedure. Animals are trained by rewarding them for pressing the left or right lever following signals of short or long duration (e.g., 2 and 8 sec), respectively. Probe trials are then conducted in which intermediate times are tested for left or right responses. Animals are not rewarded on probe trials. The PSE is the point at which the animal shows equal proportions of long and short responses. (Adapted from Church, R.M. and Deluty, M.Z., *J. Exp. Psychol. Anim. Behav. Process.*, 3, 216–228, 1977.)

a linear relationship between the duration of a stimulus, I, and the measure of uncertainty associated with that stimulus, ΔI. The relationship

$$\Delta I = kI$$

describes the amount by which a second stimulus must be changed from I in order for the two to be discriminated. It also explains the proportionality between the timed interval and the mean and standard deviations associated with that interval. This relationship breaks down for very short intervals of several seconds (discussed in detail by Cantor, 1981). This relationship also breaks down as timed intervals approach the 24-h phase of the circadian clock (Crystal, 2001).

The results from the PI procedure resemble a Gaussian distribution, as illustrated in Figure 4.2. Longer intervals between the beginning of the signal (e.g., light or sound) and the reward lead to wider distributions (Church et al., 1994; Roberts, 1981). Regardless of the length of the duration, the shape of the curve is conserved, showing that the variance scales with the length of the duration. In the temporal bisection procedure, increasing the time between signals increases the accuracy, and longer duration signals require greater disparity in length in order to achieve the same levels of discrimination (e.g., Bizo and White, 1997; Roberts, 1998).

Weber's law applies in its general form to rats (Church and Gibbon, 1982), pigeons (Cheng and Roberts, 1991), and humans (Wearden, 1991). The ecological consequences of Weber's law are discussed more specifically by Bateson (this volume) and in Section 4.5 of this chapter.

4.2.2 The Clock Can Discriminate Based on Frequency of Reward

Animals can distinguish between the reward reliability of different temporal cues. This insures that animals have the temporal acuity to rate potential rewards based on their frequency. Ecologically, this means that animals can rank foraging sites based on their density of prey.

That rats can associate different cues with different frequencies of reward was shown with the PI procedure. Light was associated with an 80% probability of food, and a tone was associated with a 20% probability of food. Animals responded to the tone at about one quarter of the peak response rate for the light, showing that they can match the memory for two different intervals with the relevant external stimuli (Roberts, 1981). This same behavior has been observed using a different procedure in pigeons (Killeen et al., 1996). In fact, studies of risk-sensitive foraging have established that animals can make the same distinctions in the wild. I will discuss this in more detail in Section 4.5.5.

4.2.3 The Clock Can Be Paused

Pigeons and rats trained in the PI procedure will respond to a 10-sec signal blackout in the middle of a trial by moving their peak response rate back by 10 sec (see Buhusi, this volume; Buhusi and Meck, 2002; Buhusi et al., in press; Hopson, this volume; Roberts, 1981, 1998). This is true for a number of different blackout times, with the interesting caveat that longer blackout times can lead to a slower resetting of the clock (de Vaca et al., 1994).

The pausing of the clock suggests higher-order control in the nervous systems of animals capable of event timing. This kind of control is unlikely to be found in more primitive event timers (see Section 4.4 and the discussion of the decay timer) or in timers like the circadian protein oscillations described in Section 4.3.

4.2.4 TEMPERATURE AFFECTS CLOCK SPEED

The first evidence that a temporal sense was not temperature compensated was provided by Hoagland (1935), who made this discovery after subjecting his sick wife to various temporal acuity tests while measuring her temperature. At hotter internal temperatures her counting rate was faster than at lower temperatures. To exclude illness, he did the same for volunteers after short periods in a freezer. His data set exhibits an exquisite linear relationship between the inverse of temperature and the log speed of counting (see Wearden and Penton-Voak, 1995).

In perhaps the only ecological study of direct interval timing and temperature, the parasitoid wasp *Trichogramma dendrolimi* was demonstrated to lay eggs in its insect host based on the duration of its walk across the host's long axis (Schmidt and Pak, 1991). The hosts are eggs of larger insects, which vary greatly in size. This lends itself to an adaptive measure of host size. Too many or too few eggs laid on the host can lead to either starvation or underutilization of resources, providing a clear evolutionary force for optimal host size assessment. The host-crossing can take between 0.5 and 20 sec, depending on the host size, but for identically sized hosts, the speed is increased at higher ambient temperatures. At higher temperatures the wasp also lays its eggs faster. However, the wasp is able to compensate for this temperature adjustment and lays the same number of eggs in identically sized hosts regardless of the ambient temperature. The wasps may have a reduced estimate of elapsed time at lower ambient temperatures to compensate for their reduced speed (Schmidt and Pak, 1991).

In circadian studies of the courtship song of *Drosophila melanogaster*, the timing pattern of the behavior is temperature compensated and directly correlated with the duration of the free-running circadian clock (Iwasaki and Thomas, 1997). Under the assumption that the wasp's internal estimate of time is affected, in the same way that Hoagland's wife was, this is our first evidence that circadian clocks are unrelated to event timing. Circadian clocks are temperature compensated, while event timers are not. I pursue this argument further in Section 4.3.

The linearity of temperature effects is limited. Severe heat stress reduces response rates, and very severe heat stress changes the motivational state of the animal (e.g., they try to escape) (Richelle and Lejeune, 1980). The problem is that temperature changes could be affecting other physiological properties with little effect on the clock. An experiment by Rozin (1965) looked at temperature effect in the goldfish *Carassius auratus*. In FI trials, goldfish at different temperatures showed similar scalloped response curves that were different in absolute, but not relative response rates over the trial interval. This suggests that effects of temperature may be operating downstream of the clock, to modify the absolute rate of behavioral processes.

4.2.5 THE CLOCK RATE CHANGES WITH REINFORCEMENT RATE

Increased rates of reinforcement increase the relative temporal rate such that, in the temporal bisection procedure, the PSE is moved toward the long response. This is supported by several studies that show altered PSEs for different between-trial durations (Bizo and White, 1997; Fetterman and Killeen, 1991; Morgan et al., 1993).

The consequences of reinforcement rate on the subjective perception of time are quite significant. To the extent that this phenomenon is real in the wild, it complicates the possibility of optimal foraging in the sense of Charnov's marginal value theorem (Charnov, 1976). In this and many optimal foraging models, the ability of the animal to measure absolute time is directly responsible for its preference for certain food types and times of patch departure. If animals are incapable of measuring absolute time, then the ability to forage optimally should be predictably changed. The consequences of rate-biased time perception have been worked out in detail for the marginal value theorem (Hills and Adler, 2002). One of the conclusions of this work is that because of the costs associated with running the nervous system, animals have probably evolved to tune their arousal (and therefore temporal accuracy) to meet the requirements of specific environments. So they pay the price of rate-biased optimal foraging to avoid the price of inappropriately tuned perceptual timing.

4.2.6 SPECIES COMPARISONS

Animals show different capacities for learning and different sensory biases (e.g., Bitterman, 1975; Dukas, 1998). The capacity for learning temporal intervals has evolved in vertebrates and is supported in at least one parasitoid wasp. Many vertebrates show scalloped responses (Richelle and Lejeune, 1980). The consensus among fish studies is that some do and some do not (Eskin and Bitterman, 1960; Richelle and Lejeune, 1980; Talton et al., 1999). The African mouthbreeder (*Tilapia macrocephala*) apparently fails to show a scalloped response, whereas goldfish (*Carassius auratus*) appear to have no problem with it (Rozin, 1965). Pigeons can show varying responses under a wide range of different reinforcement regimes (Ferster and Skinner, 1957). Surprisingly (see Section 4.5.2), honeybees (*Apis mellifera*) show break-and-run behavior in pseudonaturalistic environments with delays of 20 or 90 sec (Grossman, 1973).

Are insects in general capable of timing short intervals? It is certainly possible that we have not yet asked the question in the appropriate way. It is equally possible that the genesis of certain brain structures coincident with vertebrate evolution mark a significant dichotomy in the evolutionary history of event timing. The evolution of the vertebrate forebrain facilitates the possibility of a homologous structure with the hippocampus in all vertebrates (Colombo et al., 2001; Portavella et al., 2002). The role of the hippocampus in spatial and temporal learning and memory is widely demonstrated (Jackson et al., 1998; Ono et al., 1995; Thompson, et al., 1982). Given their capacities for spatial learning in navigation, the possibility of an analogous structure in insects should not be ruled out (Dyer, 1998). I address this relationship in more detail in Section 4.4.

4.3 CIRCADIAN AND ULTRADIAN CLOCKS

Little is know of the molecular and cellular mechanisms of event timing. Molecular geneticists have yet to isolate an event timer in animals (see Cevik, this volume). Neuroscientists have the power to isolate function of gross anatomical regions by lesion and transplant, but cell number limits functional understanding of populations of neurons almost completely to neural simulations. Neuroscientists are not readily equipped to investigate genetic and molecular function.

The point of discussing circadian and ultradian clocks is twofold. First, it will allow us to make an important distinction between mechanisms of endogenously controlled time and those of perceived time. Second, it will provide a basic description of a molecular clock as a kind of null model against which we can address future questions.

Animal timing refers to a broad class of behaviors. These include general life history events such as when to stop making sperm and the age of first reproduction. They also include rhythmic behaviors with periods on the order of a year (circannual), longer than a day (infradian), a day (circadian), or shorter than a day (ultradian). The suggestion that animals have evolved to organize their lives in time could refer to any of these levels of behavioral timekeeping. Animal event timing, as I use it here, refers to none of these.

The fundamental mechanisms of animal event timing are sensory based. Event timing requires the ability to perceive and remember the duration of an external event. Rhythmic behaviors and the scheduling of large-scale life events are based on endogenously controlled, often genetically predetermined, timetables. Circadian rhythms are 24 h long not because the animal learns a 24-h period in its lifetime, but because it is genetically predisposed to timing an interval approximating the rotation of the earth. On another planet, or in a lab where light–dark (LD) cycles are manipulated, circadian clocks are far from functional in terms of recording the length of the LD cycle. Daily activity cycles appear to fall into the same category. They are driven in well-defined ways by circadian rhythms. It is rare for an animal to learn when to be most active in the LD cycle (but see Section 4.5.4).

This section is also designed to point out that the clock used to measure event duration is not necessarily the same clock that measures time of day. Experiments designed to expose animal event timers can be confounded by this time-of-day clock. For example, intervals of 24 h do not need to be measured, because they can be posted to the circadian phase. I will provide several lines of evidence in this section that the time-of-day clock is inefficient for measuring intervals different from 24 h and show that it is quite distinct from animal event timers.

4.3.1 CIRCADIAN CLOCK GENE MRNA LEVELS OSCILLATE WITH A CIRCADIAN RHYTHM

Circadian regulation of gene expression is well established in plants (i.e., *Arabidopsis* — Millar et al., 1995), fungi (i.e., *Neurospora* — Dunlap, 1996), insects (i.e.,

Drosophila — Iwasaki and Thomas, 1997), unicellular microorganisms (Lloyd, 1998), and vertebrates (Takahashi, 1995). I will describe in some detail only the *Neurospora* and *Drosophila* system because they are typical and well-studied examples of circadian dynamics. Our understanding of the molecular details governing the circadian clock is growing rapidly (reviewed in Okamura et al., 2002; Stanewsky, 2002) and is beyond the scope of this review, but I hope to describe the basics of the system in enough detail to provide a foundation for thinking about molecular event timers.

The filamentous bread mold, *Neurospora crassa*, exhibits a circadian rhythm in its asexual spore formation (conidiation) (Edmunds, 1988). The *frq* gene encodes a central element in this circadian rhythm. Both the *frq* gene mRNA and its protein product, FRQ, cycle in amount with a period of approximately 24 h (Dunlap, 1996). In a 24-h LD cycle, *frq* mRNA and FRQ levels are at their lowest point near the middle of the dark phase. Slowly rising, they peak approximately 10 to 12 h later, with FRQ levels always lagging behind mRNA levels by about 3 h.

The vinegar fruit fly, *Drosophila melanogaster*, shows oscillations in two of its circadian clock gene transcripts (*per* and *tim*) that are exactly out of phase with its own *frq* (Hardin et al., 1992). Their periods are the same, but when *frq* mRNA is beginning to rise, *per* and *tim* mRNAs are beginning to fall. Like FRQ, PER lags behind its mRNA transcript by several hours (Iwasaki and Thomas, 1997). Interestingly, a recently identified protein in mammals, *mPer1*, sharing high sequence similarity with the *Drosophila* PER protein, oscillates in phase with *frq* (Shigeyoshi et al., 1997), and a mutation in a similar gene in humans is correlated with familial advanced sleep phase syndrome (Toh et al., 2001).

In *Drosophila*, the PER and FRQ proteins are localized to the cytoplasm. In short order, FRQ enters into a binding complex with (at least) itself and PER enters into a 1:1 heterodimeric complex with TIM, as suggested by studies in the yeast two-hybrid system (Dunlap, 1996). These relationships appear to stabilize FRQ and PER in the cytosol. They also appear to be necessary for translocation to the nucleus, where they suppress their own expression (Aronson et al., 1994; Dunlap, 1996). This is the negative feedback mechanism that generates the cyclical rise and fall of protein products, allowing the animal to keep track of time.

In the absence of light (dark–dark (DD)), all of these gene products show free-running periods of about 24 h. The phase of gene expression can also be directly entrained by light, but this response is limited to certain intervals of the cycle. For example, in *Neurospora*, when FRQ levels are low, a pulse of light leads to rapid transcription of the *frq* mRNA, moving the phase of oscillation forward in time (Crosthwaite et al., 1995). Given the swiftness of the response, it is believed that light acts directly on the *frq* promoter (Dunlap, 1996). *mPer1* shows a similar response to light (Shigeyoshi et al., 1997). In *Drosophila*, light has little effect on *per* mRNA but reduces levels of TIM. Constant light breaks down the entire rhythm (Power et al., 1995). Given this evidence, it is not surprising that manipulations by light are unable to change the period of the circadian rhythm, but are able to adjust its phase.

Light is not the only source of entrainment for the circadian clock. Temperature is also a cue (Iwasaki and Thomas, 1997). There is recent evidence that the circadian clock can also be affected by conditioned stimuli (Amir and Stewart, 1996). Some disagreement exists about the extent of this phenomenon. There is evidence that social contacts in humans can synchronize circadian pacemakers (Hastings, 1997), but recent research on blind subjects and manipulated light schedules in sighted subjects supports the necessity of light as an entrainment cue (Czeisler, 1995). Pioneering work by Winfree (1980) led him to predict strong resetting of the circadian clock by light across a wide array of species, and this appears to hold true.

4.3.2 CIRCADIAN CLOCKS ARE TEMPERATURE COMPENSATED

Although there are physiological temperature limits for the maintenance of circadian rhythms, there is no effect of stable temperatures within these limits (Iwasaki and Thomas, 1997). However, temperature changes can elicit rises or falls in gene expression (Edery et al., 1994; Rensing et al., 1995). The mechanism by which this temperature compensation operates has been worked out in detail for *Neurospora*.

Within the first 100 codons of the *frq* gene there are three methionine codons (AUG) at codons 1, 11, and 100 (Liu et al., 1997). Codon 11 is not used to initiate the sequence under normal conditions, but codons 1 and 100 are. Thus, in wild-type *Neurospora* there are two FRQ proteins, $FRQ^{100-989}$ and FRQ^{1-989}, and either of them alone is sufficient for circadian rhythms. At low temperatures, $FRQ^{100-989}$ is preferentially transcribed. At high temperatures, FRQ^{1-989} is preferentially transcribed. The ratio between the absolute levels of the two gene products is thus controlled by temperature. Removal of either form disturbs the ability of the clock to compensate for physiological temperature extremes (Liu et al., 1997).

It is possible that other mechanisms operate to compensate for temperature. Both *per* and *frq* encode an internally repetitive array of Thr-Gly codons. In *D. melanogaster* this region is polymorphic in length (Kyriacou et al., 1992). Work done by Rosato et al. (1996) shows a significant latitudinal cline in this repeat sequence ranging from Europe to North Africa. These sequences are directly related with the ability of the flies to maintain a compensated circadian rhythm at different temperatures (Sawyer et al., 1997). Furthermore, deleting the Thr-Gly region produces flies that have temperature-sensitive circadian periods (Kyriacou et al., 1992).

4.3.3 CIRCADIAN AND ULTRADIAN RHYTHMS ARE CONNECTED, BUT NOT WITH EVENT TIMING

Ultradian behaviors appear to oscillate in circadian time. They consist of such behaviors as the defecation cycle in *Caenorhabditis elegans* (Iwasaki et al., 1995) and wheel-running behavior in mice (Antoch et al., 1997). The relationship between circadian and ultradian behaviors is best understood in *D. melanogaster*. In a screen looking for mutants of the pupal–adult eclosion phenotype, Konopka and Benzer (1971) identified three mutants of the per gene: per^s showed a 19-h cycle, per^{L1} showed a 29-h cycle, and per^{01} appeared to be arrhythmic. per^s and per^{L1} mutations are due to single amino acid substitutions. per^{01} encodes a stop codon at the 460th residue (Yu et al., 1987). The amazing thing about these mutants is that their ultradian behaviors, like activity and courtship song cycles, are proportional to their circadian behaviors (Kyriacou et al., 1992). Where wild-type male flies have 60-sec courtship song cycles, per^s flies have 40-sec cycles, per^{L1} flies have 80-sec cycles, and per^{01} flies show little evidence of cycling at all.

D. melanogaster females show a preference for 55-sec songs, and *D. simulans* females prefer 35-sec songs. It is a reasonable assumption then that the various *per* mutations would show changes in female preference for male song duration. In fact, just the opposite is true (Kyriacou et al., 1992). *D. melanogaster* females, regardless of their *per* genotype, prefer 55-sec songs. To the extent that flies are not able to use other cues, this provides a fundamental difference between ultradian transmitter and receiver mechanisms. It also suggests that the event timer in *Drosophila* is not driven by an underlying circadian oscillation. This may be one of the more telling observations about circadian and event timing, explaining the peculiar inability of bees to discriminate short intervals despite their mastery of circadian time.

The *Zeitgedachtnis* (time sense) of honeybees is widely reported (Saunders, 1971; Seeley, 1995; Wilson, 1971). Bees can relocate almost anything provided it is presented at 24-h intervals (Saunders, 1971; Moore et al., 1989). As well, "marathon" dancers who return to the hive dance floor to dance for hours following a fruitful foraging trip compensate for the sun's motion as the day passes (Wilson, 1971). However, in FI trials, honeybees show no evidence that they can learn anything about a 2-min interval (Richelle and Lejeune, 1980).

Despite this evidence for a distinction between perceived time and circadian time, the relationship between ultradian and circadian behaviors is a profound one. The Syrian hamster *tau* mutant has a circadian rhythm that is shortened by 4 h from the wild type. Wild-type females have approximately 30-min periods of cortisol and luteinizing hormone fluctuations that are slightly shortened in the *tau* mutant (Loudon et al., 1994). The recently cloned circadian *Clock* gene in the mouse was isolated in a massive screen for mutants that exhibited altered wheel-running activity rhythms in constant darkness (King et al., 1997a, 1997b). Close inspection of King et al.'s (1997a) wheel-running data shows that the wheel-running activity cycle is very cyclical in its ultradian oscillations.

The pineal gland's circadian rhythm and direct control of numerous hormonally controlled behaviors is further evidence for this relationship between circadian and ultradian time. At the receiving end of the suprachiasmatic nucleus (see Section 4.3.4), the pineal gland is located at the posterior dorsal aspect of the diencephalon. The rhythmic release of melatonin is its most obvious circadian feature, as melatonin is directly responsible for transmitting the circadian LD signal to the rest of the organism (Menaker, 1997). Among other things, melatonin is known to affect human thermoregulation. Given the definitive relationship between body temperature and sleep (Wever, 1992), the pineal gland has a clear role in the sleep–wake cycle. In an experiment reported by Lavie (1992), eight young male adults experienced 20-min "days" for 48 h in a sleep–wake cycle of 7 and 13 min. In this experiment subjects were instructed to attempt to fall asleep and to resist sleep for 7 and 13 min, respectively, every 20 min. The results show a well-defined sleep–wake cycle at 24 h, despite the best efforts of the subjects to overcome this interval.

Studies of interval timing also reveal a circadian influence on attention and memory. Meck (1991) tested rats for their ability to discriminate 2- and 8-sec durations (temporal bisection procedure) over the course of the day. There was no effect of circadian phase on the clock rate — the PSE did not change. However, the overall sensitivity to time (measured by the variability of the response) was highest during the dark phase and lowest during the light phase. This relationship has also been demonstrated for honeybee arrival times, with more accuracy in the morning than later in the day (Moore et al., 1989). If there is a clear relationship between circadian clocks and event timing, this is likely to be it: circadian phase influences the accuracy of event timers. As an example, experiments on human isolation in Antarctica, submarines, and underground are typically difficult to interpret because subjects often lose their abilities to concentrate; it is as if the circadian mechanisms controlling attention are somehow lost (Harrison et al., 1989).

Unfortunately, the majority of psychophysical literature fails to mention the time of day at which experiments were performed or any salient features of the LD cycle. Ecological studies on animal behavior are seldom better, the unspoken assumption being that the behavior, once initiated, is stereotypical. Variance is regarded as "noise" in the form of genetic variance or sensitivity of the animal to subtle environmental factors. However, given Meck's (1991) results, it may be predictable circadian oscillations in the animal's sensitivity or general attention that is responsible for the behavioral variance. If multiple animals are observed over a given day, then time-of-day effect must be considered in the analysis.

A second issue with respect to the importance of recording LD cycles is that animals that are not reared in 24-h LD oscillations may not exhibit circadian rhythms (Richelle and Lejeune, 1980). For example, if the LD cycle at a particular geographical location or in a particular lab is highly unpredictable, then the activity rhythms observed in animals reared there may be quite arrhythmic or may be based on other cues, such as temperature or resource availability. These may lead to uncontrolled or unpredictable behavioral results that are not replicable under other LD cycles.

Finally, animals that exhibit circadian rhythms may be biased toward remembering 24-h event times. Honeybees are exceptionally good at following the daily peaks in pollen and nectar production by visiting only certain species of flowers at certain times of day (Saunders, 1971; Seeley, 1995). However, bees are unable to learn to return at non-24-h periods. Kestrels (*Falco tinnunculus*) and starlings (*Sturnus vulgaris*) have also been observed to follow food abundance in time, and both can be trained to return to a feeder at specific times of the day (Bell, 1991). Food-anticipatory activity rhythms have also been observed for rats fed at 24-h intervals (Rosenwasser, 1984). Rats show a clear inclination for remembering feeding times set at 24-h intervals over much shorter (3 to 14 h) or longer (34 h) times, as shown by Crystal (2001). Given the nature of temporal memory, which I discuss further in "How Event Timers Might Work" (Section 4.4), it is probably also the case that there is an annual memory bias, as animals may be able to remember more clearly their seasonal context simultaneously with important events.

More work needs to be done to verify the distinction between circadian rhythms and event timers (see Crystal, this volume). Molecular examinations of mechanisms involved in event timing require a phenotype for the isolation and cloning of relevant genes. A designer behavior would be useful in this respect. Tim Tully (DeZazzo and Tully, 1995) has made superb use of electrical stimulus to dissect memory formation in *Drosophila*. To determine if flies are capable of learning temporal durations, experimenters could train flies in a periodic shocking regime. For example, every 60 sec the experimenter runs a current through the cage. If flies have event timers, they may learn to associate flight at specific temporal intervals with the absence of shock. Reward paradigms might also be useful, in which flies are rewarded with food at consistent intervals. Once the phenotype is established, various *per* mutants could be assayed for defects in event timing (see Cevik, this volume). These mutants could then be assayed for more ecologically relevant behaviors associated with fitness. *C. elegans* might also be a useful subject for molecular study of event timers, as their nervous system and genome are amenable to neural and genetic investigation.

4.3.4 Circadian Clocks and Event Timers Are Localized to Different Areas of the Nervous System

The observation of circadian rhythms in unicellular organisms and the expression of circadian genes in many peripheral tissues in mammals suggest the possibility that all cells contain circadian rhythms (Balsalobre, 2002). While it is difficult to discount this possibility, expression patterns of circadian genes and brain lesion and transplant studies support specific localization of circadian clocks and event timers.

It is fairly well established in vertebrates that there is one circadian pacemaker: the suprachiasmatic nucleus (SCN) (Ralph and Hurd, 1995). The SCN is located over the optic chiasm rostral to the supraoptic nucleus. Cross-genotype transplants of SCN for Syrian hamster *tau* mutants have unambiguously defined the SCN as the major control mechanism of the mammalian circadian period (Ralph et al., 1990). The SCN, when moved from one hamster to another, is sufficient to alter the circadian rhythm. The SCN also shows strong expression of the *mPer1* and *Clock* genes, which are rapidly induced there by exposure to light (Hastings, 1997; Shigeyoshi et al., 1997).

SCN neurons in hypothalamic slices are observed to fire rhythmically at around 8 to 10 Hz during the day and 2 to 4 Hz at night (Hastings, 1997; Wagner et al., 1997). Isolated SCN neurons show a spontaneous rate of firing at near the same rates as in slices (Hastings, 1997). They also show a higher-order rhythm in frequency over the circadian day. The SCN reaches most of its targets via thalamus- and hypothalamus- (e.g., the pineal gland) mediated hormonal control (Hastings, 1997).

SCN function in natural environments is still poorly understood. In the laboratory, rodents rendered arrhythmic by SCN lesions live normal life spans (DeCoursey and Krulas, 1997). Besides circadian arrhythmia, the hibernation cycle is also affected by SCN lesions. SCN-lesioned female squirrels (*Spermophilus lateralis*) were observed to hibernate for almost 2 years in a laboratory setting (Ruby et al., 1996). Psychophysical experiments on event timing in SCN-lesioned animals show that, despite their inability to maintain daily rhythms, they can accurately time short intervals (Mistleberger, 1993).

An ambitious study of SCN-lesioned chipmunks returned these animals to the wild after surgery. Unfortunately, after 3000 h of fieldwork over slightly more than 2 years, there was essentially no observed effect of the SCN removal (DeCoursey and Krulas, 1998). The SCN-lesioned chipmunks did show evidence of brief arrhythmia and nighttime restlessness in the wild, but the activity cycles were largely the same for all chipmunks studied. There was also no significant effect on survivability, reproduction, or winter torpor duration.

While there is no direct relationship between the SCN and event timers, event timing does have a direct relationship with the hippocampus. The hippocampus has played a starring role in mammalian learning and memory since the lesioning of the human subject H.M.'s medial temporal lobes in the 1950s. After the surgery, H.M. was completely unable to form new declarative memories (Churchland and Sejnowski, 1994). Declarative memories are akin to semantic memory in the sense that the memory is based on the learning of semantic statement. Procedural memory, for which H.M. showed only minor deficit, is based on a kind of implicit function learning. For example, H.M. was perfectly capable of learning a motor skill, but he would be unlikely to remember that he had learned it.

Since that time, a great deal of attention has been paid to the role of the hippocampus in the formation of spatial memory. In food-storing birds, damage to the hippocampus disrupts memory for storage sites (Krebs et al., 1989). As well, bird species that store food have significantly larger hippocampal formations than those that do not. The operation of the spatial function of the hippocampus appears to work via location-coding neural cells (place fields) (e.g., Mizumori et al., 1996). When the animal returns to a specific spatial location, similar cells in the hippocampus fire. This information appears to be primarily visually coded.

The function of the hippocampus has also been established in the formation of episodic memory, but not necessarily in its retrieval (Fletcher et al., 1997). Episodic memory is typically associated with the ability to recollect past events, as originally introduced by Endel Tulving (1972). It was coined to represent any type of memory that was not strictly lexical (semantic memory), and thus it would represent essentially all temporal forms of learning. Recent evidence using positron emission tomography (PET) finds the retrieval of episodic memory events localized in the right prefrontal cortex, with a limited functional role played by the hippocampus (Fletcher et al., 1997).

Hippocampal lesions are shown to operate in the formation of temporal memory in rats (Meck et al., 1984). Once the memory is formed, however, the hippocampus becomes less important. Effects of hippocampal lesions include a lack of avoidance for previously visited maze sites, the inability to withhold or inhibit previously learned response patterns, and the foreshortening of temporal memories (Kesner, 2002; Meck, 1988; Meck et al., 1984).

4.4 HOW EVENT TIMERS MIGHT WORK

Given the hippocampal evidence, it seems likely that the hippocampus operates as a kind of ticking backdrop upon which episodic events can be hung until they are stored in a longer-term reference memory. Certainly hippocampal patterns are utilized in numerous ways, as previously assigned place fields do fire in new locations over the lifetime of a rat (Mizumori, et al., 1996).

This kind of hippocampal tagging has been modeled in detail for sequential spatial memory using a large-scale simulation of hippocampal function (Wallenstein and Hasselmo, 1997; Wallenstein et al., 1998). Multicompartmental pyramidal cells are shown to have synchronizing behavior over multitrial learning, and this is suggested as a mechanism for sequential learning. The pyramidal cells, which fire chaotically before the learning trial, use this prerandomization to settle into nonpredictable and location-specific patterns of firing. This is reminiscent of the kind of unsupervised self-organization exhibited by Kohonen networks (Haykin, 1994). It is different in that neighboring events in space share similar contextual patterns of hippocampal cell firing. In this way, when a sensory stimulus arises that is unfamiliar, the cells that fire in response to that stimulus partially stimulate the "memory" of familiar neighboring events.

Is it possible that temporal interval discrimination works by a similar mechanism? Coexisting with place fields, we would expect to find an analogous kind of "time field." By necessity, the time field could not operate exactly like the place field. In space, sensory input is constant and animals could run spatial information through the hippocampal filter persistently. This would perpetuate the synchronous cell firing in a way that time fields might be unable to do. For example, in order for a rat to learn the duration between signal and reward, it must have a sense of time. In space, the interval is exogenously applied — the space field moves and the sensory cells respond. In time, the interval must be measured by an internal clock.

What could the clock be if it is not related in some way to the circadian clock? The observation of high-frequency intracortical oscillations by electroencephalograph (EEG) provides a basis for a possible clock hand in vertebrates. It has been suggested before that composite cortical waveforms measured by EEG operate as a pacemaker in duration timing (see Artieda and Pastor, 1996; Pouthas, this volume; Sakata and Onoda, this volume; Treisman et al., 1994). Unfortunately, EEGs measure waveforms produced by populations of cells, and it is often very difficult to isolate particular areas of the brain for specific analyses. Nonetheless, evidence for the EEG pacemaker hypothesis has been provided by work showing interactions between auditory click rates, certain EEG components, and the simultaneous assessment of duration (Treisman et al., 1994).

A more specific neural pacemaker central to the hippocampus is provided by observations of theta and gamma oscillations from *in vivo* recordings of the hippocampus (Wallenstein and Hasselmo, 1997). In the sequential place field model referred to above, theta and gamma oscillations are produced by GABAergic receptor inhibition of recurrent collaterals among pyramidal cells and between pyramidal cells and nonpyramidal neurons. The effective nature of this system is to iterate and check at each time step, such that internal and external signals are integrated with the background hippocampal pattern in a meaningful way. Cells may oscillate at different or longer intervals and become associated with the duration when in specific states, such that a series of population patterns is gradually learned over progressive trials.

Iterate and check, however, may be only half of the story. Animals involved in temporal training tasks often behave in a peculiar but stereotypical way that might be further related to the spatio-temporal integration of the hippocampus. This behavior is characterized by seemingly unrelated activities between the stimulus and the reward. For example, a rat might chew its tail, a monkey might jump around its cage in a repetitive way, and a human might tap her finger or shake her head. It is also observed that animals engaging in these collateral behaviors are more efficient in their response time than animals that do not perform these behaviors (Richelle and Lejeune, 1980). These behaviors could act as a kind of context counting. Assuming the animal is unable to count (or asked not to, in the case of humans), it may, in the process of learning the interval, learn sequential behaviors associated with the particular sequence of population patterns in the hippocampus. This makes perfect sense in terms of the spatio-temporal aspects of hippocampal learning — it provides an efficient way to turn time into space, which can then be sensed continuously over the interval. It also provides a physiological basis for the behavioral theory of timing and multiple-timescale theory. Not only does the animal iterate and check, it reinforces the dynamic pattern of cellular events by engaging in context-specific behavior.

This contextual theory of timing may also explain why humans appear to have different timers for long and short durations — one temperature compensated and one not, respectively (Aschoff, 1998). The explanation may be that for short intervals, animals count either behaviors or some internally registered series, but for longer intervals, especially in humans, they do not keep track of time, but reflect on how much time should have passed given the events that have taken place in the interval. Whereas counting provides instantaneous and a more likely temperature-modulated mechanism for measuring time, reflection is initiated on a more variable schedule and is more a measurement of what the interval looked like after the fact than what it actually felt like while it was happening. That is to say, reflecting is a different kind of event timer than counting.

Evidence for an embedded, context-specific memory is supported by research on temporal memory in humans (Friedman, 1993). An appropriately reflective model for how events are recollected is the theory of reconstructive memory. Reconstruction of remembered events is based on recognition of an event with respect to extant cues during the event interval. Reconstructive memory explains otherwise anomalous characteristics of memory, like primacy (enhanced memory for initial events), scale effects (e.g., more accuracy for time of day than month or year), and facilitative effect of background temporal structure (Friedman, 1993). Subjects may be self-generating temporal structure through collateral behaviors. Reconstructive memory also supports a bias toward memory of events with more endogenous and external cues, as in 24-h and seasonal memories (see Section 4.3.4).

Because most organisms do not have a hippocampus, I would now like to discuss a smaller timer, one that is easily carried by individual cells. It shares some molecular features with that of the circadian clock, but it is linear in its response. It involves the activity of a single protein induced by a specific stimulus, which is then followed by a decay of the protein back to its original state. The protein activity could involve its production (as in circadian rhythms), its mobilization to a specific area of the cell (e.g., ion channels localized to the membrane of a neuron), or a structural change in the protein (e.g., exposing a protein binding domain). The metabolic nature of the decay timer also makes it agreeable with temperature effects on clock speed. These "clocks," for which I will use the general term decay timers, have been described in the control of countless molecular interactions (e.g., Ishijima and Yanagida, 2001; Takai et al., 2001) and are potentially our most primitive event timers.

An example of an ecological problem that is most feasibly solved by a decay timer is that of local search time in the absence of further resource acquisition. For example, if an animal in the presence of a reinforcing signal (e.g., food or mating pheromone) suddenly finds that the signal is reduced or absent, it must make a decision about how long to continue searching for the reinforcer in its present location before it decides to search elsewhere. This behavioral strategy of looking first locally and then globally is called an area-restricted search and has been observed in a wide variety of organisms (Kareiva and Odell, 1987; White et al., 1984). Underlying this strategy is a clock that keeps track of the time elapsed since the animal last encountered food. A decay timer would be appropriate for this behavior, as it could be reset by food and then directly control turning behavior by modulating proteins that control turning rates.

The run-and-tumble behavior of *Escherichia coli* bacteria follows this description, involving a phosphorylation cascade that begins with membrane proteins that bind to extracellular ligands. In the absence of a stimulating resource, these membrane proteins act through phosphorylation of downstream secondary messenger proteins, which then bind to the flagellar motor components to cause flagellar reversal and tumbly swimming (Stock and Surette, 1996). The phosphorylation schedule of these proteins is on the order of seconds, and the whole system works like clockwork to move the animals up concentration gradients.

While this is an example of a timed endogenous behavior, it is not an example of an interval timer, because the bacteria do not learn the duration of an external signal. If the animal could change the temporal dynamics of its turning in response to different resource environments, this would provide us with an understanding of how external events that are not necessarily timed with an event timer can lead to developmental changes in optimally timed behavior patterns. This developmental retiming of behaviors is likely to be a critical step in the evolution of interval timers. Promising organisms for this kind of study are *Drosophila melanogaster* and the nematode *Caenorhabditis elegans*, which are convenient organisms for molecular and genetic study that also have the wherewithal to search for food when it is no longer around. I do not believe that we are likely to find fully formed interval timers in these animals, but instead the molecular machinery from which interval timers are constructed.

Experiments designed to distinguish between decay timers and time fields, without taking into account the molecular machinery involved, are probably bound to fail. The reason is that the time field model is perfectly capable of acting like a decay timer and, in fact, undoubtedly consists of numerous decay timers that set the context for contiguous spatial and temporal phenomenon (Young and McNaughton, 2000). However, there is a rather deep distinction between these different event timers in the form of the credit-assignment problem as it is established in the psychological literature (Machado, 1997; Staddon and Higa, 1999). The credit-assignment problem is based on the animal's attention to the relevant reward cue. How, for example, does a rat learn that the onset of the red light signals food in 40 sec and that changes in air temperature are unrelated? I believe one of the premises of the decay timer must be that the credit is assigned in the evolutionary history of the animal. Bacteria, bees, and other invertebrates do not learn to assign a particular stimulus to a decay timer. That is given to them for free. On the other hand, the decay timers in the hippocampus are actually used to solve the credit-assignment problem. They do this by maintaining the firing rates of certain pyramidal cells even after the response stimulus is gone. This allows contiguous events in time and space to be contextually associated with neighboring events (Wallenstein et al., 1998). The relevant cues to which any given event timer is sensitive are intimately related to an evolutionary bias for certain environmental cues. Thus, negative results on event timing experiments may be limited to telling us about a very specific environmental stimulus (see "Navigation," Section 4.5.2).

4.5 EVENT TIMERS IN THE NATURAL WORLD

The remainder of this review is devoted to understanding why animals need event timers. One might argue that there is no need; all ecologically relevant behaviors could just as easily be performed without this faculty. I recognize this argument not because it is helpful to understanding the behavior, but because it elucidates a very real constraint on our understanding of animal event timing. In the natural world, it is very difficult to tell by what cues an animal makes its decisions. Until we can isolate the event timer or show that its properties are consistent with behavioral predictions (see Bateson, this volume), any assumptions we make about the behavioral ecology of animals using event timers run the risk of oversight. On the other hand, without understanding the ecological context to which the event timer might be adapted, claims that an animal does or does not have an event timer based on laboratory tests assume a similar risk.

With respect to the argument of clock existence, however, we are at a slight advantage. From the psychophysics, we know the behavior exists. At this point we are merely in the business of knowing why. But how can we know why if we are not sure when the behavior really exists? Pigeons in cages can learn the difference between 2 and 8 sec. Does this mean that an osprey uses time to capture prey? It does not. Does this mean that squirrels make assumptions about how much time they would have to escape given the appearance of a predator in a particular location? It does not mean this either.

The first step toward understanding the adaptive contribution of event timers is to recognize the contribution they would make if they were being used. This will provide us with some sense of the situations that might have facilitated the evolution of event timers. The following tour through event timing in the natural world will focus primarily on the domains of animal behavior that would show a positive fitness relationship in the presence of a cost-free event timer that can associate nonoverlapping events in time. I will discuss the difficulties with assessing event timer cost in "Conclusions" (Section 4.6).

4.5.1 COMMUNICATION

Communication is a distinctly temporal behavior. It requires transmitting signals in sequence and duration such that a receiving organism understands the message. There are cases when the timing of symbolic components is unimportant, for example, when one is merely trying to get another's attention. Here we are interested in the relationship between syntactical elements in time and the way in which that conveys information. More specifically, we are interested in animals that learn how to communicate.

Songbirds acquire songs by listening to other birds (Beecher et al., 1998). The primary reason for this appears to be ecological. For song sparrows, the song repertoire is usually learned after the bird leaves its birthplace and during the first season of territory establishment (Beecher, 1994). In this way, the song sparrow learns the social communication strategies of its neighbors. This is further evidenced by song sparrows showing a preference for the learning of already shared songs (Beecher et al., 1998).

The fact that birds can learn frequency and durational components of a song implies a usefulness for event timing (see MacDonald and Meck, this volume). If one bird intends to mimic the call of another, then it must be able to record that call in memory. Functionally, the mechanism that records the call is an event timer; it learns the sequence and duration of notes that constitute the local song. Whether this ability can be generalized to record the times of nonsyntactic events remains to be established.

Linguistic studies in humans recognize a temporal component, but a clear understanding of exactly how that component is manifested is far from understood (Port et al., 1995). Some models of language acquisition are distinctly similar to contemporary models of hippocampal function in their use of recurrent networks (Port et al., 1995; Wallenstein et al., 1998). There is also evidence that perceived time is shorter for familiar auditory signals than it is for unfamiliar signals, suggesting that perceived time is not absolute for auditory signals but is influenced by the content of the perceived signal (Kowal, 1984). The main difficulty with results from linguistic studies, as it is for essentially all studies on communication, is that it is extremely difficult to separate the meaning from the message.

Insect communication is understood to carry information in its gross rhythms. There is evidence that insects distinguish likely mates by the gaps between signals, the interpulse interval (Kyriacou et al., 1992; Michelsen et al., 1985). In the case of *Drosophila* the interpulse interval appears to be under genetic control with a well-defined relationship with circadian rhythms (see Section 4.3.3).

What about the receiver? Moths are useful for studying auditory transduction because they have a relatively simple ear, with one or two receptor cells attached to an accessible tympanum. In moths, specific cells can operate as frequency filters (Michelsen et al., 1985), tuned to specific sensory stimuli. Pattern-sensitive neurons have also been observed in the pyloric network of lobster (Hooper, 1998). The presence of such neurons in *D. melanogaster* would explain stable female preferences despite differences in the circadian schedule. In this way, invertebrates may bypass the need for the higher cortical functioning that goes along with vertebrate event timers.

4.5.2 Navigation

Animal navigation involves a gamut of sensory acuities to various environmental signals. As a consequence, there are countless ways to avoid noncircadian clocks:

birds, insects, and fish use celestial and magnetic compasses; honeybees use polarized light; wasps have memory for landmarks; salmon can find their breeding grounds by smell; and amphipod crustaceans use the slope of the ground (Daan, 1981; Dyer, 1998; Gould, 1998). In fact, the only evidence for an event timer in navigation is that some animals appear to know how far they have gone.

This is best exemplified by the waggle dance of honeybees. Remarkably, honeybees returning to the nest can inform other workers of the exact whereabouts of a forage site. The waggle dance transmits direction by establishing an angular relationship between the sun and the forage site in the form of a linear waggle movement at the same angle from the vertical axis of the hive. The distance to the site is transmitted by the distance of the linear waggle. The worker increases her waggle distance by about 75 msec per 100 m of foraging distance (Seeley, 1995). A decay timer would be useful in transmitting this signal, but a kind of event timer would be required for measuring the initial distance and for other bees receiving the signal at the hive. Time based on metabolic costs is probably far more prevalent than time based on the observation of events. The energy expenditure in flight could be used as the assay of flight duration. Bees may leave the hive or return to it with tuned energy stores so as to accomplish this. That another bee watches or follows the dance and then knows the distance to the forage site implies a more complicated mechanism.

The behavior of honeybees in the wild makes the FI data for honeybees particularly cumbersome (Grossman, 1973). It suggests that despite Grossman's effort to simulate a naturalistic environment in order to measure honeybee event timing, he may have been asking the question in the wrong way. There is no evidence that honeybees are confronted with anything that replenishes itself at 1-min intervals in the wild (Seeley, 1995). A honeybee may return to the hive to communicate the exact location of a new nectar source and still be unable to learn that a flower takes 20 sec to replenish its nectar stores. This entire behavior pattern appears to follow the logic of the specificity of decay timers. In general, if one expects to find event timers in insects, then one needs to accept the possibility that these event timers are very phylogenetically derived and highly specified.

The role of chemotaxis or navigation by thermal gradients is typically overlooked in the standard navigation literature. Gradient navigation requires either spatially or temporally separated samples of the environment. Which of these to choose depends critically on the size of the organism. For *Escherichia coli*, and similarly sized animals, the primary difficulty lies in the signal-to-noise ratio introduced by Brownian motion (Berg, 1983). Temporally spaced samples are of limited utility if animals are unable to find their way back to previous positions. It is for this reason that as size decreases, spatial mechanisms become more informative (Dusenberry, 1998). While there is support of spatially distributed processing at the opposite ends of *E. coli* (Grebe and Stock, 1998; Dusenberry, 1998), there is also evidence that they weight temporal experiences over time (Segall et al., 1983) (also see Section 4.4).

4.5.3 Reproduction

Reproduction involves a highly defined series of behaviors. To locate a mate, an animal must know where and when to look. I was unable to find any evidence that animals learn the temporal aspects of this behavior. In most cases, the timing of mating behaviors appears to be a possible form of sympatric speciation. For example, the temporal staggering of male mating flights in East African army ants and the nocturnal activity patterns of moths maintains a temporal separation between species in reproductive timing (Daan, 1981).

There is a relationship between event timing and circadian rhythms in the time-sharing behavior of parent doves (*Streptopelia risioria*) (Gibbon et al., 1984; Silver and Bittman, 1984). The female spends up to 18 h on the nest each day and is relieved by the male during one 6-h interval. If the male is prevented from starting his sit bout at the appropriate time, the female will return after 6 h regardless of the duration of the male's effort. The male, however, will dispute the rightful sitter with her until his 6 h are finished. Given that it is unknown whether the male learns the appropriate sitting interval, it is almost impossible to conclude that the dove has a generalized event timer. The male dove might simply have a genetically timed 6-h alarm clock in his brain that is started by sitting on the nest.

How might this alarm clock work? Presumably, in a way very similar to that of the operation of light on the circadian clock. In the case of the male dove, sitting on the nest causes a pulse of gene expression in the hypothetical "sit" gene. The SIT proteins degrade over time until they reach a minimum threshold level, at which point the dove gets up. This is, of course, the decay timer.

4.5.4 Predator Avoidance

Animals attempting to avoid predators can do so in a number of ways. They can wait for them to arrive and then try to escape. They can time their activity out of phase with predator foraging. Or they can attempt to satiate predators with fellow members of the species. All of these behaviors could gain from event timers.

There are several methods of escape in the sense that I use it above. An animal might run for cover, or it might trick the predator into thinking it is not a prey item. In either case, learning the principles of predator vigilance can increase foraging efficiency. In some cases, the presence of predators actually enhances foraging efficiency (Holtcamp et al., 1997).

For an animal foraging in the open, attention to predators is necessary for survival. But how much time should an animal devote to vigilance? Animals that economize predator vigilance strike an optimal relationship between eating and being eaten (Dukas, 1998). Potential environmental factors are the nearest possible escape and the proximity of possible predators. In the latter case, there is evidence that adult ground squirrels with obstructed views of their surroundings are more vigilant than those with clear views (Arenz and Leger, 1997). Juveniles were undeterred, suggesting the behavior is learned.

A useful trick against predators is feigning death. Vigilance is helpful here, but of equal importance is knowing how long one should stay "dead." Anxious resurrection will indubitably lead to real mortality. But staying dead until a predator unbeguiled by death arrives is an equally poor outcome. Domestic chicks perform the death-feigning behavior instinctively. The time spent inanimate is associated with the circadian phase (Richelle and Lejeune, 1980), suggesting a control mechanism possibly analogous to that found in the courtship behavior of fruit flies.

Predator avoidance can also take the form of knowing when predators are active and choosing to be active at other times. This is exemplified by the behavior of baby alligators (*Alligator* spp.), which, when heavily preyed upon by African fish eagles (*Haliatus vocifer*), move from diurnal to nocturnal activity rhythms (Curio, 1976). This suggests a phenotypic plasticity in the way behaviors are linked to circadian clocks, reminiscent of honeybees learning the daily cycles of nectar production.

By far, one of the most popular forms of predator avoidance is feeding them your neighbors. This form of predator avoidance also leads to some of the longest-known temporal periods for behavioral synchronization. Plants do this in the form of masting, which is a form of synchronized seed production that can occur over periods of many years (Silvertown, 1980). Periodical cicadas (*Magicicada* spp.) are one of the more artful exemplars of this phenomenon (Lloyd and Dybas, 1966). Having the longest-known life cycles of any insect (barring some queen ants), they emerge from the ground to mate, lay eggs, and die within weeks of one another every 13 or 17 years, depending on the species. The behavior is hypothesized to be a predator-satiating mechanism above ground and a predator avoidance mechanism below ground (Lloyd and Dybas, 1966).

The predator-satiating mechanism works on the premise that predators are limited in their maximal intake rate of prey. This can be due to simple satiation or to prey handling times. For example, when guillemot fledglings (*Uria lomvia*) jump from their breeding cliffs, the probability of death is significantly enhanced if the bird jumps alone (Daan, 1981). The usual strategy is to jump with everyone else, so that the fledglings are shielded from the predatory glaucous gulls (*Larus hyperboreus*) by other members of their cohort. If the glaucous gulls did not have a maximal intake rate, they would eat everyone as soon as they were exposed.

4.5.5 Foraging

Animals acquire resources in countless ways. Temporal perception is useful in many of them. For example, speciation mechanisms are undoubtedly related to competitive exclusion in competition for resources. This provides a force for sympatric speciation via resource partitioning in time. In this case, animals forage at different times of day but still eat the same foods, as is observed in several species of tern, lizards, crustaceans, and gastropods (Schoener, 1970, 1974). This reduces competition while simultaneously economizing resource acquisition in a kind of temporal ideal free distribution (for a description of the ideal free distribution in space, see Milinski and Parker, 1991). Whether these behaviors are learned remains to be established.

An area that seems most promising for the discovery of event timers in the wild is in the empirical testing of optimal foraging theory. A basic assumption of optimal foraging theory is that animals recognize something about resource distribution (the psychophysical evidence for this was discussed in Section 4.2.2). This recognition can be more or less behaviorally plastic, depending on the cognitive faculties of the animal. If resource distribution is relatively stable over time, a species may evolve a patch departure schedule that is based on generations of trial and error without regard for the present environmental conditions. At the other extreme, animals with event timers could measure the rate of food intake at different patches or with different foods and compare them to optimize foraging schedules in the future. Animals could also measure the time between patches and incorporate this information into the overall strategy. This updating of foraging behavior based on prior information is commonly referred to as Bayesian foraging (Getty and Krebs, 1985; Killeen et al., 1996). This is in direct contrast to the assumptions of the marginal value theorem, which assumes that animals know resource distribution, transit, and handling times even before they begin foraging (Charnov, 1976; Valone and Brown, 1989). While the marginal value theorem provides a useful null model against which to compare animal behaviors, it does not require an event timer per se, as animals may evolve to forage at optimal schedules. However, when resource distributions change over the lifetime of the animal, an event timer will be required for animals to appropriately update their Bayesian expectations. The study of Bayesian foraging behavior in organisms amenable to molecular and genetic study would provide another promising inlet into the mechanisms involved in sensing and integrating information about temporal intervals into future behavior patterns.

Bayesian foraging behavior is also likely to be ubiquitous. Many animals require patch assessment before they can make optimal foraging decisions (Valone and Brown, 1989). Constraints on forager memory and resource changes over time force the reinvestigation of patches (Belisle and Cresswell, 1997). Animals use recent information about temporal aspects of resource distribution to make decisions about patch departure. Among central place foragers, there is a positive correlation between distance traveled to the foraging site and the patch residence time (see Bateson, this volume; Kacelnik, 1984).

Studies of risk-sensitive foraging show the ability of animals to detect the variance of resource acquisition even when the mean is unchanged (Real and Caraco, 1986). For example, honeybees prefer stable rewards to unstable rewards, regardless of the mean. This requires an event timer. The adaptive explanation for stable vs. unstable preferences is described by Bateson (this volume).

The data on animal preferences do not entirely corroborate the theory of risk-sensitive foraging (Bateson and Kacelnik, 1998; Ha et al., 1990; Stephens, 1980). Explanations for this are based on cognitive constraints related to time perception and memory. Animals may discount time in different ways, depending on past experience or genetic predisposition, or they may average rate intake over different intervals. Animals also have certain constraints on their abilities to discriminate event times, as typified by Weber's law (Bateson and Kacelnik, 1998; described in Section 4.2.1). I suggested earlier that animals may suffer from distorted perceptions of time based on intake rate. The consequences for rate-biased time perception have been described by Hills and Adler (2002).

The order of experiences also appears to play a role in event timing. A kind of first impression among animals, called side bias, sometimes confounds psychophysical results (Ha et al., 1990). Side bias generally refers to some unknown force controlling the animal's behavior. Experimenters typically make an effort to remove these animals from the analyses. Nonetheless, every animal may experience this kind of bias with variable time reinforcement schedules. Large initial rewards could lead to particularly strong cognitive bias. A series of large rewards might also instill a memory of a rare event that keeps the animal coming back. Exactly how the temporal sequence of events establishes memory biases is still an open question.

Another temporal factor in foraging is the effect of time horizons (Krebs and Kacelnik, 1984). Time horizons affect the behaviors of animals that are able to anticipate the ends of foraging bouts. Late in the day an animal may choose to continue foraging in a poor patch because it does not have enough time to get to a better one. A mechanism to avoid this problem involves organizing a series of patches in time and visiting them so as to maximize resource gain over the duration.

Traplining fits the criteria of serial patch arrangement. It is a behavior seen in bats and a number of birds and frugivorous primates. It involves following a pre-specified path during the daily foraging bout (Bell, 1991). Time horizons undoubtedly affect traplining schedules, but once scheduled, traplining provides a short-term answer to the time horizon problem. A similar behavior pattern is cropping. Cropping involves visiting locations at intervals that allow for resource replenishment. Cody (1971) observed various species of finches cropping seeds in the Mohave Desert at the base of a mountain range. These birds moved their foraging sites to different distances from the mountain each day, scheduling revisitation rates to match replenishment rates. Insect-eating shore birds also appear to crop along the shore. Consistent with these observations, lab experiments on cache recovery in scrub jays (*Aphelocoma coerulescens*) show that they can learn and recall what, where, and when information about stored food items for up to 5 days (Clayton and Dickinson, 1998). In some manifestations of cropping, an event timer could help an animal know when to return to a foraging site.

4.5.6 PREY PURSUIT AND CAPTURE

For an American osprey (*Pandion haliaetus*) to intercept a fish in shallow water, it must perfectly time its descent and penetration of the water to match the location of its prey. Individual osprey have been observed catching many different kinds of fish, and this suggests that osprey learn to anticipate the position of their prey by observing something about individual fish (Bent, 1961). Numerous predators intercept moving prey (Curio, 1976), whether it is wolves (*Canis lupus*) taking down moose (*Alces alces*) in the Yukon or golden eagles (*Aquila chrysaetos*) catching rabbits (*Lepus* spp.) in the plains. The behavior seems to be a general one. But does it require an event timer?

For insects, the answer is probably no. At least in tiger beetles, the method of pursuit and capture is to constantly move where the prey is, with rapid halts to reorient its direction toward the location of the prey (Gilbert, 1997). An alternative strategy is that the insect measures the velocity of its prey and moves to where the prey will be. Evidence of the latter does not exist.

Do osprey pursue like tiger beetles? It is too early to tell. Computer imaging of predator and prey paths, like those done for the tiger beetle (Gilbert, 1997), is not yet used for larger animals. In the case of fish pursuit by birds, even simple video analysis is constrained by simultaneous water and air analyses. Still, this is likely to be the most informative method for determining the nature of larger predator pursuit and the mechanisms involved.

Another case where predator pursuit may involve timing is in group foraging efforts. Members of a concerted predatory effort must understand their duties in relation to other members. For example, observations of predatory groups breaking up to surround prey on scales at which they are not visible to one another requires an estimation of other group members in space and time (Curio, 1976). Similarly, knowing when to take over in the pursuit of prey in serial efforts necessitates an understanding of when to act. Knowing the traits of other individuals in the group, recognizing fatigue or opportunity, being at the right place at the right time — all of these things require clocks plus ample cognitive space for allocating memories and learned predatory skills. Interestingly, evolution has had no trouble solving similar problems under more predictable settings; honeybees appear to perform essentially the same feats of temporal economy within the hive (Moore et al., 1998).

4.5.7 Generalized Learning

The role of the hippocampus is well established in associative learning assays of the conditioned stimulus–unconditioned stimulus (CS-US) type (Ono et al., 1995; Thompson et al., 1982). Typically in CS-US trials an animal is trained to associate a neutral CS (e.g., an odor) with a previously meaningful US (e.g., pain or reward). The archetypical example is that of Pavlov and his bell-stimulated salivating dog. The paradigm takes advantage of the fact that the animal has some unconditioned response (UR) to the US, so that the experimenter can verify association of the CS by omitting the US while still observing the UR.

As one might expect, there is a clear relationship between the timing of various components of the CS-US and the ability of the animal to learn the association. For example, there is an optimal time of CS length that maximizes learning rate (Cooper, 1991). This is reminiscent of the credit-assignment problem and suggests that the duration of environmental signals may affect attention for those signals when they are subsequently paired with relevant stimuli.

At the level of the synapse, long-term potentiation (LTP) in the hippocampus is a form of learning based on the modulation of synaptic gain between the presynaptic and postsynaptic cell (Churchland and Sejnowski, 1994). N-methyl-D-aspartate (NMDA)–type glutamate receptors are required for LTP, and their voltage-gated properties necessitate that the postsynaptic cell fire for a certain temporal duration while the presynaptic cell fires in order for the NMDA receptors to be activated. NMDA activation operates on gene transcription (Bading et al., 1993), and this may be a mechanism for the modulation of long-term synaptic gain. Several phases exist for hippocampal LTP, and these have been compared to the various phases exhibited in vertebrate memory (DeZazzo and Tully, 1995). A recent finding suggests that this mechanism may also operate in a retrograde fashion by weakening synapses when the presynaptic cell fires after the postsynaptic cell action potential (Markram et al., 1997). This mechanism is not an event timer in the usual sense; it is more akin to a coincidence detector. But it does exhibit the flavor of event timers, and it possibly explains the contiguity of the CS-US sequence (see Matell and Meck, 2000; Matell et al., this volume).

The hippocampus appears to be required in order to get beyond coincidence-limited association. In trace conditioning assays, in which a puff of air follows a brief tone, the hippocampus is required for association of nonoverlapping stimuli (Thompson et al., 1982). Hippocampal lesions only allow learning when the CS and US are overlapping in time.

Hippocampal-type event timers offer substantial associative powers in the development of tool use and causal recollection. A Japanese macaque (*Macaca fuscata*) washing a potato must somehow learn to associate the improved potato with the action that preceded it. A more common example of this kind of temporally gapped causal association is the learned food avoidance response exhibited in a large number of vertebrates. If an animal becomes sick from eating a toxic food, it may learn to avoid that food in the future. There is a temporal relationship here because delay of adverse consequences inevitably limits the association of the food with its effects (Stephens and Krebs, 1986).

4.6 CONCLUSIONS

Time is easily one of the more slippery subjects in four dimensions. Our linear and subjective experience of it makes it rather difficult to define. Understanding how other organisms experience it is even more problematic. Although when one stops to take it all in, a considerable amount is actually known about how animals perceive time. Molecular geneticists and neurobiologists are in fact making numerous inroads into the mechanisms controlling event timing. Ecologists as well are refining their ideas to incorporate evidence from psychophysical studies of timing and time perception (see Bateson, this volume; Bateson and Kacelnik, 1998; Hills and Adler, 2002).

A hopeful contribution of this review is the distinction between circadian time and event timing. Circadian timers are temperature compensated, while event timers are most likely not. Circadian timers do not appear useful for recording event intervals that deviate from the 24-h LD cycle, whereas event timer accuracy appears to be a function of the linear increment in duration (Weber's law). Vertebrate circadian and event timers also seem to be isolated to different areas of the nervous system, but this remains to be corroborated in invertebrates. Circadian time also affects the variance in event timer responses, but not the duration. These distinctions provide us with a basis for understanding the relationship between the two mechanisms and for understanding why, in an adaptive sense, an animal is biased toward certain environmental stimuli and patterns of behavior that follow geophysical rhythms.

I have also presented possible mechanisms for event timers that are in agreement with the available evidence from psychophysics and neurophysiology. Time fields are an analogous structure to space fields and may operate by the same mechanism. Molecular decay timers are presented as an explanation for essentially all timed events (as the hippocampus contains decay timers, potentially in the form of ion channel conformation changes). The difference between decay timers and time fields may simply be in the number of cells used and the allocation of receptors to specific sensory cues. This difference may also explain failures to show evidence for event timers in invertebrates, as a consequence of sensory bias.

The problem with event timers is one that is shared by much of the literature on the adaptive value of cognitive mechanisms. We do not yet understand the costs of nervous tissue (Aiello, 1997). Costs associated with changes or new development of nervous tissue are still far from quantified. Genetic perturbations may be bringing us closer to this, but the distributed nature of nervous tissue will presumably confound our efforts for some time. In the SCN example, a portion of the brain was removed with no observable effect over a 2-year span. This suggests that a trait of nervous systems in general may be their plasticity. The costs and adaptive value of phenotypic plasticity, even though animal event timing is one of them, are still very much in the dark. Why not perceive everything, record perfect memories of it all, and be able to tell me to the minute (without looking at a clock) how long you have been reading this?

ACKNOWLEDGMENTS

This work would not have been possible without the generosity and insights of Alan Estrada, Steve Proulx, Pablo Nosa, Franz Goller, Charles Shimp, Villu Maricq, Fred Adler, Tim Brown, and especially Sara Hills.

REFERENCES

Aiello, L.C., Brains and guts in human evolution: the expensive tissue hypothesis, *Braz. J. Genet.*, 20, 141–148, 1997.

Amir, S. and Stewart, J., Resetting of the circadian clock by a conditioned stimulus, *Nature*, 379, 542–545, 1996.

Antoch, M.P., Song, E.-J., Chang, A.-M., Vitaterna, M.H., Zhao, Y., Wilsbacher, L.D., Sangoram, A.M., King, D.P., Pinto, L.H., and Takahashi, J.S., Functional identification of the mouse circadian Clock gene by transgenic BAC rescue, *Cell*, 89, 655–667, 1997.

Arenz, C.L. and Leger D.W., The antipredator vigilance of adult and juvenile thirteen-lined ground squirrels (*Sciuridae: Spermophilus tridecemlineatus*): visual obstruction and simulated hawk attacks, *Ethology*, 103, 945–953, 1997.

Aronson, B., Johnson, K., Loros, J.J., and Dunlap, J.C., Negative feedback defining a circadian clock: autoregulation in the clock gene frequency, *Science*, 263, 1578–1584, 1994.

Artieda, J. and Pastor, M.A., *Neurophysiological Mechanisms of Temporal Perception*, Elsevier, Amsterdam, 1996.

Aschoff, J., Human perception of short and long time intervals: its correlation with body temperature and the duration of wake time, *J. Biol. Rhythms*, 13, 437–442, 1998.

Bading, H., Ginty, D.D., and Greenberg, M.E., Regulation of gene expression in hippocampal neurons by distinct calcium signaling pathways, *Science*, 260, 181–186, 1993.

Balsalobre, A., Clock genes in mammalian peripheral tissues, *Cell Tissue Res.*, 309, 193–199, 2002.

Bateson, M. and Kacelnik, A., Risk-sensitive foraging: decision making in variable environments, in *Cognitive Ecology*, Dukas, R., Ed., University of Chicago Press, Chicago, 1998, pp. 297–342.

Beecher, M.D., Correlation of song learning and territory establishment strategies in the song sparrow, *Proc. Natl. Acad. Sci. U.S.A.*, 91, 1450–1454, 1994.

Beecher, M.D., Campbell, S.E., and Nordby, J.C., The cognitive ecology of song communication and song learning in the song sparrow, in *Cognitive Ecology*, Dukas, R., Ed., University of Chicago Press, Chicago, 1998, pp. 175–199.

Belisle, C. and Cresswell, J., The effects of a limited memory capacity on foraging behavior, *Theor. Popul. Biol.*, 52, 78–90, 1997.

Bell, W.J., *Search Behavior: The Behavioral Ecology of Finding Resources*, Chapman & Hall, New York, 1991.

Bent, A.C., *Life Histories of North American Birds of Prey*, Dover, New York, 1961.

Berg, H.C., *Random Walks in Biology*, Princeton University Press, Princeton, NJ, 1983.

Bitterman, M.E., Comparative analysis of learning, *Science*, 188, 699–708, 1975.

Bizo, L.A. and White, K.G., Timing with controlled reinforcer density: implications for models of timing, *J. Exp. Psychol. Anim. Behav. Process.*, 2, 44–55, 1997.

Buhusi, C.V. and Meck, W.H., Differential effects of methamphetamine and haloperidol on the control of an internal clock, *Behav. Neurosci.*, 116, 291–297, 2002.

Buhusi, C.V., Sasaki, A., and Meck, W.H., Temporal integration as a function of signal/gap intensity in rats (*Rattus norvegicus*) and pigeons (*Columba livia*), *J. Comp. Psychol.*, in press.

Cantor, M.B., Information theory: a solution to two big problems in the analysis of behavior, in *Advances in the Analysis of Behavior*, Vol. 2, *Predictability, Contiguity and Contingency*, Harzem, P. and Zeiler, M., Eds., Wiley, New York, 1981, pp. 286–320.

Charnov, E.L., Optimal foraging, the marginal value theorem, *Theor. Popul. Biol.*, 9, 129–136, 1976.

Cheng, K. and Roberts, W.A., Three psychophysical principles of timing in pigeons, *Learn. Motiv.*, 22, 112–128, 1991.

Church, R.M. and Deluty, M.Z., The bisection of temporal intervals, *J. Exp. Psychol. Anim. Behav. Process.*, 3, 216–228, 1977.

Church, R.M. and Gibbon, J., Temporal generalization, *J. Exp. Psychol. Anim. Behav. Process.*, 8, 165–186, 1982.

Church, R.M., Meck, W.H., and Gibbon, J., Application of scalar timing theory to individual trials, *J. Exp. Psychol. Anim. Behav. Process.*, 20, 135–155, 1994.

Churchland, P.S. and Sejnowski, T.J., *The Computational Brain*, MIT Press, Cambridge, MA, 1994.

Clayton, N.S. and Dickinson, A., Episodic-like memory during cache recovery by scrub jays, *Nature*, 395, 272–274, 1998.

Cody, M.L., Finch flocks in the Mohave Desert, *Theor. Popul. Biol.*, 2, 142–148, 1971.

Colombo, M., Broadbent, N.J., Taylor, C.S., and Frost, N., The role of the avian hippocampus in orientation in space and time, *Brain Res.*, 919, 292–301, 2001.

Cooper, L.D., Temporal factors in classical conditioning, *Learn. Motiv.*, 22, 129–152, 1991.

Crosthwaite, S.K., Loros, J.J., and Dunlap, J.C., Light-induced resetting of a circadian clock is mediated by a rapid increase in frequency transcript, *Cell*, 81, 1003–1012, 1995.

Crystal, J.D., Circadian time perception, *J. Exp. Psychol. Anim. Behav. Process.*, 27, 68–78, 2001.

Curio, E., *The Ethology of Predation*, Springer-Verlag, Berlin, 1976.

Czeisler, C.A., The effect of light on the human circadian pacemaker, in *Circadian Clocks and Their Adjustment*, Chadwick, D.J. and Ackrill, K., Eds., John Wiley & Sons, New York, 1995, pp. 254–290.

Daan, S., Adaptive daily strategies in behavior, in *Biological Rhythms*, Aschoff, J., Ed., Plenum Press, New York, 1981, pp. 275–298.

de Vaca, S.C., Brown, B.L., and Hemmes, N.S., Internal clock and memory processes in animal timing, *J. Exp. Psychol. Anim. Behav. Process.*, 7, 59–69, 1994.

Decoursey, P.J. and Krulas, J.R., Behavior of SCN-lesioned chipmunks in natural habitat: a pilot study, *J. Biol. Rhythms*, 13, 229–244, 1997.

DeZazzo, J. and Tully, T., Dissection of memory formation: from behavioral pharmacology to molecular genetics, *Trends Neurosci.*, 18, 212–218, 1995.

Dukas, R., Constraints on information processing and their effects on behavior, in *Cognitive Ecology*, Dukas, R., Ed., University of Chicago Press, Chicago, 1998, pp. 89–128.

Dunlap, J.C., Genetic and molecular analysis of circadian rhythms, *Ann. Rev. Genet.*, 30, 579–601, 1996.

Dusenberry, D.B., Spatial sensing of stimulus gradients can be superior to temporal sensing for free-swimming bacteria, *Biophys. J.*, 74, 2272–2277, 1998.

Dyer, F., Cognitive ecology of navigation, in *Cognitive Ecology*, Dukas R., Ed., University of Chicago Press, Chicago, 1998, pp. 201–260.

Edery, I., Rutila, J., and Rasbash, M., Phase shifting of the circadian clock by induction of the *Drosophila* period protein, *Science*, 263, 237–240, 1994.

Edmunds, L.N., *Cellular and Molecular Bases of Biological Clocks*, Springer-Verlag, New York, 1988.

Eskin, R.M. and Bitterman, M.E., Fixed-interval and fixed ratio performance in the fish as a function of prefeeding, *Am. J. Psychol.*, 73, 417–423, 1960.

Ferster, C.B. and Skinner, B.F., *Schedules of Reinforcement*, Appleton-Century-Crofts, New York, 1957.

Fetterman, J.G. and Killeen, P.R., Adjusting the pacemaker, *Learn. Motiv.*, 22, 226–252, 1991.

Fletcher, P.C., Frith, C.D., and Rugg, M.D., The functional neuroanatomy of episodic memory, *Trends Neurosci.*, 20, 213–218, 1997.

Friedman, W.J., Memory for the time of past events, *Psychol. Bull.*, 113, 44–66, 1993.

Getty, T. and Krebs, J.R., Lagging partial preferences for cryptic prey: a signal detection analysis of great tit foraging, *Am. Naturalist*, 125, 39–60, 1985.

Gibbon, J., Morrell, M., and Silver, R., Two kinds of timing in circadian incubation rhythm of ring doves, *Am. J. Physiol.*, 247, R1083–R1087, 1984.

Gilbert, C., Visual control of cursorial prey pursuit by tiger beetles (*Cicindelidae*), *J. Comp. Physiol. Sensory Neural Behav. Physiol.*, 181, 217–230, 1997.

Gould, J.L., Sensory bases of navigation, *Curr. Biol.*, 8, R731–R738, 1998.

Grebe, T.W. and Stock, J., Bacterial chemotaxis: the five sensors of bacterium, *Curr. Biol.*, 8, R154–R157, 1998.

Grossman, K.E., Continuous, fixed-ratio and fixed interval reinforcement in honey bees, *J. Exp. Anal. Behav.*, 20, 105–109, 1973.

Ha, J.C., Lehner, P.N., and Farley, S.D., Risk-prone foraging behavior in captive grey jays, Perisoreus canadensis, *Anim. Behav.*, 39, 91–96, 1990.

Hardin, P.E., Hall, J.C., and Rasbash, M., Circadian oscillations in period gene mRNA levels are transcriptionally regulated, *Proc. Natl. Acad. Sci. U.S.A.*, 89, 11711–11715, 1992.

Harrison, A.A., Clearwater, Y.A., and McKay, C.P., The human experience in Antarctica: applications to life in space, *Behav. Sci.*, 34, 253–271, 1989.

Hastings, M.H., Central clocking, *Trends Neurosci.*, 20, 459–464, 1997.

Haykin, S., *Neural Networks*, Macmillan College Publishing Company, New York, 1994.

Hills, T.T. and Adler, F.R., Time's crooked arrow: rate-biased time perception and optimal foraging theory, *Anim. Behav.*, 64, 589–597, 2002.

Hoagland, H., *Pacemakers in Relation to Aspects of Behavior*, Macmillan, New York, 1935.

Holtcamp, W.N., Grant, W.E., and Vinson, S.B., Patch use under predation hazard: effect of the red imported fire ant on deer mouse foraging behavior, *Ecology*, 78, 308–317, 1997.

Hooper, S.L., Transduction of temporal patterns by single neurons, *Nat. Neurosci.*, 1, 720–726, 1998.

Ishijima, A. and Yanagida, T., Single molecule nanobioscience, *Trends Biochem. Sci.*, 26, 438–444, 2001.

Iwasaki, K., Liu, D.W., and Thomas J.H., Genes that control a temperature-compensated ultradian clock in *Caenorhabditis elegans*, *Proc. Natl. Acad. Sci. U.S.A.*, 92, 10317–10321, 1995.

Iwasaki, K. and Thomas, J.H., Genetics in rhythm, *Trends Genet.*, 13, 111–115, 1997.

Jackson, P.A., Kesner, R.P., and Amann, K., Memory for duration: role of hippocampus and medial prefrontal cortex, *Neurobiol. Learn. Mem.*, 70, 328–348, 1998.

Kacelnik, A., Central place foraging in starlings (*Sturnus vulgaris*): I. Patch residence time, *J. Anim. Ecol.*, 53, 283–299, 1984.

Kareiva, P. and Odell, G., Swarms of predators exhibit "preytaxis" if individual predators use area-restricted search, *Am. Naturalist*, 130, 233–270, 1987.

Kesner, R.P., Neural mediation of memory for time: role of the hippocampus and medial prefrontal cortex, in *Animal Cognition and Sequential Behavior: Behavioral, Biological, and Computational Perspectives*, Fountain, S.B., Bunsey, M.D., Danks, J.H., and McBeath, M.K., Eds., Kluwer Academic, Boston, 2002, pp. 175–200.

Killeen, P.R., Palombo, G., Gottlob, L.R., and Beam, J., Bayesian analysis of foraging by pigeons (*Columba livia*), *J. Exp. Psychol. Anim. Behav. Process.*, 22, 480–496, 1996.

King, D.P., Zhao, Y., Sangoram, A.M., Wilsbacher, L.D., Tanaka, M., Aantoch, M.P., Steeves, T.D.L., Vitaterna, M.H., Kornhauser, J.M., Lowrey, P.L., Turek, F.W., and Takahashi, J.S., Functional identification of the mouse circadian Clock gene by transgenic BAC rescue, *Cell*, 89, 655–667, 1997a.

King, D.P., Zhao, Y., Sangoram, A.M., Wilsbacher, L.D., Tanaka, M., Aantoch, M.P., Steeves, T.D.L., Vitaterna, M.H., Kornhauser, J.M., Lowrey, P.L., Turek, F.W., and Takahashi, J.S., Positional cloning of the mouse circadian Clock gene, *Cell*, 89, 641–653, 1997b.

Konopka, R.J. and Benzer, S., Clock mutants of *Drosophila melanogaster*, *Proc. Natl. Acad. Sci. U.S.A.*, 68, 2112–2116, 1971.

Kowal, K., Familiar melodies seem shorter, not longer, when played backwards, *Ann. N.Y. Acad. Sci.*, 423, 610–611, 1984.

Krebs, J.R. and Kacelnik, A., Time horizons of foraging animals, *Ann. N.Y. Acad. Sci.*, 423, 278–291, 1984.

Krebs, J.R., Sherry, D.F., Healy, S.D., Perry, V.H., and Vaccarino, A.L., Hippocampal specialization of food-storing birds, *Proc. Natl. Acad. Sci. U.S.A.*, 89, 1388–1392, 1989.

Kyriacou, C.P., Greenacre, M.L., Ritchie, M.G., Peixoto, A.A., Shiels, G., and Hall, J.C., Genetic and molecular analysis of ultradian rhythms in Drosophila, in *Ultradian Rhythms in Life Processes*, Lloyd, D. and Rossi, R.L., Eds., Springer-Verlag, New York, 1992, pp. 89–105.

Lavie, P., Ultradian cycles in sleep propensity; or, Kleitman's BRAC revisited, in *Ultradian Rhythms in Life Processes*, Lloyd, D. and Rossi, R.L., Eds., Springer-Verlag, New York, 1992, pp. 284–302.

Liu, Y., Garceau, N.Y., Loros, J.J., and Dunlap, J.C., Thermally regulated translational control of FRQ mediates aspects of temperature responses in the Neurospora circadian clock, *Cell*, 89, 477–486, 1997.

Lloyd, M. and Dybas, H.S., The periodical cicada problem: I. Population ecology, *Evolution*, 20, 133–149, 1966.

Lloyd, D., Circadian and ultradian clock-controlled rhythms in unicellular microorganisms, *Advances in Microbiol Physiology*, 39, 291–338, 1998.

Loudon, A.S.I., Wayne, N.L., Krieg, R., Iranmanesh, A., Veldhuis, J.D., and Menaker, M., Ultradian endocrine rhythms are altered by a circadian mutation in the Syrian hamster, *Endocrinology*, 135, 712–718, 1994.

Machado, A., Learning the temporal dynamics of behavior, *Psychol. Rev.*, 104, 241–265, 1997.

Maricq, A.V. and Church, R.M., The differential effects of haloperidol and methamphetamine on time estimation in the rat, *Psychopharmacology*, 79, 10–15, 1983.

Markram, H., Lubke, J., Frotscher, M., and Sakmann, B., Regulation of synaptic efficacy by coincidence of postsynaptic aps and epsps, *Science*, 275, 213–215, 1997.

Matell, M.S. and Meck, W.H., Neuropsychological mechanisms of interval timing behaviour, *Bioessays*, 22, 94–103, 2000.

Meck, W.H., Hippocampal function is required for feedback control of an internal clock's criterion, *Behav. Neurosci.*, 102, 54–60, 1988.

Meck, W.H., Modality-specific circadian rhythmicities influence mechanisms of attention and memory for interval timing, *Learn. Motiv.*, 22, 153–179, 1991.

Meck, W.H., Church, R.M., and Olton, D.S., Hippocampus, time, and memory, *Behav. Neurosci.*, 98, 3–22, 1984.

Menaker, M., Commentary: what does melatonin do and how does it do it? *J. Biol. Rhythms*, 12, 532–534, 1997.

Michelsen, A., Larsen, O.N., and Sulrykke, A., Auditory processing of temporal cues in insect songs: frequency domain or time domain? in *Time Resolution in Auditory Systems*, Michelsen, A., Ed., Springer-Verlag, New York, 1985, pp. 3–27.

Milinski, M. and Parker, G.A., Competition for resources, in *Behavioral Ecology*, 3rd ed., Krebs, J.R. and Davies, N.B., Eds., Blackwell Scientific, Oxford, 1991, pp. 137–168.

Millar, A.J., Straume, M., Chory, J., Chua, N., and Kay, S.A., The regulation of circadian period by phototransduction pathways in Arabidopsis, *Science*, 267, 1163–1166, 1995.

Mistleberger, R.E., Circadian food-anticipatory activity: formal models and physiological mechanisms, *Neurosci. Biobehav. Rev.*, 18, 171–195, 1993.

Mizumori, S.J.Y., LaVoie, A.M., and Kalyani, A., Redistribution of spatial representation in the hippocampus of aged rats performing a spatial memory task, *Behav. Neurosci.*, 110, 1006–1016, 1996.

Moore, D., Angel, J.E., Cheeseman, I.M., Fahrbach, S.I., and Robinson, G.E., Timekeeping in the honeybee colony: integration of circadian rhythms and division of labor, *Behav. Ecol. Sociobiol.*, 43, 147–160, 1998.

Moore, D., Siegfried, D., Wilson, R., and Rankin, M.A., The influence of time of day on the foraging behavior of the honeybee, *Apis mellifera*, *J. Biol. Rhythms*, 4, 305–325, 1989.

Morgan, L., Killeen, P.R., and Fetterman, J.G., Changing rates of reinforcement perturbs the flow of time, *Behav. Process.*, 30, 259–272, 1993.

Okamura, H., Yamaguchi, S., and Yagita, K., Molecular machinery of the circadian clock in mammals, *Cell Tissue Res.*, 309, 47–56, 2002.

Ono, T., Nishijo, H., Eifuku, S., Kobayashi, T., and Tamura, R., Conjunctive multiple stimuli-encoding in the hippocampal formation of rats and monkeys, in *Brain Processes and Memory*, Ishikawa, K., McGaugh, J.L., and Sakata, H., Eds., Elsevier, Amsterdam, 1995, pp. 187–201.

Paule, M.G., Meck, W.H., McMillan, D.E., Bateson, M., Popke, E.J., Chelonis, J.J., and Hinton, S.C., The use of timing behaviors in animals and humans to detect drug and/or toxicant effects, *Neurotoxicol. Teratol.*, 21, 491–502, 1999.

Pavlov, I.P., *Conditioned Reflexes*, Oxford University Press, Oxford, 1927.

Port, R.F., Cummins, F., and McAuley, J.D., Naive time, temporal patterns, and human audition, in *Mind as Motion*, Port, R.F. and van Gelder, T., Eds., MIT Press, Cambridge, MA, 1995, pp. 339–372.

Portavella, M., Vargas, J.P., Torres, B., and Salas, C., The effects of telencephalic pallial lesions on spatial, temporal, and emotional learning in goldfish, *Brain Res. Bull.*, 57, 397–399, 2002.

Power, J.M., Ringo, J.M., and Dowse, H.B., The effects of period mutations and light on the activity rhythms of *Drosophila melanogaster*, *J. Biol. Rhythms*, 10, 267–280, 1995.

Ralph, M.R., Foster, R.G., Davis, F.C., and Menaker, M., Transplanted suprachiasmatic nucleus determines circadian period, *Science*, 247, 975–978, 1990.

Ralph, M.R. and Hurd, M.W., Circadian pacemakers in vertebrates, in *Circadian Clocks and Their Adjustment*, Chadwick, D.J. and Ackrill, K., Eds., John Wiley & Sons, New York, 1995, pp. 67–81.

Real, L. and Caraco, T., Risk and foraging in stochastic environments, *Ann. Rev. Ecol. Syst.*, 17, 371–390, 1986.

Rensing, L., Kallies, A., Gebauer, G., and Mohsenzadeh, S., The effects of temperature change on the circadian clock of neurospora, in *Circadian Clocks and Their Adjustment*, Chadwick, D.J. and Ackrill, K., Eds., John Wiley & Sons, New York, 1995, pp. 26–41.

Richelle, M. and Lejeune, H., *Time in Animal Behavior*, Pergamon Press, Oxford, 1980.

Roberts, S., Isolation of an internal clock, *J. Exp. Psychol. Anim. Behav. Process.*, 7, 242–268, 1981.

Roberts, S., The mental representation of time: uncovering a biological clock, in *Methods, Models, and Conceptual Issues: An Invitation to Cognitive Science*, Vol. 4, Scarborough, D. and Sternberg, S., Eds., MIT Press, Cambridge, MA, 1998, pp. 53–106.

Rosato, E., Peixoto, A.A., Gallippi, A., Kyriacou, C.P., and Costa, R., Mutational mechanisms, phylogeny, and evolution of a repetitive region within a clock gene of *Drosophila melanogaster*, *J. Mol. Evol.*, 42, 392–408, 1996.

Rosenwasser, A.M., Rats remember the circadian phase of feeding, *Ann. N.Y. Acad. Sci.*, 423, 634–635, 1984.

Rozin, P., Temperature independence of an arbitrary temporal discrimination in the goldfish, *Science*, 149, 561–564, 1965.

Ruby, N.F., Dark, J., Heller, H.C., and Zucker, I., Albation of suprachiasmatic nucleus alters timing of hibernation in ground squirrels, *Proc. Natl. Acad. Sci. U.S.A.*, 93, 9864–9868, 1996.

Saunders, D.S., *Insect Clocks*, Pergamon Press, Oxford, 1971.

Sawyer, L.A., Hennessy, M.J., Peixoto, A.A., Rosato, E., Parkinson, H., Costa, R., and Kyriacou, P.C., Natural variation in a Drosophila clock gene and temperature compensation, *Science*, 278, 2117–2120, 1997.

Schmidt, J.M. and Pak, G.A., The effect of temperature on progeny allocation and short interval timing in a parasitoid wasp, *Physiol. Entomol.*, 16, 345–353, 1991.

Schoener, T.W., Nonsynchronous spatial overlap of lizards in patchy habitats, *Ecology*, 51, 408–418, 1970.

Schoener, T.W., Resource partitioning in ecological communities, *Science*, 185, 127–185, 1974.

Seeley, T.D., *The Wisdom of the Hive: The Social Physiology of Honey Bee Colonies*, Harvard University Press, Cambridge, MA, 1995.

Segall, J.E., Block, S.M., and Berg, H.C., Temporal comparisons in bacterial chemotaxis, *Proc. Natl. Acad. Sci. U.S.A.*, 83, 8987–8991, 1983.

Shigeyoshi, Y., Taguchi, K., Yamamoto, S., Takekida, S., Yan, L., Tei, H., Moriya, T., Shibata, S., Loros, J.J., Dunlap, J.C., and Okamura, H., Light-induced resetting of a mammalian circadian clock is associated with rapid induction of the mper1 transcript, *Cell*, 91, 1043–1053, 1997.

Silver, R. and Bittman, E.L., Time sharing by parent doves, *Ann. N.Y. Acad. Sci.*, 423, 488–514, 1984.

Silvertown, J.W., The evolutionary ecology of mast seeding in trees, *Biol. J. Linnean Soc.*, 14, 235–250, 1980.

Staddon, J.E.R. and Higa, J.J., Time and memory: towards a pacemaker-free theory of interval timing, *J. Exp. Anal. Behav.*, 71, 215–251, 1999.

Stanewsky, R., Clock mechanisms in *Drosophila*, *Cell Tissue Res.*, 309, 11–26, 2002.

Stephens, D.W., The logic of risk-sensitive foraging preferences, *Anim. Behav.*, 29, 628–629, 1980.

Stephens, D.W. and Krebs, J.R., *Foraging Theory*, Princeton University Press, Princeton, NJ, 1986.

Stock, J.B. and Surette, M.G., Chemotaxis, in *Escherichia coli and Salmonella: Cellular and Molecular Biology*, Neidhardt, F.C., Ed., ASM Press, Washington, D.C., 1996, pp. 1103–1129.

Takahashi, J.S., Molecular neurobiology and genetics of circadian rhythms in mammals, *Ann. Rev. Neurosci.*, 18, 531–553, 1995.

Takai, Y., Sasaki, T., and Matozaki, T., Small GTP-binding proteins, *Physiol. Rev.*, 81, 153–208, 2001.

Talton, L.E., Higa, J.J., and Staddon, J.E.R., Interval schedule performance in the goldfish *Carassius auratus*, *Behav. Process.*, 45, 193–206, 1999.

Thompson, R.F., Berger, T.W., Berry, S.D., Clark, G.A., Kettner, R.N., LaVond, D.G., Mauk, M.D., McCormick, D.A., Solomon, P.R., and Weisz, D.J., Neuronal substrates of learning and memory: hippocampus and other structures, in *Conditioning: Representation of Involved Neural Functions*, Woody, C.D., Ed., Plenum Press, New York, 1982, pp. 1103–1129.

Toh, K.L., Jones, C.R., He, Y., Eide, E.J., Hinz, W.A., Virshup, D.M., Ptacek, L.J., and Fu, Y.H., An hPer2 phosphorylation site mutation in familial advanced sleep phase syndrome, *Science*, 9, 1040–1043, 2001.

Treisman, M., Cook, N., Naish, P.L.N., and MacCrone, J.K., The internal clock: electroencephalographic evidence for oscillatory processes underlying time perception, *Q. J. Exp. Psychol.*, 47, 241–289, 1994.

Tulving, E., Episodic and semantic memory, in *Organization of Memory*, Tulving, E. and Donaldson, W., Eds., Academic Press, New York, 1972, pp. 381–403.

Valone, T.J. and Brown, J.S., Measuring patch assessment abilities of desert granivores, *Ecology*, 70, 1800–1810, 1989.

Wagner, S., Castel, M., Gainer, H., and Yarom, Y., GABA in the mammalian suprachiasmatic nucleus and its role in diurnal rhythms, *Nature*, 387, 598–603, 1997.

Wallenstein, G.V., Eichenbaum, H., and Hasselmo, M.E., The hippocampus as an associator of discontiguous events, *Trends Neurosci.*, 21, 317–323, 1998.

Wallenstein, G.V. and Hasselmo, M.E., Gabaergic modulation of hippocampal population activity: sequence learning, place field development, and the phase precession effect, *J. Neurophysiol.*, 78, 393–408, 1997.

Wearden, J.H., Do humans possess an internal clock with scalar properties? *Learn. Motiv.*, 22, 59–83, 1991.

Wearden, J.H. and Penton-Voak, I.S., Feeling the heat: body temperature and the rate of subjective time, revisited, *Q. J. Exp. Psychol. B*, 48, 129–141, 1995.

Wever, R.A., The sleep-wake threshold in human circadian rhythms as a determinant of ultradian rhythms, in *Ultradian Rhythms in Life Processes*, Lloyd, D. and Rossi, R.L., Eds., Springer-Verlag, New York, 1992, pp. 307–322.

White, J., Tobin, T.R., and Bell, W.J., Local search in the housefly *Musca domestica* after feeding on sucrose, *J. Insect Physiol.*, 30, 477–487, 1984.

Wilson, E.O., *The Insect Societies*, The Belknap Press, Cambridge, UK, 1971.

Winfree, A.T., *The Geometry of Biological Time*, Springer-Verlag, New York, 1980.

Young, B. and McNaughton, N., Common firing patterns of hippocampal cells in a differential reinforcement of low rates of response schedule, *J. Neurosci.*, 20, 7043–7051, 2000.

Yu, Q., Jacquir, A.C., Citri, Y., and Colot, H.M., Molecular mapping of point mutations in the period gene that stop or speed up biological clocks in *Drosophila melanogaster*, *Proc. Natl. Acad. Sci. U.S.A.*, 84, 784–788, 1987.

5 Interval Timing and Optimal Foraging

Melissa Bateson

CONTENTS

5.1 Introduction 113
5.2 Interval Timing and Foraging 115
5.3 Optimal Foraging Theory 117
    5.3.1 When to Return to a Renewing Food Source 118
    5.3.2 When to Leave a Depleting Patch 119
        5.3.2.1 Sudden Patch Exhaustion 119
        5.3.2.2 Gradual Depletion 120
    5.3.3 How to Respond to Variability 120
    5.3.4 When to Sample a Changing Patch 121
    5.3.5 Constraints in Optimal Foraging Models 122
5.4 Scalar Timing Theory 123
5.5 Applications of Scalar Timing to Foraging Problems 126
    5.5.1 Scalar Timing and Patch Departure: Sudden Patch Exhaustion 127
    5.5.2 Scalar Timing and Patch Departure: The Marginal Value Theorem 130
    5.5.3 Scalar Timing and Response to Risk 133
    5.5.4 Scalar Timing and Sampling a Changing Environment 136
5.6 Conclusions 138
Acknowledgments 139
References 139

5.1 INTRODUCTION

The natural world is full of temporal regularities, and it makes sense that animals should have evolved clocks that permit them to maximize their fitness by exploiting these regularities. However, proving that animals use internal clocks to schedule their behavior is rarely possible in natural environments. To provide proof that an animal is timing, it is necessary to eliminate any external cues to the passage of time that the animal might be using in place of an internal clock (Killeen et al.,

1997). Because this is usually difficult in natural environments, data showing that animals can time intervals come from the controlled conditions of the laboratory, where the cues available to the subjects and the temporal properties of their experience can easily be manipulated. Such laboratory experiments have traditionally been the domain of operant psychologists who typically restrict their studies to the behavior of laboratory-reared rats and pigeons tested in Skinner boxes on various schedules of reinforcement. Due to the lack of a clear connection between the behavior of rats and pigeons in the artificial environment of the Skinner box and the problems faced by wild animals in their natural environments, research on interval timing has so far failed to attract widespread interest from ethologists and behavioral ecologists who usually focus on understanding behavior patterns initially identified in wild animals. Evidence for interval timing in animals has mostly been published in the psychological literature, and it is psychologists that have been responsible for setting the agenda in research on interval timing.

As a consequence of the domination of the interval timing literature by psychologists, the focus in most timing research has been on describing the psychophysics of interval timing, with the ultimate goal of elucidating the cognitive and neural mechanisms underlying the interval timing clock (e.g., Hinton and Meck, 1997; Matell and Meck, 2000; Paule et al., 1999). For example, psychologists have paid particular attention to inaccuracy and imprecision in interval timing on the grounds that imperfections in the system can be particularly important for revealing the underlying mechanisms (e.g., Gibbon and Church, 1984). It is taken for granted that the ability to time intervals is useful to animals, and the interval timing clock has consequently been conceived of as a general-purpose stopwatch-like timepiece. A result of this approach has been that questions regarding the evolution and current function of interval timing have remained largely unasked.

In contrast to the psychological approach, ethologists believe that no account of behavior is complete until both its proximate mechanisms and its ultimate evolutionary functions are understood (Tinbergen, 1963). Skinner (1989) stated that "by looking at how a clock is built, we can explain [how] it keeps good time, but not why keeping time is important." Although Skinner was correct in recognizing that questions about mechanism and function are logically distinct, the strength of Tinbergen's ethological approach lies in the belief that finding the answer to one type of question will often provide valuable insights into the answer to the other. Thus, although logically it is not necessary to know why the clock has evolved in order to understand how it works, there is reason to believe that considering both questions simultaneously could bring considerable benefits of understanding.

There are a number of reasons why it might be beneficial to study function and mechanism simultaneously. First, it is generally much easier to understand how a mechanism works if you know exactly what it is designed to do. By analogy, deciphering someone else's computer code is always facilitated if you know precisely what the function of the code is. Although we may feel we understand what clocks are for, it remains the case that a clock that functions to measure the intervals between successive prey captures in a foraging starling may have different design requirements than a clock that functions to measure the rate of the mating display in a male sage grouse. Second, interval timing clocks are likely to have evolved in response to specific types of temporal regularities that confront animals in their natural environments. As a result, the clock may behave very differently when it is probed with the natural stimuli that it has evolved to respond to than when it is probed with unnatural stimuli that are outside the range of variation it encountered during its evolutionary history. In other words, modeling the biological clock as a flexible stopwatch may be misleading; the biological clock may behave differently depending on what it is asked to do. Third, psychologists have described several ways in which the performance of the interval timing clock departs from perfect accuracy and precision. In order to understand the significance of these imperfections for the adaptive behavior of animals, we need to know whether the imperfections of the clock are true constraints of its mechanism or whether perhaps they are artifacts of studying the clock under conditions for which it was not designed. By analogy, if you rev a car on a surface such as ice or sand, the pair of wheels that drive the car will spin. This gives you some clear information about how the car works: you now know whether it has front- or rear-wheel drive. However, the behavior of the car on the sand is very different from what you would observe if you were to rev the same car on a road. Thus, studying timing in the lab may provide very useful insights into the mechanisms underlying the clock; however, it may be misleading if your aim is to understand the role of timing in the generation of adaptive behavior. The evolutionary process has resulted in a complex relationship between function and mechanism in biological systems, and it is only by considering both questions simultaneously that we can ever hope to understand animal behavior fully.

My aim for this chapter is to take an ethological approach to interval timing and attempt to integrate mechanistic and evolutionary approaches. I begin by asking the functional question of why animals need to be able to time intervals, and I present evidence suggesting that interval timing is likely to have a central role in foraging behavior. Next I introduce optimal foraging theory and show exactly how temporal information is important in making optimal foraging decisions. An important component of all optimal foraging models is the constraints that are assumed, and I go on to consider timing as a potential constraint. The most sophisticated attempts to integrate models of timing and foraging have come from studies that focus on scalar timing theory. I give a brief description of the basic scalar timing model and review how it has been adapted to model a range of foraging problems. Throughout, my aim is to highlight the benefits of understanding that have resulted from the combination of scalar timing with optimal foraging.

5.2 INTERVAL TIMING AND FORAGING

So why do animals need the ability to time intervals? Although timing seems a generally useful ability, are there specific behavioral problems faced by animals that are particularly likely to have selected for interval timing abilities? In trying to answer these questions, I am going to argue that several lines of evidence point to the likelihood that interval timing is of major importance in foraging behavior (see Hills, this volume).

Interval timing has been identified in the majority of vertebrate species in which the ability has been investigated (Lejeune and Wearden, 1991), and it is probably safe to assume that the ability is universal in the vertebrates (Bateson, 2001). This implies that interval timing is an evolutionarily ancient ability, and in searching for behavioral problems that might have specifically selected for timing, we need to identify very general problems that are faced by all vertebrate animals and are likely to have also been faced by their common ancestors. All animals need to find food, making foraging behavior a potential candidate for a general problem that might have selected for the ability to time.

Warm-blooded vertebrates such as small mammals and birds have relatively high metabolic rates, the maintenance of which requires large quantities of food on a regular basis. This need is assumed to have imposed a strong selective pressure to produce efficient foraging behavior. Birds in particular have been at the forefront of studies of optimal foraging because the high energetic demands of flight have meant that birds need to ingest particularly large quantities of food of high nutritive value. A small bird may need to eat its body weight in food per day. A blue tit (*Parus caeruleus*) weighing 11 g needs 11 kcal per day in winter, which is equivalent to around 300 small insects, and a rufous hummingbird (*Selasphorus rufus*) weighing 3.5 g will visit a nectar feeder approximately every 10 to 15 min from dawn until dusk. These high rates of foraging coupled with the fact that most birds forage only during the hours of daylight make birds attractive subjects for studies of foraging behavior.

Arthropods are the most common food consumed by small mammals and birds, and they bring with them various interesting foraging problems because the distribution of arthropod prey in the environment is usually both spatially and temporally patchy. For example, for much of the year European starlings forage on leatherjackets, the larvae of tipulid flies. Leatherjackets are hidden beneath the surface of the soil and occur in patches. Starlings search for leatherjackets by probing the soil with their bills. Given that it is impossible to visually assess the density of leatherjackets in a particular location, starlings rely on their cognitive abilities to form estimates of the rates of intake they have experienced in different locations. These estimates can then be used as a basis for making future foraging decisions.

It is possible to analyze all foraging behavior in terms of its costs and benefits to the forager. Finding, consuming, and digesting food all have both energetic and time costs associated with them, because time and energy spent foraging are time and energy taken away from other fitness-promoting activities, such as looking out for predators and reproducing. We therefore expect natural selection to have honed foraging decisions so as to optimize the trade-off between costs and benefits, and thus maximize the lifetime survival and reproductive success of the forager (e.g., Stephens and Krebs, 1986). Because the costs associated with foraging involve the length of time taken, it is likely that selection on foraging decisions has involved selection on the ability to measure these costs accurately.

The final piece of evidence linking timing with foraging comes from the observation that the majority of the comparative evidence for interval timing comes from animals performing on fixed-interval (FI) schedules of reinforcement in which food is used as the reinforcer. In a typical free-operant FI schedule, reinforcement, usually the delivery of a small amount of food, is contingent on a response made by the subject after some fixed period of time has elapsed. The interfood interval serves as the only discriminative stimulus, and the interval requirement is reset after each food reinforcement is delivered. The optimal strategy in a subject trying to maximize the frequency with which it receives food while minimizing the number of responses it has to make when faced with such a schedule is to time the fixed interval and make a single response as soon as the interval has elapsed. Although well-trained subjects never achieve this optimal strategy, they do show a postreinforcement pause that averages about two thirds of the FI value, after which they start responding at a high rate until food is delivered (e.g., Schneider, 1969). Thus, we have good evidence that animals are able to time interfood intervals and also that the delivery of food can serve to reset the animal's interval timer (Matell and Meck, 1999).

In summary, therefore, we have established that (1) all vertebrates need to forage, (2) foraging behavior is likely to have been under strong selective pressure to increase efficiency, (3) efficient foraging involves making decisions that involve timing, and (4) animals can time intervals between food deliveries and that food can reset the clock. Taken together, I suggest that the above evidence points strongly toward the possibility that interval timers may play a major role in natural foraging behavior. Of course, none of the above evidence proves that interval timers initially evolved for the purposes of foraging, or that interval timers are used solely for the purposes of foraging. However, the likely fitness consequences of inefficient foraging do suggest that the selection pressures to improve the efficiency of foraging are very likely to have been an important force in the evolution of the interval timing clock.

5.3 OPTIMAL FORAGING THEORY

Optimal foraging theory is the branch of behavioral ecology that seeks to understand how natural selection has shaped foraging behavior. The general strategy adopted is to build models of how animals should forage given various assumptions about the system, and then compare the predictions of these models with the behavior of real animals.

Classical optimal foraging models involve three kinds of assumptions: (1) those regarding the foraging decision being analyzed, (2) those regarding the currency the forager is maximizing, and (3) those regarding the constraints operating on the system (Stephens and Krebs, 1986).

In many classical optimal foraging models the currency that foraging animals are assumed to be maximizing is their long-term net rate of energy intake (Stephens and Krebs, 1986), where long-term rate is defined as the net energy intake divided by the total time spent acquiring this energy. Rate is a proximate currency that is assumed to relate closely to Darwinian fitness if it is maximized over the lifetime of the forager, because an animal that maximizes its rate of energy intake will achieve the greatest amount of energy for use in maintenance, growth, and reproduction in the least possible time, and time not spent foraging is time available for other fitness-promoting activities, such as looking out for predators and reproduction. Given that the computation of rate involves forming an estimate of the time spent foraging, interval timing is likely to be involved in many foraging decisions.

In the following sections I shall consider various foraging problems in which the ability to time intervals is potentially crucial to arriving at the optimal solution. In each case I will describe the problem faced by the forager, outline the solution to the problem that maximizes the rate of energy intake, and describe examples in which the behavior of animals has been shown to approximate the optimal foraging solution.

5.3.1 WHEN TO RETURN TO A RENEWING FOOD SOURCE

Some food sources gradually renew following depletion or depression by a forager. If this renewal process is temporally predictable, then such food sources provide a natural equivalent of a fixed-interval schedule of reinforcement. A forager that returns early to the food source will obtain less food than the maximum possible if it had delayed its return until the source had completely renewed. A forager that returns late to the food source will also obtain less than the maximum rate of food intake possible from that food source because it has waited longer than was necessary to obtain the maximum amount of food, and in situations where there is competition from other foragers, there is the added risk that the food might be lost to a competitor if the forager fails to claim it as soon as it is available. Thus, a forager that can learn the temporal predictability of the food source will be at an advantage over one that cannot, because it can schedule its visits to the source to correspond with the times at which maximum food is available and thus maximize its rate of food intake from the source.

There are several possible natural examples of renewing food sources with predictable temporal properties. For example, in some of the flower species used by nectar-feeding hummingbirds the amount of nectar available in the flower increases predictably and monotonically as a function of the time elapsed since the flower was last visited until the flower has fully refilled. In a field study of long-tailed hermit hummingbirds (*Phaethonis superciliosus*), Gill (1988) studied the responses of wild hummingbirds to 10- and 15-min FI schedules on an artificial nectar feeder. Rather than filling the feeder gradually, as would occur in a natural flower, he refilled the feeder either 10 or 15 min after the feeder had been emptied by a bird, as in a conventional FI schedule. He showed that under conditions of nearly exclusive use of a feeder by a single individual when the risk of competitive loss was low, return intervals increased to longer than the FI, thereby maximizing the probability of obtaining nectar on a visit.

Another possible example of a renewable food source with temporally predictable properties is provided by the amphipod crustacean *Corophium volutator*. *Corophium* species are the major food source of redshanks (*Tringa totanus*), wading birds that forage on tidal mudflats. The *Corophium* species live in burrows and retreat down their burrows when a redshank walks over the sediment surface to a depth where they are inaccessible to the redshanks (Goss-Custard, 1970). This is known as prey depression, and measurements of the feeding rates of redshanks have shown that it takes around 10 min for feeding rates to recover following the previous visit by a bird (Yates et al., 2000). This interval is presumed to correspond to the time it takes the *Corophium* species to resurface from their burrows. There is no information on whether redshanks are able to time this interval, but it would clearly be adaptive for them to do so, because if they return to an area within 10 min of previously feeding there, they will experience a reduced rate of prey capture.

Interval Timing and Optimal Foraging

5.3.2 WHEN TO LEAVE A DEPLETING PATCH

A decision commonly faced by foraging animals is that of when to leave a patch of food that is depleting and move on to a new patch. Because the optimal response to depleting patches varies according to whether the patch depletes gradually or suddenly, I will consider these two scenarios separately.

5.3.2.1 Sudden Patch Exhaustion

The availability of food in a patch may suddenly drop to zero. Faced with this situation, a forager needs to detect the change in status of the patch and move on to a new patch, because a forager that remains in an empty patch will be wasting time searching for food that is not there. Sometimes there may be visual or auditory cues available to the forager to indicate that a patch is exhausted, but if there are no such cues, a forager might be able to use its internal clock to detect patch exhaustion. Consider the situation in which there is a fixed intercapture interval and constant probability of sudden patch exhaustion after each prey encounter or after arrival in the patch. If the forager can learn the interfood interval, then it can use this knowledge to detect patch exhaustion. The optimal departure rule is to leave the patch as soon as exhaustion is detected, namely, as soon as the period since the last prey capture exceeds the normal intercapture interval. This optimal departure rule is independent of the time taken to travel to a new patch (Kacelnik et al., 1990).

The foraging of spotted flycatchers (*Muscicapa striata*) provides a possible example of sudden patch exhaustion (Davies, 1977). These birds hunt from a fixed perching site, making periodic forays to catch flying insects that have entered their range. Usually following a capture, the bird returns to the same or a nearby perch, but occasionally it will leave the site and travel to a new perch to continue hunting. Davies was interested in establishing what caused a bird to abandon a foraging site and move on to the next. He found that the birds would move sites when they had waited 1.5 times the average intercapture interval without making a foray, and on the basis of this finding, he suggested that the birds might be using a rule that involved estimating the average intercapture interval and moving on to a new perch when no insect had appeared within a given multiple of this time (specifically, 1.5 times). Such a moving-on rule might be optimal in an environment in which patches of prey deplete suddenly and completely, as might be the case, for instance, if a swarm of flies moves away from the immediate vicinity of the bird's perch, and where there are no external cues to indicate that the patch is now empty. Unfortunately, with the available data, it is impossible to prove that the flycatchers are really using such a time-based moving-on rule (Kacelnik et al., 1990). Proof that the birds are using an internal clock to judge their departure from a site would require eliminating the possibility that the birds do not just move on when they can no longer detect any prey items within their range. Additionally, proof that the birds wait for a fixed multiple of the intercapture interval before moving on would require manipulating intercapture intervals and demonstrating the predicted effects on moving-on times.

5.3.2.2 Gradual Depletion

The rate of food intake within a patch will often be a decelerating function of the amount of time the forager has already spent foraging in the patch. In this situation the forager is faced with the decision about when to leave the current patch and pay the cost of traveling to a new patch. The marginal value theorem (MVT) (Charnov, 1976) describes the behavior that maximizes the long-term rate of energy intake in such a situation. Long-term rate is maximized if the forager stays in each patch until its instantaneous rate within the patch falls below the background rate of gain in the environment as a whole. This background rate will be affected by the travel time between patches, and as the travel time between patches increases, the optimal patch residence time also increases. Because rates are equal to amounts per unit of time, computation of rate requires the ability to time intervals. An important prediction of the MVT is that although patch residence time is predicted to be sensitive to the average travel time, it should not be affected by variance in travel time.

A number of studies have tested the prediction that patch residence time should be positively related to the travel time between patches. Perhaps the most well known is the study by Kacelnik (1984) on European starlings (*Sturnus vulgaris*). During the breeding season starlings make regular forays from their nest to collect food for their chicks. As required by the MVT within each foraging bout, the starling suffers a decelerating rate of food acquisition. This occurs due to the starling's method of foraging, whereby it probes the ground to look for invertebrates, such as leatherjackets, hidden beneath the surface of the soil. As the starling's bill fills up with prey, the bird becomes progressively less efficient at probing the ground for further prey and its rate of prey acquisition declines. Kacelnik tested the MVT prediction that travel time should affect patch residence time by setting up feeding stations for starlings at different distances from their nests. He simulated the loading curve by delivering worms to the birds at progressively greater intervals the longer they stayed at the feeding station. He was able to show that the number of worms the starling collected before returning to its nest increased as the travel distance to the feeder increased. The observed behavior was approximately as predicted if the bird were maximizing the rate at which it delivered worms to its chicks.

5.3.3 How to Respond to Variability

Many natural food sources are variable either in the exact amount of food they provide or in the time associated with finding or extracting the food. For example, consider a forager faced with one feeding option that it knows will yield food after 5 min of searching vs. another feeding option that it knows will yield the same amount of food after the same average searching time, but the actual time taken to find the food could vary between 1 and 9 min. A rate-maximizing forager should be indifferent to such variability because the computation of the long-term rate of energy intake involves averaging the amounts and times associated with each food source, with the result that both food sources are perceived to be of equal value. However, there are circumstances where the long-term rate of energy intake may not be the currency that correlates best with fitness. Consider a small bird in winter faced with one more foraging decision before the rapidly approaching night. It is vitally important for this bird to achieve a threshold level of energy reserves before dusk in order to survive the long cold night. In this situation, it can be optimal for the forager to pay attention to the variance, or risk, as it is called in the foraging literature, in its food sources. If the bird has no chance of meeting the required threshold for survival in time by choosing the fixed option (i.e., it is on a negative energy budget), then its only chance of survival is to choose the risky option in the hope that it will be lucky and find food quickly. Conversely, if the fixed option will easily take the bird above threshold (i.e., it is on a positive energy budget), then it would be foolish to choose the variable option and risk not getting its final prey item before nightfall. These arguments are summarized in the daily energy budget rule that states that a forager on a positive energy budget should be risk averse, while one on a negative budget should be risk prone.

There is a large literature showing that animals are sensitive to risk in both the amount of food and in the delay associated with obtaining food (Kacelnik and Bateson, 1996; Bateson and Kacelnik, 1998). Unfortunately, though, there is little good empirical support for functional explanations for risk sensitivity such as the energy budget rule (but for a beautiful demonstration of the energy budget rule in yellow-eyed juncos, see Caraco et al., 1990). The overall pattern in the literature is that animals tend to be risk averse when there is variability in amount of reward, but risk prone when variability is in delay to reward. This pattern is not readily explained by any of the optimal foraging models, and there has been an ongoing discussion in the literature about whether risk sensitivity is an adaptive response to environmental variability or instead is an artifact of the cognitive mechanisms animals use to assess quantities such as amount, time, and rate (e.g., Bateson and Kacelnik, 1995b; Reboreda and Kacelnik, 1991).

5.3.4 WHEN TO SAMPLE A CHANGING PATCH

The abundance of food in the environment is seldom stable: the availability or quality of particular feeding sites or prey types may change unpredictably over the course of time, perhaps caused by depletion or depression by other foragers or by changes in the weather. As a consequence, one of the problems faced by foraging animals is keeping track of the status of their food resources. Sometimes there may be environmental cues to prey availability, but in instances where there are no such cues, a forager will be forced to sample its environment periodically. Stephens (1987) modeled the problem of how a rate-maximizing forager should track a changing environment. He considered the situation where there are two patch types available to a forager, one that fluctuates between good and bad states and one that remains stable at a value between the two states of the fluctuating patch. The fluctuating patch has a constant probability of changing state, and Stephens assumed that the forager can determine the state of the fluctuating patch as the result of a single sample. Given these assumptions, Stephens asked how a rate-maximizing forager should sample the fluctuating environment. The theoretical results of his model show that an optimal forager should sample the fluctuating patch at regular intervals. The optimal sampling frequency depends on the ratio of two kinds of cost: the lost opportunity experienced in the stable patch when the forager samples the fluctuating patch, and missing the opportunity to forage in the fluctuating site when it is in its good state.

5.3.5 Constraints in Optimal Foraging Models

Traditionally, the kinds of constraints assumed in optimal foraging models have been simple physical constraints, such as the impossibility of simultaneously handling one prey item while searching for another, or the reduction in the rate of prey acquisition as the forager's bill fills up with prey. In general, potential psychological constraints imposed by limitations of an animal's information-processing capacities have received much less attention from optimal foraging theorists. There are several reasons for this neglect. First, theoreticians usually seek to keep models as simple and as general as possible. Second, foraging theorists justify neglecting psychological constraints on the grounds that natural selection should have provided animals with near-perfect psychological mechanisms if these mechanisms have a measurable impact on optimal performance. Finally, due to the lack of integration of the psychological and ecological literatures, behavioral ecologists are often simply unaware of the wealth of data that is available regarding animals' information-processing capacities.

The ignorance of behavioral ecologists of the psychological literature has resulted in some misguided early attempts by them to introduce psychological constraints, such as imperfect timing, into optimal foraging models. For example, Yoccoz et al. (1993) presented a general theoretical framework for understanding the effects of perceptual error on optimal foraging decisions. In order to do this, they assumed explicit relationships between the real energy content and handling time of each food item and what the forager estimates these quantities to be. They incorporated these assumptions into a model of optimal diet choice (Engen and Stenseth, 1984), which is based on the assumption that animals maximize their long-term rate of energy intake, and investigated the effects of adding perceptual error on the optimal solution to the model. In their model, Yoccoz et al. (1993) represent the actual energetic gains and times as random variables G and T, and the animal's perception of these random variables as X and Y, respectively. They construct X and Y from G and T by adding unbiased normally distributed errors such that $X = G + E_G$ and $Y = T + E_T$, where E_G and E_T are the errors in gain and time, respectively. This general strategy is completely sound; however, Yoccoz et al. (1993) assume that the variances of the errors E_G and E_T are independent of the actual gains and times, G and T. This latter assumption flies in the face of over a century's worth of psychophysics showing that the error in the perception of the magnitude of a stimulus is not independent of its magnitude, but rather increases with stimulus magnitude (Bateson and Kacelnik, 1995a), a relationship known as Weber's law (see Figure 5.1).

The value of Yoccoz et al.'s (1993) model is that it demonstrates that introducing perceptual error can result in constrained rate-maximizing solutions that differ substantially from naïve, unconstrained treatments of the same problem. However, if such models are to be of use in understanding the behavior of real animals, then they need to include more realistic assumptions regarding the nature of perceptual constraints.

Interval Timing and Optimal Foraging

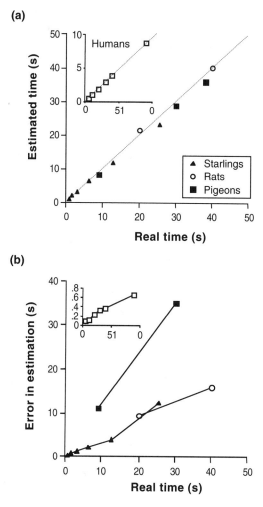

FIGURE 5.1 Examples of (a) estimated time, and (b) error in estimation of time, both vs. actual time from a variety of species. The data are taken from experiments in which subjects were required to reproduce time intervals. (Reproduced from Bateson, M. and Kacelnik, A., *Anim. Behav.*, 50, 431–443, 1995a. Copyright © 1995 by the Association for the Study of Animal Behavior. With permission.)

In the next section I introduce scalar timing theory, a sophisticated information-processing model of interval timing that has formed the basis of some of the most successful attempts to integrate realistic psychological constraints into optimal foraging models.

5.4 SCALAR TIMING THEORY

Scalar timing theory (see Church, this volume; Gibbon et al., 1984) is an information-processing account of interval timing that grew out of scalar expectancy theory

(SET) (Gibbon, 1977). All models of interval timing require three basic functions: a clock function that measures elapsed time by converting it to some physical representation, a memory function in which a recorded time interval can be represented and stored, and a decision function that uses output from the clock and the memory components to control behavior (Church, 1997). In scalar timing theory these different functions are embodied in discrete components described as follows.

The clock subsystem consists of a pacemaker, a switch and an accumulator. Time measurement is achieved by collecting pulses from the pacemaker in the accumulator when the switch is closed. The pacemaker is assumed to continuously emit pulses at a rate Λ. When the switch is closed in response to an external signal to begin timing, pulses are transmitted to the accumulator, where they accumulate until the switch is opened again. The value in the accumulator, m_t, thus represents the amount of time that has elapsed since the switch was closed and timing began. This estimate of real time grows linearly with real time, t, such that

$$m_t = \Lambda(t - T_0)$$

where T_0 is the mean latency between the external start signal and the beginning of time accumulation.

The current estimate of elapsed time, m_t, can be transferred either to working memory for immediate use or to reference memory, where it can be stored for future reference. In a fixed-interval schedule the usual cue for transferring a value to reference memory is the delivery of food. When food is delivered (i.e., $t = $ FI, the value of the fixed interval), the scalar timing theory assumes that the value of m_t, which we will now refer to as m_{FI}, is transferred to reference memory as

$$m^*_{FI} = k^* * m_{FI}$$

where k^* is a translation constant that is assumed to vary between trials. An important assumption of scalar timing theory is that m_{FI} is assumed to be represented in reference memory as a distribution of the various values of m^*_{FI} transferred to it on different trials. The form of this distribution is determined by the way in which k^* varies between trials. For the purposes of modeling timing performance on a fixed-interval procedure, the mean of k^* is usually assumed to be unity such that the mean estimate in memory is equal to the mean estimate of the current time at which reinforcement occurs. Once an animal is fully trained on a fixed-interval schedule, it is assumed to have built up a reference memory representation of the interval that has a mean of m_{FI} and a standard deviation proportional to this mean. If there is variance in the real time between reinforcements, as is the case, for instance, on a variable-interval schedule, then the memory representation formed will be equivalent to the sum of the distributions that would be formed for each of the component intervals in the variable mixture. Due to the fact that the standard deviation of the representation of an interval grows with the value of the interval, the memory distributions of variable intervals will be asymmetrical and skewed to the right. The different memory representations resulting from a range of different mixtures of intervals are illustrated in Figure 5.2.

Interval Timing and Optimal Foraging

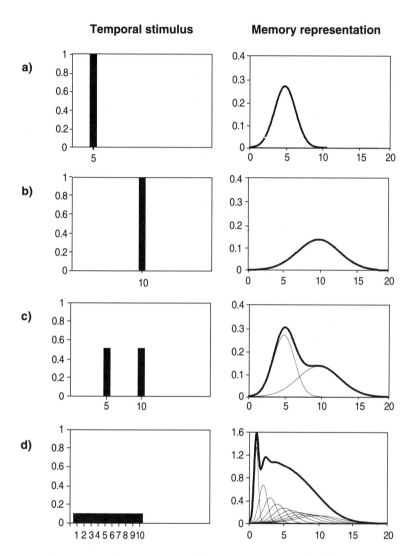

FIGURE 5.2 Examples of how a range of different temporal stimuli are represented in memory according to scalar timing theory. Temporal stimuli are shown in the left-hand panels and their associated memory representations in the right-hand panels. In all panels, probability is shown on the y-axis and the duration of the interval on the x-axis. In generating the memory representations, a coefficient of variation of 30% in the perception of the interval was assumed. (a) A fixed interval of 5 sec. (b) A fixed interval of 10 sec. (c) A variable interval that has an equal probability of being 5 or 10 sec. (d) A variable interval that has an equal probability of being 1, 2, 3, 4, 5, 6, 7, 8, 9, or 10 sec. In panels (c) and (d) the overall memory representation of the mixture is indicated by the bold line, and the representations of the individual intervals of which it is the sum are indicated by the thinner lines. Note that due to the constant coefficient of variation, the longer an interval, the less precisely it is represented; i.e., the scalar property applies. As a result of the scalar property, the memory representations for variable intervals are asymmetrical and skewed to the right.

In scalar timing theory the decision subsystem is assumed to receive input from both working and reference memories and to compare these inputs, usually using a ratio rule, to produce a behavioral output. For the purposes of modeling fixed-interval timing performance, at the beginning of each trial the decision subsystem is assumed to receive a single input from reference memory and continuously compare this with the current value of elapsed time, m_t, in working memory. Importantly, the value of m_{FI}^* used in this comparison process is assumed to be a single random sample from the distribution of m_{FI}^* stored in reference memory. The rule used to compare the values of m_t and m_{FI}^* is the discrimination ratio

$$\frac{m_t - m_{FI}^*}{m_{FI}^*}$$

When this ratio becomes greater than or equal to a threshold value, b, a decision is made to start responding. The value of b can also vary between trials, but for fixed-interval schedules it is usually assumed to have a mean value of approximately zero. Variation between trials in either k^* or b will induce the commonly observed relationship between the mean and variance of timing data known as the scalar property, whereby the coefficient of variation (i.e., the standard deviation or mean) is constant. Thus, the above scalar timing model is capable of producing the basic features of fixed-interval timing performance. The model is summarized in Figure 5.3.

One of the attractions of the scalar timing model is that the three components — clock, memory, and decision — are clearly modular, and the representation of information in memory is clearly separated from the way in which information is used in decisions. This modularity makes the scalar timing model very flexible and, as a result, particularly tractable for modeling a range of different foraging problems. The details of both the memory and decision components of the scalar timing model are altered depending on the specific task being modeled, as I shall demonstrate in the following section.

5.5 APPLICATIONS OF SCALAR TIMING TO FORAGING PROBLEMS

My aim in this section is to review examples where the scalar timing model has successfully been applied to some of the optimal foraging problems previously described. The general strategy has been to incorporate the perceptual error produced by the scalar timing model as an explicit psychological constraint in conventional optimal foraging models, and to explore whether the modified models do a better job of explaining observed behavior than their unconstrained counterparts. I shall describe how the basic scalar timing model is modified to cope with different foraging problems and highlight the benefits in understanding that have resulted. Although the following studies all tackle foraging problems that have been identified in animals in their natural habitats, the experiments described all involve analogs of these foraging problems translated to the operant laboratory. In attempting to replicate the natural foraging problems, the schedules developed for study in the lab are

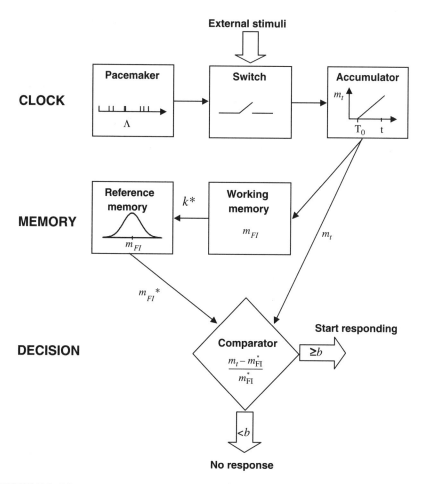

FIGURE 5.3 Diagrammatic representation of the basic scalar timing model. This version of the model predicts when a subject will start responding on a standard FI schedule. In a free-operant FI schedule the external stimulus that resets the clock and begins timing is the occurrence of food. Alternatively, in a discrete trial version of an FI schedule the external stimulus would be the stimulus that indicates the beginning of a trial, which is typically a tone or light. See the text for details of the model.

often more complex than the types of reinforcement schedules traditionally studied by operant psychologists.

5.5.1 SCALAR TIMING AND PATCH DEPARTURE: SUDDEN PATCH EXHAUSTION

Brunner and Kacelnik (Brunner et al., 1992, 1996; Kacelnik et al., 1990; Kacelnik and Brunner, 2002) set out to analyze the timing problem suggested by Davies' spotted flycatchers using European starlings foraging on an operant analog of the patch departure scenario in the lab. Their design simulated an environment in which food was distributed in patches. Each patch contained a random number of prey items (N = 0 to 4) that were delivered according to a fixed-interval schedule until the patch ended with sudden depletion. The time elapsed since the last prey item was the only cue the bird had to detect patch depletion. Once the patch had depleted, the bird could leave the patch and travel to a new patch by flying between two perches. As described above, the optimal patch departure rule, given perfect timing, is to abandon the patch as soon as exhaustion is detected, namely, when a prey is not encountered after waiting for the programmed fixed interval.

Brunner and Kacelnik tested how the patch departure of starlings was affected by the value of the fixed-interval schedule in the patch by examining the behavior of the birds tested with six different values of the fixed interval, ranging between 0.8 and 25.6 sec. As predicted by the unconstrained optimality model, the giving-in time at which a bird stopped attempting to obtain food from a patch in each trial (defined as the last peck in the patch) increased linearly as the fixed interval increased. However, in contrast to the predictions of the unconstrained model, the line relating the value of the fixed interval to the giving-in time had a slope of 1.49 rather than 1.00. Thus the starlings waited for a fixed proportion of the fixed interval before abandoning the patch. This result is particularly interesting in the light of data showing that the starlings knew accurately when food should have been expected. An analysis of the patterns of pecking in the final, unreinforced interval of each patch revealed that the birds showed a peak of pecks centered accurately on the value of the fixed interval: the line relating the value of the fixed interval to the peak time had a slope that was not significantly different from 1.00. Therefore, despite apparently knowing accurately when food should have been delivered, the birds still chose to wait 1.49 times the usual interfood interval before giving in; i.e., a multiplicative bias was introduced by the decision mechanism that controls the bird's decision to give up searching for more food in the patch.

How can we explain the multiplicative bias in the giving-in rule? It is possible to answer this question either by considering the functional implications of adopting different giving-in rules or by considering the timing processes responsible for the decision. We will start with the functional approach. If the birds knew accurately when food should have been delivered in the patch, why did they not give in and leave the patch as soon as it did not appear? The answer to this question is to be found in the behavioral variability of the birds. Although the peak times of the birds show perfect accuracy, the standard deviations of the pecking functions are linearly related to the fixed interval: in other words, the timing functions of the birds display the scalar property evident in all timing data. The giving-in times also display the scalar property because their interquartile range also increased linearly with the fixed interval. In order to understand why it is optimal to introduce a multiplicative bias in the giving-in rule, we need to consider the potential costs of giving in early (before the patch is exhausted) vs. the costs of giving in late, as was observed. It is easy to show that from the perspective of rate of energy intake, it is much more costly to give in too early than too late. This asymmetry occurs because an animal that gives in early fails to get the last food item available in the patch, and this has a much bigger impact on the rate of energy intake than the relatively small time cost imposed by giving in late. Thus we can predict that if imprecision in timing is a constraint,

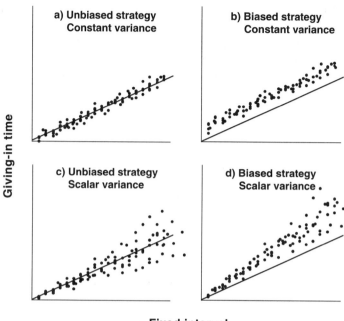

FIGURE 5.4 Variance in timing behavior and optimal giving-in time strategies. In each panel the solid line has a slope of 1.00 and indicates the optimal giving-in time predicted if timing is both perfectly accurate and precise. The top two panels (a and b) show constant variability in giving-in time, and the lower two panels (c and d) proportional or scalar variability in giving-in time. When timing is imprecise, the optimal strategy is biased in order to minimize the probability of leaving the patch before it is exhausted (panels b and d). When variability is constant, the optimal strategy is: giving-in time = fixed interval + bias (panel b); however, when variability is scalar, the optimal strategy is: giving-in time = fixed interval*bias (panel d). Brunner et al.'s (1992) starlings displayed a multiplicative bias as in panel (d) consistent with their also displaying the scalar property in their estimates of when food should have occurred. (Redrawn from Brunner, D., Kacelnik, A., and Gibbon, J., *Anim. Behav.*, 44, 597–613, 1992. Copyright © 1992 by the Association for the Study of Animal Behavior. With permission.)

then giving-in times should be biased to be longer than the fixed interval to guard against giving in too early. If, as the data revealed, the precision of timing is linearly related to the fixed interval (i.e., the scalar property applies), then the bias needs to be multiplicative, as opposed to a constant, in order to reduce the probability of giving in too early. These arguments are presented graphically in Figure 5.4.

It is also possible to analyze the behavior of the starlings from the mechanistic perspective of scalar timing theory. Scalar timing theory can be used to explain both the pecking patterns of the bird in the final unreinforced interval of the patch and the giving-in times. Both of these behavioral measures can be analyzed using the version of the scalar timing model proposed to deal with fixed-interval performance described above. The final unrewarded interval in the patch is exactly analogous to a probe trial from the peak-interval procedure. In order to use the scalar timing theory to model this performance, only a small modification to the decision mechanism is needed. Because birds start pecking at a high rate sometime prior to the FI and then stop pecking at this high rate sometime after the FI, the decision rule needs to produce both the start and stop of the high rate responding. This can be achieved by modifying the comparison made to

$$\left| \frac{m_t - m_{FI}^*}{m_{FI}^*} \right|$$

The starling responds at the high rate when this absolute ratio is greater than or equal to b. When applied over several trials, this model will result in pecking functions that are centered on the value of the fixed interval with a standard deviation proportional to the fixed interval, exactly as observed in the starlings.

The giving-in times can be modeled with another small modification of the basic model. In this case, we assume that the birds are using the same reference memory of reinforcement times used above to generate the pecking patterns, but a different decision rule. The bird leaves the patch at time g such that

$$\frac{m_g - m_{FI}^*}{m_{FI}^*} \geq b_g$$

where m_g is the perception of the current time, g, and b_g is a new threshold that is larger than b. Before Brunner et al.'s (1992) application of scalar timing theory to an optimal foraging problem, it had previously been assumed that thresholds were fixed (e.g., Gibbon 1977); however, one of the most important outcomes of Brunner et al.'s experiment is the suggestion that in fact a threshold such as b_g may be optimized by natural selection to maximize the rate of energy intake of the birds given the constraint of their imprecise timing mechanisms. The predictions of such an optimality approach are that the threshold, b_g, and consequently the giving-in times should be bigger when the cost associated with travel or the energy gain associated with the reinforcement are increased.

This study provides a beautiful example of the benefits of the ethological approach of integrating mechanistic and evolutionary analyses of the same problem. Without the scalar timing model, we would not have understood why the starlings' giving-in times are a fixed proportion of the interval between food items. Without the evolutionary approach, we would not have understood that the biases assumed in scalar timing are not arbitrarily chosen, but may actually be the outcome of an evolutionary optimization of the costs of leaving the patch too early and too late.

5.5.2 SCALAR TIMING AND PATCH DEPARTURE: THE MARGINAL VALUE THEOREM

Kacelnik and Todd (1992; Todd and Kacelnik, 1993) set up an operant analog of the MVT scenario in order to pursue the details of how foraging pigeons respond to travel time. A red flashing light signaled to the pigeon that food was available in the patch. As soon as the pigeon pecked the light, it changed to a steady red light, and a white light was also turned on. At this point the pigeon could choose to peck either the red light to obtain food in the patch or the white light to leave the patch and initiate the travel component of the schedule (that was actually a waiting time in this task). The food in the patch was delivered according to a progressive-interval schedule, such that the delay between food items increased with each successive prey delivered (as required for the MVT to apply). Using this schedule, Kacelnik and Todd (1992) could change various features of the travel component of the schedule and measure the effect that their manipulations had on the dependent variable, which was the number of prey per patch visit (PPV) taken by the bird before it pecked the white key and initiated the next travel component of the schedule.

In their first study, Kacelnik and Todd (1992) chose to test the MVT prediction that PPV should be sensitive only to mean travel time and should not be affected by variance in travel time. They tested pigeons in three treatments, all of which had mean travel times of 95 sec, but which differed in the variance in travel time. Treatment 10t consisted of a random order of ten different travel times with a coefficient of variation of 60.5%, treatment 2t consisted of a random order of two travel times with a coefficient of variation of 95%, and treatment 1t consisted of a single travel time. Contrary to the predictions of the MVT, the distribution of travel times had an effect on the mean PPV, with the birds having the highest PPV in the 1t treatment, an intermediate PPV in the 10t treatment, and the lowest PPV in the 2t treatment. Thus, PPV decreased as the coefficient of variation in travel time increased.

In order to explain the observed inverse relationship between PPV and variance in travel time, Kacelnik and Todd (1992) considered a modification of the MVT in which the representation of travel times in memory was not a perfect mean, as assumed in the unconstrained MVT, but was instead a distribution, as assumed in the scalar timing model. As shown in Figure 5.2, the travel time in the 1t treatment will be represented by a symmetrical memory representation, but the 10t and 2t distributions become progressively asymmetrical and skewed to the right. Kacelnik and Todd (1992) combined this memory representation of travel times with the MVT by assuming that in order to form an estimate of the background rate of energy intake available in the environment, the bird draws a single random sample from its reference memory holding the representation of experienced travel times. This value is then used to calculate the optimal PPV in the current patch. Due to the skew in the representations of the variable travel times, random samples drawn from the 10t representation will have an average shorter than 95 sec, and samples drawn from the 2t representation will have an average that is shorter still. Because shorter travel times will lead to a smaller optimal number of PPV, the modified MVT model is capable of explaining the observed effects of variance in travel time. In addition to explaining the effects of variance in travel time on PPV, the modified MVT model also explains another feature of the data not accommodated by the unconstrained MVT. The unconstrained MVT predicts that in the 1t treatment there should be no variance in the PPV taken by a bird; however, in reality the pigeons did have variance in their PPV in the 1t treatment. The scalar timing modification of the MVT explains this variance, because even a single travel time is represented in memory as a distribution from which random samples are taken for the purposes of decision making.

Despite the successes of the modified MVT, molecular analysis of the data from the 2t and 10t treatments revealed an important result that was not predicted by either unconstrained or modified MVT. When Kacelnik and Todd (1992) examined the PPV taken by a bird in relation to the previous travel time the bird had experienced, they found a significant positive effect, with the pigeons taking more PPV if the previous travel time had been long. Thus the pigeons appear to be not only responding to the mean and variance of the mixture of travel times, but also weighting the most recently experienced travel time more highly in their decision making. Todd and Kacelnik (1993) confirmed this finding in a subsequent study using the same paradigm that was designed to explicitly address the relative roles of the mixture of travel times and the most recently experienced travel time on PPV. They tested pigeons in two treatments that differed in mean travel time but had similar coefficients of variation (60.7 and 67.9% for the short and long mean treatments, respectively). Importantly, two of the travel times (1 and 13 sec) were contained in both mixtures. Todd and Kacelnik (1993) studied the effects of travel time by comparing the PPV after various travel times within each treatment and by comparing the PPV after equal travel times (1 and 13 sec) between treatments. The within-treatment analysis showed that in the long mean travel time treatment, PPV was correlated with the previous travel time, as they had found in the previous study. When the same travel times were compared between treatments, the pigeons were found to take higher PPV in the long mean travel time treatment than in the short mean travel time treatment. Because this effect cannot be accounted for by the within-treatment effect of the previous travel time, it implies that there is a different and independent effect on PPV of the mixture of travel times.

Because scalar timing theory is a steady-state model that makes no statements about how the reference memory representations are built up, it currently does not address short-term changes in behavior due to very recent experience. However, from a functional point of view it makes sense that the adaptive length of a memory window could vary depending on the stability of the environment. In very stable environments, it would make sense to base decisions on all of the available experience; however, in more changeable environments, it might be adaptive to weight recently acquired information more heavily than information acquired longer ago. In order to remedy this problem, Todd and Kacelnik (1993) developed a dynamic version of their previous model that involved adding an explicit learning algorithm to scalar timing theory and combining this with the MVT. Their aim was to develop a model that could handle the parallel effect of both recent and longer-term memory on foraging decisions that was suggested by their empirical data. Scalar timing theory provided a natural way to approach this problem because its parallel structure, whereby samples may be read at the same time from working and reference memory, allows a way to separate the effects of current percepts of time from remembered experience.

Todd and Kacelnik's new model contains two innovations that allow recent experience to have greater impact on decision making. First, the reference memory representation is built up and continually modified according to recent experience.

Interval Timing and Optimal Foraging

The reference memory is defined as a probability density function with bins corresponding to travel times each assigned a probability, and the total area under the function always equal to 1. Following each travel, the reference memory is updated in two steps. First, a fraction (α) of the area under the probability function is subtracted by devaluing each bin in proportion to its probability value at the time, such that the sum of the devaluations equals α. Second, an area the size of α is added back to the probability in the bin corresponding to the current travel time in working memory. Thus, following updating, the total area under the probability density function remains at unity, but the shape of the distribution is shifted toward the most recently experienced travel time, with the size of this shift controlled by the value of the parameter α. Low values of α correspond to little weight being given to recent experience, as would be predicted in a stable environment, whereas high values of α correspond to a high weight being given to recent experience, as would be predicted in changeable environments. The second innovation involves the decision subsystem. Rather than just using a random sample of travel time from reference memory in order to choose the PPV for the patch, Todd and Kacelnik (1993) assumed that value of travel time used was a weighted average of the value currently in working memory representing the most recently experienced travel time and a random sample taken from reference memory. They introduced a second parameter, β, that controlled the relative weight given to the values from working and reference memory. Just as for α, low values of β correspond to little weight being given to recent experience, whereas high values correspond to a high weighting of recent experience. As in their previous model, the weighted average of the two travel times was used as the input to the MVT to produce the optimal PPV.

Simulations of the above model run for the two different treatments experienced by the pigeons produced results that mimicked both of the main empirical results: the model produced both the observed treatment difference, with higher PPV in the high mean travel time patch than in the low mean travel time patch, and the within-treatment molecular effect of higher PPV directly following longer travel times.

Again, this study provides a clear example of the benefits of the ethological approach. Without the scalar timing model, we did not have an explanation for why pigeons should respond differently to fixed and variable travel times of the same mean. However, the scalar timing model shows that these effects occur because travel times are stored in reference memory as a distribution rather than as a mean, and as we have seen previously, the distribution representing a fixed travel time is symmetrical, whereas the distribution representing a variable travel time is asymmetrical (see Figure 5.2). The contribution of the evolutionary approach is the realization that it may not always be adaptive to weight all previous experience equally, and therefore the scalar timing model needs mechanisms that control the weighting given recent and past experience.

5.5.3 SCALAR TIMING AND RESPONSE TO RISK

There have been several experimental investigations of how foraging animals respond to variability in both delay to reward and amount of reward. For example, Bateson and Kacelnik (1995b) used an operant paradigm to study risk sensitivity in starlings. The starlings' preferences were tested in two treatments, one with variability in amount of reward and one with variability in delay to reward, and the energy budgets of the birds were maintained unchanged throughout the experiment. In the variable-amount treatment, the starlings were initially trained that a flashing red light indicated the availability of a fixed interval of 20 sec that culminated with a reward of 5 units of food, and that a flashing green light signaled the availability of a fixed interval of 20 sec culminating in a reward that was either 3 or 7 units of food with equal probability. In the variable-delay treatment, the fixed option was identical to the variable-amount treatment, but the other color signaled the availability of a variable interval of either 2.5 or 60.5 sec with equal probability, culminating in 5 units of reward. The assignment of colors to options was balanced across birds. The birds initially learned about the two options available in a treatment in forced trials in which only one option was presented. Following this training, the birds' choices were tested in trials in which both options were presented simultaneously, and the bird was required to commit to one of the options by choosing to peck one of the two flashing lights.

Bateson and Kacelnik (1995b) recorded two measures of preference in both treatments: the latency to peck the flashing lights in the training trials, and the option chosen by the birds in the choice trials. The choice data revealed that the birds were indifferent to risk in the variable-amount treatment, but strongly risk prone in the variable-delay treatment. The latency results showed that the birds had a shorter latency to begin a fixed-amount trial in the variable-amount treatment and a shorter latency to begin a variable-delay trial in the variable-delay treatment. Thus the results provide support for the general finding that animals tend to be risk averse to variability in amount, but strongly risk prone to variability in delay (Kacelnik and Bateson, 1996), because the birds preferred the variable-delay option even though the fixed delay of 20 sec was well below the mean delay of 31.5 sec in the variable-delay option.

The above results are not predicted by any purely evolutionary model; however, Reboreda and Kacelnik (1991) proposed that this pattern of preference could be explained by a modification of a version of scalar timing theory first proposed by Gibbon et al. (1988) to explain the preferences for delayed rewards seen in concurrent schedules. Gibbon et al. proposed that when animals are subjected to two alternative options between which they are required to choose, they build up a separate reference memory representation for each option. Thus, in Bateson and Kacelnik's experiment, the starlings in the variable-delay treatment would build up one reference memory for the fixed option and one for the variable option. As explained previously, the memory representation for the fixed interval would be a symmetrical distribution centered on the experienced interval, whereas the memory representation for the variable interval would be an asymmetrical distribution resulting from the sum of a symmetrical low variance distribution with a mean of 2.5 and a symmetrical high variance distribution with a mean of 60.5. The starling is assumed to choose between the two options by taking one random sample from each reference memory distribution, comparing these two samples, and choosing the option that yielded the lower sample (see Figure 5.5). Due to the skew in the representation of the variable delay (see Figure 5.2), this distribution will on average yield smaller samples than the

Interval Timing and Optimal Foraging

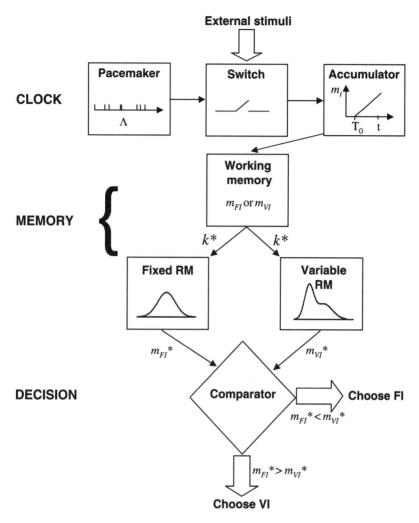

FIGURE 5.5 Diagrammatic representation of the scalar timing model applied to a choice scenario. In this case, the model is applied to the choice between a fixed interval and a variable interval (composed of two different intervals), as is found in many risk-sensitive foraging experiments.

symmetrical distribution representing the fixed option. This model can therefore explain why animals might be risk prone to variance in delay to reward. Reboreda and Kacelnik's (1991) innovation was to realize that the same argument could explain risk aversion to variance in amount. They suggested that if the birds form memory representations similar to those amounts assumed for delays, then the variable-amount option will be represented by an asymmetrical distribution formed from the sum of a symmetrical low variance distribution with a mean of 3 and a symmetrical higher variance distribution with a mean of 7. This is not an unreasonable assumption because in most experiments manipulating the amount of reward, the size of a reward is positively correlated with the time taken to consume it; thus consumption time could be used as a mechanism for measuring amount. A random sample drawn from the asymmetric distribution representing the variable-amount option will, as for the variable-delay option, on average be smaller than a random sample drawn from a symmetrical distribution with a mean of 5. However, in the case of amounts, the bird prefers the option yielding the larger sample, because although short delays to food are preferable to long delays, large amounts of food are preferable to small amounts.

One of the attractive features of the scalar timing theory account of choice is that the model not only makes predictions about the direction of preference, but also makes quantitative predictions about the exact magnitude of preference. Bateson and Kacelnik (1995b) proved that if a fixed-delay option is compared with a variable-delay option in which the delay is either short or long with equal probability, scalar timing theory predicts that the two options will become subjectively equivalent when the delay in the fixed option is equal to the geometric mean (i.e., $\sqrt{\text{short delay} \times \text{long delay}}$) of the two delays in the variable option. Bateson and Kacelnik (1996) tested this prediction in a subsequent starling experiment. They used a titration procedure to find the fixed delay equal to the variable mixture of 2.5- and 60.5-sec delays used in their previous experiment, and showed that indifference occurred when the fixed delay was equal to 5.61 sec, which is significantly below the geometric mean of 2.5 and 60.6 that equals 12.30 sec. Therefore, although scalar timing theory predicts the qualitative features of choice for variable delays, in this instance it failed to predict the quantitative detail of the results.

In the case of risk-sensitive foraging, the main outcome of considering the mechanistic account provided by scalar timing theory is that behavioral phenomena previously interpreted by behavioral ecologists as adaptive (e.g., Caraco et al., 1990) potentially now emerge as nonadaptive artifacts of the mechanisms by which animals maximize their rate of energy intake.

5.5.4 SCALAR TIMING AND SAMPLING A CHANGING ENVIRONMENT

Shettleworth et al. (1988) tested Stephens' (1987) sampling model, described previously, by setting up an operant analog of the sampling problem in the laboratory. Their experiment was carried out with pigeons in a shuttlebox in which the two feeding sites were represented by feeders and keys at either end of the box. On the stable side of the box, a random ratio schedule delivered rewards with some probability that remained unchanged throughout an experimental treatment. On the fluctuating side of the box, the schedule varied between no reward and a random ratio schedule equal to or lower (i.e., delivering more frequent rewards) than the stable side. Two different colored lights provided information about the state of the fluctuating side as soon as the bird pecked there. Following each reinforcement, the pigeon was required to return to the center of the box to initiate a new trial. Changes in the state of the fluctuating side occurred with a probability of .002 per trial. Shettleworth et al. (1988) investigated how the pigeons allocated their choices between the stable and fluctuating sides of the box in a range of treatments in which both the probability of reinforcement in the stable patch and the probability of reinforcement in the good state of the fluctuating patch were manipulated.

As predicted by Stephens' optimality model, the results showed that when the fluctuating patch was bad, pigeons chose the stable patch most of the time, only infrequently sampling the fluctuating patch. However, if on a sampling visit the fluctuating patch was in its good state, the pigeon would stay there until it reverted to its bad state. Furthermore, as predicted by the optimality model, the pigeons increased their rate of sampling the fluctuating patch as the probability of reinforcement in the constant patch was decreased. However, the behavior of the pigeons deviated in three important ways from the predictions of the unconstrained optimality model. First, sampling did not occur at the regular intervals predicted by Stephens' model, but instead occurred at random intervals. Second, sampling frequency was not sensitive to the probability of reinforcement in the good state of the fluctuating patch, and sampling still occurred when the probability of reinforcement in the good state of the fluctuating patch was equal to that in the stable patch. Finally, when the fluctuating patch was in its good state, the birds occasionally visited the constant patch, thus reducing their rate of food intake.

Because scalar timing theory was developed to model experiments in which rewards occur on the basis of time, but in this experiment Shettleworth et al. (1988) used ratio schedules in which rewards were delivered on the basis of responses, it was necessary to make the assumption that the pigeons responded at a constant rate in order to use scalar timing theory to model the data. Having made this assumption, the problem faced by the pigeon converts into a simple choice between two different distributions of delays to reward. The pigeon can simply be viewed as asking, "What is the delay to food if I stay with the current stable side vs. what is the delay to food if I sample?" According to scalar timing theory, the memory representation of the stable side will be an exponential distribution. The memory representation of the fluctuating side is more complex, and Shettleworth et al. (1988) assume that the time to food can be simplified to the sum of the time in the bad state until the good state begins and the time in the good state until food, which are both assumed to be represented by exponential distributions. As in the risky choice version of scalar timing described in the previous section, the pigeon is assumed to take a random sample from each of its two memory representations and use these to decide which side offers the shortest delay to food. Once the good state on the fluctuating side has been entered, a new reference memory for the fluctuating side applies in which the delay to reward is represented by an exponential distribution of delays as on the stable side. Thus the choice is now between two exponentially distributed delays to reward.

The scalar timing model can provide an explanation for some of the ways in which the data deviated from the unconstrained optimality model. First, sampling is not predicted to occur at regular intervals because decisions are based on random samples from memory distributions. For the same reason, both sampling of the fluctuating patch when the probability of reward is the same as the probability of reward in the stable patch, and reverse sampling of the stable patch when the fluctuating patch is currently in its good state, are predicted to occur, because occasionally the memory sample drawn from the memory for the suboptimal side will suggest a shorter delay to food than the sample drawn from the memory of the optimal side.

5.6 CONCLUSIONS

My intention has been to give some insight into how, by combining the evolutionary approach of optimal foraging theory with the hypotheses about the psychological mechanisms underlying interval timing, we can move toward a fuller understanding of both foraging behavior and timing. In all of the foraging examples described above, the basic strategy has been to take the constraints embodied in the scalar timing model, apply these to different foraging problems, and investigate how they affect optimal foraging decisions.

In all cases, the addition of scalar timing explains details of the birds' behavior that could not be accounted for with an unconstrained evolutionary approach. Three key assumptions of the scalar timing model emerge as being particularly important in producing these results. The first important assumption is the scalar property, whereby the duration of an interval is proportional to the precision with which it is represented in memory. The second assumption is that time intervals are assumed to be represented in memory as distributions of the intervals experienced rather than as a single statistic. Finally, the third crucial assumption is that reference memory is accessed via random sampling. There is extensive evidence for the scalar property; however, the latter two assumptions are more difficult to test empirically, and given the failure of the basic scalar timing model to account for some quantitative features of behavioral data (e.g., Bateson, 1993; Bateson and Kacelnik, 1996; Brunner et al., 1997), it is probably wise to maintain an open mind about the exact details of how memory is represented and accessed at this stage (e.g., Kacelnik and Brito-E-Abreu, 1998).

The examples described also highlight how our understanding of interval timing can benefit by thinking in detail about what animals use timing for. Specifically, the applications above have given insights into how the functional demands of different uses of temporal information may have shaped how this information is both represented and used in decision making. The basic scalar timing model makes no assumptions about the learning process and assumes that animals are equipped with memories that represent all possible experiences. However, the study of Todd and Kacelnik (1993) makes it clear that scalar timing models need to take account of learning, and that it may make sense for the memory to be weighted in favor of recent experience. In terms of decision making, the study of Brunner et al. (1992) shows that the biases assumed in the decision stage of scalar timing should be thought of as reflecting the optimal trade-off between responding early and late, and that the position of this trade-off will differ depending on the decision being made.

The flexibility of the scalar timing model has proved central in applying the model to a range of different foraging problems. The modularity of the model makes it easy to modify the details of how temporal information is represented in memory and how this information is used. For example, if we look first at how information is represented in the various examples, some of the applications described require a single reference memory, whereas others involving choice assume that there are two (Bateson and Kacelnik, 1995b; Reboreda and Kacelnik, 1991) or more (Shettleworth et al., 1988) memories. In most applications, experience is weighted equally in reference memory, but in some cases, it is necessary to assume that memory is biased toward recent experience (Todd and Kacelnik, 1993). If we look at how the temporal information represented in memory is subsequently used, again all of the applications make different assumptions. In some cases, a single sample from memory is compared with the current value in the accumulator (Brunner et al., 1992); in other cases, a weighted average of a sample from memory and the value in the accumulator is used in decision making (Todd and Kacelnik, 1993); and in yet other applications, the critical comparison is between two samples drawn from different reference memories (Bateson and Kacelnik, 1995b; Reboreda and Kacelnik, 1991; Shettleworth et al., 1988). An interesting point that emerges from Brunner et al.'s (1992) study is that the same reference memory may be accessed by two or more different decision-making mechanisms.

As a final thought, it is interesting to note that the flexible use of the same basic components described above has a close analogy with current thinking about the evolutionary process. The neural mechanisms responsible for producing new adaptive behavior patterns are not created from scratch, but are shaped by natural selection from small modifications of existing mechanisms. Thus, it is plausible that the tinkering necessary to apply scalar timing to a range of foraging problems is an accurate reflection of how interval timing mechanisms have evolved.

ACKNOWLEDGMENTS

I thank the Royal Society and the Natural Environment Research Council for financial support. The ideas presented in this chapter have benefited from the many discussions about timing that I have had with my mentors and colleagues, including John Gibbon, Sue Healy, Alex Kacelnik, and Warren Meck.

REFERENCES

Bateson, M., Currencies for Decision Making: The Foraging Starling as a Model Animal, unpublished doctoral dissertation, Oxford University, U.K., 1993.

Bateson, M., Interval timing, in *Frontiers of Life*, Baltimore, D., Dulbeco, R., Jacob, F., and Levi-Montalcini, R., Eds., Academic Press, New York, 2001, pp. 241–250.

Bateson, M. and Kacelnik, A., Accuracy of memory for amount in the foraging starling (*Sturnus vulgaris*), *Anim. Behav.*, 50, 431–443, 1995a.

Bateson, M. and Kacelnik, A., Preferences for fixed and variable food sources: variability in amount and delay, *J. Exp. Anal. Behav.*, 63, 313–329, 1995b.

Bateson, M. and Kacelnik, A., Rate currencies and the foraging starling: the fallacy of the averages revisited, *Behav. Ecol.*, 7, 341–352, 1996.

Bateson, M. and Kacelnik, A., Risk-sensitive foraging: decision making in variable environments, in *Cognitive Ecology*, Dukas, R., Ed., Chicago University Press, Chicago, 1998, pp. 297–341.

Brunner, D., Fairhurst, S., Stolovitzky, G., and Gibbon, J., Mnemonics for variability: remembering food delay, *J. Exp. Psychol. Anim. Behav. Process.*, 23, 68–83, 1997.

Brunner, D., Kacelnik, A., and Gibbon, J., Optimal foraging and timing processes in the starling, *Sturnus vulgaris*: effect of inter-capture interval, *Anim. Behav.*, 44, 597–613, 1992.

Brunner, D., Kacelnik, A., and Gibbon, J., Memory for inter-reinforcement interval variability and patch departure decisions in the starling, *Sturnus vulgaris*, *Anim. Behav.*, 51, 1025–1045, 1996.

Caraco, T., Blanckenhorn, W.U., Gregory, G.M., Newman, J.A., Recer, G.M., and Zwicker, S.M., Risk-sensitivity: ambient temperature affects foraging choice, *Anim. Behav.*, 39, 338–345, 1990.

Charnov, E.L., Optimal foraging: the marginal value theorem, *Theor. Popul. Biol.*, 9, 129–136, 1976.

Church, R.M., Timing and temporal search, in *Time and Behaviour: Psychological and Neurobiological Analyses*, Bradshaw, C.M. and Szabadi, E., Eds., Elsevier, Amsterdam, 1997, pp. 41–78.

Davies, N.B., Prey selection and the search strategy of the spotted flycatcher (*Muscicapa striata*), *Anim. Behav.*, 25, 1016–1033, 1977.

Engen, S. and Stenseth, N.C., A general version of optimal foraging theory: the effect of simultaneous encounters, *Theor. Popul. Biol.*, 26, 192–204, 1984.

Gibbon, J., Scalar expectancy theory and Weber's law in animal timing, *Psychol. Rev.*, 84, 279–325, 1977.

Gibbon, J. and Church, R.M., Sources of variance in an information processing theory of timing, in *Animal Cognition*, Roitblat, H.L., Bever, T.G., and Terrace, H.S., Eds., Erlbaum, Hillsdale, NJ, 1984, pp. 465–488.

Gibbon, J., Church, R.M., Fairhurst, S., and Kacelnik, A., Scalar expectancy theory and choice between delayed rewards, *Psychol. Rev.*, 95, 102–114, 1988.

Gibbon, J., Church, R.M., and Meck, W.H., Scalar timing in memory, in *Timing and Time Perception*, Gibbon, J. and Allan, L., Eds., Academic Press, New York, 1984, pp. 52–77.

Gill, F.B., Trapline foraging by hermit hummingbirds: competition for an undefended renewable resource, *Ecology*, 69, 1933–1942, 1988.

Goss-Custard, J.D., Feeding dispersion in some overwintering wading birds, in *Social Behaviour in Birds and Mammals*, Crook, J.H., Ed., Academic Press, London, 1970, pp. 3–35.

Hinton, S.H. and Meck, W.H., How time flies: functional and neural mechanisms of interval timing, in *Time and Behaviour: Psychological and Neurobiological Analyses*, Bradshaw, C.M. and Szabadi, E., Eds., Elsevier, New York, 1997, pp. 409–457.

Kacelnik, A., Central place foraging in starlings (*Sturnus vulgaris*): I. Patch residence time, *J. Anim. Ecol.*, 53, 283–299, 1984.

Kacelnik, A. and Bateson, M., Risky theories: the effects of variance on foraging decisions, *Am. Zool.*, 36, 402–434, 1996.

Kacelnik, A. and Brito-E-Abreu, F., Risky choice and Weber's law, *J. Theor. Biol.*, 194, 289–298, 1998.

Kacelnik, A. and Brunner, D., Timing and foraging: Gibbon's scalar expectancy theory and optimal patch exploitation, *Learn. Motiv.*, 33, 177–195, 2002.

Kacelnik, A., Brunner, D., and Gibbon, J., Timing mechanisms in optimal foraging: some applications of scalar expectancy theory, in *NATO ASI Series*, Series G, *Ecological Sciences*, Vol. G 20, Hughes, R.N., Ed., Springer-Verlag, Berlin, 1990, pp. 63–81.

Kacelnik, A. and Todd, I.A., Psychological mechanisms and the marginal value theorem: effect of variability in travel time on patch exploitation, *Anim. Behav.*, 43, 313–322, 1992.

Killeen, P.R., Fetterman, J.G., and Bizo, L.A., Time's causes, in *Time and Behaviour: Psychological and Neurobiological Analyses*, Bradshaw, C.M. and Szabadi, E., Eds., Elsevier, Amsterdam, 1997, pp. 79–131.

Lejeune, H. and Wearden, J.H., The comparative psychology of fixed-interval responding: some quantitative analyses, *Learn. Motiv.*, 22, 84–111, 1991.

Matell, M.S. and Meck, W.H., Reinforcement-induced within-trial resetting of an internal clock, *Behav. Process.*, 45, 159–171, 1999.

Matell, M.S. and Meck, W.H., Neuropsychological mechanisms of interval timing behaviour, *Bioessays*, 22, 94–103, 2000.

Paule, M.G., Meck, W.H., McMillan, D.E., Bateson, M., Popke, E.J., Chelonis, J.J., and Hinton, S.C., The use of timing behaviors in animals and humans to detect drug and/or toxicant effects, *Neurotoxicol. Teratol.*, 21, 491–502, 1999.

Reboreda, J.C. and Kacelnik, A., Risk sensitivity in starlings: variability in food amount and food delay, *Behav. Ecol.*, 2, 301–308, 1991.

Schneider, B.A., A two-state analysis of fixed-interval responding in the pigeon, *J. Exp. Anal. Behav.*, 12, 677–687, 1969.

Shettleworth, S.J., Krebs, J.R., Stephens, D.W., and Gibbon, J., Tracking a fluctuating environment: a study of sampling, *Anim. Behav.*, 36, 87–105, 1988.

Skinner, B.F., The origins of cognitive thought, *Am. Psychol.*, 44, 13–18, 1989.

Stephens, D., On economically tracking a variable environment, *Theor. Popul. Biol.*, 32, 15–25, 1987.

Stephens, D.W. and Krebs, J.R., *Foraging Theory*, Princeton University Press, Princeton, NJ, 1986.

Tinbergen, N., On aims and methods of ethology, *Z. Tierpsychol.*, 20, 410–433, 1963.

Todd, I.A. and Kacelnik, A., Psychological mechanisms and the marginal value theorem: dynamics of scalar memory for travel time, *Anim. Behav.*, 46, 765–775, 1993.

Yates, M.G., Stillman, R.A., and Goss-Custard, J.D., Contrasting interference functions and foraging dispersion in two species of shorebird (Charadrii), *J. Anim. Ecol.*, 69, 314–322, 2000.

Yoccoz, N.G., Engen, S., and Stenseth, N.C., Optimal foraging: the importance of environmental stochasticity and accuracy in parameter estimation, *Am. Naturalist*, 141, 139–157, 1993.

6 Nonverbal Representations of Time and Number in Animals and Human Infants

Elizabeth M. Brannon and Jamie D. Roitman

CONTENTS

6.1 Introduction 144
6.2 Number Representation in Animals 144
    6.2.1 Absolute Numerosities 145
    6.2.2 Making N Responses 146
    6.2.3 Running Speed as an Indicator of Number Representation 148
    6.2.4 Symbols and Numerosities 148
    6.2.5 Transfer between Tasks and Modalities 149
6.3 Arithmetic Reasoning in Animals 150
    6.3.1 Ordering 150
    6.3.2 Addition and Subtraction 155
6.4 Number Representation in Human Infants 155
    6.4.1 Small Number Discrimination 155
    6.4.2 Large Number Discrimination 157
    6.4.3 Cross-Modal Matching 157
    6.4.4 Continuous Dimensions Confound 158
6.5 Arithmetic Reasoning in Infancy 159
    6.5.1 Ordering 159
    6.5.2 Addition and Subtraction 160
6.6 A Brief Description of Five Models of Nonverbal Number Representation 161
    6.6.1 Arbitrary Numeron Hypothesis 161
    6.6.2 Subitizing and the Object-File Mode 162
    6.6.3 Mode-Control Model 163
    6.6.4 Connectionist Timing Model 164
    6.6.5 Dehaene and Changeux Neural Network Model 165
6.7 Differentiating the Models 166
    6.7.1 Are Animals and Infants Merely Subitizing? 166

6.7.2 Are Animals and Infants Limited to Object-Files?..........................166
6.7.3 Are Animals and Infants Counting?..167
6.7.4 How Do the Models Handle Development?...................................168
6.7.5 Weber's Law..168
6.7.6 Representation of Time and Number 169
6.8 Neural Basis of Number Representation..170
6.9 Conclusion and Directions for Future Research 174
Acknowledgments..175
References..175

6.1 INTRODUCTION

Scientists from many disciplines have been intrigued by the topic of how the mind represents number because of the question's relevance to controversial topics such as thought without language, the evolution of cognition, modularity of mind, and nature vs. nurture. Number is an abstract and emergent property of sets of discrete entities; two people and two airplanes look nothing alike, and yet the numerosity of the set is the same. Some researchers believe that the abstract nature of numerical representation makes it an unlikely candidate for a cognitive capacity held by nonhuman animals and human infants. However, a growing body of data suggests that both nonverbal animals and preverbal human infants represent number and even perform operations on these representations. In fact, a new synthesis of the data on numerical abilities in animals and infants suggests that there is an evolutionarily and developmentally primitive system for representing number as mental magnitudes with scalar variability (Gallistel and Gelman, 1992, 2000; see also Dehaene, 1997; Wynn, 1995). Furthermore, there is abundant evidence that adult humans also represent number nonverbally as analog magnitudes (Cordes et al., 2001; Dehaene, 1997; Dehaene et al., 1998; Moyer and Landauer, 1967; Whalen et al., 1999). For these reasons, numerical cognition has become an exciting area of research and an exemplary model of a cross-disciplinary field where comparative and developmental studies have influenced current conceptions of adult human numerical processing (e.g., Dehaene, 1997).

The goal of this chapter is to evaluate the evidence bearing on the nonverbal representation of number by animals and human infants. Specifically, we will (a) review the data demonstrating that nonhuman animals and human infants represent number and perform operations on these representations, (b) evaluate the proposed models of the nonverbal representation of number and the evidence for and against each model, and (c) describe what is known about the neurobiological basis of number representations. Finally, we will outline some of the outstanding questions that should guide future research aimed at understanding the evolution and development of numerical thinking.

6.2 NUMBER REPRESENTATION IN ANIMALS

Why might nonhuman animals need to represent number? In nature, many stimulus attributes correlate with number. When a set of items is simultaneously present, the total surface area, brightness, density, and cumulative perimeter of the objects are available as cues to describe the set. If the to-be-counted items are presented sequentially, then duration, rate, or rhythmic patterns may covary with number. Similarly, a food patch with a greater number of discrete items will have higher hedonic value, as measured by the volume of food ingested, the feeling of fullness, or perhaps the amount of saliva that the eating animal produces. When animals are required to discriminate endogenous stimuli, such as the number of their own responses on an operandum, the total effort expended or the duration of responding is easily confounded with number. Perhaps animals do not need to represent number to maximize food intake or function adaptively in other domains. Animals may only represent number as a last-resort strategy when there is no other way to solve a given problem (Davis and Perusse, 1988).

An alternative possibility is that animals routinely represent number and that such processing is a fundamental part of the perceptual and cognitive systems of most vertebrate taxa (Capaldi, 1993; Gallistel, 1990). In foraging decisions, many animals behave as if they calculate the rate of return in a given food patch in order to decide whether to stay or go (Krebs and Davies, 1993). In an especially compelling example, Harper (1982) dispersed food at two different rates and showed that ducks sorted themselves into groups with numbers proportional to the amount of food at each location. Importantly, sorting often occurred before every duck had had the opportunity to sample the food. The social arena is another domain where numerical competence may be useful. Wilson et al. (2001) showed that the likelihood that male chimpanzees would respond aggressively to a simulated intruder's vocalization depended on the number of potential allies present. This suggests that males determine the number of conspecifics in their party to calculate their chances at successfully repelling an intruder.

Gallistel and Gelman (1992) made the important distinction between numerical categories and numerical concepts. An organism can be said to have a numerical category insofar as it can behaviorally group all exemplars of a given numerosity regardless of irrelevant dimensions such as surface area, density, and so on. In contrast, the richness of an organism's numerical concept is dependent upon the types of operations that it can perform on representations of number. In this section, we first review the data obtained from paradigms designed to test whether animals can form purely numerical categories; subsequently, we describe the evidence that animals can manipulate these representations of number in meaningful ways and thus have numerical concepts.

6.2.1 ABSOLUTE NUMEROSITIES

A variety of paradigms have been used to assess whether animals can represent a single numerosity. In one of the earliest studies, Hicks (1956) trained eight rhesus monkeys to choose cards that contained three visual elements from among cards that contained one to five elements. Similarly, Davis (1984) trained a single raccoon to select a Plexiglas box that contained three objects from an array of boxes that contained one to five objects (see Figure 6.1). Rats have also been trained to respond selectively after three rather than two or four vibrissal strokes (Davis et al., 1989).

FIGURE 6.1 The raccoon was rewarded for selecting the Plexiglas cube that contained three objects. (From Davis, H., *Anim. Learn. Behav.*, 12, 409–413, 1984.)

Animals (e.g., Davis and Bradford, 1991; Davis, 1996) can also restrict their food intake by selecting a specific number of reinforcers from a large array (for reviews of early work on absolute numerical concepts, see Douglas and Whitty, 1941; Wesley, 1961). Davis and Bradford (1986) showed that rats can choose a specific ordinal position in a series of objects or tunnels (for a similar demonstration with monkeys, see Ruby, 1984). In that study, rats trained to run down an alley and enter the third tunnel in a row of five tunnels were successful even when the space between the tunnels varied. In sum, many animal species are adept at representing a specific numerosity or a specific ordinal position in a sequence.

6.2.2 Making N Responses

In another paradigm animals have been trained to make a certain number of responses to obtain a reward. Platt and Johnson (1971) required rats to signal when they had completed N lever presses by poking their nose into a hole equipped with a photoelectric sensor. As shown in Figure 6.2a, the number of responses the rats made before head poking was roughly normally distributed around the required number. Scalar variance was also obtained; the standard deviation of the distribution of the obtained number of responses increased linearly with the required number of responses (see also Fetterman and MacEwen, 1989; Laties, 1972; Mechner, 1958; Rilling, 1967; Wilkie et al., 1979). Whalen et al. (1999) tested adult humans with an adapted version of the Platt and Johnson task. Adults were required to rapidly press a button a certain number of times without verbally counting. Results were indistinguishable from those of the rats. As shown in Figure 6.2b and c, scalar variance was found for both species, as indicated by the linear increase in the standard deviation in the response distribution with the mean number of responses required and the constant coefficient of variation obtained for both species (see also Cordes et al., 2001).

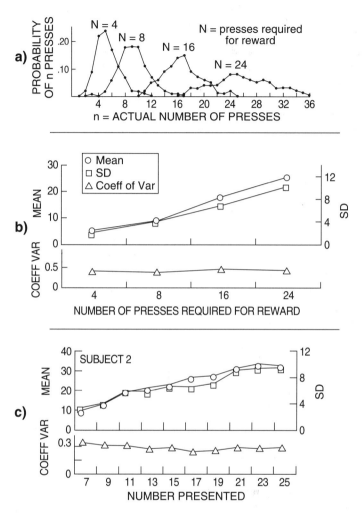

FIGURE 6.2 (a) The probability of signaling response completion as a function of the number of responses the rat made and the number that was required to obtain reward. Data from Platt and Johnson (1971). (b) The mean number of responses made (left axis, circles) and the standard deviation (right axis, squares) of the response distributions shown in panel (a), and the coefficient of variation (CV), which is the ratio of the standard deviation to the mean, as a function of the number required. (c) The mean (left axis, circles), standard deviation (right axis, squares), and CV (lower panel) as a function of the number of button responses required obtained by Whalen et al. (1999). The constant CV shown for rats in panel (b) and humans in panel (c) demonstrates that both species represent number with scalar variability.

In another version of a number production paradigm, Xia et al. (2000) trained pigeons to flexibly produce one to six responses cued by specific visual symbols. The pigeons' accuracy decreased with increasing number; however, four of six birds were above chance on all six numerosities, and nine of nine were above chance when only the numerosities 1 to 4 were tested (see also Rumbaugh and Washburn,

1993). Collectively, these studies suggest that animals can be trained to make a given number of responses; however, the very nature of these experimental designs makes it difficult to determine whether the animals were basing their judgments on the number of their own responses or instead some correlated attribute of the trials, such as the duration of responding or effort expended. To address this problem, some researchers have manipulated the rate of responding by using pharmacological manipulations or food restriction (e.g., Mechner and Guevrekian, 1962); other researchers have turned to different experimental designs altogether.

6.2.3 Running Speed as an Indicator of Number Representation

In another paradigm, Capaldi and Miller (1988) conducted a series of runway studies with rats showing that the speed with which rats run indicates that they anticipate nonreinforced runs in predictable sequences. When rats are trained with sequences of runs such as RRN and NRRN (where N denotes an unreinforced run and R a reinforced run), they run very slowly on the last run in both cases, suggesting that they can count two reinforced runs before an unreinforced run.

6.2.4 Symbols and Numerosities

Other studies have directly addressed the ability of nonhuman animals to learn the relationship between symbols and numerosities. For example, Matsuzawa (Matsuzawa, 1985; Matsuzawa et al., 1991) and Boysen (Boysen and Berntson, 1989) each trained a chimpanzee (Ai and Sheba, respectively) to choose the appropriate arabic numeral when presented with visual arrays of one to six objects. By the end of training, Ai was able to label the color, shape, and numerosity of sets of household objects. Sheba could both choose the arabic numeral that corresponded to an array of objects and choose the array of objects that corresponded to the numeral (see Figure 6.3). Pepperberg (1987) similarly trained Alex, an African gray parrot, to verbally label the numerosities 1 to 6 with the appropriate English words. Alex was even able to selectively enumerate one type of object in a visual display and ignore

FIGURE 6.3 A drawing of a chimpanzee named Sheba choosing the array of objects that depicts the numerosity of the numeral displayed on a computer monitor.

others (Pepperberg, 1994). For example, when presented with red and blue corks, Alex can answer, "How many blue cork?"

Although it is extremely impressive that chimpanzees and parrots can learn the relationship between numerosities and symbols, a striking aspect of these data is how very laborious it is to train this behavior. In fact, there is no indication of any positive transfer from one numerosity to the next (Murofushi, 1997). For example, after having been trained with the arabic numerals 1 to 3 and the corresponding numerosities 1 to 3, one might expect that when presented with a novel symbol and 3 + 1 dots that the ape would infer that the new numerosity should be paired with the new symbol. However, this is not the case. The chimpanzees required the same number of additional training trials for each new numerosity–symbol pair added to the mix. In addition, when chimpanzees were required to report the color, shape, and number of a set of objects in any order, they chose to report number last (Matsuzawa, 1985). These findings collectively suggest that the pairing of arbitrary symbols and numerosities heavily taxes the cognitive resources of chimpanzees.

6.2.5 TRANSFER BETWEEN TASKS AND MODALITIES

An abstract concept of number should not be tied to a sensory modality. Do animals appreciate that two sounds and two visual objects share "twoness"? Fernandes and Church (1982) found that rats trained to discriminate two vs. four sound bursts immediately transferred the discrimination to two vs. four light flashes. Even more impressively, rats trained to make one response after two light flashes or tones and a second response after four light flashes or tones made the four-response to a compound stimulus of two light flashes and two tones (Church and Meck, 1984). These results suggest that the rats summed over the two modalities. However, rats trained by Davis and Albert (1987) were unable to transfer an auditory numerical discrimination to the visual modality when a more complex task was used that required rats to identify the intermediate value 3 from the adjacent values, 2 and 4 (see also Pastore, 1961; Salman, 1943).

Another indication of an abstract numerical concept would be if animals could transfer a numerical discrimination learned in one task to another task (Seibt, 1982). For example, Meck (1997) trained one group of rats to behaviorally discriminate two vs. six light flashes and a second group of rats to discriminate two distinct wavelengths. Both groups were then tested in a 12-arm radial maze task where 4, 6, or 8 of the 12 arms were baited with food. Rats were required to obtain the N food items and then signal that they had completed foraging by entering a stop box. Amazingly, the rats that had been trained in the numerical discrimination task to discriminate 2 vs. 6 learned to retrieve six food items more readily than the rats that had been trained to discriminate two wavelengths. In addition, rats trained on the 2 vs. 6 discrimination performed better when they were required to retrieve 6 food items compared to 4 or 12 food items. In other words, rats trained to discriminate two vs. six light flashes exhibited savings when trained to retrieve six food items from a 12-arm maze. In contrast, when the chimpanzee Ai's ability to numerically label random dot arrays was tested, she showed no savings from her original training on numerically labeling arrays of household objects (Matsuzawa, 1985; Murofushi,

1987). More studies are needed to determine whether animals can transfer numerical discriminations learned in one task or modality to another.

6.3 ARITHMETIC REASONING IN ANIMALS

The research described above addresses the ability of nonhuman animals to form numerical categories. As eloquently articulated by Gallistel and Gelman (2000), animals might be expected to reason arithmetically because this would aid the many calculations for which continuous and discrete variables need to be combined. For example, it could be useful to multiply item size by number when deciding between a patch of three large food items and six tiny food items. Similarly, when faced with a decision between two food patches, it would be profitable to calculate the rate of return as defined by the number of food items divided by some unit of time. These calculations are greatly facilitated if discrete quantities are represented in the same currency as continuous quantities such as time and surface area. Below we review the evidence that animals reason arithmetically.

6.3.1 ORDERING

Davis and Perusse (1988) argued that making a relative numerosity judgment is simpler than the ability to form an absolute representation of number. The implicit idea here is that it is not necessary to represent cardinal values when making relative numerosity judgments. However, an alternative possibility is that ordering is an operation that is performed on absolute numerical representations. It is likely that there is a continuum of complexity for both cardinal and ordinal judgments.

Honig and colleagues (Honig, 1991; Honig and Matheson, 1995; Honig and Stewart, 1989, 1993) have conducted a series of studies showing that pigeons are sensitive to the relative number of icons in visual matrices. In one study, pigeons were trained to respond to a homogeneous array of X's and to avoid responding to a homogeneous array of O's (Honig and Stewart, 1989). Pigeons were then tested with arrays that had different proportions of X's and O's, and the proportion of responses the pigeons made was a systematic function of the proportion of X's in the test arrays. These data suggest that the pigeons were responding on the basis of the relative numerosity of the X's compared to the O's. Similarly, Meck and Church (1983) trained rats to discriminate two from eight sounds or successively presented visual stimuli and found a generalization gradient for intermediate values (for a similar demonstration in pigeons, see Roberts, 1995). More recently, Emmerton (1998) used a similar paradigm and demonstrated that pigeons trained to discriminate one and two from six and seven simultaneously presented visual stimuli subsequently showed a generalization gradient to the intermediate values.

Many primate species are adept at choosing the larger of two quantities of food (e.g., Anderson et al., 2000; Boysen and Berntson, 1995; Boysen et al., 2001; Call, 2000; Dooley and Gill, 1977; Rumbaugh et al., 1987, 1988). For example, Rumbaugh et al. (1987, 1988) presented chimpanzees with two food wells that contained different numbers of discrete food items and then measured the percentage of trials in which the chimps chose the larger quantity. They found that with values 0 through

4 the chimpanzees chose the larger quantity on approximately 95% of trials. Beran (2001) has also shown that chimpanzees can track sequential accumulations into two separate containers and subsequently compare the two nonvisible quantities to choose the larger quantity. In that study, chimpanzees watched as each of two buckets received up to three smaller sets in alternation (see also Call, 2000). Accuracy in these comparison tasks is controlled by the ratio of the two values being compared (Dooley and Gill, 1977). Interestingly, chimpanzees seem completely unable to learn to choose the smaller of two food quantities in order to obtain the larger quantity, suggesting that chimpanzees may be unable to inhibit their desire to choose the larger quantity (Boysen et al., 1999). Orangutans may differ in this respect (Shumaker et al., 2001).

Thomas et al. (1980) presented squirrel monkeys with two random dot patterns and reinforced them for choosing the display with the smaller number. The monkeys were trained until they reached a performance criterion on successive numerical pairs beginning with 1 vs. 2. One monkey reached criterion with 7 vs. 8, and a second monkey reached criterion with 8 vs. 9. These data suggest that monkeys may appreciate ordinal relations between numerosities; however, because each numerical pair was trained successively, it remains unclear whether the monkeys learned a series of absolute discriminations or were using a more abstract ordering operation (see also Thomas and Chase, 1980).

Washburn and Rumbaugh (1991) trained rhesus monkeys to use a joystick to select one of two arabic numerals presented on a computer monitor. Whichever numeral the monkey selected resulted in the immediate delivery of the corresponding number of food pellets. The monkeys learned to choose the symbol that produced the larger number of pellets and succeeded on tests of transitive inference. In addition, the monkeys' accuracy was dependent on the numerical distance between the quantities the symbols represented. The monkeys may have represented the discrete number of pellets associated with each numeral; alternatively, they may have represented the total amount of food each numeral yielded. Regardless, these results show an impressive capacity to order symbols based on underlying magnitudes (for a similar demonstration with squirrel monkeys, see Olthof et al., 1997; with dolphins, Mitchell et al., 1985).

In a series of studies, Brannon and Terrace (1998, 1999, 2000, 2002; Brannon et al., unpublished data) have examined the ability of rhesus macaques to discriminate the numerosities 1 to 9 and represent their ordinal relations. In these studies rhesus monkeys were first trained to respond in ascending or descending order to a small set of numerical values (1-2-3-4; 4-3-2-1). Nonnumerical confounds were controlled by varying relative element size over a large number of stimulus sets (see Figure 6.4). After extensive training, the monkeys were tested on their ability to order novel exemplars of these same numerosities where nonnumerical dimensions were varied randomly across sets of exemplars. The monkeys learned the response rule and performed well above chance when tested with novel exemplars of the numerosities, regardless of the ordinal direction used in training.

However, were the monkeys using an ordinal numerical rule? Did their behavior reveal any appreciation of the fact that 2 is more than 1, or that 3 is less than 4? To address this question, the monkeys were then tested on their ability to order novel

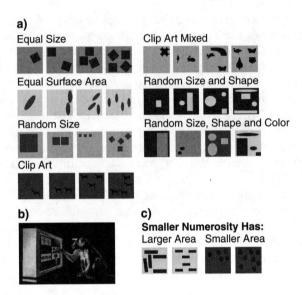

FIGURE 6.4 (a) Exemplars of the seven different types of stimulus sets used by Brannon and Terrace (1998). *Equal size*: Elements were of same size and shape. *Equal area*: Cumulative area of elements was equal. *Random size*: Element size varied randomly across stimuli. *Clip art*: Identical nongeometric elements selected from clip art software. *Clip art mixed*: Clip art elements of variable shape. *Random size and shape*: Elements within a stimulus were varied randomly in size and shape. *Random size, shape, and color*: Same as previous, with background and foreground colors varied between stimuli. (b) Examples of stimulus sets used in the pair-wise numerosity test. (c) Examples of pair-wise tests where the smaller numerosity has the larger and smaller surface areas.

numerical values that were outside the range of the training values (e.g., 5 vs. 8). The size of the elements was varied such that the smaller numerosity had the larger surface area on 50% of trials. If the monkeys had learned a specific arbitrary ordering of a set of numerical values, then they should have had no basis for ordering novel values that exceed the training range. To illustrate that point, imagine having learned the beginning of an alphabet and then being asked to order the letters at the end of the alphabet. This would be an impossible task given that the ordering of an alphabet is arbitrary. If, on the other hand, the monkeys learned an ordinal rule such as "respond to the stimulus with the smallest number of elements and continue to do so without replacement," then they should have had no trouble ordering pairs of novel values.

Rosencrantz and Macduff, trained to respond in ascending order to the numerosities 1 to 4, performed at above-chance levels when tested with pairs of the novel values 5 to 9. These results show that monkeys do not require training on each specific numerical pair to appreciate their ordinal relations. In fact, the monkeys were equally good at this task when the smaller number had a larger surface area compared to when it had a smaller surface area. These data have recently been replicated in a baboon, a squirrel monkey, and two cebus monkeys (Evans et al., 2002; Smith et al., 2002). In contrast to Rosencrantz and Macduff's success in extrapolating the ascending numerical rule to novel values, Benedict, the monkey trained to respond in descending order to the values 1 to 4 (Brannon and Terrace, 2000), was at chance when tested with novel numerosities.

We have recently conducted a new experiment that sheds light on Benedict's failure to generalize to novel values after descending training on the values 1 to 4 (Brannon et al., unpublished data). In this experiment we trained monkeys to respond 4-5-6 or 6-5-4 and then tested their ability to order pairs of all the values 1 to 9. The monkeys learned the 4-5-6 and 6-5-4 rule and were able to appropriately order novel exemplars of 4 to 6 with above-chance performance. However, when tested with pairs of the values 1 to 9, the monkeys showed an interaction whereby they were above chance when tested with values that continued the training sequence (e.g., 4-5-6 training yielded good 7-8-9 pairwise performance, and 6-5-4 training yielded good 3-2-1 pairwise performance), but below chance when tested with values that preceded the training range (Brannon et al., unpublished data). These results suggest two important conclusions. First, familiarity and novelty are not the most important factors in monkeys' numerical discriminations. In fact, the monkeys were in some cases able to order novel pairs with higher accuracy than familiar pairs. Second, in addition to learning about the ordinal direction in which they are expected to respond, monkeys seem to learn something more specific about the anchor point in the training series. The first value learned in the training sequence seems to act as an anchor and figure heavily into the monkeys subsequent comparison calculations. To illustrate, consider the 4-5-6 training series and the test pairs 1 vs. 3 and 7 vs. 9. The value 3 is closer than 1 to the anchor value 4, whereas the value 7 is closer than 9 to the anchor value 4. Thus, the monkey will successfully order 7 and then 9, but unsuccessfully order 3 and then 1.

Moyer and Landauer (1967) first showed that when adults are presented with two arabic numerals and required to indicate the symbol that represents the larger magnitude, their accuracy and latency are systematically influenced by the numerical disparity between the magnitudes. Specifically, reaction time decreases and accuracy increases with increasing numerical disparity. Furthermore, when distance is held constant, performance decreases with increasing numerical magnitude. Since this pioneering work, the distance and magnitude effects have been replicated in many languages, and representational formats with adult humans and have also been found in children as young as 5 years old (e.g., Buckley and Gilman, 1974; Temple and Posner, 1998).

Moyer and Landauer (1967) interpreted their results as evidence that numbers are represented as analog magnitudes. This easily replicated finding is also a signature property of animal number discrimination. For example, Brannon and Terrace (1998, 2000) showed that when rhesus monkeys choose the larger or smaller of two visual quantities, their reaction time and accuracy are systematically affected by the numerical disparity between the response alternatives and their magnitudes. Thus, number discrimination follows Weber's law such that as absolute magnitude increases, a larger disparity is needed to obtain the same level of discrimination. Brannon and Terrace then tested college students with the same task and stimuli and instructed the participants to respond as quickly as they could while still getting the majority of problems correct. Figure 6.5 shows the results for college students and monkeys. The striking similarity in the latency and accuracy functions obtained with

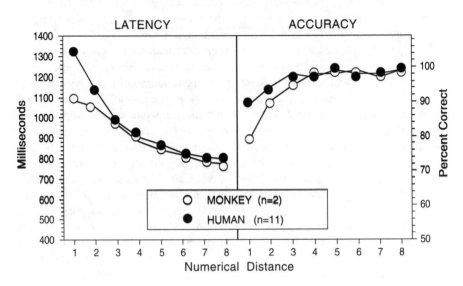

FIGURE 6.5 Latency to the first response (left) and accuracy (right) in a pair-wise numerical comparison task as a function of numerical disparity. Monkeys (open circles) and humans (closed circles) were required to respond first to the stimulus with the fewer number of elements.

college students and monkeys suggests that a similar representational format and comparison process is used.

In another approach, Hauser et al. (2000) conducted a series of studies on the spontaneous numerical abilities of rhesus macaques free-ranging on the island of Cayo Santiago. In all of these studies Hauser et al. tested a given monkey for a single trial and compared large groups of monkeys in each experimental condition (between-subjects design). This approach is notable because it assesses spontaneous cognition without the extensive training typically required in number discrimination tasks with animals. In one paradigm experimenters successively dropped apple slices into two distinct buckets and determined the percentage of trials on which the monkeys chose the bucket with the larger number of apples (Hauser et al., 2000). The monkeys successfully chose the bucket with the larger number of food items with the comparisons 2 vs. 3 and 3 vs. 4, but were at chance with the comparisons 4 vs. 5 and 5 vs. 6. Surprisingly, the monkeys also failed when presented with the contrast of 4 vs. 8, even though this involved a large ratio. These results differ from those obtained by Brannon and Terrace where rhesus monkeys discriminated values as large as 8 vs. 9 and where the ratio of the values largely controlled performance.

One possible explanation for this discrepancy is that Hauser's paradigm assesses the spontaneous number discrimination abilities of rhesus macaques, whereas Brannon and Terrace's paradigm shows what a monkey is capable of after extensive training. This explanation may be partly correct; however, it is important to note that the monkeys in Brannon and Terrace's experiments performed above chance on large number contrasts despite the fact that they had no prior training on those particular values. There are two other major differences between these paradigms. First, Brannon and Terrace presented simultaneous visual arrays, whereas Hauser presented items sequentially and sets were not visible. Second, food items were used by Hauser rather than the abstract shapes used by Brannon and Terrace. More work is needed to parse the relative contribution of these three factors to the different performance of rhesus monkeys in these two paradigms.

6.3.2 ADDITION AND SUBTRACTION

Experiments on addition and subtraction have been done with New World monkeys, Old World monkeys, and apes. In perhaps the most impressive demonstration of animal arithmetic, a chimpanzee named Sheba was led around a room to three different hiding places that contained a total of one to four food items. Subsequently, Sheba was led to a platform where she was required to choose the arabic numeral that corresponded to the total number of items. Sheba chose the correct numeral on 75% of trials without any explicit training. Amazingly, Sheba was still successful when the oranges were replaced by arabic numerals (Boysen and Berntson, 1989).

Summation may also be seen in tasks where primates are required to choose the larger of two food quantities. Rumbaugh et al. (1987, 1988) presented chimpanzees with two sets of two food wells, each of which contained some number of chocolates, to determine whether the chimps could choose the set of wells that had the largest cumulative quantity. To choose the set with the largest quantity, the chimpanzees would have to sum the chocolates in each of the two sets and then compare the two summed values. On the critical trials where the largest set of wells did not contain the largest single value, the chimps chose the larger quantity on approximately 90% of trials (see also Beran, 2001; Call, 2000). Olthof et al. (1997) conducted a similar experiment with squirrel monkeys and arabic numerals. The monkeys were first trained to choose the symbol that indicated the larger quantity and then presented with two sets of one to three arabic numerals each. Again, both monkeys reliably chose the set that contained the largest sum, even when this set did not contain the largest individual value.

Hauser et al. (1996) adapted the violation-of-expectation paradigm developed by Wynn (1992) to test arithmetic reasoning in human infants. In this paradigm human infants or monkeys view dolls or food items. A screen is then raised to obscure the items on the stage. A hand then adds or removes an object behind the screen. Finally, the screen is lowered to reveal the expected or unexpected number of objects and looking time is measured. Figure 6.6 illustrates a $2 - 1 = 1$ or 2 trial. Hauser et al. (1996) found that monkeys looked longer when the unexpected outcome was revealed for $1 + 1 = 1$ or 2 and $2 - 1 = 1$ or 2 operations (see also Sulkowski and Hauser, 2001). Uller et al. (1999) also found that cotton-top tamarins look longer at the unexpected outcome of 3 or 1 compared to the expected outcome of 2 when they witness a $1 + 1$ operation.

6.4 NUMBER REPRESENTATION IN HUMAN INFANTS

6.4.1 SMALL NUMBER DISCRIMINATION

The habituation paradigm has provided abundant evidence that infants, even newborns, represent number. In the habituation paradigm, infants are presented with many exemplars of one numerosity. The time the infant looks at each stimulus is

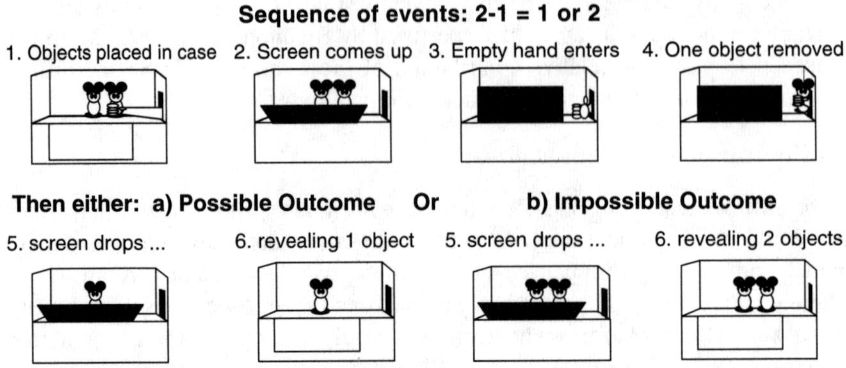

Sequence of events: 2-1 = 1 or 2

1. Objects placed in case 2. Screen comes up 3. Empty hand enters 4. One object removed

Then either: a) Possible Outcome   Or   b) Impossible Outcome

5. screen drops ...   6. revealing 1 object   5. screen drops ...   6. revealing 2 objects

FIGURE 6.6 Illustration of a 2 − 1 = 1 or 2 trial. Two Mickey Mouse® dolls (Wynn, 1992) or eggplants (not shown; Hauser et al., 1996) were placed on a stage. A screen was then raised to hide the objects. A hand entered the scene and removed one object. The screen was then lowered to reveal one (expected) or two (unexpected) objects. The time a monkey or human infant spent looking at the final outcome was measured.

recorded, and stimuli are presented repeatedly until a substantial reduction in looking time occurs (habituation criterion). Infants are then shown new exemplars of the same number and a new number. Infants tend to look longer at the new number, demonstrating that they discriminate the novel from the familiar numerosity. In the first study of this sort, Starkey and Cooper (1980) tested infants between 16 and 30 weeks on their ability to discriminate two vs. three or four vs. six dots. They found 2 vs. 3 but not 4 vs. 6 discrimination (for a similar demonstration with slides of household objects, see also Strauss and Curtis, 1981). Antell and Keating (1983) tested newborn infants and also found evidence of 2 vs. 3 but not 4 vs. 6 discrimination. Van Loosbroek and Smitsman (1990) found 2 vs. 3 and 3 vs. 4, but not 4 vs. 5 discrimination in 5-month-old infants tested with dynamically changing displays rather than static images. Bijeljac-Babic et al. (1993), using the high-amplitude sucking procedure, found evidence that 4-day-old infants discriminate two vs. three consonant–vowel syllables. Collectively these studies suggest that infants' ability to make numerical discriminations may be limited to small set sizes.

In a recent study, Wynn et al. (2002) habituated infants to two or four randomly moving dots (see Figure 6.7) and then tested both groups of infants with two groups of four dots or four groups of two dots. Infants habituated to two dots looked longer at the four groups of two dots, and infants habituated to four dots looked longer at the two groups of four dots. This is a particularly powerful finding in that the test displays contained the exact same eight dots and were thus precisely controlled for surface area and perimeter. In another study from Wynn's (1996) research group, infants were habituated to two or three puppet jumps and then tested with both numbers of jumps. Infants' looking time was measured to the static puppet after the jumps were completed, and the timing of the jumps was carefully controlled so that total duration was not available as a cue. Infants looked longer after the novel number of puppet jumps, showing that they discriminated two from three puppet jumps even when total duration and tempo were controlled.

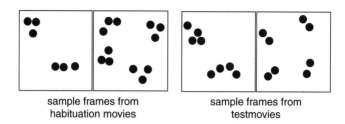

sample frames from
habituation movies

sample frames from
testmovies

FIGURE 6.7 Sample frames of habituation and test movies used by Wynn et al. (2002) to test 5-month-old infants in a 2 vs. 4 discrimination. (From Wynn, K., Bloom, P., and Chiang, W., *Cognition*, 83, B55–B62, 2002.)

A new method makes use of infants' reaching behavior and shows that both the duration and number of reaches an infant makes into an opaque box are correlated with the number of objects that the infant observed being placed into the box (Van de Walle et al., 2000). Infants were shown one or two objects being placed into a box. The experimenter surreptitiously removed one of the objects from the box and then allowed the infant to reach in and withdraw one object. After the experimenter took the first object away from the infant, the infant was then allowed to continue searching the box. The time infants spent searching the empty box was then measured. Infants who observed two objects being placed into the box were more likely to reach for a second object than infants who observed one object.

6.4.2 LARGE NUMBER DISCRIMINATION

Contrary to the view that infants can only discriminate small values, Xu and Spelke (2000) found that 6-month-old infants discriminate 8- vs. 16-element displays (see Figure 6.8). Infants were habituated to 8- or 16-element displays. The 8- and 16-element displays contained the same cumulative surface area of dots, and stimulus size was constant in habituation such that density covaried with number. In test, the continuous variables that had been held constant in habituation now covaried with number, whereas the variables that covaried with number in habituation were held constant in test. Specifically, element size was held constant in the 8- and 16-element test displays, whereas the stimulus size varied such that density was constant. The particular element size used in test was chosen so that the total surface area in test differed from that of habituation for both the 8- and 16-element displays by a constant amount. Infants looked longer at the novel numerosity. Supporting these findings, Brannon (2000b) replicated this experiment and Lipton and Spelke (2001, 2002), using the head-turn procedure, have also found large number discrimination in 6- and 9-month-old infants with tones rather than visual displays.

6.4.3 CROSS-MODAL MATCHING

Starkey et al. (1983, 1990) reported that when 6- to 8-month-old infants heard two or three drumbeats, they looked longer at a visual display that contained the matching number of elements than at visual displays of the other numerosity. This finding, however, was not replicated by other researchers (Mix et al., 1997; Moore et al.,

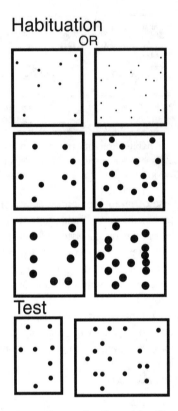

FIGURE 6.8 Sample habituation and test stimuli used by Xu and Spelke (2000) to test 6-month-old infants in an 8 vs. 16 discrimination. (From Xu, F. and Spelke, E.S., *Cognition*, 74, B1–B11, 2000.)

1987). Therefore, more definitive evidence that infants form numerical representations that are modality independent is still needed.

6.4.4 CONTINUOUS DIMENSIONS CONFOUND

In experiments with animals it is possible to use large sets of very diverse stimuli to ensure that only numerical discriminations would allow the animal to be successful. With human infants, it is often the case that sessions are limited to 10 to 15 min, and each infant generally participates in one or two sessions, each of which contains a handful of stimuli rather than the endless trials that can be conducted with nonhuman animals. For these reasons, it is a more challenging enterprise to design studies that conclusively demonstrate numerical competence in infants.

Recent studies suggest that under some circumstances infants may keep track of continuous dimensions such as total contour length rather than number (Clearfield and Mix, 1999; Feigenson et al., 2002b; Mix et al., 2002; Newcombe, 2002). For example, Clearfield and Mix (1999) habituated 6- to 8-month-old infants to stimuli with a constant number (2 or 3) and contour length and then tested the infants with stimuli of a new number with an unchanged contour length or a new contour length with an unchanged number. Infants looked longer to the test stimuli that changed in contour length, compared to the last few habituation trials, and did not show a comparable elevation in looking time to the test stimuli that changed in number. Such results at the very least suggest that continuous dimensions must be more carefully controlled in studies of numerical cognition. These studies also suggest a need for psychophysical studies of the perception of both number and continuous dimensions in infancy. However, it is important to remember that there are two distinct questions that should not be confused. One question is whether infants are capable of purely numerical discriminations and computations. The second question is whether continuous dimensions are more salient than number in infancy. The Clearfield and Mix study is more pertinent to the second question.

6.5 ARITHMETIC REASONING IN INFANCY

The studies described above suggest that infants may be able to discriminate the number of elements in a visual array or in a sequence of events. We now turn to the question of what types of operations infants' numerical representations might support. Do infants appreciate ordinal relationships between numerosities? Can infants add or subtract?

6.5.1 ORDERING

Although previous work suggested that infants do not represent ordinal relationship between numerosities in the first year of life (Cooper, 1984; Strauss and Curtis, 1984), Brannon (2002a) recently found evidence that infants as young as 11 months old can make ordinal numerical judgments. In that study, infants were habituated to ascending or descending numerical sequences (see Figure 6.9). For example, infants habituated to the ascending ordinal direction were shown a small numerosity followed by a larger numerosity followed by an even larger numerosity. Between trials the absolute values changed but the ordinal direction was constant. Infants were then tested with new numerical values where the ordinal direction was maintained or reversed. Eleven-month-old infants looked for significantly longer when the ordinal direction was reversed.

In related work, Feigenson et al. (2002a) tested 10- and 12-month-old infants on their ability to choose the larger of two food quantities. The paradigm directly paralleled that used by Hauser et al. (2000) described above. The experimenters successively dropped graham crackers into each of two buckets and then allowed the infants to crawl to either bucket. Infants at both ages spontaneously chose the bucket with the larger number of crackers when tested with 1 vs. 2 and 2 vs. 3 comparisons, but not with 2 vs. 4, 3 vs. 4, and 3 vs. 6. These results suggest that the numerical ratio was not what controlled performance, but instead that infants were limited by the absolute set size of the values being compared. However, the infants also seemed to represent the absolute amount of food. When food amount was equated (e.g., one large and two small crackers), the infants chose randomly. This is not surprising given that the infants should be motivated to maximize food

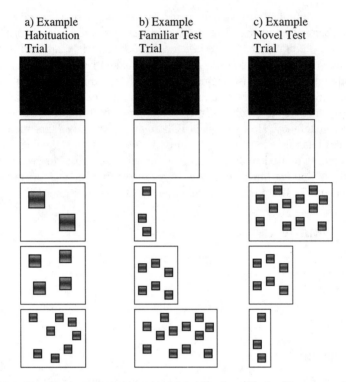

FIGURE 6.9 The five frames of a sample (A) habituation, (B) familiar test, and (C) novel test trial in Brannon (2002a). Infants were habituated to one ordinal direction of numerosities and then tested with the same direction and the reversed direction. Achromatic elements are shown here; however, elements were rainbow colored in the actual experiment. (From Brannon, E.M., *Cognition*, 83, 223, 2002.)

intake, not number of items. Again, more research is needed to explain why different paradigms yield different results. One possibility is that these paradigms involve different representational systems.

6.5.2 Addition and Subtraction

As described above, the violation-of-expectancy paradigm exploits the fact that infants look longer at unexpected outcomes than expected ones. Wynn (1992) first showed that infants form expectations about the number of objects that are successively placed behind an opaque barrier. Infants look longer at the unexpected outcome when tested with $2 - 1 = 1$ or 2 (as illustrated in Figure 6.6), $1 + 1 = 1$ or 2, and $1 + 1 = 2$ or 3. These findings have been replicated and extended in other laboratories (Baillargeon, 1994; Feigenson et al., 2002b; Koechlin et al., 1998; Simon et al., 1995; Wynn, 1995). For example, Simon et al. (1995) replicated the Wynn paradigm but used Elmo dolls in the initial phase of the trials and then surreptitiously replaced Elmo dolls with Ernie dolls. The infants' expectations were not violated by this identity switch; they looked longer only when the result was an unexpected number. This suggests that in some sense infants represent the number of objects stripped of their particular nonnumerical features. In another study, the Wynn paradigm was replicated; however, the dolls were placed on a revolving surface such that the location of the objects changed during a trial (Koechlin et al., 1998). Infants still looked longer at the unexpected outcome. Thus, infants' representations in these tasks do not preserve object identity or location.

In another test of the Wynn paradigm, Feigenson et al. (2002b) tested 7-month-old infants and pitted violations in surface area against violations in number. Infants saw two small objects placed behind a screen. The screen was then lowered to reveal one or two large objects. The outcome of one large object was expected in area and unexpected in number, while the outcome of two large objects was expected in number and unexpected in area. Infants looked longer at the unexpected area outcome than at the unexpected number outcome. These results suggest that information about surface area is preserved and used in the comparison process.

Despite the many replications of the Wynn paradigm described above, Wakely et al. (2000) obtained negative evidence in a $1 + 1 = 1$ or 2 and $2 - 1 = 1$ or 2 study with 5-month-old infants. The main differences in the methods were the use of an automated display and rigid intertrial intervals. Wynn (2000) suggested that the failure to replicate might result from subtle details in the procedure that did not optimize infants' attention to the displays. More research is needed to resolve this conflict (See also Cohen and Marks, 2002).

6.6 A BRIEF DESCRIPTION OF FIVE MODELS OF NONVERBAL NUMBER REPRESENTATION

The sections above describe what nonhuman animals and human infants are capable of in the numerical domain. But how are they accomplishing these mathematical feats? What is the process by which they form numerical representations, and how can we best describe the format of their numerical representations? In this section we will describe five proposals for how number is represented and then review the evidence that supports each model.

6.6.1 ARBITRARY NUMERON HYPOTHESIS

The first proposal is that animals and preverbal children possess an ordered set of arbitrary nonverbal neural representations of numerosities (Gelman and Gallistel, 1978). In the original formulation of this model, numerons were described as *indirect* symbols with an arbitrary relationship to the numerosities they represent (see Figure 6.10a). Like words or arabic numerals, the proposal was that numerons did not in any way resemble the numerosities they served to represent. Numerons were applied to the set of objects or events through a process conforming to Gelman and Gallistel's five counting principles (Gelman and Gallistel, 1978). First, the *one-to-one* principle states that each element in the to-be-counted set must be mapped to one and only one numeron (e.g., the verbal label three can be applied to only one of the items in a set). Second, the *stable-order* principle requires that numerons be applied in a consistent order (e.g., one cannot count a five-element sequence as 1-

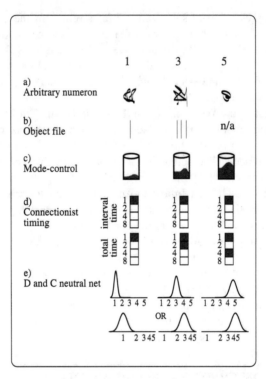

FIGURE 6.10 Format of number representations for the values 1, 3, and 5. (a) *Arbitrary numeron* model: Each number is represented by an abstract, arbitrary symbol. (b) *Object-file* model: Each object is represented by an object-file. There is no symbol that represents the set of objects. System is limited by the small number of available object-files. (c) *Mode-control* model: Number is represented as the accrual of pulses from a pacemaker into an accumulator. (d) *Connectionist timing* model: Memory for intervals is represented as a storage vector that contains information about the half phases of multiple oscillators. Number is derived by dividing the total time that has elapsed by the duration of the intervals between events. (e) *Dehaene and Changeux neural network* model: Number is represented by a map of numerosity detectors.

2-3-4-5 on one occasion and 1-3-4-5-2 on the next). The third principle is *cardinality*; it states that the last numeron assigned to an element of a set must also serve to represent the whole set. The final two principles are not essential to the counting process, but describe counting in its most abstract sense. The fourth principle is the *abstraction* principle; it states that anything can be counted (apples, people, buildings, sounds, etc.). The final principle is the *order irrelevance* principle; it states that objects can be counted in any order (e.g., left to right, circular pattern, etc.).

6.6.2 Subitizing and the Object-File Mode

A popular distinction in the literature of human and animal numerical abilities is subitizing vs. counting. Subitizing is defined as a fast, effortless, parallel perceptual process that is limited to the apprehension of the small values 1 to 4 (Kaufman et al., 1949; Mandler and Shebo, 1982). In contrast, counting is a slower, more effortful sequential process that is used for sets of five or more items.

More recently, the subitizing hypothesis has been further specified in the object-file model, which posits that an object-file is opened for each element in a visual array (Carey, 1998; Hauser and Carey, 1998; Leslie et al., 1998; Simon, 1997; Uller et al., 1999). In this model there is no symbol that represents the numerosity of a set; instead, as shown in Figure 6.10b, each element is represented by a file stripped of object features such as color, shape, and size (see also Trick and Pylyshyn's (1993, 1994) application of the FINST model). The model posits that the visual system contains a limited number of object-files that can be assigned to an object. A given set can only be represented if there are a sufficient number of object-files available (Pylyshyn and Storm, 1988). Thus, if an organism's only means of representing number was the object-file model, it could only represent small sets of objects.

6.6.3 MODE-CONTROL MODEL

The third proposal is the mode-control model, or accumulator model; it shares with the arbitrary numeron hypothesis the idea that animals use a serial process that conforms to the counting definition proposed by Gelman and Gallistel (1978). However, the mode-control model (Meck and Church, 1983; Meck et al., 1985) posits that number is represented as continuous magnitudes that directly reflect the magnitude of the discrete quantities they serve to represent (see Figure 6.10c). Thus the mode-control model contends that the nervous system inverts the representational convention whereby numbers are used to represent linear magnitudes. It is proposed that instead of using number to represent magnitude, the nervous system uses magnitude to represent number (see Gallistel and Gelman, 2000; Meck, 1997). Furthermore, the mode-control model posits that a single mechanism serves to represent time and number.

The mode-control model was originally developed as an adaptation of the information-processing model of animal timing behavior by Gibbon and Church (1984). Like the pure timing model, the mode-control model is composed of a pacemaker, accumulator, working memory buffer, reference memory, and comparator (see Figure 6.11a). At the onset of a relevant stimulus, pulses are gated into an accumulator, which then integrates the number of pulses over time. The critical innovation for the mode-control model is the addition of a mode switch, which allows the system to work like a timer or a counter. Pulses are gated into the accumulator by one of three different modes, depending on the nature of the stimulus. These three modes provide the mechanism with the ability to act as both a counter and a timer. In the *run mode*, the initial stimulus starts an accumulation process that continues until the end of the signal or trial; in the *stop mode*, the process occurs whenever the stimulus is physically present; in the *event mode*, each onset of the stimulus produces a relatively fixed duration of the process regardless of stimulus duration. This mechanism thus provides a way to estimate duration (the run and stop modes) or number (the event mode). The accumulator value is transferred to working memory and sometimes to reference memory. The animal then compares the current value in the

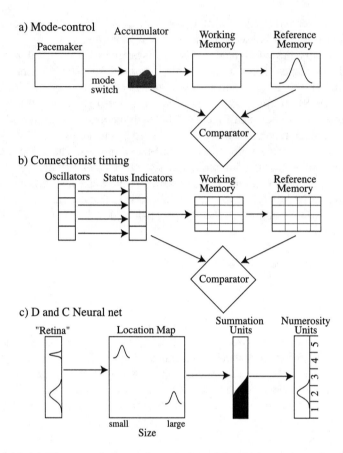

FIGURE 6.11 (a) Diagram of the mode-control model, which consists of a pacemaker, switch, accumulator, working memory, reference memory, and comparator. (b) Diagram of the connectionist timing model, which consists of multiple oscillators rather than a single pacemaker, status indicators that provide information about the phase of the oscillators rather than an accumulator value, a working memory, a reference memory, and a comparator. (c) Diagram of the workings of the Dehaene and Changeux neural net model, which consists of an input retina, a topographical map of object locations that normalizes for size of the objects, and a map of numerosity detectors that sums the outputs from the location maps and provides a representation of number.

accumulator with the value(s) in reference memory to determine what type of response to make. Although the mode-control model was developed to explain timing and counting in rats, it has since been used to explain numerical behavior in human infants (e.g., Wynn, 1995).

6.6.4 CONNECTIONIST TIMING MODEL

A fourth proposal is the connectionist timing model (Church and Broadbent, 1990, 1991) (see Figure 6.11b). This model posits a set of oscillators with different periods ranging from milliseconds to hundreds of seconds. The oscillators are reset (entrained) at the onset of a stimulus. When the stimulus terminates, the duration of the interval is represented by a *status indicator*, which is a vector that provides information about the phases of multiple oscillators. Memory for intervals is represented by a matrix of connection strengths between elements on the status indicators. Thus, temporal information is represented in a distributed fashion. Like the mode-control model, the connectionist timing model posits a comparison process between working and reference memory. In this model, number is derived by dividing the cumulative duration of a sequence of events by the duration of one interval (see Figure 6.10d).

Both the connectionist timing model (Church and Broadbent, 1990, 1991) and the mode-control model (Meck and Church, 1983) propose that the representation of time and number are inextricably linked; however, there are some important differences. The mode-control model uses a single pacemaker as the basis for interval timing and counting. The connectionist timing model instead proposes that multiple oscillators with a wide range of periodicities (from hundreds of milliseconds to hundreds of seconds) serve as the basis for timing. Number is represented in the mode-control model as the magnitude of an accumulation process. In the connectionist timing model, the relationship between durations represents number. Therefore, the connectionist timing model, in contrast to the mode-control model, cannot uniquely identify number for arrhythmic presentations (e.g., Breukelaar and Dalrymple-Alford, 1998; Davis et al., 1975). Finally, both the connectionist timing and mode-control models process information serially from a clock to working memory to reference memory to a comparison. The variables that represent time and number in the mode-control model are scalars, which are associated with some uncertainty. In the connectionist timing model, temporal variables are encoded by vectors and matrices, resulting in parallel distributed representations of time, and therefore number.

6.6.5 DEHAENE AND CHANGEUX NEURAL NETWORK MODEL

A fifth model is the Dehaene and Changeux (1993) neural network model. This model posits that numerosity detectors represent the abstract number of objects independently of the size and configuration of the stimuli. Figure 6.11c shows the three layers to this model: an input "retina," a map of object locations, and an array of numerosity detectors. The map of object locations converts stimuli from the retina to a representation of each stimulus irrespective of object size (an echoic auditory memory also allows the system to enumerate sounds as well as objects). The location map sends its output to numerosity detectors, which consist of summation units and numerosity units. Each summation unit has a set threshold of a particular value. When the total activity from the output of the location map (which is proportional to numerosity) exceeds the summation unit's threshold, it will be activated. These units differ from the event mode of the mode-control model, as they are active only when the number of events exceeds some level, for example, greater than or equal to 3. Finally, the summation clusters project to numerosity clusters, which represent the numerosities 1 through 5. A given numerosity cluster will be activated if the corresponding summation cluster is active (e.g., 3), but those representing higher values (e.g., 4 and 5) are not. Therefore presentation of stimuli with the same numerosity, despite differences in size, location, and modality, results in the activation of the same numerosity detectors (see Figure 6.10e).

6.7 DIFFERENTIATING THE MODELS

6.7.1 Are Animals and Infants Merely Subitizing?

When adult humans name the number of elements in a visual display, the slope in the function relating reaction time to numerosity is approximately 57 msec per item for values 1 to 4 and 300 msec per item for values 5 to 20 (Klahr, 1973). This apparent elbow in the slope of the reaction time function for labeling the number of items in visual displays is typically offered as evidence that two distinct processes are engaged for small and large numbers. However, there is controversy over whether a true discontinuity actually exists (e.g., Gallistel and Gelman, 1991; Trick and Pylyshyn, 1994). Balakrishnan and Ashby (1992) conducted a quantitative analysis of reaction time distributions and found no evidence for a discontinuity.

Although there are few data that can directly address whether animals show a discontinuity in the reaction time function for estimating numerosity, there have nevertheless been suggestions in the literature that animals are limited to subitizing (e.g., Davis and Perusse, 1988; Rumbaugh et al., 1987). Given that animals are capable of discriminating the large values 8 vs. 9, or even double-digit values when the ratio is large enough, the idea that animals are limited to a subitizing mechanism seems untenable. It is possible that animals use subitizing for small values and a counting-like process for larger values (Matsuzawa et al., 1991; Murofushi, 1997). If separate processes support small and large number representations, we might expect that rules learned in one numerical range would not be extrapolated to the whole spectrum of values. However, as discussed previously, monkeys trained to respond in descending order to 6-5-4 performed at higher accuracies on pairs of the small values 1, 2, and 3 (1 vs. 2, 2 vs. 3, and 1 vs. 3) than they did on pairs of familiar values such as 4 vs. 6 (Brannon et al., unpublished data).

6.7.2 Are Animals and Infants Limited to Object-Files?

The object-file model can in some sense be thought of as a more specified version of the subitizing hypothesis. Both seek to explain why small numbers are treated differently from large numbers. However, the object-file model proposes that the ability to perceive small numbers is a by-product of the visual system. If animals or infants were limited to the object-file model as a means of representing number, then they should be limited by absolute set size. In keeping with this idea, Starkey and Cooper (1980) found that 4- to 7-month-old infants could discriminate 2 vs. 3 in a visual habituation paradigm, but failed to discriminate 4 vs. 6. More recently, Feigenson et al. (2002a) found that 10- and 12-month-old infants could reliably choose the larger set in a 2 vs. 3 comparison, but not a 2 vs. 4 comparison. Both of these studies suggest that infants' numerical abilities are limited by absolute set size.

However, the object-file model is not sufficient to explain all of the available data on infants' numerical abilities. The object-file model cannot explain how infants could discriminate sounds or events because object-files are limited to the visual system (e.g., Bijeljac-Babic et al., 1993; Lipton and Spelke, 2001, 2002; Wynn, 1996). In addition, if limited to object files, infants should not be able to discriminate large values such as 8 vs. 16 (Xu and Spelke, 2000). Also, the object-file and subitizing models have no built-in mechanism for comparing ordinal relations (Brannon, 2002a). The available data suggest that in at least some contexts, infants represent number as mental magnitudes and are not limited to object-file representations. Carey (1998, 2001) and Spelke (2000) have suggested that infants may possess both an object-file and an accumulator mechanism, and that these processes may operate over different numerical ranges. Future work should investigate whether there are really two distinct processes and what conditions invoke each system.

There is much stronger evidence that animals possess a mechanism for representing number that is more powerful than subitizing or the object-file model. The main lines of evidence are that (a) the values that animals can discriminate greatly exceed the range than can be handled by object-files; (b) animals can perform computations with their numerical representations, such as ordering; and (c) animal number discrimination follows Weber's law, implicating an analog representational format. Thus animals and human infants both appear to have a mechanism for representing number that is more powerful than subitizing or the object-file model. The remaining models are attempts to specify a mechanism for representing number nonverbally and predict the distance and magnitude effects found in animals.

6.7.3 ARE ANIMALS AND INFANTS COUNTING?

The arbitrary numeron model and the mode-control model are the only models described above that conform to Gelman and Gallistel's (1978) counting principles. In the event mode, each to-be-counted event results in a fixed increment (200 msec for rats) in the accumulator and thus abides by the *one-to-one correspondence* principle, which states that only one numeron can be assigned to each event. The *stable-order* principle is upheld because the states of the accumulator have a fixed order (you cannot get four pulses without first having three pulses). Finally, the resulting value in the accumulator represents the numerosity of events or objects and thus conforms to the *cardinal* principle. A more detailed description of why the mode-control model of temporal integration is functionally equivalent to counting has been provided by Broadbent et al. (1993) and Meck (1997).

Thus the mode-control model predicts that infants and animals serially enumerate. In contrast, the Dehaene and Changeux neural network model and the connectionist timing model predict that infants and animals perceive number in parallel (i.e., all at once). Unfortunately, there are few or no data that can help determine whether animals or infants use an iterative or parallel process. However, it is interesting to note that given multiple switches and a single accumulator, the mode-control model could readily be modified to accumulate multiple simultaneously presented events or objects in parallel. This idea has not been explicitly described; however, the assumption that there are multiple switches is inherent to the model in that animals are thought to count (event mode) and time (run or stop mode) simultaneously (e.g., Meck and Church, 1983; Roberts et al., 2000).

6.7.4 How Do the Models Handle Development?

A comprehensive model of nonverbal numerical abilities should address the development of cardinal and ordinal numerical knowledge and the development of numerical abilities in relation to nonnumerical abilities. The Dehaene and Changeux neural network model posits that the initial state of the organism consists solely of numerosity detectors and that the system learns the relationship between quantities and motor outputs from external reward input or from an autoevaluation loop. Thus, the ability to make ordinal numerical judgments arises after the ability to make cardinal discriminations. Furthermore, a specific prediction made by the Dehaene and Changeux neural network model is that ordinal judgments come on line with the maturation of the frontal lobes at about 10 months of age in humans. This prediction is supported by the finding that 11- but not 9-month-old infants detect a reversal in the ordinal direction of a numerical sequence (Brannon, 2002a).

6.7.5 Weber's Law

The mode-control model, the connectionist timing model, and the Dehaene and Changeux neural network model all predict that numerical discriminations will follow Weber's law. The mode-control model specifically posits that the variability in the memory distributions for each numerical value increases proportionally to the mean of that value. Confusion results from increasing overlap in the memory distributions, which increases the probability that a value pulled randomly from two distributions with means X and Y will be equal or reversed in their ordinal relations (e.g., $Y_1 < X_1$). Thus, the mode-control model posits that the internal number line is linearly spaced, but that the variability in the representations of each number is proportional to the value represented. An alternative hypothesis is that the internal number line might be logarithmically scaled (Dehaene, 1992).

The linear-with-scalar-variability subjective number line hypothesis and the logarithmic subjective number line hypothesis have been treated as functionally equivalent because of their similar empirical predictions (Dehaene, 1992). However, Gibbon and Church (1981) developed a clever experimental paradigm to address whether time is subjectively represented in linear or logarithmic coordinates (time-left). This paradigm makes use of the simple fact that if the subjective time scale is logarithmically scaled, behavior based on the difference between two values will be the same whenever the two values have the same ratio, no matter how far apart these values are objectively. In other words, subtraction in a logarithmic scale is equal to division in a linear scale. Brannon et al. (2001) adapted the time-left paradigm to investigate the subjective scaling of number. The task required pigeons to compare a numerical difference that varied from trial to trial to a constant value (e.g., 8 – 2 vs. 4, or 8 – 6 vs. 4). The pigeons should choose whichever value seemed smaller, the difference or the constant value. If the number scale is logarithmically compressed, then the comparison of the difference with the constant value will depend not on the objective difference between the two numbers, but rather on their objective ratio. Thus, the number that is judged as equal to a given subjective difference will not increase if the two numbers being compared are made objectively bigger, as long as the same ratio is maintained between them. In contrast, when the two numbers are scaled up proportionally, the objective difference between them increases on a linear scale. If number is represented in linear coordinates, the number judged as equal to the subjective difference should increase linearly with objective number.

Results indicated that changing the absolute values of the numerical constant and the numerical difference had a linear effect on the number that was subjectively equal to a given subjective difference, even though the ratio between the constant and the difference was maintained. The reward schedule insured that number rather than time controlled the pigeons' behavior. These results are contrary to the predictions of the logarithmic compression hypothesis and indicate that subjective number is linearly related to objective number. However, the Dehaene and Changeux neural network model can be implemented with linear spacing of the mental number line and scalar variability (Dehaene, 2001).

6.7.6 REPRESENTATION OF TIME AND NUMBER

A prediction of the mode-control and connectionist timing, but not the Dehaene and Changeux neural network model, is that time and number are represented by the same mechanism. A great deal of evidence supports this claim in animals (Church and Meck, 1984; Fetterman, 1993; Meck and Church, 1983; Meck et al., 1985; Roberts, 1995; Roberts and Boisvert, 1998; Roberts et al., 1995, 2000, 2002; Roberts and Mitchell, 1994; Santi and Hope, 2001). Here we will review four relevant findings.

First, Meck and Church (1983) trained rats in the duration bisection procedure to make one response to a 2-sec two-cycle stimulus and another distinct response to an 8-sec eight-cycle stimulus. Rats were then tested with duration held constant at 4 sec and number varied, or number held constant at 4 and duration varied. In both cases, the rats' behavior was modulated by the stimulus dimension that varied, showing that the rats had encoded both number and time when the two were confounded. Second, when the probability of making a "long" or "many" response was plotted against stimulus duration or number, the point of subjective equality (PSE) was equivalent for time and number and the functions were virtually identical. The PSE is the value at which the animals were equally likely to categorize the stimulus as long or short, or many or few. Third, when the rats were administered methamphetamine, they showed the same leftward shift in the psychophysical curve that relates the probability of a "long" and "many" response to the actual times or counts with which the animal was presented (Meck and Church, 1983). This suggests that the drug manipulation may function to increase the speed of the pacemaker and implicates a single pacemaker for timing and counting in the seconds range (Church and Meck, 1984). Fourth, studies that evaluate working memory for time and number also suggest that a single mechanism is used to time and count. Spetch and Wilkie (1983) trained pigeons in a delayed-match-to-sample procedure to make one response after a short stimulus and a second response after a long stimulus. As the retention interval between the sample presentation and the choice was increased, the 2-sec sample retention curve was unaffected while the 8-sec retention curve suffered substantially. Thus, as the retention interval increased, the pigeons' bias to choose the small response increased. A parallel effect was found when pigeons discriminated small and large numbers of responses (Fetterman and MacEwen, 1989) or sequences of flashes (Roberts et al., 1995).

In the mode-control model, animals are thought to encode time and number automatically. However, recent studies question whether time may be more salient than number for animals under some situations. In one study, a key color cued pigeons of reward after a fixed number of flashes and a second color cued reward after a fixed duration. Midway through a trial, the experimenters changed the cue and found that when the pigeons had originally been cued to count, they had also kept track of time; however, when cued to time, the pigeons did not count (Roberts et al., 2000). It is thus possible that there is an asymmetry in the mode-control model whereby timing can occur automatically and counting does less so (see also Breukelaar and Dalrymple-Alford, 1998; Roberts et al., 2002).

Unfortunately, there are no data that can address whether infants use a single mechanism to time and count. In fact, in contrast to the large number of studies that have investigated what infants know about number, relatively few studies have asked about infants' ability to represent time intervals (for a review, see Pouthas et al., 1993). Wynn (2000) recently tested 6-month-old infants in a habituation paradigm and found that they discriminated intervals with a 1:2 ratio but not a 2:3 ratio. This suggests that infants can represent time and that their discriminations may be controlled by the ratio of the intervals juxtaposed. Similarly, in preliminary studies in our lab (Brannon et al., 2002), 10-month-old infants showed a mismatch negativity (MMN) to a time deviation when 500-msec interstimulus intervals (ISIs) were occasionally and unexpectedly intermixed with 1500-msec ISIs. The MMN is elicited by deviant auditory stimuli in a train of standard stimuli and is a negative deflection, between 130 and 200 msec poststimulus, in the difference wave derived from subtracting the event-related potential (ERP) elicited to the standard from the ERP elicited to the deviant. Collectively, the behavioral data from Wynn's lab and our ERP results suggest that human infants are capable of interval timing in the seconds range. Future studies must test whether time discrimination in human infants follows Weber's law and what the relationship is between timing and number discrimination in infancy.

In summary, there is a great deal of evidence that shows that many animal species possess a mechanism for representing number that is more powerful than a subitizing process or the object-file model. A handful of studies also implicate analog magnitude representations of number in human infancy. In support of the mode-control and connectionist timing models, a large literature substantiates the idea that animals represent number and time with a single currency, but no data test this hypothesis for human infants. More research is needed to test predictions of these three hypotheses in both nonhuman animals and human infants and to investigate how any of these models might be implemented in the nervous system.

6.8 NEURAL BASIS OF NUMBER REPRESENTATION

What area of the brain is involved in representing number and making arithmetic calculations? The neural basis of numerical cognition has been studied in adult humans using ERPs, functional magnetic resonance imaging (fMRI), positron emission tomography (PET), and patient populations (for a review, see Dehaene, 2000). Generally, this literature implicates the parietal cortex in number processing (inferior or superior parietal lobule). However, the majority of these studies have employed tasks that require the recognition and manipulation of arabic numerals and therefore engage symbolic numerical processing and not enumeration per se (but see Fink et al., 2001; Sathian et al., 1999).

Comparably little is known about the neural basis of number representation in animals. For many years, the only report of number-related neural activity was that in the association cortex of the anesthetized cat (Thompson et al., 1970). In that study, neural activity was recorded in response to a series of ten auditory or visual stimuli. A small number of neurons recorded (5 of 500) discharged more to a particular position in the sequence of lights or tones (see Figure 6.12a). These five number cells responded to the values 2, 5, 6, 6, and 7 in the series, regardless of stimulus modality (auditory or visual) and frequency (interstimulus intervals varied from 1 to 5 sec). A similar, but more broadly tuned, cell was found in an 8-day-old kitten. These results are intriguing in that they suggest that single cells may be selective for particular numerical values and that tuning for number may increase over development. However, replication and extension of these results are sorely needed.

More recently, number-related activity was found in the parietal cortex of monkeys as they performed a series of repetitive arm movements (Sawamura et al., 2002). Monkeys were trained to repeat a movement (pushing or turning a handle) for five trials and then switch to the other movement for five trials, and so on. There were no external cues as to how many movements the monkey had made; thus the monkey needed to keep track of the number of repetitions he had performed. The duration of each trial was varied to prevent monkeys from switching to the other movement on the basis of elapsed time, and electromyograms were recorded to insure that the five movements within a series did not differ in any systematic fashion.

The investigators selected neurons from the superior lobule of the parietal cortex with somatosensory receptive fields in the proximal forelimb and trunk of the arm performing the movement. However, they found that the neurons' responses also depended on the number of movements the monkey made. This number-modulated activity was observed during a period in which the monkeys waited for a "go" signal to execute the movement. As shown in Figure 6.12b, many of the cells were selective for only one of the positions in the sequence (e.g., third). In the remaining neurons, firing rate increased during more than one period, such as the first and second. Across the population of neurons studied, there were neurons that were selective for each of the ordinal positions in the sequence.

Although the cumulative amount of motor effort and the cumulative amount of juice within a sequence were both confounded with number behaviorally, the fact that there were neurons selective for each of the ordinal positions within the sequence suggests that these neurons may function as cardinal rather than ordinal number detectors. It was not reported, but it would be interesting to determine whether variability in activity increased with number.

An exciting new study has isolated neural activity in the prefrontal cortex of awake, behaving macaque monkeys that is associated with the number of

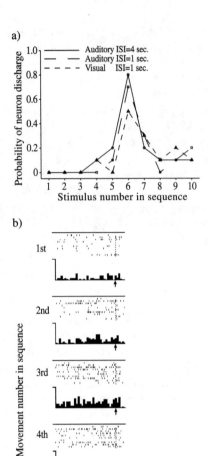

FIGURE 6.12 Neural basis of number representation. (a) Number 6 neuron from cat association cortex. Action potentials were recorded as a sequence of ten tones or lights with different interstimulus intervals presented to an anesthetized cat. Each sequence was repeated ten times. The probability that the neuron discharged to a stimulus is plotted as a function of the stimulus's position in the sequence. This neuron was most active to the sixth stimulus in the sequence, regardless of modality or ISI. (b) Number 5 neuron from monkey parietal cortex. Data are presented from ten blocks of trials in which a monkey repeated an arm movement (push) five times. Each of the panels depicts the trials from a particular movement in the sequence. On the top half of each panel, each row represents one trial. The vertical tick marks indicate the time of each action potential relative to the time of the "go" signal to start the movement (arrows). On the bottom of each panel, the average activity from all trials was calculated in 50-msec time bins. Activity of this neuron was elevated before the fifth movement of the sequence, despite the fact that the movement itself did not differ from any others in the sequence.

simultaneously presented elements in a visual display (Nieder et al., 2002). In this study, monkeys were trained to perform a delayed same–different task in which a "same" answer was rewarded if the second stimulus was an exemplar of the equivalent numerosity as the first stimulus. Monkeys initiated each trial by grasping a lever, and subsequently a sample stimulus was presented, which contained between one and five items. Following a delay period, a test stimulus containing either the same number of elements or a set that differed by one element was presented. Monkeys reported that the sample and test stimuli were the same by releasing the lever. The area, circumference, arrangement, density, and shapes of the items in the stimuli were systematically varied to ensure that number alone was the basis for judgment.

Consistent with previous behavioral observations, accuracy in responding declined as the number of items in the sample increased. In an additional set of behavioral experiments where the sample and test stimuli differed by more than one element, performance improved as the difference between the sample and test stimuli increased. The performance of monkeys in this task thus demonstrated distance and magnitude effects typically associated with numerical discrimination in humans and other animals in a wide variety of tasks.

While monkeys performed the delayed same–different task, the activity of randomly selected neurons in the prefrontal cortex was measured. For approximately one third of neurons studied, activity measured during stimulus presentation or the delay period was maximal for one quantity and declined as distance from that quantity increased. Many neurons in the prefrontal cortex were selective for a single numerosity, but the largest proportion of these number-selective neurons preferred the numerosity 1. The authors showed that numerical selectivity was broader for larger numbers, suggesting a possible mechanism for numerical distance and magnitude effects. They also reported that the onset of prefrontal activity was approximately 120 msec following sample presentation, possibly implicating a parallel model of numerical processing (Dehaene, 2002). This conclusion must be viewed with caution, however, since this measure actually indicates the point in time at which neuronal activity increased above baseline and does not indicate the time at which neurons differentially responded to a given quantity.

In contrast with previous studies, the authors found that neurons in the prefrontal cortex were much more likely to be selective for number than neurons in the parietal cortex (about one third compared with 7% of those sampled). This discrepancy may be due to a difference in the spatial response properties of neurons in the two areas. Parietal neurons are much more spatially selective than prefrontal neurons, but stimuli in this study were presented in random spatial locations. Future studies should probe number-related activity in the parietal cortex using stimuli tailored to the spatially selective response properties of each neuron.

Activity in the parietal lobe has also been shown to correlate with monkeys' performance on an interval time discrimination task (Leon and Shadlen, 2000). Monkeys were trained to compare the duration of two sequentially presented colored lights. On each trial, a short or long standard stimulus (316 or 800 msec) was presented, followed by a test stimulus of varying duration (126 to 800 msec for short and 400 to 1600 msec for long). The monkey indicated whether the test stimulus was longer or shorter than the standard with a saccadic eye movement directed to one of two corresponding choice targets. The monkeys' accuracy increased with the ratio of the standard to the target value, as predicted by Weber's law.

Leon and Shadlen (2000) recorded the activity of neurons in the lateral intraparietal area. These neurons discharge before saccades to a particular location; however, their activity also appeared to represent the duration of the test stimulus compared to the standard. The investigators placed one of the targets within the neuron's receptive field and the other outside it. When the test stimulus was first presented, the neurons behaved as if the monkey intended to choose the shorter target; however, if the test stimulus was in fact longer than the standard, the neuron's firing rate began to indicate that the longer target would be chosen. Future studies should investigate the relationship between the representation of time and number in the parietal cortex.

6.9 CONCLUSION AND DIRECTIONS FOR FUTURE RESEARCH

The data reviewed in this chapter suggest that language is not a necessary prerequisite for representing number and reasoning arithmetically. Many animal species represent numerosities independently of continuous variables such as time, density, and area and represent the ordinal relations between numerical values. Similarly, when surface area and density are well controlled, infants too are capable of representing number. A key question is whether human infants, adult humans, and nonhuman animals use the same system to represent number nonverbally. Do all three populations use analog magnitude representations of number? If so, are these representations formed by the same mechanism? Does the process conform to the counting principles defined by Gelman and Gallistel (1978)?

As reviewed above, animals and adult humans exhibit distance and magnitude effects when making numerical comparisons (Brannon and Terrace, 1998, 2000; Moyer and Landaeur, 1967) and exhibit scalar variability in their memory for specific magnitudes (Platt and Johnson, 1971; Whalen et al., 1999). Such data provide strong evidence that both nonhuman animals and adult humans represent number as continuous mental magnitudes. However, as of yet there are no estimates of the variability in infants' numerical representations (Gelman and Cordes, 2001). Methods have not been established that would allow one to assess whether infants show distance and magnitude effects. Such data would help delineate the similarities and differences between animal and human infant numerical representations.

A great deal of research has been done to characterize young children's early understanding of number (for a review, see Gelman and Meck, 1983). While there is abundant evidence that children have rich numerical concepts before they enter school, it is also apparent that the verbal counting system is mastered slowly. A challenge for the field is to determine how young children map the language-specific count words onto nonverbal representations of number. Do young children map count words onto analog magnitude representations of number, as suggested by Gelman and Gallistel (1978; Gallistel and Gelman, 1992), or are the first mappings accomplished by mapping count words onto plural–singular distinctions in language (Bloom and Wynn, 1997) or object-file representations (Carey, 1998; Spelke, 2000)?

It will be important to determine whether there are multiple formats for representing number and the developmental time course of these representations. Other questions that remain unanswered are whether the representations of number held by animals and infants are independent of sensory modality. What is the relative salience of number vs. continuous dimensions such as surface area, density, and time? Does the ability to discriminate and operate upon representations of time or surface area develop before the ability to discriminate number? Is one system used to represent time and number in infancy? Is the enumeration process parallel or serial?

It is ultimately essential to map the neural substrates of numerical processing and determine whether the same substrates underlie numerical abilities in animals, adult humans, and human infants. Are numerical mental magnitudes represented by the parietal cortex in human infants, nonhuman animals, and adult humans? If so, how specific is this neural substrate? Are all analog magnitude comparisons performed by this brain area, or only those that are numerical in nature? Do single cells represent cardinal and ordinal aspects of number, or is there a more distributed representation of number?

In conclusion, while we have outlined some of the many unanswered questions in the study of the nonverbal representation of number, it should be clear that there is an abundance of evidence that the sophisticated and dazzling numerical abilities seen in modern humans have both evolutionary and developmental precursors. Like most of complex cognition, we can map the seeds of mathematical prowess by studying babies and nonhuman animals. It is at the intersection between these two fields that we can begin to understand the nature of thought without language.

ACKNOWLEDGMENTS

We thank members of the Brannon Lab and Michael Platt for helpful comments on this chapter. E.M.B is especially grateful to Susan Carey, Randy Gallistel, Rochel Gelman, John Gibbon, Warren Meck, and Herb Terrace for conversations that inspired many experiments and greatly influenced the perspective offered in this chapter.

REFERENCES

Anderson, J.R., Awazu, S., and Fujita, K., Can squirrel monkeys learn self-control? A study using food array selection tests and reverse-reward contingency, *J. Exp. Psychol. Anim. Behav. Process.*, 26, 87–97, 2000.

Antell, S.E. and Keating, D.P., Perception of numerical invariances in neonates, *Child Dev.*, 54, 695–701, 1983.

Baillargeon, R., How do infants learn about the physical world? *Curr. Directions Psychol. Sci.*, 3, 133–140, 1994.

Balakrishnan, J.D. and Ashby, F.G., Subitizing: magical numbers or mere superstition? *Psychol. Res.*, 54, 80–90, 1992.

Beran, M., Summation and numerousness judgments of sequentially presented sets of items by chimpanzees (*Pan troglodytes*), *J. Comp. Psychol.*, 115, 181–191, 2001.

Bijeljac-Babic, R., Bertoncini, J., and Mehler, J., How do four-day-old infants categorize multisyllabic utterances? *Dev. Psychol.*, 29, 711–721, 1993.

Bloom, P. and Wynn, K., Linguistic cues in the acquisition of number words, *J. Child Lang.*, 24, 511–533, 1997.

Boysen, S.T. and Berntson, G.G., Numerical competence in a chimpanzee (*Pan troglodytes*), *J. Comp. Psychol.*, 103, 23–31, 1989.

Boysen, S.T. and Berntson, G.G., Responses to quantity: perceptual versus cognitive mechanisms in chimpanzees (*Pan troglodytes*), *J. Exp. Psychol. Anim. Behav. Process.*, 21, 82–86, 1995.

Boysen, S.T., Berntson, G.G., and Mukobi, K.L., Size matters: impact of item size and quantity on array choice by chimpanzees (*Pan troglodytes*), *J. Comp. Psychol.*, 115, 106–110, 2001.

Boysen, S.T., Mukobi, K.L., and Berntson, G.G., Overcoming response bias using symbolic representations of number by chimpanzees (*Pan troglodytes*), *Anim. Learn. Behav.*, 27, 229–235, 1999.

Brannon, E.M., The development of ordinal numerical knowledge in infancy, *Cognition*, 83, 223–240, 2002a.

Brannon, E.M., The Development of Ordinal Numerical Knowledge in Infants, paper presented at ICIS, Toronto, Canada, 2002b.

Brannon, E.M., Kovary, I., Prasankumar, R., and Terrace, H.S., Asymmetrical Extrapolation of Ordinal Numerical Rules by Rhesus Macaques, unpublished.

Brannon, E.M. and Terrace, H.S., Ordering of the numerosities 1–9 by monkeys, *Science*, 282, 746–749, 1998.

Brannon, E.M. and Terrace, H.S., Corrections and clarifications, *Science*, 283, 1852, 1999.

Brannon, E.M. and Terrace, H.S., Representation of the numerosities 1–9 by rhesus monkeys (*Macaca mulatta*), *J. Exp. Psychol. Anim. Behav. Process.*, 26, 31–49, 2000.

Brannon, E.M. and Terrace H.S., The Evolution and Ontogeny of Ordinal Numerical Ability, in *The Cognitive Animal,* Bekoff, M., Allen, C., and Burghardt, G.M., Eds., The MIT Press, Cambridge, MA, 2002.

Brannon, E.M., Wolfe, L., Meck, W.H., and Woldorff, M., Electrophysiological Correlates of Timing in Human Infants, poster presented at Society for Neuroscience, Orlando, FL, 2002.

Brannon, E.M., Wusthoff, C.J., Gallistel, C.R., and Gibbon, J., Subtraction in the pigeon: evidence for a linear subjective number scale, *Psychol. Sci.*, 12, 238–243, 2001 (poster presented at Society for Neuroscience 589.11).

Breukelaar, J.W.C. and Dalrymple-Alford, J.C., Timing ability and numerical competence in rats, *J. Exp. Psychol. Anim. Behav. Process.*, 24, 84–97, 1998.

Broadbent, H.A., Rakitin, B.C., Church, R.M., and Meck, W.H., Quantitative relationships between timing and counting, in *Numerical Skills in Animals*, Boysen, S. and Capaldi, E.J., Eds., Erlbaum, Hillsdale, NJ, 1993, pp. 171–187.

Buckley, P.B. and Gillman, C.B., Comparisons of digit and dot patterns, *J. Exp. Psychol.*, 103, 1131–1136, 1974.

Call, J., Estimating and operating on discrete quantities in orangutans (*Pongo pygmaeus*), *J. Comp. Psychol.*, 114, 136–147, 2000.

Capaldi, E.J., Animal number abilities: implications for a hierarchical approach to instrumental learning, in *The Development of Numerical Competence; Animal and Human Models,* Boysen, S.T. and Capaldi, E.J., Eds., Lawrence Erlbaum Associates, Hillsdale, NJ, 1993, pg. 191–210.

Capaldi, E.J. and Miller, D.J., Counting in rats: its functional significance and the independent cognitive processes which comprise it, *J. Exp. Psychol. Anim. Behav. Process.*, 14, 3–17, 1988.

Carey, S., Knowledge of number: it's evolution and ontology, *Science*, 282, 641–642, 1998.

Carey, S., On the very possibility of discontinuities in conceptual development, in *Language, Brain, and Cognitive Development: Essays in Honor of Jacques Mehler*, Dupoux, E., Ed., MIT Press, Cambridge, MA, 2001, pp. 303–324.
Church, R.M. and Broadbent, H.A., Alternative representations of time, number, and rate, *Cognition*, 37, 55–81, 1990.
Church, R.M. and Broadbent, H.A., A connectionist model of timing, in *Neural Network Models of Conditioning and Action: Quantitative Analyses of Behavior Series*, Commons, M.L. and Grossberg, S., Eds., Erlbaum, Hillsdale, NJ, 1991, pp. 225–240.
Church, R.M. and Meck, W.H., The numerical attribute of stimuli, in *Animal Cognition*, Roitblat, H.L., Bever, T.G., and Terrace, H.S., Eds., Erlbaum, Hillsdale, NJ, 1984, pp. 445–464.
Clearfield, M.W. and Mix, K.S., Number versus contour length in infants' discrimination of small visual sets, *Psychol. Sci.*, 10, 408–411, 1999.
Cohen, L.B. and Marks, K.S., How infants process addition and subtraction events, *Developmental Science*, 5, 186–201, 2002.
Cooper, R.G.J., Early number development: discovering number space and addition and subtraction, in *Origins of Cognitive Skills*, Sophian, C., Ed., Erlbaum, Hillsdale, NJ, 1984, pp. 157–192.
Cordes, S., Gelman, R., Gallistel, C.R., and Whalen, J., Variability signatures distinguish verbal from nonverbal counting for both large and small numbers, *Psychonomic Bull. Rev.*, 8, 698–707, 2001.
Davis, H., Discrimination of the number three by a raccoon (*Procyon lotor*), *Anim. Learn. Behav.*, 12, 409–413, 1984.
Davis, H., Numerical competence in ferrets (*Mustela putorius furo*), *Int. J. Comp. Psychol.*, 9, 51–64, 1996.
Davis, H. and Albert, M., Failure to transfer or train a numerical discrimination using sequential visual stimuli in rats, *Bull. Psychonomic Soc.*, 25, 472–474, 1987.
Davis, H. and Bradford, S.A., Counting behavior by rats in a simulated natural environment, *Ethology*, 73, 265–280, 1986.
Davis, H. and Bradford, S.A., Numerically restricted food intake in the rat in a free-feeding situation, *Anim. Learn. Behav.*, 19, 215–222, 1991.
Davis, H., MacKenzie, K.A., and Morrison, S., Numerical discrimination by rats (*Rattus norvegicus*) using body and vibrissal touch, *J. Comp. Psychol.*, 103, 45–53, 1989.
Davis, H., Memmott, J., and Hurwitz, H.M., Autocontingencies: a model for subtle behavioral control, *J. Exp. Psychol. Gen.*, 104, 169–188, 1975.
Davis, H. and Perusse, R., Numerical competence: from backwater to mainstream of comparative psychology, *Behav. Brain Sci.*, 11, 602–615, 1988.
Dehaene, S., Varieties of numerical abilities, *Cognition*, 44, 1–42, 1992.
Dehaene, S., *The Number Sense: How the Mind Creates Mathematics*, Oxford University Press, New York, 1997.
Dehaene, S., Cerebral bases of number processing and calculation, in *The New Cognitive Neurosciences*, 2nd ed., Gazzaniga, M.S., Ed., MIT Press, Cambridge, MA, 2000.
Dehaene, S., Subtracting pigeons: logarithmic or linear?, *Psychological Sci.*, 12, 244–246, 2001.
Dehaene, S., Single-neuron arithmetic, *Science*, 297, 1652–1653, 2002.
Dehaene, S. and Changeux, J., Development of elementary numerical abilities: a neuronal model, *J. Cognit. Neurosci.*, 5, 390–407, 1993.
Dehaene, S., Dehaene-Lambertz, G., and Cohen, L., Abstract representation of numbers in the animal and human brain, *Nat. Neurosci.*, 21, 355–361, 1998.

Dooley, G.B. and Gill, T.V., Acquisition and use of mathematical skills by a linguistic chimpanzee, in *Language Learning by a Chimpanzee: The Lana Project*, Rumbaugh, D.M., Ed., Academic Press, New York, 1977, pp. 247–260.

Douglas, J.W.B. and Whitty, W.M., An investigation of number appreciation in some subhuman primates, *J. Comp. Psychol.*, 31, 129–143, 1941.

Emmerton, J., Numerosity differences and effects of stimulus density on pigeons' discrimination performance, *Anim. Learn. Behav.*, 26, 243–256, 1998.

Emmerton, J., Lohmann, A., and Niemann, J., Pigeons' serial ordering of numerosity with visual arrays, *Anim. Learn. Behav.*, 25, 234–244, 1997.

Evans, T.A., Judge, P.G., Holzworth, C.L., and Vyas, D.K., Judging Relative Numerousness in Captive Capuchin Monkeys (Cebus apella), paper presented at American Society of Primatologists, Oklahoma City, OK, 2002.

Feigenson, L., Carey, S., and Hauser, M., The representations underlying infants' choice of more: object files versus analog magnitudes, *Psychol. Sci.*, 13, 150–156, 2002a.

Feigenson, L., Carey, S., and Spelke, E., Infants' discrimination of number vs. continuous extent, *Cognit. Psychol.*, 44, 33–66, 2002b.

Fernandes, D.M. and Church, R.M., Discrimination of the number of sequential events by rats, *Anim. Learn. Behav.*, 10, 171–176, 1982.

Fetterman, J.G., Numerosity discrimination: both time and number matter, *J. Exp. Psychol. Anim. Behav. Process.*, 19, 149–164, 1993.

Fetterman, G. and MacEwen, D., Short-term memory for responses: the "choose small" effect, *J. Exp. Anal. Behav.*, 52, 311–324, 1989.

Fink, G.R., Marshall, J.C., Gurd, J., Weiss, P.H., Zafiris, O., Shah, N.J., and Zilles, K., Deriving numerosity and shape from identical visual displays, *Neuroimage*, 13, 46–55, 2001.

Gallistel, C.R. and Gelman, R., The what and how of counting, *Cognition*, 34, 197–199, 1990.

Gallistel, C.R. and Gelman., R., Subitizing: the preverbal counting process, in *Memories, Thoughts and Emotions: Essays in Honor of George Mandler*, Kessen, W., Ortony, A., and Craik, F., Eds., Erlbaum, Hillsdale, NJ, 1991, pp. 65–81.

Gallistel, C.R. and Gelman., R., Preverbal and verbal counting and computation, *Cognition*, 44, 43–74, 1992.

Gallistel, R. and Gelman, R., Non-verbal numerical cognition: from reals to integers, *Trends Cognit. Sci.*, 4, 59–65, 2000.

Gelman, R. and Cordes, S., Counting in animals and humans, in *Language, Brain, and Cognitive Development: Essays in Honor of Jacques Mehler*, Dupoux, E., Ed., MIT Press, Cambridge, MA, 2001, pp. 279–301.

Gelman, R. and Gallistel, C.R., *The Child's Concept of Number*, Harvard University Press, Cambridge, MA, 1978.

Gelman, R. and Meck, W., Preschoolers' counting: principles before skill, *Cognition*, 13, 434–459, 1983.

Gibbon, J. and Church, R.M., Time left: linear versus logarithmic subjective time, *J. Exp. Anal. Behav.*, 7, 87–107, 1981.

Gibbon, J. and Church, R.M., Sources of variability in an information processing theory of timing, in *Animal Cognition*, Roitblat, H.L., Bever, T.G., and Terrace, H.S., Eds., Erlbaum, Hillsdale, NJ, 1984, pp. 465–488.

Harper, D.G., Competitive foraging in mallards: "ideal free" ducks, *Anim. Behav.*, 30, 2, 575–584, 1982.

Hauser, M. and Carey, S., Building a cognitive creature from a set of primitives: evolutionary and developmental insights, in *The Evolution of Mind*, Cummins-Dellarosa, D. and Allen, C., Eds., Oxford University Press, New York, 1998, pp. 51–106.

Hauser, M.D., Carey, S., and Hauser, L.B., Spontaneous number representation in semi-free-ranging rhesus monkeys, *Proc. R. Soc. Lond.*, 267, 829–833, 2000.

Hauser, M.D., MacNeilage, P., and Ware, M., Numerical representations in primates, *Proc. Natl. Acad. Sci. U.S.A.*, 93, 1514–1517, 1996.

Hicks, L.H., An analysis of number-concept formation in the rhesus monkey, *J. Comp. Physiol. Psychol.*, 49, 212–218, 1956.

Honig, W.K., Discrimination by pigeons of mixture and uniformity in arrays of stimulus elements, *J. Exp. Psychol. Anim. Behav. Process.*, 17, 68–80, 1991.

Honig, W.K. and Matheson, R., Discrimination of relative numerosity and stimulus mixture by pigeons with comparable tasks, *J. Exp. Psychol. Anim. Behav. Process.*, 21, 348–363, 1995.

Honig, W.K. and Stewart, K.E., Discrimination of relative numerosity by pigeons, *Anim. Learn. Behav.*, 17, 134–146, 1989.

Honig, W.K. and Stewart, K.E., Relative numerosity as a dimension of stimulus control: the peak shift, *Anim. Learn. Behav.*, 21, 346–354, 1993.

Kaufman, E., Lord, M., Reese, T., and Volkmann, J., The discrimination of visual number, *Am. J. Psychol.*, 62, 498–525, 1949.

Klahr, D., A production system for counting, subitizing and adding, in *Visual Information Processing,* Chase, W.G., Ed., Academic Press, New York, 1973.

Koechlin, E., Dehaene, S., and Mehler, J., Numerical transformations in five-month-old human infants, *Math. Cognit.*, 3, 89–104, 1998.

Krebs, J.R. and Davies, R., *Introduction to Behavioral Ecology*, Blackwell Scientific, Oxford, 1993.

Laties, V., The modification of drug effects on behavior by external discriminative stimuli, *J. Pharmacol. Exp. Ther.*, 183, 1–13, 1972.

Leon, M.I. and Shadlen, M.N., Representation of elapsed time by neurons in area LIP of the macaque, *Soc. Neurosci. Abstr.*, 26, 668, 2000.

Leslie, A., Xu, F., Tremoulet, P., and Scholl, B., Indexing and the object concept: developing 'what' and 'where' systems, *Trends Cognit. Sci.*, 2, 10–18, 1998.

Lipton, J.S. and Spelke, E.S., Large Number Discrimination of Auditory Sequences in 6- and 9-Month-Old Infants, poster presented at ICIS, Toronto, Canada, 2001.

Lipton, J.S. and Spelke, E.S., Large Number Discrimination in 6-Month-Olds, poster presented at SRCD, Minneapolis, MN, 2002.

Mandler, G. and Shebo, B., Subitizing: an analysis of its component processes, *J. Exp. Psychol. Gen.*, 111, 1–22, 1982.

Matsuzawa, T., Use of numbers by a chimpanzee, *Nature*, 315, 57–59, 1985.

Matsuzawa, T., Itakura, S., and Tomonaga, M., Use of numbers by a chimpanzee: a further study, in *Primatology Today*, Ehara, A., Kimura, T., Takenaka, O., and Iwamoto, M., Eds., Elsevier, Amsterdam, 1991, pp. 317–320.

Mechner, F., Probability relations within response sequences under ratio reinforcement, *J. Exp. Anal. Behav.*, 1, 109–122, 1958.

Mechner, F. and Guevrekian, L., Effects of deprivation on counting and timing in rats, *J. Experimental Anal. Behav.*, 5, 463–466, 1962.

Meck, W.H., Application of a mode-control model of temporal integration to counting and timing behaviour, in *Time and Behaviour: Psychological and Neurobehavioural Analyses*, Vol. 120, *Advances in Psychology*, Bradshaw, C.M. and Szabadi, E., Eds., Elsevier, Amsterdam, 1997, pp. 133–184.

Meck, W.H. and Church, R.M., A mode control model of counting and timing processes, *J. Exp. Psychol. Anim. Behav. Process.*, 9, 320–334, 1983.

Meck, W.H., Church, R.M., and Gibbon, J., Temporal integration in duration and number discrimination, *J. Exp. Psychol. Anim. Behav. Process.*, 11, 591–597, 1985.

Murofushi, K., Numerical matching behavior by a chimpanzee (*Pan troglodytes*): subitizing and analogue magnitude estimation, *Jpn. Psychol. Res.*, 39, 140–153, 1997.

Mitchell, R.W., Yao, P., Sherman, P., and O'Regan, M., Discriminative responding of a dolphin (*Tursiops truncatus*) to differentially rewarded stimuli, *J. Comp. Psychol.*, 99, 218–225, 1985.

Mix, K.S., Huttenlocher, J., and Levine, S.C., Multiple cues for quantification in infancy: is number one of them?, *Psychological Bull.*, 128, 278–294, 2002.

Mix, K.S., Levine, S.C., and Huttenlocher, J., Numerical abstraction in infants: another look, *Developmental Psychol.*, 33, 423–428, 1997.

Moore, D., Benenson, J., Reznick, J.S., and Peterson, M., Effect of auditory numerical information on infants' looking behavior: contradictory evidence, *Dev. Psychol.*, 23, 665–670, 1987.

Moyer, R.S. and Landauer, T.K., Time required for judgments of numerical inequality, *Nature*, 215, 1519–1520, 1967.

Newcombe, N.S., The nativist-empiricist controversy in the context of recent research on spatial and quantitative development, *Psychological Sci.*, 13, 395–401, 2002.

Nieder, A., Freedman, D.J., and Miller, E.K., Representation of the quantity of visual items in the primate prefrontal cortex, *Science*, 297, 1708–1711, 2002.

Olthof, A., Iden, C.M., and Roberts, W.A., Judgments of ordinality and summation of number symbols by squirrel monkeys (*Saimiri sciureus*), *J. Exp. Psychol. Anim. Behav. Process.*, 23, 325–339, 1997.

Pastore, N., Number sense and "counting" ability in the canary, *Z. Tierpsychol.*, 18, 561–573, 1961.

Pepperberg, I.M., Evidence for conceptual quantitative abilities in an African grey parrot: labeling of cardinal sets, *Ethology*, 75, 37–61, 1987.

Pepperberg, I.M., Numerical competence in an African gray parrot (*Psittacus erithacus*), *J. Comp. Psychol.*, 108, 36–44, 1994.

Piaget, J., *The Child's Conception of Number*, W.W. Norton, New York, 1965 (original work published in 1941).

Platt, J.R. and Johnson, D.M., Localization of position within a homogeneous behavior chain: effects of error contingencies, *Learn. Motiv.*, 2, 386–414, 1971.

Pouthas, V., Droit, S., and Jacquet, A.Y., Temporal experience and time knowledge in infancy and early childhood, *Time Soc.*, 2, 199–218, 1993.

Pylyshyn, Z.W. and Storm, R.W., Tracking multiple independent targets: evidence for a parallel tracking mechanism, *Spat. Vis.*, 3, 179–197, 1988.

Rilling, M., Number of responses as a stimulus in fixed interval and fixed ratio schedules, *J. Comp. Physiol. Psychol.*, 63, 60–65, 1967.

Roberts, W.A., Simultaneous numerical and temporal processing in the pigeon, *Curr. Directions Psychol. Sci.*, 4, 47–51, 1995.

Roberts, W.A. and Boisvert, M.J., Using the peak procedure to measure timing and counting processes in pigeons, *J. Exp. Psychol. Anim. Behav. Process.*, 24, 416–430, 1998.

Roberts, W.A., Coughlin, R., and Roberts, S., Pigeons flexibly time or count on cue, *Psychol. Sci.*, 11, 218–222, 2000.

Roberts, W.A., Macuda, T., and Brodbeck, D.R., Memory for number of light flashes in the pigeon, *Anim. Learn. Behav.*, 23, 182–188, 1995.

Roberts, W.A. and Mitchell, S., Can a pigeon simultaneously process temporal and numerical information? *J. Exp. Psychol. Anim. Behav. Process.*, 20, 66–78, 1994.

Roberts, W., Roberts, S., and Kit, K.A., Pigeons presented with sequences of false flashes use behavior to count but not to time, *J. Exp. Psychol. Anim. Behav. Process.*, 28, 137–150, 2002.

Ruby, L.M., An investigation of number-concept appreciation in a rhesus monkey, *Primates*, 25, 236–242, 1984.

Rumbaugh, D.M., Savage-Rumbaugh, S., and Hegel, M.T., Summation in the chimpanzee (*Pan troglodytes*), *J. Exp. Psychol. Anim. Behav. Process.*, 13, 107–115, 1987.

Rumbaugh, D.M., Savage-Rumbaugh, E.S., and Pate, J.L., Addendum to summation in the chimpanzee (*Pan troglodytes*), *J. Exp. Psychol. Anim. Behav. Process.*, 14, 118–120, 1988.

Rumbaugh, D.M. and Washburn, D.A., Counting by chimpanzees and ordinality judgments by macaques in video-formatted tasks, in *The Development of Numerical Competence; Animal and Human Models,* Boysen, S.T. and Capaldi, E.J., Eds., Lawrence Erlbaum Associates, Hillsdale, NJ, 1993, pg. 87–108.

Salman, D.H., Note on the number conception in animal psychology, *Br. J. Psychol.*, 33, 209–219, 1943.

Santi, A. and Hope, C., Errors in pigeons' memory for number of events, *Anim. Learning Behav.*, 29, 208–220, 2001.

Sawamura, H., Shima, K., and Tanji, J., Numerical representation for action in the parietal cortex of the monkey, *Nature*, 415, 918–922, 2002.

Sathian, K., Simon, T.J., Peterson, S., Patel, G.A., Hoffman, J.M., and Grafton, S.T., Neural evidence linking visual object enumeration and attention, *J. Cognit. Neurosci.*, 11, 36–51, 1999.

Seibt, U., Zahlbegriff und Zahlverhalten bei Tieren: Neu Versuche und Deutungen, *Z. Tierpsychol.*, 60, 325–341, 1982.

Shumaker, R.W., Palkovich, A.M., Beck, B.B., Guagnano, G.A., and Morowitz, H., Spontaneous use of magnitude discriminations and ordination by the orangutan (*Pongo pygmaeus*), *J. Comp. Psychol.*, 115, 385–391, 2001.

Simon, T.J., Reconceptualizing the origins of number knowledge: a "non-numerical account," *Cognit. Dev.*, 12, 349–372, 1997.

Simon, T.J., Hespos, S.J., and Rochat, P., Do infants understand simple arithmetic? A replication of Wynn (1992), *Cognit. Dev.*, 10, 253–269, 1995.

Smith, B.R., Piel, A.K., and Candland, D.K., The Numerical Abilities of a Socially-Housed Hamadryas Baboon (*Papio hamadryas*) and Squirrel Monkey (*Saimiri sciureaus*), Abstracts of Psychonomic Society, 545, 81, 2002.

Spelke, E.S., Core knowledge, *Am. Psychol.*, 55, 1233–1243, 2000.

Spetch, M. and Wilkie, D.M., Subjective shortening: a model of pigeons' memory for event duration, *J. Exp. Psychol. Anim. Behav. Process.*, 9, 14–30, 1983.

Starkey, P. and Cooper, R.G., Jr., Perception of numbers by human infants, *Science*, 210, 1033–1035, 1980.

Starkey, P., Spelke, E.S., and Gelman, R., Detection of intermodal numerical correspondences by human infants, *Science*, 222, 179–181, 1983.

Starkey, P., Spelke, E.S., and Gelman, R., Numerical abstraction by human infants, *Cognition*, 36, 97–128, 1990.

Strauss, M.S. and Curtis, L.E., Infants perception of numerosity, *Child Dev.*, 52, 1146–1152, 1981.

Strauss, M.S. and Curtis, L.E., Development of numerical concepts in infancy, in *Origins of Cognitive Skills*, Erlbaum, Hillsdale, NJ, 1984.

Sulkowski, G.M. and Hauser, M.D., Can rhesus monkeys spontaneously subtract?, *Cognition*, 79, 239–262, 2001.

Temple, E. and Posner, M., Brain mechanisms of quantity are similar in 5-year-olds and adults, *Proc. Natl. Acad. Sci. U.S.A.*, 95, 7836–7841, 1998.

Thomas, R.K. and Chase, L., Relative numerousness judgments by squirrel monkeys, *Bull. Psychonomic Soc.*, 16, 79–82, 1980.

Thomas, R.K., Fowlkes, D., and Vickery, J.D., Conceptual numerousness judgments by squirrel monkeys, *Am. J. Psychol.*, 93, 247–257, 1980.

Thompson, R.F., Mayers, K.S., Robertson, R.T., and Patterson, C.J., Number coding in association cortex of the cat, *Science*, 168, 271–273, 1970.

Trick, L.M. and Pylyshyn, Z.W., What enumeration studies can show us about spatial attention: evidence for limited capacity preattentive processing, *J. Exp. Psychol. Hum. Percept. Perform.*, 19, 331–351, 1993.

Trick, L.M. and Pylyshyn, Z.W., Why are small and large numbers enumerated differently? A limited capacity preattentive stage in vision, *Psychol. Rev.*, 101, 80–102, 1994.

Uller, C., Carey, S., Huntley-Fenner, G., and Klatt, L., What representations might underlie infant numerical knowledge, *Cognit. Dev.*, 14, 1–36, 1999.

Van de Walle, G.A., Carey, S., and Prevor, M., Bases for object individuation in infancy: evidence from manual search, *J. Cognit. Dev.*, 1, 249–280, 2000.

Van Loosbroek, E. and Smitsman, A.W., Visual perception of numerosity in infancy, *Dev. Psychol.*, 26, 916–922, 1990.

Wakely, A., Rivera, S., and Langer, J., Can young infants add and subtract? *Child Dev.*, 71, 1525–1534, 2000.

Washburn, D. and Rumbaugh, D.M., Ordinal judgments of numerical symbols by macaques (*Macaca mulatta*), *Psychol. Sci.*, 2, 190–193, 1991.

Wesley, F., The number concept: a phylogenetic review, *Psychol. Bull.*, 58, 420–428, 1961.

Whalen, J., Gelman, R., and Gallistel, C.R., Non-verbal counting in humans: the psychophysics of number representation, *Psychol. Sci.*, 10, 130–137, 1999.

Wilkie, D.M., Webster, J.B., and Leader, L.G., Unconfounding time and number discrimination in a Mechner counting schedule, *Bull. Psychonomic Soc.*, 13, 390–392, 1979.

Wilson, M.L. Hauser, M.D., and Wrangham, R.W., Does participation in intergroup conflict depend on numerical assessment, range location, or rank for wild chimpanzees? *Anim. Behav.*, 61, 1203–1216, 2001.

Wynn, K., Addition and subtraction by human infants, *Nature*, 358, 749–750, 1992.

Wynn, K., Origins of numerical knowledge, *Math. Cognit.*, 1, 35–60, 1995.

Wynn, K., Infants' individuation and enumeration of actions, *Psychol. Sci.*, 7, 164–169, 1996.

Wynn, K., Psychological foundations of number: numerical competence in human infants, *Trends Cognit. Sci.*, 2, 296–303, 1998.

Wynn, K., Findings of addition and subtraction in infants are robust and consistent: reply to Wakeley, Rivera, and Langer, *Child Dev.*, 71, 1535–1536, 2000.

Wynn, K., Bloom, P., and Chiang, W., Enumeration of collective entities by 5-month-old infants, *Cognition*, 83, B55–B62, 2002.

Xia, L., Siemann, M., and Delius, J.D., Matching of numerical symbols with number of responses by pigeons, *Anim. Cognit.*, 3, 35–43, 2000.

Xu, F. and Spelke, E.S., Large number discrimination in 6-month-old infants, *Cognition*, 74, B1–B11, 2000.

form
7 Temporal Experience and Timing in Children

Sylvie Droit-Volet

CONTENTS

7.1 Introduction 183
7.2 Interval Timing in Children: Procedures and Modeling of Data 184
    7.2.1 Temporal Bisection in Children 184
    7.2.2 Temporal Generalization in Children 188
7.3 Is an Internal Clock Functional at an Early Age? 194
7.4 The Effect of Sensory Modality in Children 196
7.5 Sources of Developmental Changes in Interval Timing 200
    7.5.1 Long-Term Memory Sources of Developmental Changes in Timing 200
    7.5.2 Attentional Sources of Developmental Changes in Timing 202
7.6 Conclusion 204
Acknowledgments 205
References 205

7.1 INTRODUCTION

In 1928, at the first International Congress of Physics and Philosophy in Davos, Switzerland, Piaget heard Einstein presenting his theory of relativity on physical time. From then, he tried to demonstrate that psychological time is intrinsically related to the physical time defined by Einstein: "The hypothesis that I want to defend is that psychological time depends on the speed or the movements with their speed" (Piaget, 1946). Thus, during several decades, Piaget and a whole generation of psychologists investigated, with sophisticated methods, the development of the ability to judge time in children as a result of the development of the capacity to logically reason about time.

    In parallel, researchers accumulated data from animals and human adults showing their abilities to accurately estimate time, and that the subjective time is linear with the objective time. Then they proposed models of temporal information processing, according to which the raw material for time judgments comes from an internal mechanism similar to that of a clock. Recently, developmental psychologists have become aware of the utility of these models to better understand the interval timing abilities in children, and have decided to break with the dominant Piagetian theory. Thus, they made an effort to reconsider interval timing in children within the framework of the most completely developed model of interval timing, the scalar timing theory (e.g., Gibbon et al., 1984). First, they submitted children to the tasks related to this theory, in order to provide empirical data on age-related changes in timing. Then they examined different potential sources of developmental changes at each level of the temporal information processing.

7.2 INTERVAL TIMING IN CHILDREN: PROCEDURES AND MODELING OF DATA

7.2.1 TEMPORAL BISECTION IN CHILDREN

Recent studies have adapted for the temporal bisection procedure, initially used with animals (e.g., Church and Deluty, 1977; Meck, 1983), and later modified for human adults (e.g., Allan and Gibbon, 1991; Wearden, 1991) [young children (e.g., Droit-Volet and Wearden, 2001, 2002; Gautier and Droit-Volet, 2002a; McCormack et al., 1999; Rattat and Droit-Volet, 2001)]. In the temporal bisection task, the children generally received three experimental phases: pretraining, training, and testing. In the pretraining, they were presented a short and a long standard duration in the form of a sound or a visual stimulus (e.g., blue circle). Then they were trained on successive blocks of trials to press one button after the short standard duration and another one after the long standard duration. In this training phase, a correct response resulted in the appearance of a smiling clown, and an incorrect one of a frowning clown, as illustrated in Figure 7.1. The training terminated when the child made no errors during a block of trials (i.e., eight trials, four for each standard). The number of the training blocks required to discriminate the short from the long standard decreased with age; the 3-year-olds took at most three or four training blocks, the 5-year-olds up to two or three, and the 8-year-olds, as the adults, one or two training blocks. Thus, the simple learning of two standard durations took longer in the younger children. After a successful training block, the children were given a testing

FIGURE 7.1 The smiling clown (positive feedback) and the frowning clown (negative feedback) presented on the center of the computer screen after a correct response and an incorrect response, respectively, in both the temporal bisection and temporal generalization procedures.

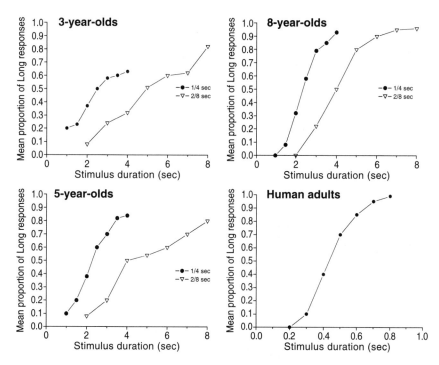

FIGURE 7.2 Psychophysical functions obtained from the 3-, 5-, and 8-year-olds in a temporal bisection task in two anchor duration conditions (1/4 and 2/8 sec) (Droit-Volet and Wearden, 2001), with an example of temporal bisection function obtained from the human adults derived from Wearden (1991). (From Droit-Volet, S. and Wearden, J., *J. Exp. Child Psychol.*, 80, 142–159, 2001.)

phase that maintained the conditions of training, except that the feedback was discontinuous. Furthermore, they received eight or ten blocks of seven trials, two for the two standard durations and five for the intermediate nonstandard durations.

A typical psychophysical function found in human adults in a temporal bisection task, with the mean proportion of long responses (i.e., comparison stimulus durations identified as similar to the long standard duration), plotted against a comparison stimulus duration, is shown in the bottom right panel of Figure 7.2. The slope of this temporal bisection function is steep, indicating a high sensitivity to duration. The Weber ratio (difference limen divided by the bisection point), that is, an index of the steepness of the temporal bisection function, is therefore particularly small (e.g., 0.17 in Wearden's (1991) study). In light of adult temporal bisection performance, several questions can be raised from a developmental perspective. Are children in different age groups capable of duration discrimination in a temporal bisection task? Does the temporal bisection performance in children differ from that in adults? If yes, what might developmental changes in temporal bisection performance look like? And, finally, what are the main causes of these developmental changes?

Until now, the youngest children who have been tested on a temporal bisection task are 3 years old. Figure 7.2 shows their temporal bisection functions (top left), as well as those of older children aged 5 years old (bottom left) and 8 years old (top right). In this figure extracted from Droit-Volet and Wearden's (2001) study, there are two ranges of durations tested, each longer than 1 sec (i.e., 1 to 4 sec and 2 to 8 sec), similar to those generally used in experiments with rats and pigeons. Using these time values without any concurrent distracting task is possible in young children because, unlike adults, they do not spontaneously use counting strategies to time durations up to 10 years (Friedman, 1990a; Wilkening et al., 1987). However, as in studies preventing counting strategies in human adults, Droit-Volet and Wearden (2002) and McCormack et al. (1999) have also used time values shorter than 1 sec with children and have obtained nearly identical temporal bisection functions to those reported in Figure 7.2.

Thus, Figure 7.2 reveals that the temporal bisection task produced orderly data from children as young as 3 years old. At this age, the proportion of long responses increased as the stimulus duration value increased. Therefore, young children are able to discriminate different durations in a temporal bisection task. Furthermore, the use of two ranges of durations allows us to test whether interval timing in children conforms to the scalar property of variance, characteristic of interval timing in animals and human adults. This property is a form of Weber's law (Gibbon, 1977; Gibbon et al., 1984), in which the standard deviation of time judgments grows as a constant fraction of the mean time judgment when different time values are judged. A way to test this scalar property is to examine how well the temporal bisection functions for different time values superimpose when plotted on the same relative time scale (Allan and Gibbon, 1991). Following up on this test in children with both short and long durations, Droit-Volet and Wearden (2001, 2002) found that children's temporal bisection behavior conforms well to the principle of superimposition at all age groups tested. Thus, the scalar property of variance common to the interval timing behavior of animals and human adults is characteristic of interval timing behavior at different levels of the ontogenetic scale, or at least from the age of 3 years in children.

Beyond the similarities, there were also differences in temporal bisection performance between children and adults. In particular, the slope of the psychophysical functions increased with age, being flatter in the 3- and 5-year-olds than in the 8-year-olds, the age at which it was close to the adults'. As a result, the Weber ratio was on average higher in the 3- and 5-year-olds (0.39 and 0.37, respectively) than in the 8-year-olds (0.21), with adults typically showing a Weber ratio of 0.17 (e.g., Wearden, 1991). As the bisection slope and the Weber ratio are two indexes of temporal sensitivity, these data indicate that there was an increase with age in sensitivity for duration in the temporal bisection task.

The question now is: How can we explain these age-related changes in temporal sensitivity in a temporal bisection task? One way is to use the models based on the scalar timing theory accounting for temporal bisection performance in human adults (e.g., Allan and Gibbon, 1991; Wearden, 1991; for more details, see Droit-Volet and Wearden, 2001). More specifically, we used the modified difference model of Wearden (1991). The scalar timing theory proposes that the raw material for time judgments comes from a pacemaker–accumulator system. However, the model also involves memory and decision processes. In the case of bisection, the comparison stimulus duration, t, is assumed to be stored in working memory and reflects the amount of pulses counted in the accumulator of the clock. By contrast, the short and the long standard durations are represented in long-term memory. According to the scalar timing model, the decision to classify the comparison stimulus duration as more similar to the short or the long standard is governed by the difference between the comparison stimulus duration and samples drawn from the long-term memory of the short and the long standard (s^*, l^*). The memory representations of the short and the long standard are stored as distributions rather than single values, with means equal to the standard values and some coefficient of variation c. The higher this coefficient of variation, the greater is the variability of standard representation in memory, and the flatter the psychophysical function will be. The coefficient of variation of the remembered time is thus a kind of sensitivity parameter controlling the slope of the psychophysical function. The first explanation that we can provide for our developmental data is that the increase in timing sensitivity with age would be related to the decrease in the coefficient of variation of the remembered time. However, psychophysical functions can also differ because of random responses, given regardless of stimulus duration values. Random responding is rare in human adults, but this has been observed in animals. Bridging the gap between animal and human adult timing behavior, we can propose a second hypothesis that the amount of random responses in children would decrease with the increasing age. Interestingly, the probability of random responses is the only parameter that we need to add to the scalar timing model in order to obtain an excellent correspondence between the simulation and our temporal bisection data from young children.

In addition, according to the scalar timing model, responding "short" or "long" depends on a decision threshold adopted by the subject. The model predicts that if the difference between $D(s^*, t)$ and $D(l^*, t)$ is greater than some threshold value, b, the subject responds short if $D(s^*, t) < D(l^*, t)$ and long if $D(s^*, t) > D(l^*, t)$, where D equals the difference. However, facing ambiguous cases, if the differences are not clearly discriminated and are less than the threshold value, b, the model chooses to respond long. The b parameter is thus a kind of bias toward responding long. This bias toward long responses does not affect the slope of the temporal bisection function, but it shifts the psychophysical function laterally. Thus, the different bisection patterns observed at different ages could also be in part due to a developmental change in the bias toward long responses.

This scalar timing model adapted to children's timing behavior fits our data very well (Droit-Volet and Wearden, 2001; Rattat and Droit-Volet, 2001) and has allowed us to identify two main sources of developmental changes in temporal bisection performance. First, the coefficient of variation of the representation of standard durations was the highest in the 3-year-olds and the lowest in the 8-year-olds. Thus, the increase in the sensitivity to time was related to the decrease in the variability of the representation of standard durations in long-term memory. Second, the probability of random responding was between 10 and 20% of responses in the 3- and 5-year-olds and near zero in the 8-year-olds (0.01), as in the adults. Thus, the probability of random responses contaminating timing behavior in temporal bisection decreased with the increasing age of the children. By contrast, the b values were small and constant across the different age groups. In other words, this bias toward responding long did not vary with age. To sum up, temporal sensitivity in temporal bisection is lower in younger children, and the developmental version of the scalar timing model suggests that this is due to the decrease with increasing age in the variability of the long-term memory representation of the standard durations, and the probability of random responding.

7.2.2 Temporal Generalization in Children

Interval timing behavior in children has also been tested with the temporal generalization procedure. In our studies, the children were first presented a standard duration. Then they received a series of comparison stimulus durations shorter than, longer than, or equal to the standard duration. Their task was to indicate whether the just-presented comparison duration was the standard (yes or no response). A correct response resulted in the appearance of a smiling clown, and an incorrect response in a frowning clown (Figure 7.1). Under similar experimental conditions, McCormack et al. (1999) evaluated temporal generalization performance in children from 5 to 10 years using one standard duration that was shorter than 1 sec, and Droit-Volet et al. have evaluated children from 3 to 8 years of age using two ranges of durations, in order to test conformity to the scalar property of variance. In a first study (Droit-Volet et al., 2001), they used two standard durations longer than 1 sec, and in a second study (Droit-Volet, 2002), a standard duration of 400 msec and another ten times as long (i.e., 4.0 sec). The performance obtained from children in this latter study, with an example of generalization gradients obtained from human adults, is shown in Figure 7.3.

As one can see in Figure 7.3, for human adults, the temporal generalization task produces a gradient that peaks at the standard duration, with the proportion of yes responses declining as the deviation from the standard increases. In the present situation, the same orderly gradients were observed in children at the ages of 3, 5, and 8 years old. At the age of 3 years, the children were able to discriminate different durations in the temporal generalization task, as they did in the temporal bisection task. In addition, Droit-Volet et al. found neither a significant effect of the standard durations used nor a significant interaction involving this variable. This lack of effect with the standard duration is consistent with the scalar property of interval timing, according to which the overall proportion of yes responses was not affected by the standard duration. Supporting these results, in all age groups, the generalization gradients for different time values superimposed reasonably well when plotted on the same relative scale, as shown in Figure 7.3. Therefore, the scalar property of interval timing is obtained in children at different ages, as well as in animals and human adults, and this occurred in different temporal discrimination procedures: not only for the temporal bisection task, but also for the temporal generalization task. In short, the ability to exhibit interval timing is a general behavioral property observed at different levels of the ontogenetic scale.

However, in the temporal generalization procedure, there were also clear differences in performance between the children and the adults. Notably, the generalization gradients were flatter in the 3-year-olds than in the 5-year-olds and in the 5-year-olds than in the 8-year-olds. At this last age, the gradients were nearly identical to

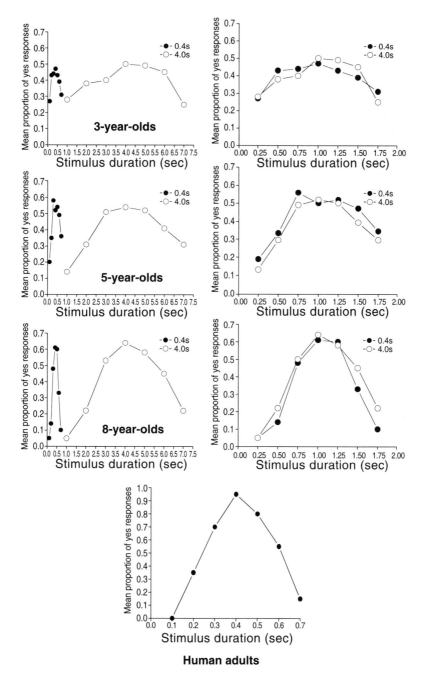

FIGURE 7.3 Temporal gradients obtained from the 3-, 5-, and 8-year-olds in a temporal generalization task in two standard duration conditions (4 msec and 4 sec) (Droit-Volet, 2002), with an example of temporal gradient obtained from the human adults derived from Wearden (1992). (From Droit-Volet, S., *Q. J. Exp. Psychol.*, A, 55A, 1193–1209.)

those found in the adults. Furthermore, there was an age-related change in the shape of the temporal generalization gradient. As shown in Figure 7.3, the human adults usually exhibit a right-skewed gradient, so that stimuli longer that the standard are more likely to be confused with the standard than are stimuli shorter by the same amount. However, in our studies, the adult-like gradient has been obtained only at the age of 8 years old; the youngest children, aged 3 and 5, produced a more or less symmetrical gradient, typical to those found in animals. Thus, there is an age-related shift from animal-like symmetrical gradients to adult-like rightward-skewed gradients in children (Droit-Volet, 2002; Droit-Volet et al., 2001). However, McCormack et al., (1999) did not find this developmental shift in the shape of their temporal generalization gradients. Gradients for their 5- and 8-year-olds were significantly skewed to the left, symmetrical gradients being observed at the age of 10 years, and rightward skewed gradients only at adulthood.

In the framework of a systematic analysis of developmental changes in temporal generalization performance, the question is always how to explain these differences between adults and children. Figure 7.4 shows adult-like gradients (black curves)

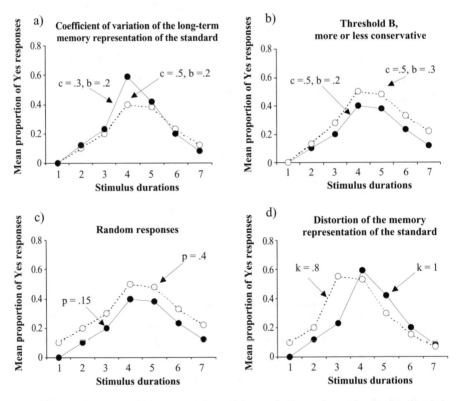

FIGURE 7.4 Illustration of four potential developmental changes in temporal generalization gradients. Results come from simulations using a computer model based on the developmental version of scalar timing theory (for details, see text and Droit-Volet et al., 2001). (From Droit-Volet, S., Clément, A., and Wearden, J., *J. Exp. Child Psychol.*, 80, 271–288, 2001.)

and several theoretical cases of departure (open curves) from these gradients produced by different sources of variation. Of course, these different sources of variation can cumulate, resulting in a particular pattern of timing behavior. Figure 7.4a shows gradients that have the characteristics of those found in adults, but that differ in relative steepness. One gradient is significantly flatter than the other. Flatter gradients indicate lower temporal sensitivity. The first potential source of developmental change is in the temporal sensitivity manifested in the steepness of the gradient. Figure 7.4b also shows adult-like gradients, but one gradient shows an asymmetrical increase in the probability of yes responses for comparison stimulus durations. This is the second developmental possibility: identical shape at different ages, but different threshold levels for responding. Figure 7.4c shows a similar possibility, but now a higher proportion of yes responses occurs also at the shortest comparison stimulus durations. A high proportion of yes responses for the shortest durations is virtually unknown in human adults, who produce a proportion of yes responses near zero for these durations. However, Church and Gibbon (1982) have observed this pattern of responses in rats and have interpreted it as the result of responses emitted at random, and not controlled by signal duration. Therefore, the third possibility is a higher tendency in young children to produce random responses. Figure 7.4d shows the last developmental possibility, with an adult-like rightward asymmetrical gradient and a gradient that is shifted to the left. In this case, stimuli shorter than the standard are confused more with the standard than stimuli longer than it. So, the last developmental possibility is a change in the shape of the generalization gradient.

Our developmental model based on scalar timing theory — a developmental version of the modified Church and Gibbon model (Wearden, 1992) — allows us to take into account these different potential developmental changes in temporal generalization performance. As described above, this model supposes that the amount of pulses accumulated for the just-presented duration is stored in short-term memory. As regards the standard duration, it is stored in long-term memory as a distribution with a coefficient of variation c. At the decision level, the subject evaluates the difference between the just-presented duration and the memory of the standard duration. When this difference is judged small, the subject responds, "Yes, it's the standard." Furthermore, the final judgment is controlled by a decision threshold b, which can be more or less conservative.

Some properties of this model are illustrated in Figure 7.4. The effect of varying the coefficient of variation of the long-term memory representation of the standard duration is shown in Figure 7.4a. Increasing the coefficient of variation makes the memory representation of the standard fuzzy, which decreases the time sensitivity, and flattens the gradient. The effect of varying the threshold b, when the coefficient of variation is constant, is shown in Figure 7.4b. Increasing b makes the decision as to whether to respond yes less conservative, and the overall proportion of yes responses increases. Figure 7.4c shows the effect of varying the proportion of random responses. Increasing p increases the proportion of random responses for each stimulus duration, even the shortest ones. Figure 7.4d shows the temporal generalization gradient shifted leftward relative to the standard, that is, related to a change in the memory representation of the standard, modeled in scalar timing theory as a change in the multiplicative memory translation constant, k* (see Gibbon et al.,

1984; Meck, this volume). Therefore, McCormack et al. (1999) suggested that the leftward shifted gradient observed in their young children is due to a distortion of the representation of the standard. Thus, they added a distortion parameter k. If k was equal to 1.0, the standard value was remembered veridically; if k was less than 1.0, the standard was remembered as shorter than it really was; and if k was more than 1.0, it was remembered as longer than it really was.

In sum, our model proposes four parameters to take into account the age-related differences in the temporal generalization gradients obtained from children: (1) the variability of the long-term memory representation of the standard duration, (2) the decision threshold more or less conservative, (3) the proportion of random responses, and (4) the memory distortion of the standard duration. This developmental model fits our data very well and suggests that three of these four parameters were involved in the developmental changes observed in temporal generalization gradients (for more details, see Droit-Volet, 2002). The first is the coefficient of variation for the memory representation of the standard, which was higher in the 3- and 5-year-olds than in the 8-year-olds. Therefore, the increase with age in the steepness of the generalization gradient, namely, the increase in temporal sensitivity, was related to the decrease in the variability of the memory representation of the standard. The second relevant parameter is the proportion of random responses; the younger children emitted a greater number of random responses than did the older children. The third contributing parameter is the threshold value, b. Our model suggested that there was a small, but significant increase of the threshold value with increasing age. Thus, in their temporal judgments, the 8-year-olds were less conservative than the 3- or 5-year-olds. The real role of this decision threshold is still unclear (see Wearden, 1999). However, some studies have shown that subjects who are more confident in their knowledge are also less conservative, i.e., they take more risks and produce more errors when facing ambiguous cases (Nelson, 1996). We can therefore suppose that the 8-year-olds are less conservative because they were more confident in their temporal knowledge. We believe this for three main reasons. First, the 8-year-olds' memory representation of the standard was more precise than that of the younger children. Second, they have a better mastery of the concept of time (for reviews, see Droit-Volet, 2000a, 2000b; Droit and Pouthas, 1992; Pouthas et al., 1993; Friedman, 1990a). Consequently, the temporal generalization task may appear for them relatively easy. Third, they obtained more positive feedback than the younger children, who made more errors due to a higher level of random responding and a more variable memory representation.

In contrast, the last parameter, the distortion value applied to the memory representation of the standard, did not vary with age; it was stable throughout the three age groups at a value of 1.0. Therefore, in our studies of children as young as 3 years of age, no multiplicative distortions were observed in their memory representation of the standard (Droit-Volet, 2002). The problem now is to explain the developmental shift from a symmetrical to a rightward asymmetrical response gradient. A view consistent with scalar timing theory would postulate that the main difference in temporal generalization gradients between animals and human adults lies in the type of decision rule employed. The discrepancy between the just-presented duration and the standard memory sample is normalized by the standard

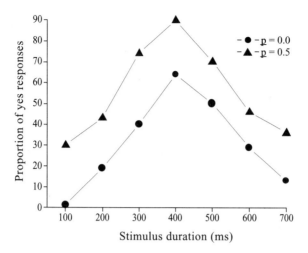

FIGURE 7.5 Illustration of the effect on temporal generalization gradients of an increase in the proportion of random responses (parameter p). Data derived from the computer simulation of individual subjects based on the developmental version of scalar timing theory. (From Droit-Volet, S., *Q. J. Exp. Psychol.*, A, 55A, 1193–1209.)

sample in animals, that is, $(s^* - t)/s^*$, and by the duration value in human adults, that is, $(s^* - t)/t$ (i.e., Church and Gibbon (1982) model vs. Wearden (1992) model). However, the reason why animals and human adults would use different decision rules is not entirely clear (Allan, 1998). Moreover, in the course of ontogenesis, there is no particular reason to suppose such changes in the decision rules. In contrast, there is a decrease in the number of random responses with increasing age in children. As suggested earlier and illustrated in Figure 7.4c, the increase in random responses increases the proportion of yes responses for each stimulus duration, but the proportion of yes responses for the shortest duration increases relatively more. In this case, the greater proportion of yes responses at the shortest durations can balance the greater proportion of yes responses at the longest durations, usually obtained in the adult-like right-skewed gradient. Consequently, the temporal generalization gradient that is rightward asymmetrical can become symmetrical when the number of random responses increases. This is precisely what we found when we simulated individual data from subjects by changing the proportion of random responses while holding the other parameters constant (Figure 7.5). When the proportion of random responses is fixed at zero percent, the gradient is rightward asymmetrical, and when the proportion of random responses is higher, the gradient significantly shifts toward symmetry. Therefore, the decrease in the proportion of random responses can, in part, explain this age-related shift from symmetrical to adult-like asymmetrical gradients.

In summary, our data on temporal generalization are consistent with those found in temporal bisection. There is an increase in the sensitivity to signal duration with increasing age, due to the decrease both in the variability of remembered time and in the probability of random responses. However, in temporal generalization, the decision threshold also changed, with the older children becoming more confident in their temporal knowledge and less conservative. In the temporal bisection procedure, this effect of age on the decision threshold was not observed. This is probably linked to the temporal bisection task, which is relatively less difficult than the temporal generalization task (McCormack et al., 1999; Wearden et al., 1997). Supporting this idea, other studies have shown that interval timing performance was worse in elderly adults than in young adults in a temporal generalization task, but not in a temporal bisection task (Wearden et al., 1997; but see Lustig and Meck, 2001).

7.3 IS AN INTERNAL CLOCK FUNCTIONAL AT AN EARLY AGE?

Beyond age differences in temporal performance, the bisection and generalization methods used in animals and human adults produced orderly data from all children. Furthermore, children's timing behavior conformed well to the scalar property of interval timing. On the whole, these findings suggest that the clock-based system underlying time perception in animals and human adults is functional at an early age (see Brannon and Roitman, this volume).

The success of the scalar timing theory comes precisely from the revival of the idea that animals and human adults might possess a sort of internal clock, and from studies derived from this theory demonstrating its existence. The clearest evidence comes from animal studies that have succeeded in speeding up or slowing down the pacemaker of the internal clock by administering drugs (Maricq et al., 1981; Meck, 1983). However, Treisman et al. (1990, 1994) have recently invented a new method that is able to alter the speed of the internal clock and is easier to use in human subjects. This method consists of accompanying or preceding an event to be timed by a short period of repetitive sensory stimulation that clicks or flashes. Using this new method, several studies have shown that adults judge stimuli to be longer if they are preceded by repetitive auditory clicks (e.g., Burle and Casini, 2001; Penton-Voak et al., 1996). In order to solve the issue of the functionality of an early internal clock, we conducted a series of experiments in children aged 3 to 8 years, based on the repetitive stimulation procedure first used by Treisman (Droit-Volet and Wearden, 2002).

In this experiment, the children were given a temporal bisection task in two different duration conditions. In the first condition, the short standard was 200 msec and the long standard 800 msec; in the second condition, the anchor durations were 400 and 1600 msec. In the training phase, the children were presented with a white circle for 5 sec, followed immediately by a blue circle, which was the stimulus to be timed (i.e., standard duration). In the testing phase, they received comparison stimulus durations either with the flicking of the white circle (flicker condition) or without flicking of the white circle (no flicker condition). As illustrated in Figure 7.6, the results showed that the flicker shifted the psychophysical functions toward the left, whatever the age group or the duration condition tested. Consequently, the bisection points were lower with flicker than without flicker, and to the same extent in the three age groups. As of the age of 3 years, the stimulus durations were thus judged longer with than without repetitive stimulation produced by the flicker.

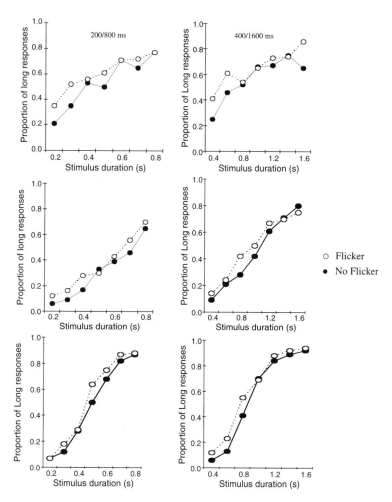

FIGURE 7.6 Psychophysical functions obtained with repetitive stimulation (flicker condition) and without repetitive stimulation (no flicker condition) in a temporal bisection task in two anchor duration conditions (200/800 and 400/1600 msec) (Droit-Volet and Wearden, 2002). Top two panels: Data from the 3-year-olds. Center two panels: Data from the 5-year-olds. Lower two panels: Data from the 8-year-olds. (From Droit-Volet, S. and Wearden, J., *Q. J. Exp. Psychol.*, B, 55B, 193–211.)

According to most internal clock models, the overestimation of times obtained in our flicker condition can be explained either by an increase in clock speed or by a decrease in the latency to start the clock. Indeed, in each case, more pulses would be accumulated in the counter and the subjective time would be longer with the repetitive stimulation than without. However, the mathematics of internal clock models predict distinct effects for the clock speed and latency interpretations. For the clock speed explanation, the models predict that the effect should be proportional to the duration being timed, i.e., the overestimation of duration in absolute numbers of seconds would be greater at longer times than at shorter times. For the latency-to-start-the-clock explanation, the models predict an additive effect that is independent of the duration judged; in other words, a constant number of seconds increase in the duration judgment whatever the duration values used. As indicated before, the clearest demonstration that some manipulations can affect the speed of an internal clock comes from pharmacological studies in which stimulant drugs (e.g., methamphetamine) are administrated to rats performing in interval timing tasks (e.g., Maricq et al., 1981; Meck, 1983, 1996). In particular, Maricq et al. (1981) reported data from a temporal bisection procedure in which three different short–long standard duration conditions were used: 1/4, 2/8, and 4/16 sec. In all cases, the administration of methamphetamine shifted psychophysical functions to the left and lowered the bisection points. Furthermore, these shifts were proportional to the intervals being timed with the dose of the drug held constant. This proportional horizontal shift is precisely that required by changes in pacemaker speed, and that we obtained in our experiment with children using Treisman's repetitive stimulation technique. Indeed, in our study, the magnitude of the leftward shift of the bisection points was greater in the 400- to 1600-msec than in the 200- to 800-msec duration condition, and this was observed for all age groups. Thus, the flicker increased clock speed in children in a temporal bisection task just as a stimulant drug (methamphetamine) did when administered to rats (Maricq et al., 1981). Our study was then a clear demonstration that the pacemaker–accumulator system is functional at an early age. However, as we will see, another demonstration of changes in clock speed is provided by tests in children of the effect of sensory modality on time perception.

7.4 THE EFFECT OF SENSORY MODALITY IN CHILDREN

There is ample evidence that auditory stimuli are judged longer than equivalent visual stimuli, and visual stimuli shorter than auditory ones. For example, by varying contextual factors (e.g., intensity, color of lights, level of practice) as well as the timing procedures (e.g., production, reproduction, verbal estimation, pair comparison), Goldstone and Lhamon (1972, 1974) observed that subjects systematically judged the auditory stimuli longer than the visual ones, when they shared a common duration. The robustness of this phenomenon led Goldstone and Lhamon (1972, p. 626) to conclude that the auditory–visual difference is a "fundamental property of human temporal processing" (see Penney, this volume).

Recently, in the context of scalar timing theory, Penney et al. (1998, 2000) have succeeded in explaining the effect of sensory modality in time judgments by using a variant of the temporal bisection procedure in which auditory and visual stimuli of the same physical durations were presented during the same sessions. The result was that psychophysical functions produced by human adults were shifted toward the left for auditory stimulus durations, compared to visual ones, showing that the former are judged longer than the latter. Furthermore, this leftward shift of the temporal bisection functions was proportional to the stimulus duration value to be estimated. That is to say, the auditory–visual difference in the bisection functions increased as the duration values increased. This proportional effect with the auditory–visual duration values supports the assumption that differences in clock speed were the main cause of the modality effect on temporal bisection judgments. Notably, the internal clock runs faster for auditory than for visual stimuli. Thus, for the same objective duration, more pulses are accumulated for auditory signals than for visual signals, and the subjective time seems longer. Using other interval timing tasks (e.g., temporal generalization, verbal time estimation), Wearden et al. (1998) obtained similar results, consistent with this clock speed interpretation.

Within a developmental perspective, we might assume that if the pacemaker–accumulator system is functional at an early age, the modality difference observed in human adults in temporal bisection would also be found in children, whatever the age groups tested. Therefore, the following experiment was designed to test the modality effect on temporal bisection performance in children aged 5 and 8 years, as well as in adults (Droit-Volet et al., submitted). The subjects were given two sessions of temporal bisection, one per day. Each session was composed of pretraining, training, and testing. In the first session, the subjects were presented the stimulus in the same sensory modality for both the pretraining, training, and testing phases, either visual or auditory (visu/visu and audi/audi). In the second session, they were presented stimuli in the same modality as in session 1 for the pretraining and training phases, but in another modality for the testing phase (visu/audi and audi/visu). Therefore, session 1 required a within-modal comparison and session 2 a cross-modal comparison. In session 2, the subjects were thus required to make temporal judgments relating an auditory comparison stimulus to a visual standard for one group (visu/visu-visu/audi group), and a visual comparison stimulus to an auditory standard for the other group (audi/audi-audi/visu group).

The temporal bisection functions obtained for the 5-year-olds, 8-year-olds, and adults for both session 1 (visu/visu or audi/audi) and session 2 (visu/audi or audi/visu) are shown in Figure 7.7. Consistent with our previous results, the psychophysical functions were orderly in all age groups, although flatter in the 5-year-olds than in the 8-year-olds and the adults. In addition, in each age group, the temporal bisection functions shifted to the left for the auditory stimuli, compared to the visual stimuli; hence, bisection points were lower. Thus, the subjects judged the auditory stimuli as being longer than the visual stimuli. This was observed for each transfer modality order, both from auditory toward visual (audi/audi–audi/visu) and from visual toward auditory (visu/visu–visu/audi), despite a less marked auditory–visual difference in the last condition for the two oldest age groups (for more details, see Droit-Volet et al., submitted). Both the statistical analyses and the modeling of our data attributed this modality effect to differences in the speed of the internal clock, with the clock running faster for the auditory than for the visual stimuli. The developmental version of the scalar timing model presented above, and adapted for this experiment, found similar differences in the clock rate for the adults and the children aged 8 years. In these two age groups, the visual clock rate was on average 10.5% slower than the auditory one. By contrast, the magnitude of this difference in clock speed was relatively larger in the 5-year-olds than in the 8-year-

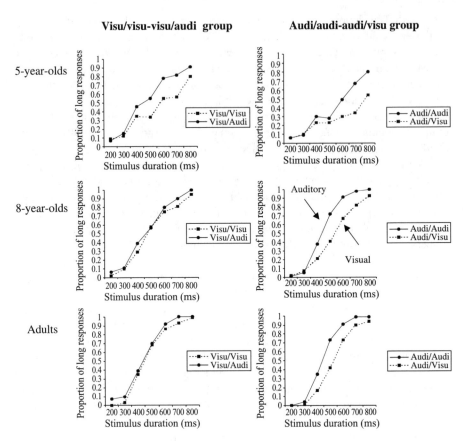

FIGURE 7.7 Psychophysical functions obtained from the 5- and 8-year-olds and the adults in a bisection task in the visu/visu–visu/audi group and the audi/audi–audi/visu group for the within-modal comparison sessions (visu/visu, audi/audi) and the cross-modal comparison sessions (visu/audi, audi/visu). (From Droit-Volet, S., Tourret, S., and Wearden, J., submitted.)

olds and the adults. It is therefore reasonable to suppose that clock speed runs even slower for the visual than for the auditory modality in the youngest children.

In the present study, other interesting developmental trends were found. The 5-year-olds judged the visual stimuli as shorter than the auditory stimuli, but their temporal judgments were also more variable for the visual stimuli, contrary to the 8-year-olds and the adults. Indeed, in the 5-year-olds, the psychophysical functions were flatter and the Weber ratio greater for the visual than for the auditory stimuli. According to scalar timing theory, clock speed could be a cause of this variability difference in modalities for the youngest children. Indeed, this theory suggests that the temporal judgments depend on a count of pulses generated by a Poisson source that emits pulses at random, but at some constant mean rate (Gibbon, 1977, p. 284). On the basis of the mathematics of a Poisson pacemaker, Gibbon (1977) explained how a slower pacemaker rate produces more variable temporal estimates. Consequently, whether the clock speed was slower for the visual stimuli than for the auditory stimuli in the 5-year-olds, their temporal discrimination was more variable and their bisection functions flatter, with a larger Weber ratio.

However, it has been argued that differences in pacemaker speed are not the main source of variability in a timing system (Gibbon et al., 1984). In agreement with this, Penney et al. (1998, 2000) assumed in their explanation of the modality effect a low level of variability in temporal integration. In fact, according to scalar timing theory, the main source of variance is in the memory-encoding process, which we are going to discuss below. In other interval timing procedures (e.g., temporal generalization and verbal estimation), Wearden et al. (1998) observed that adult's duration judgments, like those of the youngest children in temporal bisection, were also more variable with visual than with auditory stimuli. These researchers suggested that variation in clock speed is not the only difference affecting the processing of temporal information when subjects compare auditory and visual stimuli. Notably, the modality effect on the variability of duration judgments can be caused by greater variance of the switch latency for the visual than for the auditory stimuli. In order to dissociate these two effects (clock speed and switch latency) on the variability of duration judgments, Wearden et al. (1998) conducted an experiment in which they combined both the sensory modality of the presented stimulus durations (visual or auditory) and the presence or absence of a 5-sec train of clicks, presumed to increase the clock speed. In this experimental condition, they observed that the presence or absence of clicks changed the length of duration estimates, as did the modality of the stimulus, but only the modality effect produced significant differences in the coefficient of variation of duration judgments. Thus, Wearden et al. (1998) concluded that the greater variance observed in the latency to open and close the switch, used to start and stop the clock when timing signals, explained the increased variability in duration judgments for visual signals compared to auditory signals.

In our temporal bisection task, only the youngest children were more variable in their duration judgments for the visual than for the auditory signals. We can therefore assume that the processing of the duration of visual stimuli requires relatively more attentional effort from younger children. Indeed, the visual stimuli require a continuous focus of attention on the physical source of presentation of visual information (i.e., computer screen) in order to begin the processing of duration at the onset of the stimulus. In contrast, the processing of the duration of auditory stimuli requires less attentional effort, these stimuli being more directly perceived (e.g., Meck, 1984). As the younger children have limited attentional capacity and are easily distracted (for a review, see Dempster and Brainerd, 1995), the variability of the latency to close the switch would be greater for the visual than for the auditory stimuli. The modeling of our data provided support for this idea by showing that the memory representation of the standard durations in the youngest children was relatively less precise for the visual than for the auditory stimuli.

The greater sensitivity to duration for auditory than for visual stimuli in young children suggests a sort of primacy of audition over vision in the processing of temporal information. This is an old idea, already put forward by studies in infants on the perception of temporal characteristics of speech sound and rhythms (for a review, see Pouthas et al., 1993; Droit-Volet, 2000b). For example, Eimas et al. (1971) have shown that infants can distinguish elementary speech segments (i.e.,

Pa, Ta, Ba, Da) only on the basis of their temporal acoustic characteristics. Lewkowicz (1989, 1992) has also shown that infants are able to detect changes in frequency of rhythm, but that this ability emerges in the auditory modality before it emerges in the visual modality. He stated that audition is specialized in the processing of temporal information, whereas vision is specialized to detect motion. Friedman (1990a, p. 87) added that the "early sensitivity to the temporal characteristic of sound may be a special biological adaptation that allows infants to process information about speech."

Although there is a form of primacy of audition over vision in the processing of temporal information, the present study, as the previous one, demonstrates that the clock mechanism underlying time perception is functional at an early age. Indeed, as in human adults, the internal clock of children appears to run faster for auditory than for visual signals. However, this does not preclude the possibility of developmental changes at this level, such as an increase in pacemaker speed. Nevertheless, the results of our studies using the temporal bisection and generalization tasks suggest that the main sources of developmental changes in timing behavior are probably elsewhere than in the basic functioning of the internal clock.

7.5 SOURCES OF DEVELOPMENTAL CHANGES IN INTERVAL TIMING

7.5.1 LONG-TERM MEMORY SOURCES OF DEVELOPMENTAL CHANGES IN TIMING

Our developmental models of interval timing suggest that the increase with age in the sensitivity to time in the temporal bisection and generalization tasks is due to a decrease in the variability of the long-term memory representations of the standard durations. In short, young children have a "fuzzier" memory of durations. The issue is: Why do they have a fuzzier memory representation of time? A first working hypothesis is to consider that the standard durations have been correctly encoded and stored in long-term memory, in the form of a distribution with means equal to the standard values and some given coefficient of variation. In this case, the higher variability of time representation in memory obtained from young children could be attributed to a greater degradation over time, in other words, to a larger degree of memory decay for durations. This decay would erase memory traces of the standard durations and would increase the coefficient of variation of their representation, without alteration of their mean representation. In sum, with this sort of decay or forgetting, the long-term memory of standard durations would get fuzzier, which would flatten the psychophysical functions.

In order to test whether the degradation over time of the long-term memory representation of the standard durations is greater in younger children, we submitted children aged 5 and 8 years old to a temporal bisection task with a 2-sec short standard and an 8-sec long standard. In addition, the children were given an interfering task lasting for 15 min, either between the training and the testing phase, for one experiment, or between two testing phases, for a second experiment (Rattat, in preparation). The results from Rattat and Droit-Volet (2001) showing the temporal

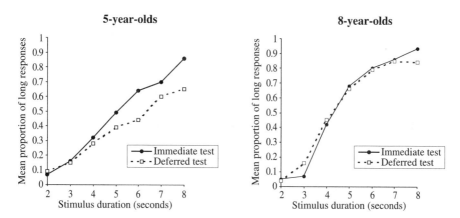

FIGURE 7.8 Psychophysical functions for the 5- and 8-year-olds in a bisection task before an interfering task (immediate test) and after (deferred test). (From Rattat, A.-C. and Droit-Volet, S., *Behav. Process.*, 55, 81–91, 2001.)

bisection functions for the 5- and 8-year-olds in the testing phase before and after the interfering task (immediate vs. deferred test) are shown in Figure 7.8. In these two age groups, the psychophysical functions were flatter and the Weber ratio higher after the interfering task than before, but more particularly in the 5-year-olds. Our developmental version of the scalar timing model applied to these bisection data attributed the lower temporal sensitivity with the interfering task to an increase in the coefficient of variation for the remembered time. Furthermore, the magnitude of this increase was greater for the 5-year-olds (before, 0.65; after, 0.95) than for the 8-year-olds (before, 0.30; after, 0.50). Thus, the forgetting of the standard durations was more significant in the younger children, which produced a greater variability in the memory representation of durations. In the field of psychology, the development of long-term memory in children is still not well understood (for reviews, see Cowan, 1997; Gathercole, 1998). However, some studies suggest that the memory of events in children is influenced by the repeated experience of events and how well the mental representation of these events is organized. Indeed, a well-organized representation of events in memory allows a better recall of event memories and their preservation over a longer period of time. In this case, the variability of the memory representation of the standard durations at the outset of the temporal bisection testing phase can explain, in part, the greater effect on the bisection performance of the youngest children.

In any event, with regard to familiar events or actions for which the temporal characteristics have been reactivated sufficiently in memory to strengthen their memory traces, the observed forgetting is greatly reduced. Friedman (1990b) showed that children aged 3.6 years old are able to correctly recall, on a verbal estimation scale, the relative duration of familiar activities, such as drinking a glass of milk and viewing a TV cartoon. In a recent study, we showed that 5-year-olds are able to keep in memory the duration of a learned action for up to 6 weeks. More precisely, the 3- and 5-year-old children were trained to produce a given action duration of 5 sec (button press) with the procedural learning methods used by Droit-Volet (1998) and

Droit-Volet and Rattat (1999). In this method, the children were forced to repetitively produce a 5-sec action duration by simultaneously imitating the experimenter's action. Each correct response duration (between 4 and 6 sec) resulted in positive feedback (smiling clown), and incorrect response duration (shorter than 4 sec or longer than 6 sec) in negative feedback (frowning clown). Then, after an immediate test of the learning of the required response duration, the 3- and 5-year-olds were randomly assigned to a retention interval condition given 15 min, 1 h, 24 h, or 48 h later. The results showed that the proportion of correct responses decreased for the 3-year-olds as soon as the 15-min retention interval. However, the 5-year-olds maintained their level of performance until 48 h. Increasing the duration of the retention interval up to 6 days or even 6 weeks did not alter their performance (Rattat, in preparation). Therefore, with a repeated experience (i.e., procedural learning) the memory for duration is not or slightly affected by forgetting. Finally, the most critical process in children's abilities to time events is probably the encoding of duration.

7.5.2 Attentional Sources of Developmental Changes in Timing

Difficulties in the encoding of duration can also account for the high variability in the memory representation of the standard durations in young children. Indeed, the memory representation is the result of how it has been encoded. Among the cognitive processes involved in the encoding of duration, we have specifically investigated the role of attention. Indeed, the inefficiency of attentional processes contributes to increases in the trial-by-trial variance and, as a consequence, makes the memory of event durations fuzzier. Furthermore, it is well known that the prefrontal cortex plays a central role in attention and that its maturation is not complete until the end of adolescence. More precisely, its maturation is relatively fast until 2 years, but it continues after, with an important evolution between 4 and 7 years, followed by a more progressive evolution up to adulthood (Luria, 1961, 1973; see Pang and McAuley, this volume). Thus, the maturation of the prefrontal cortex influences the efficiency of different cognitive operations relative to attention, such as the mobilization of attentional resources in a dual task, the resistance to distraction, or the selective orientation of attention for the simultaneous processing of information.

Within a developmental framework for the mobilization of attentional resources, there is a consensus according to which the amount of attentional resources available to process a given task increases with age. Indeed, most psychologists believe that the total capacity of the pool of attentional resources remains constant throughout development, or at least after the age of 2 years. However, thanks to cognitive development, the processing demands of any particular task decrease, thus releasing attentional resources. For example, the more the processing of information becomes automatic, the less it consumes attentional resources and the more these resources are available for the processing of other information. However, psychologists do not agree on the causes of this releasing of resources through cognitive development. For Case (1985), this release of resources is caused by a better chunking of information related to the development of knowledge; for Pascual-Leone and Baillargeon (1994), it is caused by the increase in automatic cognitive operations as a result of practice; and for Bjorklund and Harnishfeger (1995), it comes about from an improvement in inhibitory efficiency. Recently, others have suggested that an increase with age of the speed of general cognitive processing is involved: the faster the processing of information, the more resources are available for the processing of other information. In any event, all these psychologists agree that the amount of attentional resources increases with age during early development.

According to attentional models of time perception in human adults (e.g., Thomas and Weaver, 1975; Zakay, 1989), subjective time is directly related to the amount of attentional resources allocated to the processing of temporal information: the fewer the attentional resources that are allocated to time, the shorter the time estimate. This shortening effect is interpreted as the consequence of a loss of pulses by the flickering of the switch connecting the pacemaker to the accumulator (for reviews, see Buhusi, this volume; Fortin; this volume). The predictions of the attentional models have been extensively validated in human adults by studies using the dual-task paradigm (see Fortin, this volume). In contrast, few studies have investigated children's timing performance with this paradigm (Arlin, 1986a, 1986b; Gautier and Droit-Volet, 2002b). If the amount of attentional resources is less in younger children, then we might expect a greater interference effect of a concurrent nontemporal task on duration estimates, in terms of underestimation. This is exactly what was found by Gautier and Droit-Volet (2002b). In their study, the 5- and 8-year-olds were required to reproduce a given duration (6 or 12 sec) in both a single- and dual-task condition with a concurrent nontemporal task. The concurrent nontemporal task was to name the pictures presented in the center of the stimulus to be timed. These pictures were easy to name (low attentional demand) or difficult to name (high attentional demand). The results showed that the 5- and 8-year-olds, as well as adults, reproduced shorter durations in the dual task than in the single task (Figure 7.9). This provides support to the predictions of the attentional models and extends them to children as young as 5 years old, and even as young as 3 years, as reported by other studies (Gautier, 2002). This shortening effect was also greater in the 5-year-olds than in the 8-year-olds. Moreover, in the 5-year-olds, temporal reproductions were significantly shorter in both dual tasks (low and high attentional demands) than in the single task, whereas in the 8-year-olds, differences reached significance only between the higher attentional demand dual task and the single task. These findings indicated a greater interference effect on duration judgments in the 5-year-olds than in the 8-year-olds. The amount of available attentional resources is therefore a potential contributor to the developmental changes we have observed in the timing behavior of children.

Nevertheless, as we can easily imagine on the basis of our results, the overall explanation of the age-related changes in the development of interval timing is particularly complex. There is not one, but several sources of developmental changes. Probably related to the limited pool of attentional resources in young children, we found that children also have difficulties in resisting distraction or are susceptible to interference from nontemporal information. Using an attentional distracter in a temporal bisection task, we showed that a distracter disrupted the bisection performance more in the 5-year-olds than in the 8-year-olds (Gautier and Droit-Volet, 2002a). In the same vein, when the subjects were instructed to process the duration of a sequence

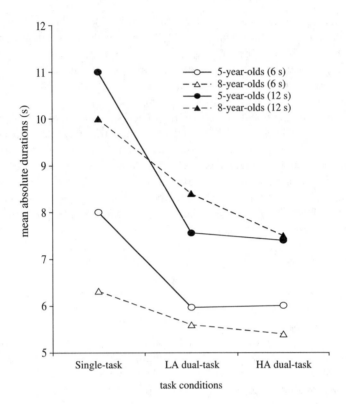

FIGURE 7.9 Mean durations (in seconds) reproduced by the 5- and 8-year-old children for the 6- and 12-sec stimulus durations in the single task and the low (LA) and high (HA) attentional dual tasks. (From Gautier, T. and Droit-Volet, S., *Behav. Process.*, 58, 57–66, 2002b.)

of stimuli and to ignore the varying number of stimuli in this sequence, the number of stimuli interfered more with temporal bisection performance in the 5-year-olds than in the 8-year-olds and the adults (Droit-Volet, Clément, and Fayol, in press). In contrast, the duration dimension did not interfere with the numerical bisection performance when the subjects were instructed to process the number of stimuli in the sequence while ignoring the varying duration. The younger children therefore had more difficulty in inhibiting the processing of nontemporal information (i.e., number) that altered their duration judgments and masked their fundamental competence to time events. On the other hand, it was relatively easy for them to selectively ignore time, for which the processing was more attentionally demanding (for a discussion of counting and timing in infants, see Brannon and Roitman, this volume).

7.6 CONCLUSION

In conclusion, the results of our different studies on interval timing in children suggest that, whatever their age, children are able to estimate duration. In accordance with the scalar timing theory, this fundamental temporal competence is possible because they possess, as animals and human adults do, a functional internal clock. Clear evidence for the early functionality of this clock comes from our repetitive stimulation studies designed to speed up the clock. However, despite an early temporal competence, the sensitivity for duration increases with age in the temporal bisection and temporal generalization tasks. Our developmental version of scalar timing theory (Gibbon et al., 1984) suggests that this is due to a greater number of responses emitted at random by younger children and to a greater variability in their memory representation of durations. Some studies have suggested that this greater variability in the memory representation of duration in young children can be explained both by a more rapid forgetting of durations and by a poorer encoding of durations related to their limited attentional capacity. Future research will be aimed at understanding both the psychological and neurobiological processes involved in enhancing the sensitivity to time as a function of early development.

ACKNOWLEDGMENTS

The author extends her gratitude to Warren Meck for his invitation to publish a chapter in this book and for his correction of the English language used in the manuscript.

REFERENCES

Allan, L.G., The influence of the scalar timing model on human timing research, *Behav. Process.*, 44, 101–117, 1998.

Allan, L.G. and Gibbon, J., Human bisection at the geometric mean, *Learn. Motiv.*, 22, 39–58, 1991.

Arlin, M., The effect of quantity, complexity, and attentional demand on children's time perception, *Percept. Psychophys.*, 40, 177–182, 1986a.

Arlin, M., The effects of quantity and depth of processing on children's time perception, *J. Exp. Child Psychol.*, 42, 84–98, 1986b.

Bjorklund, D. and Harnishfeger, K., The evolution of inhibition mechanisms and their role in human cognition and behavior, in *Interference and Inhibition in Cognition*, Dempster, F. and Brainerd, C., Eds., Academic Press, New York, 1995.

Burle, B. and Casini, L., Dissociation between activation and attention effects in time estimation: implication for internal clock models, *J. Exp. Psychol. Hum. Percept. Perform.*, 27, 195–205, 2001.

Case, R., *Intellectual Development: Birth to Adulthood*, Academic Press, New York, 1985.

Church, R.M. and Deluty, M.Z., Bisection of temporal intervals, *J. Exp. Psychol. Anim. Behav. Process.*, 3, 216–228, 1977.

Church, R.M. and Gibbon, J., Temporal generalization, *J. Exp. Psychol. Anim. Behav. Process.*, 8, 165–186, 1982.

Cowan, N., *Development of Memory in Childhood*, Psychology Press, Hove, U.K., 1997.

Dempster, F.N. and Brainerd, C.J., *Interference and Inhibition in Cognition*, Academic Press, New York, 1995.

Droit, S. and Pouthas, V., Changes in temporal regulation of behavior in young children: from action to representation, in *Time, Action and Cognition: Towards Bridging the Gap*, Macar, F., Pouthas, V., and Friedman, W.J., Eds., Kluwer Academic Publishers, Dordrecht, Netherlands, 1992, pp. 46–53.

Droit-Volet, S., Time estimation in young children: an initial force rule governing time production, *J. Exp. Child Psychol.*, 68, 236–249, 1998.

Droit-Volet, S., L'estimation du temps: perspective développementale, *L'Année Psychol.*, 100, 443–464, 2000a.

Droit-Volet, S., L'enfant et les différentes facettes du temps, *Enfances Psy.*, 13, 26–40, 2000b.

Droit-Volet, S., Scalar timing in temporal generalization in children with short and long stimulus durations, *Q. J. Exp. Psychol.*, 55A, 1193–1209, 2002.

Droit-Volet, S., Clément, A., and Fayol, M., Time and number discrimination in a bisection task with a sequence of stimuli: a developmental approach, *J. Exp. Child Psychol.*, in press.

Droit-Volet, S., Clément, A., and Wearden, J., Temporal generalization in 3- to 8-year-old children, *J. Exp. Child Psychol.*, 80, 271–288, 2001.

Droit-Volet, S. and Rattat, A.-C., Are time and action dissociated in young children's time estimation? *Cognit. Dev.*, 14, 573–595, 1999.

Droit-Volet, S., Tourret, S., and Wearden, J., Perception of the duration of auditory and visual stimuli in children and adults, submitted.

Droit-Volet, S. and Wearden, J., Temporal bisection in children, *J. Exp. Child Psychol.*, 80, 142–159, 2001.

Droit-Volet, S. and Wearden, J., Speeding up an internal clock in children? Effects of visual flicker on subjective duration, *Q. J. Exp. Psychol.*, 55B, 193–211, 2002.

Eimas, P., Siqueland, E., Jusczyk, P., and Vigorito, J., Speech perception in infants, *Science*, 171, 303–306, 1971.

Friedman, W., *About Time: Inventing the Fourth Dimension*, MIT Press, Cambridge, MA, 1990a.

Friedman, W., Children's representations to the daily activities, *Child Dev.*, 61, 1399–1412, 1990b.

Gathercole, S.E., The development of memory, *J. Child Psychol. Psychiatry*, 39, 3–27, 1998.

Gautier, T., Estimation du temps et attention chez le jeune enfant, Doctoral thesis, University Blaise Pascal, France, 2002.

Gautier, T. and Droit-Volet, S., Attention and young children's time perception in temporal bisection task, *Int. J. Psychol.*, 37, 27–35, 2002a.

Gautier, T. and Droit-Volet, S., Attention and time estimation in 5- and 8-year-old children: a dual-task procedure, *Behav. Process.*, 58, 57–66, 2002b.

Gibbon, J., Scalar expectancy theory and Weber's law in animal timing, *Psychol. Rev.*, 84, 279–325, 1977.

Gibbon, J., Church, R.M., and Meck, W.H., Scalar timing in memory, in *Annals of the New York Academy of Sciences: Timing and Time Perception*, Vol. 423, Gibbon, J. and Allan, L., Eds., New York Academy of Sciences, New York, 1984, pp. 52–77.

Goldstone, S. and Lhamon, W.T., Auditory-visual differences in human temporal judgment, *Percept. Motor Skills*, 34, 623–633, 1972.

Goldstone, S. and Lhamon, W.T., Studies of auditory-visual differences in human timing judgment: I. Sounds are judged longer than lights, *Percept. Motor Skills*, 39, 63–82, 1974.

Lewkowicz, D.J., The development of temporally-based intersensory perception in human infants, in *Time, Action and Cognition: Towards Bridging the Gap*, Macar, F., Pouthas, V., and Friedman W.J., Eds., Kluwer Academic Publishers, Dordrecht, Netherlands, 1992, pp. 33–44.

Lewkowicz, D.J., The role of temporal factors in infant behavior and development, in *Time and Human Cognition: A Life Span Perspective*, Elsevier, Amsterdam, 1989, pp. 9–58.

Luria, A., *The Role of Speech in the Regulation of Normal and Abnormal Behavior*, Pergamon, New York, 1961.

Luria, A., *The Working Brain: An Introduction to Neuropsychology*, Basic Books, New York, 1973.

Lustig, C. and Meck, W.H., Paying attention to time as one gets older, *Psychol. Sci.*, 12, 478–484, 2001.

Maricq, A.V., Roberts, S., and Church, R.M., Metamphetamine and time estimation, *J. Exp. Psychol. Anim. Behav. Process.*, 7, 18–30, 1981.

McCormack, T., Brown, G.D.A., Maylor, E.A., Darby, R.J., and Green, D., Developmental changes in time estimation: comparing childhood and old age, *Dev. Psychol.*, 35, 1143–1155, 1999.

Meck, W.H., Selective adjustment of the speed of the clock and memory processes, *J. Exp. Psychol. Anim. Behav. Process.*, 9, 171–201, 1983.

Meck, W.H., Attentional bias between modalities: effect on the internal clock, memory, and decision stages used in animal time discrimination, in *Annals of the New York Academy of Sciences: Timing and Time Perception*, Vol. 423, Gibbon, J. and Allan, L., Eds., New York Academy of Sciences, New York, 1984, pp. 528–541.

Meck, W.H., Neuropharmacology of timing and time percpetion, *Cognit. Brain Res.*, 3, 227–242, 1996.

Nelson, T.O., Consciousness and metacognition, *Am. Psychol.*, 51, 102–116, 1996.

Pascual-Leone, J. and Baillargeon, R., Developmental measurement of mental attention, *Int. J. Behav. Dev.*, 17, 161–200, 1994.

Penney, T.B., Allan, L.G., Meck, W.H., and Gibbon, J., Memory mixing in duration bisection, in *Timing of Behavior: Neural, Computational, and Psychophysical Perspectives*, Rosenbaum, D.A. and Collyer, C.E., Eds., MIT Press, Cambridge, MA, 1998, pp. 165–193.

Penney, T.B., Gibbon, J., and Meck, W.H., Differential effects of auditory and visual signals on clock speed and temporal memory, *J. Exp. Psychol. Hum. Percept. Perform.*, 26, 1770–1787, 2000.

Penton-Voak, I.S., Edwards, H., Percival, A., and Wearden, J.H., Speeding up an internal clock in humans? Effects of click trains on subjective duration, *J. Exp. Psychol. Anim. Behav. Process.*, 22, 307–320, 1996.

Piaget, J., *Le développement de la notion de temps chez l'enfant*, P.U.F., Paris, 1946.

Pouthas, V., Droit, S., and Jacquet, A.Y., Temporal experiences and time knowledge in infancy and early childhood, *Time Soc.*, 2, 199–218, 1993.

Rattat, A.-C., Estimation du temps et attention chez le jeune enfant: le rô le de la mémoire de référence, Doctoral thesis, University Blaise Pascal, France, in preparation.

Rattat, A. C. and Droit-Volet, S., Variability in children's memory for duration, *Behav. Process.*, 55, 81–91, 2001.

Thomas, E. and Weaver, W., Cognitive processing and time perception, *Percept. Psychophys.*, 17, 363–367, 1975.

Treisman, M., Cook, N., Naish, P., and McCrone, J., The internal clock: electroencephalographic evidence for oscillatory processes underlying time perception, *Q. J. Exp. Psychol.*, 47A, 241–289, 1994.

Treisman, M., Faulkner, A., Naish, P., and Brogan, D., The internal clock: evidence for a temporal oscillator underlying time perception with some estimates of its characteristic frequency, *Perception*, 19, 705–743, 1990.

Wearden, J.H., Human performance on an analogue of an interval bisection task, *Q. J. Exp. Psychol.*, 43B, 59–81, 1991.

Wearden, J.H., Temporal generalization in humans, *J. Exp. Psychol. Anim. Behav. Process.*, 18, 134–144, 1992.

Wearden, J.H., Beyond the fields we know ... : exploring and developing scalar timing theory, *Behav. Process.*, 45, 3–21, 1999.

Wearden, J.H., Edwards, H., Fakhri, M., and Percival, A., Why sounds are judged longer than lights: application of a model of the internal clock in humans, *Q. J. Exp. Psychol.*, 51B, 97–120, 1998.

Wearden, J.H., Wearden, A., and Rabitt, P., Age and QI effects on stimulus and response timing, *J. Exp. Psychol. Hum. Percept. Perform.*, 23, 962–979, 1997.

Wilkening, F., Levin, I., and Druyan, S., Children's counting strategies for time quantification and integration, *Dev. Psychol.*, 23, 822–883, 1987.

Zakay, D., Subjective and attentional resource allocation: an integrated model of time estimation, in *Time and Human Cognition*, Levin, I. and Zakay, D., Eds., North-Holland, Amsterdam, 1989, pp. 365–397.

8 Modality Differences in Interval Timing: Attention, Clock Speed, and Memory

Trevor B. Penney

CONTENTS

8.1 Introduction 209
8.2 Modality Effects on Time Perception..212
    8.2.1 Filled vs. Empty Duration Discrimination 212
    8.2.2 Are Auditory Signals Subjectively Longer than Visual Signals?...213
8.3 Mechanisms Underlying Modality Effects in Time Perception..................220
8.4 Stimulus Range and Modality Effects..224
8.5 Modality Effects as a Methodological Tool 225
    8.5.1 Circadian Modulation of Interval Timing 225
    8.5.2 Interval Timing Sensitivity in Persons at Risk for Schizophrenia 225
8.6 Summary and Conclusions 227
References..228

8.1 INTRODUCTION

The study of timing and time perception bridges durations ranging from milliseconds to days. Very long durations, such as circadian rhythms, appear to be governed by a periodic oscillatory process, sensitive to an external zeitgeber for reset and entrainment, that has extremely low variability and is used to time each single 24-h duration (e.g., Aschoff, 1984). In contrast, interval timing in the seconds-to-minutes range shows much greater variability, but is also highly flexible in terms of the durations that can be accurately timed (e.g., Gibbon et al., 1997; Hinton and Meck, 1997).

Many of the conceptualizations of interval timing proposed during the past 40 years characterized the ability to perceive the passage of time as the function of a specialized chronometric mechanism (e.g., Creelman, 1962; Fraisse, 1963; Frankenhauser, 1960; Gibbon et al., 1984; Killeen, 1984; Michon, 1967; Treisman, 1963). Most models of this type suggest five basic components (time base, gate, counter, memory, and comparator), but the attributes of these components vary across timing models. For example, the specification of the time base ranges from the mean number of events perceived in a given time interval (Frankenhauser, 1960) to pulses emitted by an internal pacemaker (Gibbon and Church, 1984; Treisman, 1963). Moreover, in some pacemaker models, the pacemaker–accumulator system is continuous in that distinct times, limited only by the base rate of the pacemaker, are represented (Gibbon et al., 1984), whereas in other models, the timer is categorical (Kristofferson, 1984), meaning that all durations falling into a particular range are equivalent and therefore indistinguishable.

One feature the clock-counter models proposed to date share, however, is that they do not explicitly account for the effects of signal characteristics on duration perception. Whether nontemporal signal characteristics influence timing is an important question because it is intimately related to the cognitive and physiological nature of the timing process itself. For example, whether the internal clock is modality specific, i.e., whether there are different clocks for different modalities or whether the clock is amodal, tells us something very basic about the timing system and its instantiation in both cognitive and neurophysiological terms. In addition, whether nontemporal signal characteristic effects are consistent across ranges of stimulus durations may be suggestive of whether durations in the milliseconds range access the same timing system as durations in the seconds-to-minutes range. There are arguments in the literature that both milliseconds and seconds range timing have the same underlying neural substrate (Gibbon et al., 1997), as well as arguments that the neural substrates for these two duration ranges are different (Ivry, 1996). In a more general vein, the influence of signal characteristics on duration perception may have diagnostic value for discriminating among models of timing. For example, Grondin (2001) suggested that if one accepts the claim that independence of temporal judgments from sensory characteristics supports the idea of a central timing mechanism, then the evidence that nontemporal marker characteristics influence duration discrimination might lead one to question whether there is a central timing mechanism. Clearly then, understanding the influence of nontemporal stimulus characteristics is important for developing accurate models of timing behavior.

It should be noted, however, that there are models of time perception and production that inherently account for the influence of nontemporal stimulus characteristics. For example, Ornstein (1969) rejected the idea of a specialized chronometric mechanism and suggested instead that the appreciation of elapsed duration is dependent on both the amount of information available during the time interval and the perceptual and memory processes involved in processing that information. Within this framework, time is an emergent property of information processing. One advantage of such a purely cognitive approach to temporal processing is that the effect of nontemporal attributes of the signal on perceived duration is explicitly incorporated into the model. Time is merely the amount of information processed, and because nontemporal characteristics are also processed, they must influence perceived duration. Based on a meta-analytic review, however, Block and Zakay

(1997) concluded that memory-based models such as Ornstein's were better suited for explaining retrospective temporal judgments, whereas attentional models are needed to explain prospective temporal judgments. Given this conclusion, the focus here is on prospective temporal judgments, and a specialized system of interval timing is assumed. More specifically, the literature is approached from the perspective of the scalar timing theory. The scalar timing theory (also referred to as the scalar expectancy theory), originally developed by Gibbon (1977) and subsequently modified and extended by Gibbon and his colleagues, is perhaps the most highly cited clock-counter model in the timing literature (for a discussion of this model's impact, see Allan, 1998), although more recent work suggests that oscillatory or coincidence detection systems may be the appropriate biological implementation of timing (see Malapani and Rakitin, this volume; Matell and Meck, 2000; Matell et al., this volume; Miall, 1989). Included with a precise mathematical description of the psychophysical nature of the time sense is a companion information-processing model (see Figure 8.1) that specifies a distinct multistage timing mechanism consistent with the components of internal clock models outlined above (Gibbon, 1977, 1991; Gibbon and Church, 1984; Gibbon et al., 1984).

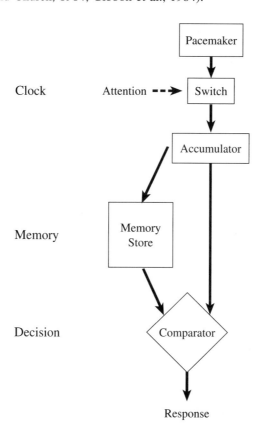

FIGURE 8.1 Three-stage information-processing model of scalar timing theory. (Adapted from Meck, W.H., *Cognit. Brain Res.*, 3, 227–242, 1996.)

In its original form, scalar timing theory does not include an explicit account of the influence of signal characteristics on the perception of time. One reason for this omission may be that the literature is somewhat contradictory with regard to the effects on temporal experience of nontemporal stimulus characteristics such as whether the signal is filled or empty; its intensity, modality, and pitch; and the cognitive load experienced during the task. For example, increasing signal intensity has usually, but not always been found to increase the perceived duration of the signal (e.g., Berglund et al., 1969; Goldstone and Lhamon, 1974; Goldstone et al., 1978; Steiner, 1968; Zelkind, 1973). Cognitive variables have also been shown to influence judgment of duration, but not entirely systematically. The presentation time of nonsense words is judged longer than equivalent duration real words (Avant et al., 1975), but more familiar words are judged as longer than unfamiliar words (Devane, 1974; Warm et al., 1964; Warm and McCray, 1969). Allan (1979, 1992) provides an excellent summary of the effects of a number of nontemporal signal characteristics on perceived duration. In the present chapter, the focus is on the effects of stimulus modality — specifically the auditory and visual modalities — on temporal experience.

8.2 MODALITY EFFECTS ON TIME PERCEPTION

8.2.1 FILLED VS. EMPTY DURATION DISCRIMINATION

The durations used in studies of filled and empty interval discrimination are generally on the order of tens to hundreds of milliseconds, and the empty intervals are usually demarcated by very brief stimuli on the order of tens of milliseconds in duration. Although, Rammsayer and Lima (1991) reported superior discrimination for 50-msec filled auditory intervals, compared to empty intervals, subsequent work by Grondin (1993) showed that whether filled or empty intervals support superior discrimination is modulated both by the modality of the signals used to demarcate the temporal interval and the length of the temporal interval (for a review, see Grondin, 2001). For example, although Grondin (1993) found superior performance for empty, as opposed to filled, intervals for both auditory and visual stimuli using the method of single stimuli and a test duration of 250 msec, with 50-msec target intervals, empty intervals demarcated by visual signals showed superior performance, and there was no difference for the auditory modality. Grondin (1993) proposed an *internal marker* hypothesis to account for these results. The basic idea is that it takes longer for the internal trace of a timing signal to disappear and timing to stop than it does to initiate timing when a signal is presented. For filled signals, timing is initiated with signal onset and terminated when the trace of the physical signal is eliminated. In contrast, for empty intervals timing might begin when the trace of the physical signal has disappeared and ends as soon as the second signal is detected. The end result is that the duration actually timed for an empty signal is smaller than that for a filled interval. Easier discrimination for brief empty intervals follows from this because, according to Weber's law, the size of the change necessary for a detectable difference to be perceived increases with the size of the stimulus, in this case, the amount of time perceived for a filled or empty interval. With long durations, the effect of the timing onset–offset difference washes out and the discrimination advantage for empty intervals disappears. Finally, this hypothesis can account for the result that the difference between empty and filled intervals is larger in the visual than in the auditory modality if one assumes visual signals persist longer than auditory signals.

8.2.2 ARE AUDITORY SIGNALS SUBJECTIVELY LONGER THAN VISUAL SIGNALS?

The question of whether signal modality influences the subjective experience of time has a long history in psychology. For example, Gridley (1932) traced the suggestion to compare the senses on accuracy in judging the same time intervals back as far as Czermak in 1857. Most subsequent work on signal modality and the subjective experience of time has been on the visual and auditory modalities. Sometimes modality effects have been examined directly, whereas in other cases, the experiment allowed the question to be addressed, but it was not the primary focus of the research. Although, the literature includes research with both human (e.g., Behar and Bevan, 1961) and nonhuman animals (e.g., Meck, 1991; Meck and Church, 1982), the review that follows is primarily restricted to experiments with human participants.

A number of investigators have examined modality differences in perceived duration (e.g., Behar and Bevan, 1961; Bobko et al., 1977; Brown and Hitchcock, 1965; Goldstone and Goldfarb, 1964; Hawkes et al., 1961; Walker and Scott, 1981; Wearden et al., 1998). Some of this work, utilizing a number of different paradigms, has indicated that auditory signals are judged longer than equivalent duration visual signals (e.g., Behar and Bevan, 1961; Goldstone and Goldfarb, 1964; Walker and Scott, 1981; Wearden et al., 1998).

Goldstone and Goldfarb (1964) conducted one of the most extensive analyses of modality differences. In a series of experiments, participants classified auditory and visual signal durations relative to either a social standard or a subjective standard. For the former case, the standard was 1 clock second on a scale ranging from very much less than 1 sec to very much more than 1 sec, whereas in the latter case, the signal duration was judged on a scale ranging from very, very short to very, very long. The presented durations ranged from 0.15 to 1.95 sec. For both the social standard and the subjective standard conditions the visual signals were judged shorter than equivalent duration auditory signals. This effect was obtained both when signal modality was a between-subjects factor and when it was a within-subjects factor. Although the authors also found that low-intensity lights seemed shorter than equivalent duration high-intensity lights, they claimed that intensity differences could not explain the modality effect because a separate experiment had failed to reveal significant effects of pitch or intensity on the perceived duration of auditory signals.

Behar and Bevan (1961) also found differences in the perceived duration of auditory and visual intervals. Their participants classified a series of probe durations (1, 2, 3, 4, and 5 sec) using an 11-category scale ranging from very, very, very short to very, very, very long. In two experiments there was an additional probe duration that was an outlier relative to the other probe durations (9 or 0.2 sec). They found that auditory durations were judged longer than objectively equivalent visual durations, although it is possible this difference may have been due to the extreme outliers influencing visual signal classification more than auditory signal classification. A subsequent within-subjects experiment, however, using the same basic design, but without the extreme outlier manipulation, indicated that auditory signals were judged about 20% longer than equivalent duration visual signals.

Walker and Scott (1981) asked participants to reproduce the duration of a presented stimuli by pressing and holding down a response button. They used intervals of 500, 1000, and 1500 msec demarcated by either a 600-Hz tone, illumination of a 15 × 20 mm light, or both. They found that 1000- and 1500-msec tones were perceived as longer than separately presented lights of equal duration, but that there was no difference for 500-msec stimuli. When both modalities were presented together, the duration reproduced was longer than that for light presented alone, but there was no difference between the combined stimulus and tone alone. In this experiment, the tone stimulus remained on during the light gap period and vice versa, so the conditions are not comparable to those described above for empty and filled duration comparisons. A second experiment used the same basic design as experiment 1, but with two exceptions. First, the time intervals were gaps in otherwise continuous tone and light presentations. Second, the intensity of the auditory signal was reduced relative to the first experiment. They found that a silent gap in an otherwise continuous tone was perceived as longer than an equivalent gap in an otherwise continuous light. If the light and tone turned off together, however, the gap duration was judged equivalent to that of the tone-alone gap duration. Apparently, when both signals were presented together, the auditory signal dominated because the combined signal was perceived like an auditory signal. Interestingly, a third experiment demonstrated that a 500-msec light was judged longer than a 500-msec tone if the intensity of the tone was sufficiently reduced. However, they did not obtain such a difference for either the 1000- or the 1500-msec stimulus. They suggested that the difference in auditory and visual duration perception might have been due to differences in the perception of the onset and offset of visual and auditory signals. If the auditory signal had a faster perceptual onset and a slower or equal offset to the visual signal, then the auditory signal would seem longer. However, they rejected this possibility based on the finding that offset detection is faster for auditory than for visual signals (Goldstone, 1968). Unfortunately, they did not offer an alternative explanation for the auditory–visual difference, and they did not explain why they obtained different effects for the 1000- and 1500-msec stimuli than for the 500-msec stimulus.

Wearden et al. (1998) addressed the issue of modality effects in duration perception using both temporal generalization and verbal estimation methods in combination with within-subjects designs. For both paradigms, the auditory stimulus was a 500-Hz tone and the visual stimulus was a 4 × 4 cm blue square. In the temporal generalization experiment, the target duration selected on each test block came from a range of 400 to 600 msec, whereas in the verbal estimation experiment, duration length ranged from 77 to 1183 msec. They found that auditory stimuli were judged as longer than equivalent duration visual stimuli in both tasks, and that the auditory judgments were less variable than those for the visual stimuli. They interpreted the results as indicating that an internal pacemaker accumulating subjective time ran faster for auditory signals than for visual signals. The difference in variability they attributed to a variability difference in the operation of the mode "switch," which allows the internal pacemaker counts to accumulate in memory.

Other work, however, has failed to find evidence of modality differences (e.g., Bobko et al., 1977; Brown and Hitchcock, 1965; Hawkes et al., 1961; Kagerer et al., 2002; Szelag et al., 2002). Bobko et al. (1977) failed to find a statistically significant modality effect in four experiments, although they did report a clear trend for auditory signals to be judged longer than visual signals. Verbal estimation and magnitude estimation with and without a standard comparison stimulus were used, and the durations ranged from 0.25 to 5 sec in steps of 0.25 sec. The design was between subjects in that each participant experienced only a visual signal, the letter x presented tachistoscopically, or an auditory signal, a 1500-Hz tone.

Brown and Hitchcock (1965) used a reproduction method in which participants were initially trained on an interval of one signal modality, auditory (A) or visual (V), and were subsequently asked to reproduce the interval when it was demarcated by a signal of the same or a different modality. The four resulting conditions were AA, AV, VV, and VA. A group of ten participants was tested in each experimental condition, but all participants reproduced a complete set of intervals that ranged from 1 to 17 sec in increments of 2 sec. Comparisons of the AA and VV conditions failed to reveal any effects of signal modality on reproduced duration. In addition, the comparisons of the VA vs. VV, AA vs. AV, AA vs. VA, and AV vs. VV conditions at all test intervals revealed only two significant effects: those for AA vs. AV at 3 and 7 sec. Moreover, these effects were in opposite directions, with AV eliciting longer reproductions at 3 sec and AA eliciting longer reproductions at 7 sec.

Hawkes et al. (1961) used production, reproduction, and verbal estimation methods and durations of 0.5, 1, 2 and 4 sec. Each individual test session used a single method and modality, meaning that each participant served in nine test sessions total. They failed to obtain any significant differences in duration judgments between auditory and visual signals across the range of durations and methodologies examined.

Szelag et al. (2002) examined duration processing in children. The child's task was to reproduce the duration of a standard auditory or visual stimulus. Specifically, a standard duration ranging from 1 to 5.5 sec or 1 to 3 sec, depending on the condition, was presented, and following a brief delay, the stimulus reappeared and the participant had to respond on the keyboard when the correct duration had elapsed. They found that auditory standards were more accurately reproduced than visual standards for 1-, 1.5-, and 2-sec stimuli out of a stimulus range of 1 to 5.5 sec, but they did not obtain any evidence that auditory stimuli were experienced as perceptually longer than equivalent duration visual stimuli. In any case, the authors interpreted the accuracy difference as due to the nature of stimulus presentation rather than to a modality effect per se. The light stimulus was presented on a screen 1.7 m from the participant, whereas the tone was presented over headphones. They suggested that the auditory stimuli were better attended than the visual stimuli and therefore were reproduced more accurately. Interestingly, both modalities were presented in the same test session, but in a blocked fashion. Moreover, as noted above, on each trial the participant was presented with a target duration that then had to be reproduced. The significance of this aspect of the design is discussed below.

Kagerer et al. (2002) examined reproduction of durations in a range from 1000 to 5500 msec in a group of patients with unilateral focal brain injuries and a group of normal controls. The specific stimuli used in the experiment, rather than the comparison of brain-injured and control participant groups, are of interest here. They used four stimulus categories, and participants were tested on a different stimulus category on each of four test days. Two of the stimulus types were auditory (simple, 500-Hz tone; complex, sound of running water) and two were visual (simple, green oblong shape projected onto the wall of the test room; complex, 19th-century painting). They failed to find any differences across the different stimulus modalities and degrees of complexity.

At first glance, the conflicting results among studies that obtained and those that failed to obtain modality differences appear difficult to reconcile because of the variety of experimental procedures used. However, closer examination reveals some commonalities in the pattern of results across experiments. In general, the design of the studies that obtained an effect of modality allowed or encouraged the comparison of both the auditory and the visual probe durations with a common representation of the standard interval, be it a preexisting representation such as 1 clock second or a standard comparison interval presented in the experiment. This could occur through the use of a within-subjects design in which both signal modalities were presented in the same test session and participants were likely to use the same representation of the standard duration for both the auditory and the visual probe durations (e.g., Walker and Scott, 1981; Wearden et al., 1998). In addition, comparison to a prior representation, such as 1 clock second, or on a scale from very short to very long means, in effect, that the auditory and visual signals are being compared to the same internal standard, a situation that is similar to that of a within-subjects design (e.g., Behar and Bevan, 1961; Goldstone and Goldfarb, 1964). Of course, even if a participant experiences both modalities in the same test session, the design could allow or require the participant to compare an auditory probe with a representation of an auditory standard and a visual probe with a representation of a visual standard (e.g., Szelag et al., 2002). Furthermore, if a blocked design is used, then it is unlikely that the participant would use the same representation of the standard for both auditory and visual signals (e.g., Hawkes et al., 1961; Kagerer et al., 2002). Obviously, a between-subjects design, depending on its specific details, often absolutely prevents the use of a common standard representation for both modalities (e.g., Bobko et al., 1977). The possible exception to this may be when participants are asked to make comparisons to an explicit social standard like 1 clock second or on a subjective scale from short to long. Assuming the average objective duration of the social standard is approximately equivalent across groups of participants, then the groups are effectively using a common memory representation to which both auditory and visual signal durations are compared (Behar and Bevan, 1961; Goldstone and Goldfarb, 1964). As noted above, one possible explanation of the modality effect reasons that the internal clock runs at a faster rate for auditory than for visual signals. Therefore, the accumulated clock value for a given duration will be larger when the signal is auditory than when it is visual. As a consequence, if the auditory and visual accumulations are compared to a common memory representation, or to each other, then an auditory signal will seem longer than an equivalent duration visual signal.

Recently, Penney et al. (2000) addressed this possibility directly by comparing timing performance when the participants experienced both auditory and visual signals in the same test session with timing performance when the participants experienced only one or the other of the signal modalities in a test session. In the latter case, the experiment was a between-subjects design because each group of participants experienced a single signal modality. They used a duration bisection task in which participants were trained with examples of the short and long anchor durations prior to the test session. Three possible anchor duration ranges (2 to 8 sec, 3 to 4 sec, and 4 to 12 sec) were used across the series of experiments. An additional feature of some of the conditions, not addressed in detail here, was the simultaneous presentation of auditory and visual signals on the same trial, but with asynchronous onset and offset. In other conditions, single auditory and visual signals were presented sequentially in the same test session. The major finding was that when auditory and visual signals appeared in the same test session and had the same objective duration anchor values, auditory signals appeared subjectively longer than equivalent duration visual signals, as illustrated in Figure 8.2. This was true for both the simultaneous and the sequential presentation conditions. However, when auditory and visual signals appeared in separate test sessions, there were not any differences

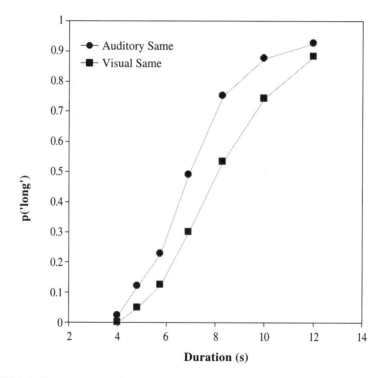

FIGURE 8.2 Group response functions averaged across participants for visual and auditory timing signals presented in the same test session. The anchor durations were 4 and 12 sec. p('long') = probability of a long response. (Data replotted from Penney, T.B., Gibbon, J., and Meck, W.H., *J. Exp. Psychol. Hum. Percept. Perform.*, 26, 1770–1787, 2000.)

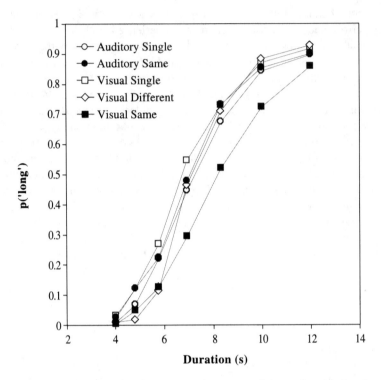

FIGURE 8.3 Group response functions averaged across participants for visual and auditory timing signals. The auditory single and visual single response functions are from single-modality test sessions. The auditory same and visual same response functions are from the same two-modality test session. The visual different response function is from a two-modality test session in which a different anchor duration pair was used for the auditory signals. The anchor durations were 4 and 12 sec. p('long') = probability of a long response. (Data replotted from Penney, T.B., Gibbon, J., and Meck, W.H., *J. Exp. Psychol. Hum. Percept. Perform.*, 26, 1770–1787, 2000.)

in the classification of equivalent duration auditory and visual signals. Crucially, an additional experiment demonstrated that the appearance of auditory and visual signals in the same test session was not sufficient to elicit the modality effect. It was also necessary for the anchor durations of the signals to be objectively equivalent. For example, when a series of auditory signals with an anchor duration pair of 3 to 6 sec appeared in the same test session as visual signals with an anchor duration pair of 4 to 12 sec, then there was no modality effect. That is to say, the response function for visual signals 4 to 12 sec in this condition was indistinguishable from that generated by participants who experienced visual signals only in the test session, auditory signals only in the test session, or the auditory response function generated by participants who experienced both auditory and visual signals in the test session, as illustrated in Figure 8.3.

In accounting for their results, Penney et al. (2000) claimed that the auditory and visual signals drove an internal clock at different rates. They located this rate difference at the level of the mode switch within the scalar timing model (see Figure 8.1), rather than suggesting that the pacemaker "pulsed" at a faster rate for auditory than for visual signals. In their view, the pacemaker rate was the same for both auditory and visual signals, but the probability that the mode switch would be in the closed state was different for the auditory and visual signals. They suggested that for visual signals the mode switch could oscillate or "flicker" between an open and a closed state with the result that some number of pacemaker pulses or counts would be lost during the presentation of the visual signal. The degree of count loss would be proportional to the duration of the timing signal with the end result that the size of the modality effect would be proportional to the signal duration. In principle, the state of the mode switch would be determined by the ability of the signal to capture attention because the mode switch requires attention to be allocated in order to enter the closed state (see Droit-Volet, this volume; Fortin, this volume). Assuming auditory signals more easily capture attention than visual signals, and there is some evidence that auditory signals are processed automatically, whereas visual signals are subject to controlled processing (Meck, 1984; Posner, 1978), then they are better able to maintain the switch in the closed state. Therefore, for visual signals, many clock ticks would fail to accumulate, and consequently, the visual duration would seem subjectively shorter than an equivalent auditory duration if both signals were compared to a common memory representation. In truth, the experimental evidence available from this series of experiments does not allow for a differentiation between a change in pacemaker rate and a change in the efficacy of the mode switch, but there is evidence in the literature, discussed in detail below, that auditory and visual signals share a common pacemaker but have separate switch–accumulator modules (Rousseau and Rousseau, 1996). However, no matter what the underlying cause of the different temporal accumulations for auditory and visual signals, it is clear that the modality effect of auditory signals being experienced as subjectively longer than equivalent duration visual signals requires more than just differential clock rates for both the visual and auditory signal modalities. The modality effect also requires that the auditory and visual signals be compared to a common memory representation in order for a clock rate difference to manifest as a difference in perceived duration. As noted above, the common memory representation could be due to a common memory pool to which both auditory and visual training signals contribute to varying degrees (cf. Penney et al., 2000) or due to the use of a common social standard. In the conditions that revealed a modality effect in the bisection experiments described above, the subjects failed to maintain separate memory representations for the visual and auditory anchor durations. In other tasks, subjects may compare the current passage of time to some preexisting standard, such as 1 clock second. Clearly in this case also, a fast auditory accumulation and a relatively slower visual accumulation would seem different relative to the common comparison duration; therefore, auditory signals would seem longer than visual signals. Even in cases where the participants are not told explicitly what the comparison should be (e.g., when they are told to use a scale from very, very short to very, very long), it is probable that they would make the comparison relative to some preexisting mental representation even as they built up a sense of the range of durations presented in the specific experimental context (see Lustig, this volume).

The present model, however, does not easily account for the failure, described earlier, of Brown and Hitchcock (1965) to find consistent modality effects when comparing participants asked to reproduce signal durations using a different modality from that of the sample duration. For example, based on the preceding, one would expect the group of participants presented with an auditory sample and a visual test to reproduce longer durations than the group of participants presented with a visual sample and an auditory test; yet this did not happen. Interestingly, Droit-Volet et al. (submitted; Droit-Volet, this volume) recently found modality effects in a version of the temporal bisection task that shares critical features with the Brown and Hitchcock (1965) design. Specifically, participants were trained and tested with one signal modality (e.g., visual) and, on a subsequent test day, experienced the same signal (e.g., visual) in training, but a different modality signal (e.g., auditory) in the test. The results indicated that participants experienced auditory signals as longer than equivalent duration visual signals. As participants would have compared an auditory signal with a visual memory representation of the anchors or a visual signal with an auditory memory representation of the anchors, these results are consistent with the model outlined above.

Note, however, that a common memory representation combined with clock rate differences is not a requirement for differences in discriminability between auditory and visual signals, because clock rate differences alone will allow one signal to be more easily discriminated than another based upon the total number of pacemaker pulses accumulated within a given duration (Gibbon, 1977). Rammsayer and Lima (1991) raised this possibility as an account of differences in discriminability between empty and filled durations. In brief, the higher the pulse rate, the finer the discrimination between intervals that can be made because the delay between pulses is shorter.

8.3 MECHANISMS UNDERLYING MODALITY EFFECTS IN TIME PERCEPTION

The following section outlines some of the evidence suggesting that there are latency differences in signal detection or differences in clock speed between the visual and auditory modalities. As noted above, both timing onset–offset latency and clock speed have been proposed as putative mechanisms underlying modality differences.

Evidence that the response time (RT) for an auditory stimulus is shorter than that for a visual stimulus, by about 40 msec, comes from experiments showing that in order for an auditory and visual stimulus to be considered simultaneous in a temporal order task, the onset of the visual signal must precede the onset of the auditory signal by an amount slightly larger than this RT difference (Jaskowski et al., 1990). The evidence for this modality difference is somewhat ambiguous, however, because Rutschman and Link (1964) also found a 40-msec RT difference, but in their experiment the auditory signal had to precede the visual signal for judgments of simultaneity on the temporal order task. Interestingly, Hirsh and Sherrick (1961) found simultaneity with no onset difference. In both the peak-interval procedure and the temporal bisection task (Meck, 1984; Penney et al., 1996), presentation of a brief stimulus (e.g., 0.5-sec visual cue) indicating the modality of the upcoming timing signal (e.g., continuous visual signal) reduced the latency to begin timing in rats. Conversely, if the cue incorrectly predicted the modality of the timing signal (e.g., visual cue followed by an auditory signal), the latency to begin timing was increased. Moreover, Penney et al. (1996) found that clonidine, a noradrenergic antagonist, increased the latency to initiate timing in rats. This effect was localized to the switch process of the internal clock model.

A large body of work in rats, and other animals, much of it interpreted within the framework of scalar timing theory, has addressed both the neural and neurochemical substrates of the pacemaker–accumulator component of the internal clock (for a detailed review of the pharmacological basis of the internal clock, see Meck, 1996). For example, Maricq and Church (1983) reported that the dopaminergic agonist methamphetamine increased the rate of the internal clock, whereas the dopaminergic antagonist haloperidol decreased the rate of the internal clock when rats timed intervals in the seconds range. The importance of the dopaminergic system for timing has also been extended to human participants (Malapani et al., 1998; Malapani and Rakitin, this volume; Raamsayer, 1997a, 1997b). Although there is strong experimental evidence that both the rate of the internal clock and the latency with which timing processes are initiated are affected by pharmacological manipulations, the question of whether behavioral manipulations are also effective in modulating pacemaker speed is more important for the present discussion. Current evidence indicates that they are. For example, introducing rapid stimulus modulations, i.e., tone clicks, within a time interval demarcated by asterisks presented on a computer screen increases the apparent duration of the interval (Burle and Casini, 2001; Treisman et al., 1990). The verbal estimation task employed by Treisman et al. used stimulus durations in the hundreds of milliseconds range and a click train frequency that varied from 2.5 to 27.5 Hz across experiments. The primary purpose of the experiments was to obtain support for the existence of an internal temporal oscillator — an oscillator that could be perturbed by external stimuli via an influence on the arousal levels of the participants.

Perhaps most relevant here, however, is the work by Wearden and colleagues. Penton-Voak et al. (1996) followed up the findings of Treisman et al. (1990) by demonstrating that preceding a timing stimulus by an auditory click train influences the perceived duration of the timed stimulus. In four experiments using a variety of methods (temporal generalization, pair comparisons, verbal estimation, and short-duration production) and probe durations ranging from 200 to 1000 msec across experiments, they found that preceding the timing signal with a click train, which ranged from 1 to 5 sec in duration and had a frequency of 5 or 25 Hz, changed the expected target time by about 10% in human subjects in the temporal generalization, verbal estimation, and production paradigms. Droit-Volet and Wearden (2002; Droit-Volet, this volume) examined the influence of visual flicker stimuli prior to the training signals in a visual temporal bisection task with 3- to 8-year-old children and anchor durations of 400 to 800 msec and 400 to 1600 msec. Compared to the no-flicker condition, visual flicker caused a leftward shift in the response functions, and this shift was proportional to the duration being timed, a result that is consistent with a pacemaker speed effect. Earlier work by Wearden et al. (1999) also showed that auditory click trains relatively sped up or slowed down the internal clock in a temporal bisection task with short and long standards of 200 and 800 msec. In one condition, the intermediate probe durations were preceded by clicks, but the standards were not, whereas in another condition, the standards were preceded by clicks and the probe durations were not. Preceding the probes by clicks shifted the response function to the left, whereas preceding the standards by clicks shifted the response function to the right. The results from these click train studies were interpreted as indicating an effect on the speed of an internal pacemaker. Specifically, the click trains caused the pacemaker to emit pulses at a faster-than-normal rate, possibly because of an increase in arousal level or the amount of attentional resources dedicated to timing. Moreover, auditory click trains elicited effects for both auditory and visual timing signals. This outcome is significant because it suggests that there is a common amodal timing mechanism that is influenced by the click train.

Rousseau and Rousseau (1996) obtained strong evidence that a common amodal central pacemaker underlies timing of visual and auditory signals by using a task in which participants were not explicitly asked to time. In the most basic version of their stop reaction time task, participants experienced a unimodal sequence of brief auditory or visual stimuli that had a constant stimulus onset asynchrony (SOA). Participants were instructed to respond as quickly as possible when they thought the sequence had ended, but because the sequence length varied, counting stimuli did not allow participants to complete the task accurately. Therefore, although participants were not explicitly instructed to time, and the task was very different from the explicit timing tasks mentioned in this chapter, the task did require them to build up a representation of the SOA and compare the interval following each stimulus with this representation to determine whether to respond that the sequence had ended.

With simple unimodal sequences, Rousseau and Rousseau (1996) found that stop reaction times for visual sequences were 45 msec slower than those for auditory sequences across all SOA durations tested (250 to 1000 msec). This result is consistent with a modality-dependent switch-latency effect rather than a clock speed difference between auditory and visual signals, and suggests that both auditory and visual timing access a common pacemaker.

Some of their most interesting stop reaction time results, however, were obtained from a series of experiments that used bimodal single SOA sequences, bimodal multiple SOA (polyrhythmic) sequences, and bimodal polyrhythmic sequences. In the bimodal experiments, each sequence consisted of alternating auditory and visual stimuli. The results from the bimodal sequences indicated that participants processed these bimodal sequences as two concurrent unimodal subsequences; i.e., they timed the unimodal subsequences in parallel. In the unimodal polyrhythmic sequences, there were two SOAs, but all signals were identical (i.e., the same duration, frequency, and amplitude for all the auditory signals in the sequence, and the same central source and luminance for all the visual signals in the sequence). The unimodal polyrhythmic sequence experiments demonstrated that participants were able to time the auditory subsequences in parallel, but were unable to time the visual subsequences in parallel.

Rousseau and Rousseau (1996) interpreted their results within a multiple switch–accumulator framework developed by Church and Meck (Church, 1984; Meck and Church, 1984) to account for the ability of rats to simultaneously time

Modality Differences in Interval Timing

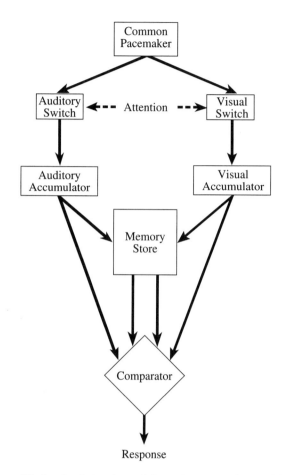

FIGURE 8.4 A modified scalar timing theory information-processing model that incorporates the modality-specific switch–accumulator modules proposed by Rousseau and Rousseau (1994, 1996). In this model a common amodal pacemaker is assumed, and inputs from modality-specific switch–accumulator modules feed into common memory and decision mechanisms.

more than one signal. According to Rousseau and Rousseau (1996), there is a central pacemaker that subserves multiple switch–accumulator timing modules. The parallel timing ideas developed by Rousseau and Rousseau (1994, 1996) can be incorporated into a modified scalar timing information-processing (IP) model, as illustrated in Figure 8.4. In this modified IP model a common pacemaker is assumed so that modality effects may be manifested through either differences in switch closure latency or differences in the ease of maintaining the switch in the closed state only, and not via differences in pacemaker speed. Inputs from both modality-specific switch–accumulator modules are shown feeding into common memory and decision mechanisms. This is a requirement of the model given the claim made earlier that some modality effects are a consequence of comparing both visual and auditory temporal accumulations to a common comparison interval.

8.4 STIMULUS RANGE AND MODALITY EFFECTS

Although, as reviewed above, there are numerous reports in the literature of modality effects in the milliseconds range, recently Melgire et al. (submitted) directly compared the influence of signal modality on temporal bisection performance in the seconds and milliseconds ranges. Each participant completed two test sessions on different days. For half of the participants, the first session anchor durations were 2 and 8 sec, and the second session anchor durations were 200 and 800 msec. The order of duration assignment was reversed for the remaining participants. Both auditory and visual probe durations were presented sequentially in random order in the same test session. Interestingly, although there was a robust effect of signal modality when 2 and 8 sec were used as the anchor durations, thereby replicating the results of Penney et al. (2000) showing that sounds are judged longer than lights, there was no effect of signal modality when 200 and 800 msec were used as anchor durations. One possible explanation is that the durations used in the milliseconds version of the task were too brief to allow a modality effect mediated by a flickering mode switch to be revealed. It is possible that when a timing signal initially captures attention, the switch is maintained in the closed state for some time before "flickering" develops. If this is the case, then for short-duration signals, i.e., milliseconds range, there may not be sufficient opportunity for the development of differential levels of flicker during the stimulus presentation. Consequently, the auditory and visual signals might not seem any different in the temporal bisection task when durations in the short milliseconds range are used. For longer signals (e.g., 2 vs. 8 sec), mode switch flicker develops and the modality effect is revealed as a shift in subjective duration that is proportional to the length of the probe duration. The major shortcoming with this explanation is that there is extensive evidence of robust modality effects with milliseconds range stimuli from a number of other timing paradigms (e.g., Goldstone and Goldfarb, 1964; Wearden et al., 1998). Perhaps there is something unique about the temporal bisection procedure in the milliseconds range, but this seems unlikely given that the data obtained by Melgire et al. (submitted) were typical of such procedures in other respects; i.e., there was an orderly increase in the percent "long" signal classification that assumed a sigmoidal shape (cf. Allan and Gerhardt, 2001; Allan and Gibbon, 1991; Penney et al., 1998, 2000; Wearden, 1991; Wearden and Ferrara, 1995, 1996); the point of subjective equality was observed to be close to the geometric mean; and functions superimposed on a relative time scale (Allan and Gibbon, 1991). Alternatively, the primary mechanism underlying the modality effect in the seconds range may be distinct from that involved in the milliseconds range. For example, milliseconds range modality effects may be primarily due to differential latencies to initiate and stop timing for the various signal modalities. Because onset–offset effects are absolute amounts independent of the duration of the timing signal, they would make a larger contribution to the size of a modality effect for a milliseconds range signal than they would for a seconds range signal. Indeed, as Grondin (1993) noted for filled vs. empty duration comparisons, timing onset–offset differences may wash out with longer durations. Of course, even if this is true, it does not explain why a modality effect was not obtained in the 200- vs. 800-msec condition of Melgire et al. (submitted).

8.5 MODALITY EFFECTS AS A METHODOLOGICAL TOOL

Modality effects, specifically auditory and visual differences, can be used as a methodological tool to advance our understanding of normal timing. Application of interval timing paradigms that include a modality component to particular populations, such as healthy aged persons, can provide a window to the nature of the timing system, its function over the life span, and how it interacts with other cognitive and physiological variables. For example, there are numerous studies of the effects of divided attention on temporal processing. Although the majority of these studies used a timing task in combination with some other nontiming task, or instructed subjects to differentially allocate attention (e.g., Brown, 1985; Fortin, this volume; Macar et al., 1994; Sawyer et al., 1994), there are a few studies of simultaneous temporal processing (see Pang and McAuley, this volume). Most of these simultaneous temporal processing studies in humans have used two timing signals taken from different modalities (e.g., Penney et al., 2000). Therefore, it is important to consider the influence of stimulus modality in addition to the attentional demands of the specific task when interpreting the results. Indeed, the argument presented here is that the modality effect is, in essence, an attentional effect that relates to issues of automatic and controlled processing and switch closure efficacies.

8.5.1 CIRCADIAN MODULATION OF INTERVAL TIMING

Lustig and Meck (2001) used a modality manipulation in a temporal bisection task to examine the interaction of interval and circadian timing processes. In brief, they found that the timing sensitivity of older participants was modulated by the time of day at which testing occurred. Older participants were better timers during the morning than during the afternoon, whereas younger participants were better timers during the afternoon than in the morning. Although both groups perceived auditory signals as subjectively longer than visual signals of the same duration, the older adults were more sensitive to visual durations than auditory durations when tested in the morning. In contrast, the younger adults were equally sensitive to auditory and visual signals in the morning and in the afternoon. The authors suggested that the older participants concentrated their controlled attentional resources on the single visual trials when tested in the morning (for additional details about aging and time perception, see Lustig, this volume).

8.5.2 INTERVAL TIMING SENSITIVITY IN PERSONS AT RISK FOR SCHIZOPHRENIA

Given the findings of different sensitivities to modality effects in young and aged populations, it is worthwhile to consider whether the existence of modality effects may prove a useful tool for examining the cognitive and neurophysiological bases of timing via studies of patient populations. For example, a number of studies have reported temporal processing deficits in patients diagnosed with schizophrenia (Sz). Overestimation of duration in the seconds range has been reported for verbal estimation tasks (Clausen, 1950; Densen, 1977; Johnson and Petzel, 1971; Lhamon and Goldstone, 1956; Wahl and Sieg, 1980; Weinstein et al., 1958), and underestimation in the seconds range has been reported for reproduction and production tasks (Clausen, 1950; Johnson and Petzel, 1971; Tysk, 1983, 1990; Wahl and Sieg, 1980). Overestimation in verbal estimation tasks and underestimation in production tasks by patients diagnosed with Sz may be consistent with either an increased internal clock speed or an increased memory storage speed, depending on the training and test procedures used and the pattern of obtained results (e.g., Meck, 1983, 1996; Meck and Benson, 2002).

One difficulty with studying cognitive processing in a psychiatric population, however, is that performance deficits may be a consequence of a lack of motivation or task comprehension rather than a specific cognitive deficit. In addition, psychiatric patients are often medicated, and most psychoactive drugs have varied effects on cognitive processing. These difficulties may be circumvented through studying unmedicated individuals at genetic risk for the psychiatric disorder in question. The risk to offspring of one schizophrenic parent, the relatives most commonly studied in high-risk research, has been estimated to be 12% (Gottesman et al., 1987; Holzman and Matthysse, 1990). Moreover, the pattern of empiric risks for schizophrenia in relatives of schizophrenic probands is not Mendelian, but suggests instead the action of multiple genes (e.g., Gottesman et al., 1987). It appears that the clinical illness is not expressed in all individuals who carry the full genetic liability because the concordance rate for the disorder is only about 45% in monozygotic twins (e.g., Gottesman et al., 1982; Kendler, 1988). Some clinically unaffected individuals, however, may show cognitive or other neurobehavioral deficits associated with schizophrenia. Thus, the rate of such deficits in high-risk individuals may be greater than the expected rate of the illness itself (e.g., Erlenmeyer-Kimling, 1996). Importantly, participants at high risk for developing schizophrenia (i.e., first-degree relatives of diagnosed schizophrenic patients) sometimes exhibit structural and functional abnormalities of the brain that are similar to those exhibited by schizophrenia patients (for a review, see Cannon, 1996).

Recently, Penney et al. (submitted) examined differences in temporal processing between individuals at high risk for schizophrenia (HrSz) and normal controls (NC) using the duration bisection task. Both groups of participants showed a robust difference between auditory and visual signal classification. Interestingly, however, the HrSz group showed a statistically larger difference between the auditory and visual response functions than did the NC group. Within the framework outlined above, this result suggests a greater auditory–visual clock speed difference for the HrSz participants than for the NC participants. One possibility is that the HrSz participants had a slower visual temporal accumulation than normal because they were less able to maintain the mode switch in a closed state, indicating an attentional dysfunction.

As mentioned earlier, studies of seconds-to-minutes range timing in schizophrenic patients have usually found overestimation on estimation tasks and underestimation on production or reproduction tasks. These studies have been interpreted as indicative of an increased clock rate in schizophrenic patients, an interpretation that has been assumed consistent with animal studies of the effects of dopamine agonists on interval timing (Rammsayer, 1990). However, as described above, a clock rate difference by itself will not always result in a behavioral effect. To obtain underestimation, the information-processing model requires that the stored memory values be distorted short. As a consequence, accumulated time on the test trial reaches the stored subjective time target value before the objective time target duration has elapsed. Therefore, the timing dysfunction in diagnosed schizophrenic patients may be due to disruption of frontal lobe structures involved in memory storage. In the Penney et al. (2002) study, the HrSz auditory response functions were equal to those of the NC group, whereas the visual functions were shifted farther to the right. Our interpretation does not necessarily require a faster overall clock rate for the HrSz participants, although we do require a larger relative difference between auditory and visual clock rates for the HrSz participants. Given that a number of studies using a variety of tasks have found attentional deficits in both schizophrenic patients and individuals at high risk for schizophrenia (e.g., Erlenmeyer-Kimling and Cornblatt, 1992; Erlenmeyer-Kimling et al., 1979; Mirsky et al., 1995; Nuechterlein, 1977), it is not unreasonable to interpret the auditory–visual difference for HrSz participants as the result of an attentional effect at the level of the mode switch.

8.6 SUMMARY AND CONCLUSIONS

This chapter presents an extensive, although nonexhaustive, survey of the effects of auditory and visual signal modality on interval timing. It is clear from the literature described here that the modality of a timing signal can have a significant influence on the discriminability of a timing signal, its subjective duration, or both. It is also clear that visual–auditory modality effects on timing take at least two forms.

In some cases, the magnitude of the modality effect is an absolute constant amount in that the duration of the timing signal does not influence the size of the modality effect. Results of this form are consistent with modality-dependent latency differences in initiating and terminating timing. Within the scalar timing theory, this type of effect is localized to differences in switch closure and opening efficacy. Importantly, such effects can manifest themselves either as differences in the subjective duration of equivalent duration auditory and visual signals or as differences in duration discriminability (cf. Grondin, 2001).

In other cases, the magnitude of the modality effect is a constant proportion of the duration of the signal being timed. Results of this form are consistent with differences in clock speed. Interestingly, work in a number of labs (e.g., Droit-Volet, this volume; Penney et al., 2000; Wearden et al., 1998) has shown that although switch and clock speed modality effects can manifest themselves in the same experiment, they are separable effects. As described above, there are at least two plausible accounts of the source of clock speed differences within the scalar timing theory framework. One possibility is that the pacemaker runs at a faster rate for auditory than for visual signals (e.g., Wearden et al., 1998). The alternative, and the position promoted here, is that auditory and visual signals are differentially efficient at maintaining the switch in a closed state. The probability that the mode switch will oscillate between a closed and an open state is greater for visual than for auditory signals, with the result that more pacemaker counts are lost when the signal is visual than when it is auditory. As is the case for timing latency effects, clock speed effects can manifest themselves either as differences in the subjective duration of equivalent duration auditory and visual signals or as differences in duration discriminability.

In some cases, whether a modality effect is obtained also depends on the specific nature of the timing task used. For example, if the task requires participants to compare auditory and visual signals to a common memory representation, then auditory signals will seem subjectively longer than equivalent duration visual signals. In the case of tasks that measure duration discrimination acuity, then either timing onset–offset latency or clock speed effects can result in better discrimination for auditory than for visual signals, regardless of whether the auditory and visual signals are directly or indirectly compared to one another (cf. Grondin, 2001; Raamsayer and Lima, 1991).

As noted in the introduction, the influence of signal characteristics on timing is intimately related to the cognitive and physiological nature of the timing process itself. A strong claim has been made here that auditory and visual signals are not equally efficient in either initiating switch closure or maintaining the switch in a closed state. The obvious next question is whether a single pacemaker–switch–accumulator system is activated by both auditory and visual signals or whether there are modality-specific clocks. Following Rousseau and Rousseau (1996), we propose that distinct switch–accumulator modules process auditory and visual timing signals, although these modules receive input from a common amodal pacemaker.

Of course, if modality-specific switch–accumulator modules are assumed, then it should be possible to isolate these timing modules in the brain. The evidence from the study with persons at high risk for developing schizophrenia is valuable in this regard because it indicates that it is possible for modality differences to be exacerbated in some populations. Significantly, the exaggerated modality effect appeared to be mediated by the visual modality because differences in auditory response functions between the HrSz and NC groups failed to obtain. This outcome provides additional support for the idea that there are multiple modality-specific switch–accumulator modules and that these modules can be independently modulated.

In conclusion, modality effects raise a number of fundamental questions about the nature of interval timing in their own right and also provide a window for examining basic aspects of the interval timing system. Although modality effects on timing have been investigated since the beginning days of experimental psychology, much remains to be learned.

REFERENCES

Allan, L.G., The perception of time, *Percept. Psychophys.*, 26, 340–354, 1979.
Allan, L.G., The internal clock revisited, in *Time, Action and Cognition: Towards Bridging the Gap*, Macar, F., Pouthas, V., and Friedman, W.J., Eds., Kluwer Academic Publishers, Dordrecht, Netherlands, 1992, pp. 191–202.
Allan, L.G., The influence of the scalar timing model on human timing research, *Behav. Process.*, 44, 101–117, 1998.
Allan, L.G. and Gerhardt, K., Temporal bisection with trial referents, *Percept. Psychophys.*, 63, 524–540, 2001.

Allan, L.G. and Gibbon, J., Human bisection at the geometric mean, *Learn. Motiv.*, 22, 39–58, 1991.
Aschoff, J., Circadian timing, in *Timing and Time Perception*, Gibbon, J. and Allan, L., Eds., New York Academy of Sciences, New York, 1984, pp. 442–468.
Avant, L.L., Lyman, P.J., and Antes, J.R., Effects of stimulus familiarity upon judged visual duration, *Percept. Psychophys.*, 17, 253–262, 1975.
Behar, I. and Bevan, W., The perceived duration of auditory and visual intervals: cross modal comparison and interaction, *Am. J. Psychol.*, 74, 17–26, 1961.
Berglund, B., Berglund, U., Ekman, G., and Frankenhauser, M., The influence of auditory stimulus intensity on apparent duration, *Scand. J. Psychol.*, 10, 21–26, 1969.
Block, R.A. and Zakay, D., Prospective and retrospective duration judgments: a meta-analytic review, *Psychonomic Bull. Rev.*, 4, 184–197, 1997.
Bobko, D.J., Thompson, J.G., and Schiffman, H.R., The perception of brief temporal intervals: power functions for auditory and visual stimulus intervals, *Perception*, 6, 703–709, 1977.
Brown, D.R. and Hitchcock, L., Time estimation: dependence and independence of modality specific effects, *Percept. Motor Skills*, 21, 727–734, 1965.
Brown, S.W., Time perception and attention: the effects of prospective versus retrospective paradigms and task demands on perceived duration, *Percept. Psychophys.*, 38, 115–124, 1985.
Burle, B. and Casini, L., Dissociation between activation and attention effects in time estimation: implication for internal clock models, *J. Exp. Psychol. Hum. Percept. Perform.*, 27, 195–205, 2001.
Cannon, T.D., Abnormalities of brain structure and function in schizophrenia: implications for aetiology and pathophysiology, *Ann. Med.*, 28, 533–539, 1996.
Church, R.M., Properties of the internal clock, in *Annals of the New York Academy of Sciences*, Vol. 423, *Timing and Time Perception*, Gibbon, J. and Allan, L.G., Eds., New York Academy of Sciences, New York, 1984, pp. 566–582.
Clausen, J., An evaluation of experimental methods of time judgment, *J. Exp. Psychol.*, 40, 76–761, 1950.
Creelman, C.D., Human discrimination of auditory duration, *J. Acoust. Soc. Am.*, 34, 582–593, 1962.
Densen, M.E., Time perception and schizophrenia, *Percept. Motor Skills*, 44, 436–438, 1977.
Devane, J.R., Word characteristics and judged duration for two response sequences, *Percept. Motor Skills*, 38, 525–526, 1974.
Droit-Volet, S., Tourret, S., and Wearden, J., Perception of the duration of auditory and visual stimuli in children and adults, submitted.
Droit-Volet, S. and Wearden, J., Speeding up an internal clock in children? Effects of visual flicker on subjective duration, *Q. J. Exp. Psychol., Comp. Phys. Psych.*, 55B, 193–211, 2002.
Erlenmeyer-Kimling, L., A look at the evolution of developmental models of schizophrenia, in *Psychopathology: The Evolving Science of Mental Disorder*, Matthysse, S., Levy, D., Kagan, J., and Benes, F.M., Eds., Cambridge University Press, Cambridge, U.K., 1996, pp. 229–252.
Erlenmeyer-Kimling, L. and Cornblatt, B.A., A summary of attentional findings in the New York high-risk project, *J. Psychiatry Res.*, 26, 405–426, 1992.
Erlenmeyer-Kimling, L., Cornblatt, B., and Fleiss, J., High-risk research in schizophrenia, *Psychiatr. Ann.*, 9, 79–111, 1979.
Fraisse, P., *The Psychology of Time*, Harper & Row, New York, 1963.
Frankenhauser, M., Subjective time as affected by gravitational stress, *Scand. J. Psychol.*, 1, 1–6, 1960.

Gibbon, J., Scalar expectancy and Weber's law in animal timing, *Psychol. Rev.*, 84, 279–325, 1977.

Gibbon, J., Ubiquity of scalar timing with a Poisson clock, *J. Math. Psychol.*, 35, 1–11, 1991.

Gibbon, J. and Church, R.M., Sources of variance in an information processing theory of timing, in *Animal Cognition*, Roitblat, H.L., Bever, T.G., and Terrace, H.S., Eds., Lawrence Erlbaum Associates, Hillsdale, NJ, 1984, pp. 465–488.

Gibbon, J., Church, R.M., and Meck, W.H., Scalar timing in memory, in *Annals of the New York Academy of Sciences*, Vol. 423, *Timing and Time Perception*, Gibbon, J. and Allan, L.G., Eds., New York Academy of Sciences, New York, 1984, pp. 52–77.

Gibbon, J., Malapani, C., Dale, C., and Gallistel, C.R., Toward a neurobiology of temporal cognition: advances and challenges, *Curr. Opin. Neurobiol.*, 7, 170–184, 1997.

Goldstone, S., Production and reproduction of duration: intersensory comparisons, *Percept. Motor Skills*, 26, 755–760, 1968.

Goldstone, S. and Goldfarb, J.L., Auditory and visual time judgment, *J. Gen. Psychol.*, 70, 369–387, 1964.

Goldstone, S. and Lhamon, W.T., Studies of the auditory-visual differences in human time judgments: I. Sounds are judged longer than lights, *Percept. Motor Skills*, 39, 63–82, 1974.

Goldstone, S., Lhamon, W.T., and Sechzer, J., Light intensity and judged duration, *Bull. Psychonomic Soc.*, 12, 83–84, 1978.

Gottesman, I.I., McGuffin, P., and Farmer, A.E., Clinical genetics as clues to the "real" genetics of schizophrenia (a decade of modest gains while playing for time), *Schizophr. Bull.*, 13, 23–47, 1987.

Gottesman, I.I., Shields, J., and Hanson, D.R., *Schizophrenia: The Epigenetic Puzzle*, Cambridge University Press, Cambridge, U.K., 1982.

Gridley, P.F., The discrimination of short intervals of time by finger-tip and by ear, *Am. J. Psychol.*, 44, 18–43, 1932.

Grondin, S., Duration discrimination of empty and filled intervals marked by auditory and visual signals, *Percept. Psychophys.*, 54, 383–394, 1993.

Grondin, S., From physical time to the first and second moments of psychological time, *Psychol. Bull.*, 127, 22–44, 2001.

Hawkes, G.R., Bailey, R.W., and Warm, J.S., Method and modality in judgments of brief stimulus duration, *J. Aud. Res.*, 1, 133–144, 1961.

Hinton, S.H. and Meck, W.H., The "internal clocks" of circadian and interval timing, *Endeavour*, 21, 82–87, 1997.

Hirsh, I.J. and Sherrick, C.E., Jr., Perceived order in different sense modalities, *J. Exp. Psychol.*, 62, 423–432, 1961.

Holzman, P.S. and Matthysse, S., The genetics of schizophrenia: a review, *Psychol. Sci.*, 1, 279–286, 1990.

Ivry, R.B., The representation of temporal information in perception and motor control, *Curr. Opin. Neurobiol.*, 6, 851–857, 1996.

Jaskowski, P., Jaroszyk, F., and Hojan-Jezierska, D., Temporal-order judgments and reaction time for stimuli of different modalities, *Psychol. Res.*, 52, 35–38, 1990.

Johnson, J.E. and Petzel, T.P., Temporal orientation and time estimation in chronic schizophrenics, *J. Clin. Psychol.*, 27, 194–196, 1971.

Kagerer, F.A., Wittmann, M., Szelag, E., and Steinbüchel, V.N., Cortical involvement in temporal reproduction: evidence for differential roles of the hemispheres, *Neuropsychologica*, 40, 357–366, 2002.

Kendler, K.S., The genetics of schizophrenia and related disorders, in *Relatives at Risk for Mental Disorder*, Dunner, D.L., Gershon, E.S., and Barrett, J.E., Eds., Raven Press, New York, 1988.

Killeen, P., Incentive theory III: adaptive clocks, in *Annals of the New York Academy of Sciences*, Vol. 423, *Timing and Time Perception*, Gibbon, J. and Allan, L.G., Eds., New York Academy of Sciences, New York, 1984, pp. 515–527.

Kristofferson, A.B., Quantal and deterministic timing in human duration discrimination, in *Annals of the New York Academy of Sciences*, Vol. 423, *Timing and Time Perception*, Gibbon, J. and Allan, L.G., Eds., New York Academy of Sciences, New York, 1984, pp. 3–15.

Lhamon, L.T. and Goldstone, S., The time sense: estimation of one second durations by schizophrenic patients, *Arch. Neurol. Psychiatry*, 76, 625–629, 1956.

Lustig, C. and Meck, W.H., Paying attention to time as one gets older, *Psychol. Sci.*, 12, 478–484, 2001.

Macar, F., Grondin, S., and Casini, L., Controlled attention sharing influences time estimation, *Mem. Cognit.*, 22, 673–686, 1994.

Malapani, C., Rakitin, B., Levy, R., Meck, W.H., Deweer, B., Dubois, B., and Gibbon, J., Coupled temporal memories in Parkinson's disease: a dopamine-related dysfunction, *J. Cognit. Neurosci.*, 10, 316–331, 1998.

Maricq, A.V. and Church, R.M., The differential effects of haloperidol and methamphetamine on time estimation in the rat, *Psychopharmacology*, 79, 10–15, 1983.

Matell, M.S. and Meck, W.H., Neuropsychological mechanisms of interval timing behaviour, *Bioessays*, 22, 94–103, 2000.

Meck, W.H., Selective adjustment of the speed of internal clock and memory storage processes, *J. Exp. Psychol. Anim. Behav. Process.*, 9, 171–201, 1983.

Meck, W.H., Attentional bias between modalities: effects on the internal clock, memory, and decision stages used in animal time discrimination, in *Annals of the New York Academy of Sciences*, Vol. 423, *Timing and Time Perception*, Gibbon, J. and Allan, L., Eds., New York Academy of Sciences, New York, 1984, pp. 528–541.

Meck, W.H., Modality-specific circadian rhythmicities influence mechanisms of attention and memory for interval timing, *Learn. Motiv.*, 22, 153–179, 1991.

Meck, W.H., Neuropharmacology of timing and time perception, *Cognit. Brain Res.*, 3, 227–242, 1996.

Meck, W.H. and Benson, A.M., Dissecting the brain's internal clock: how frontal-striatal circuitry keeps time and shifts attention, *Brain Cognit.*, 48, 195–211, 2002.

Meck, W.H. and Church, R.M., Abstraction of temporal attributes, *J. Exp. Psychol. Anim. Behav. Process.*, 8, 226–243, 1982.

Meck, W.H. and Church, R.M., Simultaneous temporal processing, *J. Exp. Psychol. Anim. Behav. Process.*, 10, 1–29, 1984.

Meck, W.H., Church, R.M., and Olton, D.S., Hippocampus, time, and memory, *Behav. Neurosci.*, 98, 3–22, 1984.

Melgire, M., Ragot, R., Penney, T.B., Meck, W.H., and Pouthas, V., Timing mechanisms for auditory and visual stimuli depend on duration range, submitted.

Miall, C., The storage of time intervals using oscillating neurons, *Neural Comput.*, 1, 359–371, 1989.

Michon, J.A., Magnitude scaling of short durations with closely spaced stimuli, *Psychonomic Sci.*, 9, 359–360, 1967.

Mirsky, A.F., Ingraham, L.J., and Kugelmass, S., Neuropsychological assessment of attention and its pathology in the Israeli cohort, *Schizophr. Bull.*, 21, 193–204, 1995.

Nuechterlein, K.H., Refocusing on attentional dysfunctions in schizophrenia, *Schizophr. Bull.*, 3, 457–468, 1977.

Ornstein, R.E., *On the Experience of Time*, Penguin, Harmondsworth, U.K., 1969.

Penney, T.B., Allan, L.G., Meck, W.H., and Gibbon, J., Memory mixing in duration bisection, in *Timing of Behavior: Neural, Psychological, and Computational Perspectives*, Rosenbaum, D.A. and Collyer, C.E., Eds., MIT Press, Cambridge, MA, 1998, pp. 165–193.

Penney, T.B., Gibbon, J., and Meck, W.H., Differential effects of auditory and visual signals on clock speed and temporal memory, *J. Exp. Psychol. Hum. Percept. Perform.*, 26, 1770–1787, 2000.

Penney, T.B., Holder, M.D., and Meck, W.H., Clonidine-induced antagonism of norepinephrine modulates the attentional processes involved in peak-interval timing, *Exp. Clin. Psychopharmacol.*, 4, 82–92, 1996.

Penney, T.B., Roberts, S., Meck, W.H., Gibbon, J., and Erlenmeyer-Kimling, L., Attention mediated temporal processing deficits in individuals at high risk for schizophrenia, submitted.

Penton-Voak, I.S., Edwards, H., Percival, A., and Wearden, J.H., Speeding up an internal clock in humans? Effects of click trains on subjective duration, *J. Exp. Psychol. Anim. Behav. Process.*, 22, 307–320, 1996.

Posner, M.I., *Chronometric Explorations of Mind*, Erlbaum, Hillsdale, NJ, 1978.

Rammsayer, T., Temporal discrimination in schizophrenic and affective disorders: evidence for a dopamine-dependent internal clock, *Int. J. Neurosci.*, 53, 111–120, 1990.

Rammsayer, T.H., Effects of body core temperature and brain dopamine activity on timing processes in humans, *Biol. Psychol.*, 46, 169–192, 1997a.

Rammsayer, T.H., Are there dissociable roles of the mesostriatal and mesolimbocortical dopamine systems on temporal information processing in humans? *Neuropsychobiology*, 35, 36–45, 1997b.

Rammsayer, T.H. and Lima, S.D., Duration discrimination of filled and empty auditory intervals: cognitive and perceptual factors, *Percept. Psychophys.*, 50, 565–574, 1991.

Rousseau, L. and Rousseau, R., Stop-reaction time and the internal clock, Les Cahiers de Recherche de l'École de Psychologie, Université Laval, Montreal, Quebec, Canada, March 1994.

Rousseau, L. and Rousseau, R., Stop-reaction time and the internal clock, *Percept. Psychophys.*, 58, 434–448, 1996.

Rutschman, J. and Link, R., Perception of temporal order of stimuli differing in sense mode and simple reaction time, *Percept. Motor Skills*, 18, 345–352, 1964.

Sawyer, T.F., Meyers, P.J., and Huser, S.J., Contrasting task demands alter the perceived duration of brief time intervals, *Percept. Psychophys.*, 56, 649–657, 1994.

Steiner, S., Apparent duration of auditory stimuli, *J. Aud. Res.*, 8, 195–205, 1968.

Szelag, E., Kowalska, J., Rymarczyk, K., and Pöppel, E., Duration processing in children as determined by time reproduction: implications for a few seconds temporal window, *Acta Psychol.*, 110, 1–19, 2002.

Treisman, M., Temporal discrimination and the indifference interval: implications for a model of the "internal clock," *Psychol. Monogr.*, 77, 1–31, 1963.

Treisman, M., Faulkner, A., Naish, P.L., and Brogan, D., The internal clock: evidence for a temporal oscillator underlying time perception with some estimates of its characterisitic frequency, *Perception*, 19, 705–743, 1990.

Tysk, L., Estimation of time and the subclassification of schizophrenic disorders, *Percept. Motor Skills*, 57, 911–918, 1983.

Tysk, L., Estimation of time by patients with positive and negative schizophrenia, *Percept. Motor Skills*, 71, 826, 1990.

Wahl, O.F. and Sieg, D., Time estimation among schizophrenics, *Percept. Motor Skills*, 50, 535–541, 1980.

Walker, J.T. and Scott, K.J., Auditory-visual conflicts in the perceived duration of lights, tones, and gaps, *J. Exp. Psychol. Hum. Percept. Perform.*, 7, 1327–1339, 1981.

Warm, J.S., Greenberg, L.F., and Dube, C.S., II, Stimulus and motivational determinants in temporal perception, *J. Psychol.*, 58, 243–248, 1964.

Warm, J.S. and McCray, R.E., Influence of word frequency and length on the apparent duration of tachistoscopic presentations, *J. Exp. Psychol.*, 79, 56–58, 1969.

Wearden, J.H., Human performance on an analogue of an interval bisection task, *Q. J. Exp. Psychol.*, 43, 59–81, 1991.

Wearden, J.H., Edwards, H., Fakhri, M., and Percival, A., Why "sounds are judged longer than lights": application of a model of an internal clock in humans, *Q. J. Exp. Psychol.*, 51, 97–120, 1998.

Wearden, J.H. and Ferrara, A., Stimulus spacing effects in temporal bisection by humans, *Q. J. Exp. Psychol.*, 48, 289–310, 1995.

Wearden, J.H. and Ferrara, A., Stimulus range effects in temporal bisection by humans, *Q. J. Exp. Psychol.*, 49, 24–44, 1996.

Wearden, J.H., Philpott, K., and Win, T., Speeding up and (... relatively ...) slowing down an internal clock in humans, *Behav. Process.*, 46, 63–73, 1999.

Weinstein, A.D., Goldstone, S., and Boardman, W.R., The effect of recent and remote frames of reference on temporal judgments of schizophrenic patients, *J. Abnorm. Soc. Psychol.*, 57, 241–244, 1958.

Zelkind, I., Factors in time estimation and a case for the internal clock, *J. Gen. Psychol.*, 88, 295–301, 1973.

9 Attentional Time-Sharing in Interval Timing

Claudette Fortin

CONTENTS

9.1 Introduction 236
9.2 The Break Location Effect 237
9.3 Interpretation of the Break Location Effect: Attentional Shifts during Pulse Accumulation 238
9.4 Expecting — or Not — a Break in the Absence of a Break 241
9.5 Expectation of a Break and Rate of Accumulation 242
9.6 Uncertainty, Level of Expectancy, and Rate of Accumulation 243
9.7 Experiments 1 and 2: General Method 243
    9.7.1 Participants 243
    9.7.2 Apparatus and Stimuli 243
    9.7.3 Procedure: Training Sessions 244
9.8 Experiment 1: Manipulating Uncertainty about Break Occurrence with Frequency of Trials with Breaks 244
    9.8.1 Procedure 244
    9.8.2 Results and Discussion 245
9.9 Experiment 2: Manipulating Uncertainty about the Time of Break Occurrence with a Forewarning Cue 249
    9.9.1 Procedure 249
    9.9.2 Results and Discussion 250
9.10 General Discussion 252
    9.10.1 Break Expectancy and Uncertainty 252
    9.10.2 The Slope of Production Functions: An Index of Accumulation Rate Determined by Frequency or Duration of Attentional Shifts 252
9.11 The Effect of Expecting a Break: Generality across Stimulus Conditions and Time Estimation Methods 253
9.12 Conclusion 256
Acknowledgments 257
References 258

9.1 INTRODUCTION

Diverting attention from ongoing temporal events shortens perceived duration. This claim was actually made more than a hundred years ago: "Tracts of time ... shorten in passing whenever we are so fully occupied with their content as not to note the actual time itself" (James, 1890, p. 626). This might be why "time flies when we are having fun"; i.e., if we are enjoying some experience, we might not be attending to the time dimension very much and relatively few clock "pulses" are accumulated. In contrast, when we are bored by a lecture or some other occurrence and time seems to drag on and on, we may be attending quite a bit to the time dimension of events, and hence seem to feel that the experience is lasting quite a long time. Impressive experimental support for these statements has been gathered over the last century (e.g., for a recent review, see Brown, 1997). Accordingly, most current models of interval timing consider some form of attentional control in prospective time estimation, when one knows in advance that a time interval must be estimated. One common assumption is that a period of time is estimated through an accumulation of temporal information from an internal source, a process that requires attention (e.g., Hicks et al., 1976; Lejeune, 1998; Meck, 1984; Thomas and Weaver, 1975; Zakay and Block, 1996).

In internal clock models, one source of attentional control is in a pacemaker–accumulator system, where some perceptual representation of duration is built. For example, in Gibbon and Church's (1984) information-processing model of interval timing, pulses continuously emitted from an internal source of temporal information, a pacemaker, are transmitted to an accumulator. Pulses are not continuously transferred; a switch, located between the pacemaker and the accumulator, is closed when an organism is estimating time, which allows pulse transfer during timing. In general, the switch provides a constant source of variability (as opposed to the Poisson variability of the pacemaker and the scalar variability of memory and response thresholds) by closing at the beginning of a duration to be estimated and opening when it ends, interrupting pulse accumulation (Gibbon et al., 1984). Activation of the switch would be under attentional control (Meck, 1984), so that attention is necessary for accumulation to proceed. If attention is diverted from timing, opening of the switch interrupts pulse accumulation. This mechanism results in temporal information still being emitted but not accumulated, which will not contribute to the estimated duration. The net effect of interrupting pulse accumulation during timing of an interval is a shortening of its perceived duration.

Interruption in the accumulation process was inferred from effects of nontemporal processing on concurrent time estimation. Using dual tasks, numerous studies showed that perceived duration shortens as the duration of nontemporal processing increases (e.g., Brown, 1985, 1997; Fortin and Couture, 2002; Fortin and Massé, 1999; Zakay et al., 1983). In time production tasks, shortening of perceived duration results in produced time intervals lengthening with increasing duration of nontemporal processing. In many of these studies, pulse accumulation was assumed to be interrupted because attention was diverted from timing to execute some concurrent task. Interruption has never been manipulated directly and independently from concurrent processing, however, so its effects could not be dissociated from effects of concurrent tasks.

On the other hand, interruption itself was systematically investigated in nonhuman animal research using the peak-interval procedure, with temporary breaks or gaps in stimuli that animals learn to time. In a recent study, we modified an experimental task where visual or memory processing was interpolated in human time production (e.g., Fortin et al., 1993) by replacing the nontemporal task with an empty break (Fortin and Massé, 2000). As in peak-interval experiments with breaks, participants had to interrupt timing during the break in order to correctly produce the target interval. This modification allowed the effects due to interruption to be independent from those due to executing some nontemporal task. The results showed effects related to break location, which seemed actually to be more specifically related to expectation of the break occurrence. The objectives of this chapter are (1) to review data on effects of break location, and their interpretation in terms of attentional time-sharing; (2) to present new data from two experiments where further predictions derived from this interpretation — predictions relating expectation, accumulation of temporal information, and uncertainty — are tested; and (3) to summarize data showing that the effect of break anticipation is general and independent of specific stimulus conditions or time estimation methods.

9.2 THE BREAK LOCATION EFFECT

Fortin and Massé (2000) used a time production task in which participants were trained to produce a given target duration (e.g., 2.5 sec) by pressing a key twice in practice sessions. A tone presentation started on the first key press and ended on the second key press. Feedback on time production accuracy was provided in practice trials, allowing participants to develop a stable and precise performance.

In probe trials with breaks, the task was the same, but with a silent break in tone presentation, which occurred at a predetermined duration after the first key press, as illustrated in Figure 9.1. A typical production trial with a break can be divided in three periods: (1) the prebreak period, which took place between the first key press and the break onset; (b) the silent break period; and (c) the postbreak period, defined as the time between the break offset and the second key press. The task was to execute the second key press when the sum of the prebreak and postbreak periods (a + c) corresponded to the target duration. Participants had therefore to estimate time after the first key press until the break onset, to interrupt timing during the

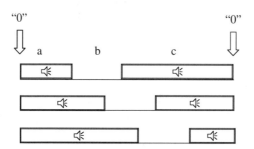

FIGURE 9.1 Time interval production with break in Fortin and Massé's (2000) experiments.

break, and to resume timing at the break offset until they judged that the total tone duration corresponded to the target interval. For example, in experiment 2 of Fortin and Massé (2000) the target duration was 2.5 sec. The prebreak duration (a in Figure 9.1) could be 800, 1300, or 1800 msec, and break duration could be 2, 3, or 4 sec. In a trial, the participant started the temporal production by pressing a 0 key on a keyboard, which triggered a tone lasting for 800, 1300, or 1800 msec. The tone was then interrupted for 2, 3, or 4 sec, and then resumed until the participant ended the temporal production by pressing 0. The participant was instructed to execute this second key press when the total tone duration was equal to the practiced target duration. So in principle, if prebreak duration was 800 msec, the participant had to wait 1700 msec after the break in tone presentation to produce a perfect 2.5-sec target interval.

Location (which corresponds to prebreak duration) of breaks and their duration were varied in these experiments, as illustrated in Figure 9.1. The procedure is similar to that in animal break or gap experiments (e.g., Meck et al., 1984; Roberts and Church, 1978; Roberts, 1981). In those experiments, normal rats seem to be able to time a stimulus before and after a gap in stimulus presentation, while interrupting timing during the gap. This defines a *stop* mode where the switch controlling pulse accumulation from the pacemaker to the accumulator is opened during the break, thus stopping the accumulation process for this period. In human experiments, a gap in stimulus presentation was used similarly to induce a break in time estimation. Produced intervals were defined as the sum of the prebreak and postbreak durations (a + c). In this task, a stop mode would result in produced intervals corresponding to the practiced target duration (e.g., 2.5 sec) and should not be affected by break location or duration.

The mean produced intervals in Fortin and Massé's (2000) experiment 2, averaged over the three break durations, are shown in Figure 9.2 as a function of break location. Perfect production in a stop mode should result in 2500-msec intervals and should not vary with break location. Actually, the main finding was that mean produced intervals lengthened proportionally with increasing value of location, that is, of prebreak duration. In fact, even though mean produced intervals were generally close to the 2500-msec target duration, they were clearly positively related to prebreak duration. This lengthening is called the *break location effect*.

9.3 INTERPRETATION OF THE BREAK LOCATION EFFECT: ATTENTIONAL SHIFTS DURING PULSE ACCUMULATION

The proposed interpretation of the break location effect referred to an internal clock framework where accumulation of temporal information is assumed to require attention. During the prebreak period, attentional time-sharing between pulse accumulation and monitoring the source from which the break signal will come caused some relative loss in accumulation. Because of that loss, the criterion number of pulses corresponding to the target duration would be reached later, hence postponing the end of temporal production. That interpretation had been proposed to explain similar

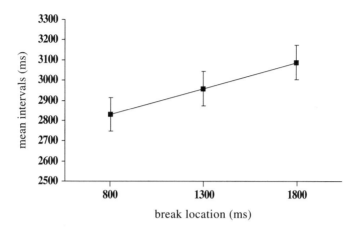

FIGURE 9.2 Results from Fortin and Massé's (2000) experiment 2. Mean produced intervals, not including breaks (±SE), as a function of break location.

effects in time production, where increasing the delay between the start of an interval and presentation of a tone to be discriminated lengthened produced intervals (Rousseau et al., 1984). More precisely, Rousseau et al. (1984) proposed that pulses are emitted and transferred to a counter. Two possible states are assumed in pulse transfer, ON or OFF, so accumulation proceeds in an all-or-nothing mode. Pulse accumulation is under attentional control, so this process is ON when attention is selectively allocated to timing. During an interval production, the proportion of time that pulse transfer is ON is represented by a p parameter. $(1 - p)$ is the proportion of time OFF, when attention is not allocated to timing. An increase of $(1 - p)$ accounts for longer intervals when the tone to be discriminated was delayed, because during this delay, attention was diverted from timing to monitor the auditory channel for tone detection.

Note that effects of location of intervening elements on timing are not specific to temporal production. They have also been observed with time discrimination tasks (e.g., Buffardi, 1971; Casini and Macar, 1997; Israeli, 1930), and an interpretation of attentional time-sharing during pulse accumulation has been considered recently by Casini and Macar (1997) to explain such effects.

Fortin and Massé's (2000) study showed effects of location with no concurrent task by varying the time of occurrence of empty breaks during time production. The time-sharing interpretation of break location in a production task is illustrated in Figure 9.3. In this figure, interval production starts with a first key press (kp1). When no break is expected, as in practice trials, pulse accumulation proceeds until a criterion number of pulses corresponding to the target duration is reached (A), which triggers the second key press, ending the interval (kp2A). When the prebreak duration is 800 msec, accumulation is slower until the break onset (dashed part of the accumulation function) because attentional shifts increase the proportion of time where pulse transfer is OFF. Accumulation is completely interrupted during the break (flat part of the function) and resumes after the break at its usual rate until the criterion is reached (B), which triggers the end of production (kp2B). The time

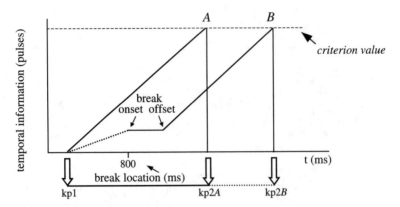

FIGURE 9.3 Pulse accumulation during a time interval production. Produced interval is the time between kp1 and kp2A when no break is expected, in a trial with no break. The time between kp1 and kp2B is the temporal production in a trial with a break, which occurs 800 msec after the beginning of the interval production. In a trial with a break, productions are lengthened by pulse loss in the prebreak duration (dotted line) and by the interruption during the break period (solid horizontal line between break onset and offset).

between the two key presses in production with the break (kp1 and kp2B) is longer than with no break (kp1 and kp2A). Two factors contribute to longer productions in trials with breaks. The most obvious is the duration of the break itself, during which time is elapsing, but timing is interrupted in accordance with instructions. The second factor is the rate of accumulation before break onset, which is slowed down by attentional shifts during the prebreak period. In summary, accumulation is interrupted (1) during the experiment-controlled break itself, and (2) during brief attentional shifts caused by anticipating the experiment-controlled break.

The time-sharing interpretation of the break location effect is illustrated in Figure 9.4. Note that for the sake of simplicity, in that figure, as in the following figures, the break period is no longer included. The two main periods of interest during which the accumulation process operates in these experiments, the prebreak and postbreak periods, are represented with dotted and solid lines respectively. In Figure 9.4, the abscissa represents time, and white arrows represent key presses executed by the participants. The time between key presses corresponds to produced intervals as defined previously: time between the first and second key presses from which the break period (time between break onset and offset in Figure 9.3) has been removed. For example, when a break occurs 800 msec after the first key press, the produced interval, defined as the sum of prebreak and postbreak durations, will be the time between kp1 and $kp2B_1$. When a break occurs 1300 msec after the first key press, the produced interval will be the time between kp1 and $kp2B_2$. Productions are longer when break location is 1300 msec than when it is 800 msec because the longer period of time-sharing (dotted line) before the break causes an extra loss in pulse accumulation. After the break, accumulation resumes at its normal rate in both location conditions (solid lines of accumulation functions). The criterion is reached later when the break occurs later because a larger amount of temporal information

Attentional Time-Sharing in Interval Timing

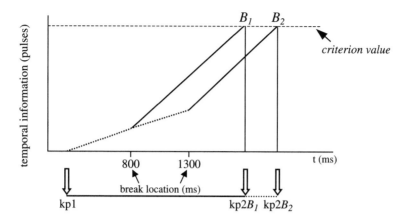

FIGURE 9.4 Pulse accumulation in time production during the prebreak and postbreak periods. Note that the break period itself is *not* represented. Accumulation in time production starts at the first key press (kp1), is slower during the prebreak period (parts of accumulation function shown with dotted lines), and resumes at its usual rate (parts of accumulation functions shown with solid lines) after the break. Production is shown when a break is expected for 800 msec (time between kp1 and kp2B_1) or 1300 msec (time between kp1 and kp2B_2). Production is longer when a break is expected for 1300 than 800 msec because if accumulation is slower for a longer period, the criterion would be reached later.

was lost in the prebreak period. Consequently, produced intervals are longer at a later break location.

While the time-sharing interpretation can account for the break location effect, it is based on an expectancy assumption. It is not the break itself, but its expectation, that induces time-sharing. Experiments 3 and 4 from Fortin and Massé's (2000) study provided a test of the expectancy assumption. In these experiments, trials with and without breaks were mixed within blocks of trials.

9.4 EXPECTING — OR NOT — A BREAK IN THE ABSENCE OF A BREAK

In trials with breaks from Fortin and Massé's (2000) experiment 3, production of a 2.6-sec duration was interrupted for 1, 2, or 3 sec after 600, 1100, 1600, and 2100 msec. There were no breaks in 20% of the trials. Therefore, a break was presumably expected until its occurrence in trials with breaks, and for the most part of target duration in trials with no breaks. The expectation assumption of the time-sharing hypothesis leads to a prediction that produced intervals would be positively related to prebreak duration in trials with breaks, with longest intervals being produced in trials with no breaks. The data clearly supported the prediction: overall, productions were positively related to prebreak duration, and longest productions were observed in trials with no breaks.

The most conclusive support for the attentional time-sharing interpretation was obtained when expectancy was manipulated in trials with no breaks (Fortin and

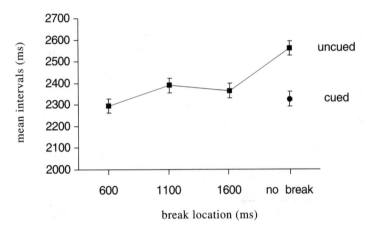

FIGURE 9.5 Results from Fortin and Massé's (2000) experiment 4. Mean produced intervals, not including breaks (±SE), as a function of break location in trials with and without breaks, in the cued and uncued conditions.

Massé, 2000, experiment 4). In that experiment, trials with and without breaks were mixed, but there were two types of no-break trials, cued and uncued. In cued trials, a short message, presented on a computer screen at the start of the trial, forewarned participants of the break absence. Participants were therefore not expecting a break in these cued trials in principle, whereas a break was expected in the other half of the no-break trials where no cue was presented. A strong prediction of the expectancy interpretation was that production in identical trial conditions without breaks should differ depending on a break being uncued or cued, thus expected or not during the interval production. Intervals should be longer in uncued trials, where participants are expecting a break, than in cued trials, where participants, being forewarned of the break absence, are not expecting a break. As shown in Figure 9.5, this pattern of results was actually observed: in trials with no breaks, productions were significantly longer in uncued trials than in cued trials, that is, when participants were expecting a break than when no break was expected. Overall, longest productions were obtained in uncued trials with no breaks.

9.5 EXPECTATION OF A BREAK AND RATE OF ACCUMULATION

Expecting a break in timing is assumed to cause attentional shifts, which would slow down the rate of pulse accumulation. In Figures 9.3 and 9.4, two rates of accumulation corresponding to two extreme levels of expectancy for a break are represented: accumulation is relatively slow or fast depending on a break being expected or not. Accumulation is fast when a break is not expected for the whole duration of a temporal production (A), or when it is no more expected after the break offset in trials with breaks (B in Figure 9.3, B_1 and B_2 in Figure 9.4). Accumulation is slow when a break is expected, for example, during the prebreak period in trials with breaks.

The inferred relationship between the rate of accumulation and break expectancy suggests that the rate of accumulation itself could be manipulated by varying the *level* of expectancy for a break. One way of varying level of expectancy is to manipulate the degree of certainty about some event.

9.6 UNCERTAINTY, LEVEL OF EXPECTANCY, AND RATE OF ACCUMULATION

Uncertainty is a factor classically related to expectation in psychological research (e.g., Kahneman and Tversky, 1982; Sanders, 1966; for a discussion related to timing, see also Barnes and Jones, 2000). According to a probabilistic view of expectancy, the more certain you are about some event, the greater your expectation is about this event. Consequently, increasing the degree of certainty about break occurrence should increase the level of expectancy for this occurrence. The increase in expectancy level should intensify attentional shifting between accumulation and monitoring for the break signal, which should in turn reduce the rate of pulse accumulation. Two new experiments are reported here where evidence supporting this hypothesis was found.

In experiment 1, uncertainty was manipulated by varying the relative frequency of trials with and without breaks. Two main conditions were compared: a break was highly probable in the *high-frequency condition* and less probable in the *low-frequency condition*. The expectancy level about break occurrence should be higher in the high-frequency condition, thus leading to stronger effects of break location.

In experiment 2, all trials included a break. There was therefore no uncertainty concerning the break occurrence itself, but uncertainty concerning the time of break occurrence was manipulated. Two main conditions were compared: a cued condition, where a cue, presented at a fixed 500-msec duration before the break onset, informed participants that the break was imminent, and a no-cue condition, where this information was not provided. By eliminating uncertainty about the time of its occurrence, the cue should induce a maximum level of expectancy for the break, thus interrupting pulse accumulation.

9.7 EXPERIMENTS 1 AND 2: GENERAL METHOD

The methods were very similar in experiments 1 and 2. The apparatus, training procedures, and general method also shared many features with those of Fortin and Massé (2000), where additional details are provided.

9.7.1 PARTICIPANTS

Twenty participants, 12 women and 8 men, aged between 20 and 38 years old ($M = 22.7$), received a small honorarium for their participation.

9.7.2 APPARATUS AND STIMULI

A PC-compatible computer running Micro Experimental Laboratory software controlled stimulus and feedback presentations. The preferred hand rested on the numerical keyboard, the 0 key being used to produce time intervals. Intervals were recorded to the nearest millisecond.

9.7.3 Procedure: Training Sessions

Production of target duration, 2900 msec, was practiced in two sessions comprising five 48-trial blocks. In the first four blocks, feedback was provided after the temporal production, informing the participant whether the interval was too short, too long, or correct, within a temporal window of 10% (±145 msec) centered on the target interval. No feedback was provided in the last block. Practice sessions were followed by experimental sessions. All sessions were completed on successive days.

9.8 EXPERIMENT 1: MANIPULATING UNCERTAINTY ABOUT BREAK OCCURRENCE WITH FREQUENCY OF TRIALS WITH BREAKS

This experiment was designed to test the effect of frequency of trials with breaks in two groups of participants. The main hypothesis was that the effect of expecting a break would be stronger in the high-frequency group than in the low-frequency group.

A secondary objective of experiment 1 was to use new stimulus conditions, mainly to enhance generality of results. In Fortin and Massé's (2000) experiments, break periods were delimited by a silent interruption in tone presentation: the tone was interrupted and resumed at the beginning and the end of the break, respectively. These stimulus conditions are similar to those in the standard gap procedure used in animal studies, where a gap in stimulus presentation is used to induce a break in timing. In the present experiment, the stimulus conditions were reversed: a tone onset indicated the beginning of the break period, and a tone offset, the end of the break. Performance in a similar reversed-gap procedure seemed to differ from that in a standard gap procedure in recent animal experiments (Buhusi and Meck, 2000, 2002).

9.8.1 Procedure

Twenty participants were randomly assigned in equal number to one of two frequency conditions. In the low-frequency condition, 46% of trials in a block included a break, whereas 92% of trials included a break in the high-frequency condition. For all participants, break duration and location were varied within blocks, and trials with no breaks were mixed among trials with breaks. In trials with breaks, break location was 700, 1200, 1700, or 2200 msec and break duration was 1, 2, or 3 sec.

All participants were tested in two break-signal conditions in separate sessions. A tone (500 Hz) was presented continuously during the break period in a continuous break-signal condition. This condition was compared with a discontinuous condition, where the break period was bounded by two 50-msec tones (500 Hz), the break beginning with the first tone onset and ending with the second tone offset. Experimental trials in both conditions are illustrated in Figure 9.6.

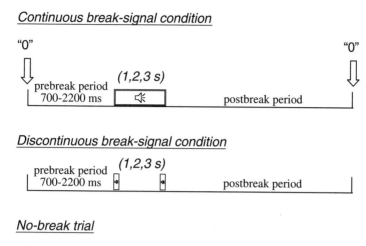

FIGURE 9.6 Experimental trials in the continuous break-signal condition. Participants start the temporal production by pressing the 0 key. After a silent prebreak duration, a tone is presented for 1, 2, or 3 sec in trials with breaks. The task is to end the temporal production when the total silent duration corresponds to the target duration to be produced, 2900 msec.

Therefore, a 2 (frequency of trials with breaks) × 4 (break location) × 3 (break duration) × 2 (discontinuous or continuous break signal) mixed design was used, with repeated measures on the latter three factors.

The exact proportion of trials with and without breaks was not mentioned to the participant. In the high-frequency condition, the instructions were that "Most trials will include a break, during which time estimation must be interrupted." In the low-frequency condition, "most trials" was replaced by "some trials." At the beginning of a session (continuous or discontinuous), the experimenter informed the participant whether a continuous or a discontinuous break signal was used.

There were four experimental sessions: two successive sessions in the continuous and two successive in the discontinuous break-signal conditions, the continuous and discontinuous conditions being presented in counterbalanced order. Each session included one 48-trial block of practice trials with feedback, followed by four 26-trial experimental blocks with no feedback.

9.8.2 RESULTS AND DISCUSSION

The data submitted to analyses were time intervals produced in trials with breaks, not including break duration. For example, if in a trial with a 3-sec break, the time between the first and the second key press was 7400 msec, the datum used in the analysis was 4400 msec. For each participant, means and standard deviations of produced intervals were computed and data more than 3 SDs from the means were discarded, which represented 70 of 8320 observations. In both low-frequency and high-frequency groups, the data were averaged over trials with breaks to obtain a

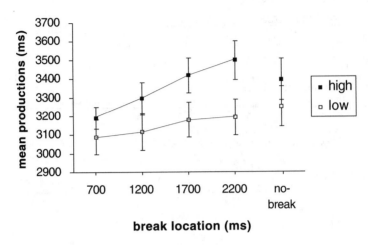

FIGURE 9.7 Results from experiment 1. Mean produced intervals, not including breaks (± SE), as a function of break location in trials with and without breaks, in conditions of high and low frequency of trials with breaks.

mean produced interval at each combination of break location, duration, and type of break signal. A mean produced interval was also computed in trials with no breaks.

Mean intervals not including break duration are shown as a function of break location in the high- and low-frequency conditions in Figure 9.7. In trials with breaks, mean intervals increased linearly with increasing break location in the high-frequency condition, $R^2 = .99$, with a mean slope of 0.21 msec. The relation was strongly linear in the low-frequency condition as well, $R^2 = .94$, but with a much lower slope, 0.08 msec. Slopes of productions as a function of break location in trials with breaks were computed for each participant. A t-test for independent samples showed that the mean slope was significantly greater in the high-frequency condition ($M = 0.21$, $SD = 0.13$) than in the low-frequency condition ($M = 0.08$, $SD = 0.11$), $t(18) = 2.46$, $P < .05$. These results show generally that the break location effect, attributed to expectation of a break, is stronger when trials with breaks are more frequent. This observation supports our main hypothesis: increasing the degree of certainty about break occurrence increases the level of expectancy for a break, hence strengthening the break location effect.

An analysis of variance (ANOVA) was performed on mean produced intervals not including breaks, in trials with breaks only, with one nonrepeated (break frequency) and two repeated (break location and type of break signal) factors. The interaction between break location and frequency was significant: $F(3, 54) = 3.46$, $P < .05$. The lengthening of produced intervals as a function of break location was stronger in the high- than in the low-frequency condition. The effect of break location was also significant: $F(3, 54) = 15.91$, $P < .0001$. Productions lengthened with increasing prebreak duration. Mean intervals did not differ significantly in low- and high-frequency conditions: $F(1, 18) = 2.82$, $P > .05$. No other effects were significant.

As shown in Figure 9.7, in the low-frequency condition, productions in no-break trials seem to follow the general linear trend present in trials with breaks. This is not the case in the high-frequency condition, where mean productions were even shorter in trials with no breaks than at the longest prebreak duration, 2200 msec, although a t-test for dependent samples showed that the difference was not significant ($M = 3498.1$, $SD = 329.74$ vs. $M = 3393.9$, $SD = 349.97$; $t(9) = 2.00$).

The results of experiment 1 confirm the break location effect found in previous experiments: participants produced longer intervals when they were expecting a break for a longer duration. This supports Fortin and Massé's (2000) conclusion that expecting the interruption of a signal being timed perturbs pulse accumulation, as illustrated in Figures 9.3 and 9.4.

The main finding here, however, is the interaction between break location and frequency of trials with breaks. The lengthening of produced intervals as a function of break location was more pronounced when the degree of certainty about the break occurrence was high. These results suggest a slower rate of accumulation during the prebreak period when participants are relatively certain that a break will occur. The slower rate of accumulation could result from an increase in frequency of attentional shifts. An illustration of the attentional-shift interpretation, showing pulse accumulation in both frequency conditions, is shown in Figure 9.8.

Accumulation starts on the first key press and proceeds with brief interruptions due to attentional shifts until the break occurs. Figure 9.8 shows the process for break location of 2200 msec. In the low-frequency condition, attentional shifts are assumed to be less frequent so that when the break occurs, the pulse count would be higher than in the high-frequency condition. After the break, accumulation resumes at its normal rate in both conditions until the criterion is reached, which

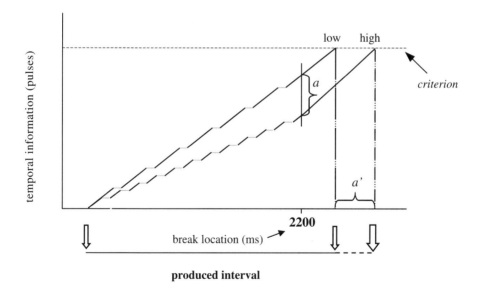

FIGURE 9.8 Pulse accumulation in conditions of low and high frequency of trials with breaks. Accumulation is faster in the low- than in the high-frequency condition, because attentional shifts are less frequent. The criterion is reached earlier when accumulation is faster, resulting in relatively short produced intervals.

triggers the end of interval production. The difference in accumulated counts at the end of prebreak duration (a in Figure 9.8 when break location is 2200 msec) would determine the difference between produced intervals (a') in the high- and low-frequency conditions. Because the temporal criterion is reached earlier in the low- than in the high-frequency condition, produced intervals are shorter. The difference would be proportional to prebreak duration: we can deduce from Figure 9.8 that the difference in accumulated counts between the low- and high-frequency conditions would be smaller if the break occurred earlier, for example, 700 msec after the first key press rather than 2200 msec. Differences between produced intervals in the low- and high-frequency conditions are thus proportional to prebreak duration, which results in the observed interaction between prebreak duration and frequency of trials with breaks.

The relation between produced intervals and prebreak duration is linear in both high- and low-frequency conditions, suggesting that the duration and frequency of shifts are relatively stable throughout the prebreak duration. Of course, even though shifts are equally spaced and of the same duration for illustrative purposes in Figure 9.8, we do not assume this to be the case.

In this interpretation, accumulation proceeds according to an ON–OFF operation mode, depending on whether time is selected as the focus of attention. Accumulation would be under control of a mechanism compatible with a flickering switch, discussed in recent articles on the role of selective attention in timing (Lejeune, 1998, 2000; Penney et al., 2000; Zakay, 2000). In the information-processing model of interval timing (Church, 1984; Gibbon and Church, 1984; Gibbon et al., 1984), pulses are accumulated when the switch is closed, and accumulation is interrupted when the switch is opened. Given the constraints of a production task, the switch is normally closed during the interval production, allowing pulse accumulation. The state of the switch is modified in two periods of interval production with breaks. First, it is opened during the break itself, interrupting accumulation (e.g., Meck et al., 1984; Roberts and Church, 1978; Roberts, 1981). Second, it alternates between closed and opened states during the prebreak period (Fortin and Massé, 2000). The results from the present experiment suggest that the level of expectancy for a break controls the frequency of changes in the state of the switch during the prebreak period: the switch would be opened more frequently when a break is strongly expected. Note, however, that longer openings of the switch would produce the same results.

The slope of mean productions as a function of prebreak duration is almost three times greater in the high-frequency condition than in the low-frequency condition (0.21 vs. 0.08). This could be taken as an estimate of the relative frequency (or duration) of OFF periods in the two conditions. On average, OFF periods would be three times more frequent (or longer) in the high- than in the low-frequency condition.

Mean intervals in trials with no breaks, shown in Figure 9.7, seem to follow the linear trend of the production function in the low-frequency condition, but not in the high-frequency condition. This may be related to the difference in proportion of trials with and without breaks. Because almost all trials included a break in the high-frequency condition, participants could develop a precise representation of the four possible break locations so that when the last possible location was past, they stopped expecting a break. That would explain why productions are not longer than they were at the last break location in the high-frequency condition. In contrast, a minority of trials included a break in the low-frequency condition. Representation of the possible break locations was therefore probably much less accurate, making it more likely that participants were still expecting a break after the last possible break location. In these conditions, productions should be longer in the no-break trials than in trials at the longest prebreak duration, which was observed in the low-frequency condition.

Given that the primary interest here was to identify the attentional mechanisms involved in break expectancy, the focus of data analysis in experiment 1 was on the break location effect. Break duration was therefore not included in our analysis.

Overall, the results of experiment 1 replicate and extend those obtained in Fortin and Massé's (2000) study. As in this study, we noted in postexperimental interviews that participants were unaware of the effect of prebreak duration. This suggests that time-sharing during break expectancy is not under voluntary control.

9.9 EXPERIMENT 2: MANIPULATING UNCERTAINTY ABOUT THE TIME OF BREAK OCCURRENCE WITH A FOREWARNING CUE

In experiment 2, uncertainty concerning the time of break presentation was manipulated. Two main conditions were compared: a cued condition, in which a cue presented 500 msec before the break onset informed participants that the break was imminent, and a no-cue condition, in which this information was not provided. In research on visual attention, cues for visual targets orient attention by reducing uncertainty about its spatial location (e.g., Posner, 1980). We applied the same reasoning to the temporal domain here by presenting a forewarning cue at a fixed time, 500 msec, before the break signal. By reducing uncertainty about temporal location of the break, the cue should have a significant impact on orienting attention toward the break occurrence. On the other hand, because the cue is perfectly valid and is presented on every trial in the cued group, another possible result is that, knowing they will be forewarned of the break occurrence, participants will stop shifting their attention. This implies, however, that attentional time-sharing in these conditions is under voluntary control and may be prevented.

The task was basically the same as that in experiment 1 in the continuous break-signal condition, with all trials including a break.

9.9.1 PROCEDURE

Twenty participants were randomly assigned in equal number to one of the two cue conditions. In the cued condition, a row of asterisks was presented on a screen 500 msec before the onset of break signal. This condition was compared with a no-cue condition, where no asterisks were presented. Location and duration of breaks were varied from trial to trial. Break location had four levels, 700, 1200, 1700, and 2200 msec; break duration had three levels, 1, 2, and 3 sec. Thus, a 2 (cue or no cue) × 4 (break location) × 3 (break duration) mixed design was used, with repeated measures on the latter two factors.

The two training sessions were followed by two experimental sessions, either in the cue or in the no-cue condition, depending on the group of participants. The data submitted to analyses were collected in the two experimental sessions, which comprised four 24-trial experimental blocks.

Participants were informed that a tone was presented in all trials. In both cue and no-cue groups, the instructions were to interrupt timing at the beginning of tone presentation and to resume timing at the tone offset. In the cued group, participants were told that they would be forewarned of the break location with a row of asterisks presented on the screen at a constant duration before the tone onset.

9.9.2 Results and Discussion

As in experiment 1, the analyses were carried out on produced intervals not including break duration, that is, on the sum of prebreak and postbreak durations. For each participant, means and standard deviations of temporal intervals not including break duration were computed and data more than 3 SDs from the means were discarded, which represented 30 of 3840 observations. Data were averaged in the cued and uncued groups of participants to obtain a mean produced interval at each combination of levels of break location and duration. An ANOVA with one repeated factor, break location, and one nonrepeated factor, cue or no cue, was carried out on these means.

Mean intervals not including break duration are shown as a function of break location in the cue and no-cue conditions in Figure 9.9. The usual break location effect was found: mean intervals increased with increasing prebreak duration $F(3, 54) = 19.70$, $P < .0001$. More importantly, they were generally longer in the cue ($M = 3657$ msec) than in the no-cue ($M = 3184$ msec) condition $F(1, 18) = 5.22$, $P < .05$. The interaction between break location and the cue or no-cue condition was not significant $F(3, 54) = 1.28$, $P = .29$. Note that the cue and no-cue groups

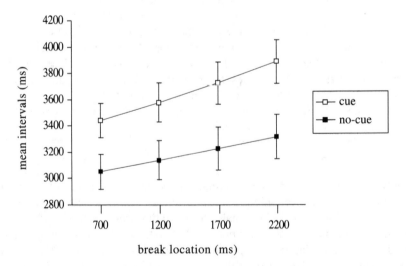

FIGURE 9.9 Results from experiment 2. Mean produced intervals, not including breaks (± SE), as a function of break location in the cue and no-cue conditions.

Attentional Time-Sharing in Interval Timing

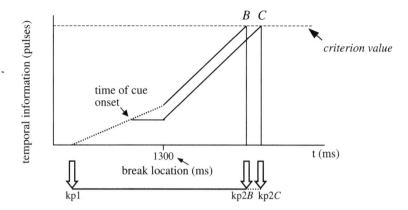

FIGURE 9.10 Pulse accumulation in the cue and no-cue conditions of experiment 2. In both conditions, accumulation is slower when a break is expected. When a forewarning cue is presented, however, attention is oriented toward anticipation of the break, thus interrupting completely (or almost completely) the accumulation process. Premature interruption in the cue condition results in longer produced intervals.

did not produce different mean intervals in practice trials with no feedback: $F(1, 18) = 1.62$, $P = .22$. Thus, the two groups could be considered as equivalent as to their temporal performance before introduction of the cue manipulation.

The key finding in this experiment is that presenting a forewarning cue 500 msec before the break signal lengthened produced intervals of about the same duration, 473 msec, a significant difference. This may be explained by an interruption in pulse accumulation as soon as the cue is presented, in the cued condition. This interpretation is illustrated in Figure 9.10.

In the no-cue condition, accumulation starts on the first key press and proceeds with brief interruptions due to attentional shifts (dotted line in the accumulation function) until the break occurs, for example, at 1300 msec. After the break, accumulation resumes at its usual rate until the criterion is reached (B), which triggers the end of production (kp2B). In the cue condition, accumulation also begins at the first key press. When the cue is presented, for example, 500 msec before the break signal, attention is totally oriented to anticipation of the break. This could result in complete (illustrated in Figure 9.10) or almost complete interruption in pulse accumulation on cue presentation. The criterion would be reached later with this premature interruption (C), postponing the end of interval production (kp2C). Because the cue is presented at a constant duration before the break signal, the net result would be mean intervals of relatively constant longer duration in the cue condition than in the no-cue condition, at all values of break duration, as observed in experiment 2.

These data support the hypothesized link between the degree of certainty, level of expectancy, and attentional shifts from accumulation. Expectancy implies some uncertainty; when subjective probability for one event approaches certainty, expectancy for an event becomes a "set" toward it. Attentional set is usually defined as directing attention toward one object or dimension with exclusion of other possibilities (Sanders, 1966). Because it was perfectly reliable, the cue, when it was presented, established a set toward anticipation of the break signal, while excluding temporal processing.

9.10 GENERAL DISCUSSION

9.10.1 BREAK EXPECTANCY AND UNCERTAINTY

Expectation of a break in timing seems strongly influenced by uncertainty about the break occurrence. The two experiments reported here suggest this to be the case whether uncertainty is varied by manipulating information on a global or a local level. Thus, in experiment 1, information was manipulated on a global level, because subjective probability about break occurrence developed over trials. In experiment 2, local information, a cue, was provided within each trial to inform participants of the break arrival. In both cases, longer intervals were produced when participants were more certain about the break, suggesting that pulse accumulation was affected.

Results from Fortin and Massé's (2000) study also support the hypothesized relationship between break expectancy and uncertainty. In that study, the strongest break location effect was obtained when all trials included a break, that is, when participants were certain that a break would occur (experiment 2; see results in Figure 9.2 of this chapter). The effect tended to weaken when trials with no breaks were included, hence reducing certainty about the break (experiment 4; see results in Figure 9.5 of this chapter). Furthermore, the effect was almost abolished when participants knew before starting the interval production that there would be no break in the current trial (cued condition of experiment 4; shown in Figure 9.5 of this chapter). In that condition where participants were informed that no break would occur, productions shortened drastically, suggesting that accumulation was relatively undisturbed. The opposite was obtained in the cued condition of experiment 2 reported here, where the cue had the exact opposite function, that is, made participants certain of the break occurrence. This presumably induced a complete or almost complete interruption in pulse accumulation (see Figure 9.10), hence longer productions in the cue condition (see Figure 9.9).

9.10.2 THE SLOPE OF PRODUCTION FUNCTIONS: AN INDEX OF ACCUMULATION RATE DETERMINED BY FREQUENCY OR DURATION OF ATTENTIONAL SHIFTS

In experiment 1, the slope was 0.08 msec in the low-frequency condition (46% of trials with breaks), which suggests that about 8% of prebreak duration was lost in this condition because of attention sharing. In the high-frequency condition (92% of trials with breaks) the slope is 0.21 msec, suggesting that the loss was more than doubled. In experiment 2, where all trials included a break, the slope (averaged over the cued and uncued conditions) is 0.24 msec. Interestingly, this value is very close to the slope obtained in a similar experiment, where 100% of the trials included a break, 0.26 msec (Fortin and Massé, 2000, experiment 2; see results in Figure 9.2 of this chapter).

In Rousseau et al.'s (1984) experiment 2, where in all trials a tone to be discriminated was presented at different locations during a 2-sec production, the corresponding slope was 0.38 msec. This value was taken as an estimate of $(1-p)$, the probability of pulse transfer to be in an OFF state. The slope of production functions is interpreted in a comparable way here, as an index of accumulation rate and, consequently, of the loss in accumulation when it operates in a flickering switch mode.

9.11 THE EFFECT OF EXPECTING A BREAK: GENERALITY ACROSS STIMULUS CONDITIONS AND TIME ESTIMATION METHODS

The effect of expecting a break appeared in the two experiments of the present report in spite of differences in stimulus conditions relative to Fortin and Massé's (2000) experiments, where the effect was first observed. Whereas the break period was indicated by an interruption in tone presentation in previous experiments (see Figure 9.1), new stimulus conditions were used here, with the break period marked by a tone presentation (see Figure 9.6) in the continuous break-signal condition of experiment 1, and by two tones bounding the break period in the discontinuous signal condition. The break location effect did not vary across stimulus conditions, which was also observed when different stimulus conditions (visual vs. auditory, filled vs. empty breaks) were manipulated within experimental sessions, in a within-subject design (Fortin and Bédard, 1999).

We have also observed effects of break expectation in time discrimination studies (Tremblay et al., 2001; Tremblay and Fortin, submitted). In one experiment, two tones of 2.5 and 3.0 sec were used as short and long durations, respectively. Participants were first familiarized with the two target durations. In each of the following experimental trials, one of the two tones was presented with a break in tone presentation. Break location was varied: it could occur when 30, 50, or 70% of the tone duration had elapsed. Break duration, as well as break location, was varied from trial to trial. The stimuli used in that experiment are illustrated in Figure 9.11. The task was to classify the total tone duration (prebreak plus postbreak) as corresponding to the short or long target duration by providing a "short" or "long" response.

A shortening of perceived duration with increasing prebreak duration was suggested in responding to both short and long durations: the proportion of correct responses increased with increasing prebreak duration in "short" trials, whereas it decreased in "long" trials. Overall, participants responded short more often as prebreak duration increased, a result compatible with previous data obtained in discrimination experiments when location of a nontemporal task was manipulated (Buffardi, 1971; Casini and Macar, 1997).

The break location effect was large and reliable: substantial differences in proportion of short responses were observed with varying break location. More importantly, when the effect was tested over a wider range of location values, the function relating the proportion of short responses to prebreak duration was not linear, but sigmoid. Thus, differences in proportions of short responses were reduced at extreme values of break location, that is, when two values of break location placed near the

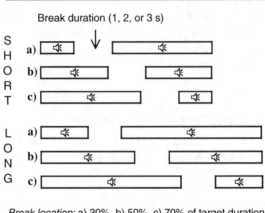

FIGURE 9.11 Trials in Tremblay et al.'s (2001) and in Tremblay and Fortin's (submitted) experiments. A short or long tone was presented, with a break in tone presentation. Break location and duration were varied.

beginning (or near the end) of tone presentation were compared. For example, in one experiment, the proportion of short responses was higher when the break occurred when 65% of tone presentation had elapsed than when it occurred at 50% of tone presentation (0.62 vs. 0.55, respectively), but was similar whether the break occurred at 65 or 80% of tone presentation (0.62 vs. 0.63, respectively).

These results from discrimination experiments may be interpreted in part as an effect of attentional time-sharing due to break expectation, as illustrated in Figures 9.3 and 9.4. In that respect, the interpretation is essentially the same as that used to explain the effect of varying location of empty breaks (Fortin and Massé, 2000) and of a concurrent task (Rousseau et al., 1984) in a production paradigm, as well as of a concurrent task in time discrimination (Casini and Macar, 1997). This interpretation could not account for the whole pattern of results in time discrimination with breaks, however, especially for the flattening of the function relating the proportion of short responses to break location at extreme values of break location. In fact, slower accumulation due to break expectation would predict strictly linear functions, with a regular increase in proportion of short responses with increasing prebreak duration at any value of break location.

A more accurate account of the results in time discrimination with empty breaks is provided by coupling the attentional time-sharing hypothesis illustrated in Figures 9.3 and 9.4 to a real-time criterion hypothesis (Kristofferson, 1977). According to this hypothesis, the response would be determined by the outcome of a race between a criterion value, possibly based on the short duration, and the presented stimulus. For example, on presentation of one of the two target durations, accumulation would start. As illustrated in Figure 9.12, it would stop if either one of the two following conditions is met: (a) the tone offset indicates the end of the duration to be estimated, or (b) the criterion is met. If (a) occurs before (b), the tone will be perceived as

Attentional Time-Sharing in Interval Timing

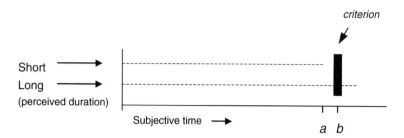

FIGURE 9.12 Perceived duration as a result of an accumulation process in time discrimination. In a two-choice time discrimination task, a tone will be perceived as short if (a) its presentation ends before (b) a criterion is met. (a) is more likely to occur before (b) when a short tone is presented or when there is subjective shortening caused by slower accumulation during break expectancy. When tone presentation ends well before (or well after) the criterion is crossed, as at extreme values of break location, a small increase in duration of break expectancy will not change the decision to classify the tone as short or long.

short. If (b) occurs before (a), the tone will be perceived as long. Of course, the end of accumulation because of tone offset (a) is more probable when the short duration is presented, whereas reaching the criterion (b) first is more likely in trials where a long duration is presented.

As in time production, expecting a break induces attentional shifts that cause brief interruptions in accumulation, hence slowing down pulse accumulation. For example, if a break is located at 75% of the target duration, accumulation would be slower during the prebreak period and would resume at its usual rate after the break. At the end of the target duration, the criterion might not be reached on these trials because of the relatively important loss in accumulation due to attentional time-sharing in the prebreak period. In comparison with a trial where accumulation proceeds with no time-sharing, expecting a break would change the outcome: a decision for a short rather than a long response would be made.

Generally, increasing prebreak duration makes the relative loss more important, thus making the end of tone presentation (Figure 9.12a) occurring before the criterion is met (Figure 9.12b) more probable, hence increasing the probability of a short response. At extreme values of break location, however, increasing prebreak duration may not change the proportion of short responses, because increasing the duration of expectancy would not change the outcome of accumulation relative to the criterion. For example, if a break occurs when 75% of the target duration has elapsed, perceived duration will be quite short because a break was expected for the most part of tone presentation. In this case, as shown in Figure 9.12, (a) will occur well before (b), and a short response will be provided. Increasing further prebreak duration, for example, to 85% of tone duration, will shorten perceived duration even more, because of additional loss in pulse accumulation due to the break being expected for a longer period of time. This extra loss will not change the decision in classifying the tone as short or long, however, because (a) would still occur before (b) and, in fact, would even occur earlier, in subjective time, than when the break is located at 75% of tone presentation. At a break location of 85%, as at a break location of 75%, a short response would therefore be decided. Similarly, changing the break location from 15 to 25% of target duration could have no effect on the outcome of accumulation relative to the criterion. For example, when a long duration is presented, the criterion could be crossed in both cases before the end of target duration, leading to a long response at both break locations.

Finally, as with time production, the shortening of perceived duration may not be attributed to some weighting or comparison of prebreak and postbreak durations. In effect, there is a clear shortening of perceived duration in no-break trials, when a break is expected for the most part of the stimulus duration, but does not occur. Thus, in one experiment of Tremblay and Fortin's (submitted) study, the proportion of short responses was higher when a break was expected but did not happen than in most trials with breaks. Dividing tone presentation in prebreak and postbreak periods is not necessary for the subjective shortening of target duration to occur because the tone is also perceived shorter if a break is expected during its uninterrupted presentation.

9.12 CONCLUSION

Expecting a break in time estimation shortens perceived duration. In time production, the effect is directly proportional to the duration for which the break is expected. Moreover, perceived duration is shortened further by increasing certainty about the break occurrence: higher levels of expectancy would increase the time allocated to monitoring the source from which the break signal will come, leading to greater loss in accumulation of temporal information. Taken together, the data presented here confirm the generality of the phenomenon. They also support its interpretation in terms of expectancy because it appears to be modulated by certainty, as are other expectancy-related effects in psychology.

The empirical work reviewed in this chapter is interpreted within an internal clock framework, where accumulation of temporal information is under attentional control (Gibbon et al., 1984; Meck, 1984; Zakay and Block, 1996). Expecting an interruption in accumulation perturbs this process, possibly because of attentional shifts from accumulation to monitor for the break signal. Linear functions in our time production experiments with breaks suggest that attentional shifts are regularly distributed during the expectancy period, as illustrated in Figure 9.8. The production data reported in the current chapter suggest that the total duration of shifts seems longer when certainty about the break is higher, which would result from more frequent or longer shifts from accumulation.

In time discrimination, the effect seems functionally similar, although a detailed analysis of performance suggested that the response is determined by a decisional criterion in that paradigm.

The role of attention in time estimation is a fundamental issue that may be analyzed with a variety of experimental strategies. Expectancy effects in temporally structured sequences of events have been thoroughly analyzed (e.g., Jones and Boltz, 1989; Jones et al., 1993). Dual tasks, where participants execute some nontemporal task and estimate its duration simultaneously, have been widely used to study attention in human timing research (e.g., Brown, 1985, 1997; Hicks et al., 1976; Macar, 1996; McClain, 1983; Zakay et al., 1983). Interpolating nontemporal tasks in time production (e.g., Burle and Casini, 2001; Fortin and Rousseau, 1987; Rousseau et al., 1984; for a review of studies using this paradigm, see also Fortin, 1999) or in time discrimination (e.g., Casini and Macar, 1997) has also been used in the same purpose. However, even though disruption or interruption in accumulation of temporal information was often inferred to interpret results from dual-task studies, none of them addressed the problem of interruption in timing itself. In contrast, effects of breaks in stimuli to be timed were systematically investigated using the peak-interval procedure in animal timing research (e.g., Meck et al., 1984; Roberts, 1981; Roberts and Church, 1978), which provided reference data and a useful theoretical framework to interpret interruption-related effects. In this chapter we reviewed recent experiments with human participants inspired from these break or gap experiments. The results show strong attentional effects related to expecting a break in human timing, in the absence of any concurrent nontemporal task during the interruption. We showed that these effects are reliable and may be generalized to diverse timing conditions. An internal clock framework — where brief closures of a flickering switch under attentional control during pulse accumulation are induced by break expectation — may account for these data. One major point made by these experiments is that attentional effects related to expecting an interruption in timing must be distinguished and considered independently from any additional cost of concurrency between temporal and nontemporal processes usually assumed to interrupt or disrupt timing.

Although results with human participants are often similar to those in animal experiments, they differ in some respects. For example, effects of break location, consistently observed with humans, as reviewed in this chapter, seem weaker and are not always reliable with rats (see, for example, Roberts, 1981, experiment 2; Meck et al., 1984). In gap experiments, rats and pigeons seem to use different processing rules when stimulus conditions are changed (Buhusi, this volume; Buhusi and Meck, 2000, 2002; Buhusi et al., 2002), which is not the case with humans, as shown in the two experiments reported here. Some discrepancies in results may be explained by differences in procedures, for example, by training with no-break trials, which is much more intensive in animal experiments than in human ones. Data from human and animal experiments with breaks have still to be compared systematically, however. In the future, such comparison should provide invaluable information on the role of attention in timing and, more generally, on rules common to animal and human timing.

ACKNOWLEDGMENTS

The work reported here was supported by a grant from the Natural Sciences and Engineering Research Council of Canada (NSERC) to C. Fortin. The author thanks Robert Rousseau and Sébastien Tremblay for useful comments on an earlier draft of the chapter.

REFERENCES

Barnes, R. and Jones, M.R., Expectancy, attention, and time, *Cognit. Psychol.*, 41, 254–311, 2000.

Brown, S.W., Time perception and attention: the effects of prospective versus retrospective paradigms and task demands on perceived duration, *Percept. Psychophys.*, 38, 115–124, 1985.

Brown, S.W., Attentional resources in timing: interference effects in concurrent temporal and nontemporal working memory tasks, *Percept. Psychophys.*, 59, 1118–1140, 1997.

Buffardi, L., Factors affecting the filled-duration illusion in the auditory, tactual, and visual modalities, *Percept. Psychophys.*, 10, 292–294, 1971.

Buhusi, C.V. and Meck, W.H., Timing for the absence of a stimulus: the gap paradigm reversed, *J. Exp. Psychol. Anim. Behav. Process.*, 26, 305–322, 2000.

Buhusi, C.V. and Meck, W.H., Differential effects of methamphetamine and haloperidol on the control of an internal clock, *Behav. Neurosci.*, 116, 292–297, 2002.

Buhusi, C.V., Sasaki, A., and Meck, W.H., Temporal integration as a function of signal/gap intensity in rats (*Rattus norvegicus*) and pigeons (*Columba livia*), *J. Comp. Psychol.*, 116, 381–390, 2002.

Burle, B. and Casini, L., Dissociation between activation and attention effects in time estimation: implications for internal clock models, *J. Exp. Psychol. Hum. Percept. Perform.*, 27, 195–205, 2001.

Casini, L. and Macar, F., Effects of attention manipulation on judgments of duration and of intensity in the visual modality, *Mem. Cognit.*, 25, 812–818, 1997.

Church, R.M., Properties of the internal clock, in *Annals of the New York Academy of Sciences*, Vol. 423, *Timing and Time Perception*, Gibbon, J. and Allan, L., Eds., New York Academy of Sciences, New York, 1984, pp. 566–582.

Fortin, C., Short-term memory in time interval production, in *Short-Term/Working Memory: A Special Issue of the International Journal of Psychology*, Neath, I., Brown, G.D.A., Poirier, M., and Fortin, C., Eds., Psychology Press, Hove, U.K., 1999, pp. 308–316.

Fortin, C. and Bédard, M.-C., Stimulus Compatibility in Interrupted Time Interval Production, poster presented at the annual meeting of the Psychonomic Society, Los Angeles, CA, 1999.

Fortin, C. and Couture, E., Short-term memory and time estimation: beyond the 2-second "critical" value, *Can. J. Exp. Psychol.*, 56, 120–127, 2002.

Fortin, C. and Massé, N., Order information in short-term memory and time estimation, *Mem. Cognit.*, 27, 54–62, 1999.

Fortin, C. and Massé, N., Expecting a break in time estimation: attentional time-sharing without concurrent processing, *J. Exp. Psychol. Hum. Percept. Perform.*, 26, 1788–1796, 2000.

Fortin, C. and Rousseau, R., Time estimation as an index of processing demand in memory search, *Percept. Psychophys.*, 42, 377–382, 1987.

Fortin, C., Rousseau, R., Bourque, P., and Kirouac, E., Time estimation and concurrent nontemporal processing: specific interference from short-term-memory demands, *Percept. Psychophys.*, 53, 536–548, 1993.

Gibbon, J. and Church, R.M., Sources of variance in an information processing theory of timing, in *Animal Cognition*, Roitblat, H.L., Bever, T.G., and Terrace, H.S., Eds., Erlbaum, Hillsdale, NJ, 1984, pp. 465–488.

Gibbon, J., Church, R.M., and Meck, W.H., Scalar timing in memory, in *Annals of the New York Academy of Sciences*, Vol. 423, *Timing and Time Perception*, Gibbon, J. and Allan, L., Eds., New York Academy of Sciences, New York, 1984, pp. 52–77.

Hicks, R.E., Miller, G.W., and Kinsbourne, M., Prospective and retrospective judgments of time as a function of amount of information processed, *Am. J. Psychol.*, 89, 719–730, 1976.

Israeli, N., Illusions in the perception of short time intervals, *Arch. Psychol.*, 18, 1930.

James, W., *The Principles of Psychology*, Vol. 1, Macmillan, London, 1890.

Jones, M.R. and Boltz, M., Dynamic attending and responses to time, *Psychol. Rev.*, 96, 459–491, 1989.

Jones, M.R., Boltz, M., and Klein, J.M., Expected endings and judged duration, *Mem. Cognit.*, 21, 646–665, 1993.

Kahneman, D. and Tversky, A., Variants of uncertainty, *Cognition*, 11, 143–157, 1982.

Kristofferson, A.B., A real-time criterion theory of duration discrimination, *Perception & Psychophysics*, 21, 105–117, 1977.

Lejeune, H., Switching or gating? The attentional challenge in cognitive models of psychological time, *Behav. Process.*, 44, 127–145, 1998.

Lejeune, H., Prospective timing, attention and the switch: a response to "Gating or switching? Gating is a better model of prospective timing" by Zakay, *Behav. Process.*, 52, 71–76, 2000.

Macar, T., Temporal judgments on intervals containing stimuli of varying quantity, complexity and periodicity, *Acta Psychologica*, 92, 297–308, 1996.

McClain, L., Interval estimation: effect of processing demands on prospection and retrospection reports, *Percept. Psychophys.*, 34, 185–189, 1983.

Meck, W.H., Attentional bias between modalities: effect on the internal clock, memory, and decision stages used in animal time discrimination, in *Annals of the New York Academy of Sciences*, Vol. 423, *Timing and Time Perception*, Gibbon, J. and Allan, L., Eds., New York Academy of Sciences, New York, 1984, pp. 528–541.

Meck, W.H., Church, R.M., and Olton, D.S., Hippocampus, time, and memory, *Behav. Neurosci.*, 98, 3–22, 1984.

Penney, T.B., Gibbon, J., and Meck, W.H., Differential effects of auditory and visual signals on clock speed and temporal memory, *J. Exp. Psychol. Hum. Percept. Perform.*, 26, 1770–1787, 2000.

Posner, M.I., Orienting of attention, *Q. J. Exp. Psychol.*, 32A, 3–25, 1980.

Roberts, S., Isolation of an internal clock, *J. Exp. Psychol. Anim. Behav. Process.*, 7, 242–268, 1981.

Roberts, S. and Church, R.M., Control of an internal clock, *J. Exp. Psychol. Anim. Behav. Process.*, 4, 318–337, 1978.

Rousseau, R., Picard, D., and Pitre, E., An adaptive counter model for time estimation, in *Annals of the New York Academy of Sciences*, Vol. 423, *Timing and Time Perception*, Gibbon, J. and Allan, L., Eds., New York Academy of Sciences, New York, 1984, pp. 639–642.

Sanders, A.F., Expectancy: application and measurement, *Acta Psychol.*, 25, 293–313, 1966.

Thomas, E. and Weaver, W., Cognitive processing and time perception, *Percept. Psychophys.*, 17, 363–367, 1975.

Tremblay, S. and Fortin, C., Expecting a break in timing increases proportion of "short" responses in time discrimination, submitted.

Tremblay, S., Fortin, C., and Jones, D., Expecting a Break in Timing Increases Proportion of "Short" Responses in Time Discrimination, poster presented at the annual meeting of the Psychonomic Society, Orlando, FL, 2001.

Zakay, D., Gating or switching? Gating is a better model of prospective timing (a response to 'switching or gating?' by Lejeune), *Behav. Process.*, 52, 63–69, 2000.

Zakay, D. and Block, R.A., The role of attention in time estimation processes, in *Time, Internal Clocks and Movement*, Pastor, M.A. and Artieda, J., Eds., Elsevier, Amsterdam, 1996.

Zakay, D., Nitzan, D., and Glickson, J., The influence of task and external tempo on subjective time estimation, *Percept. Psychophys.*, 34, 451–456, 1983.

10 Grandfather's Clock: Attention and Interval Timing in Older Adults

Cindy Lustig

CONTENTS

10.1 Introduction 261
    10.1.1 An Information-Processing Model of Interval Timing 262
    10.1.2 Age Differences in Attention Relevant to Interval Timing.............264
10.2 How Do Age Differences in Attention Affect Older Adults' Interval Timing? 266
    10.2.1 Absolute Time Judgments and Subjective/Objective Duration Ratios: Are Older Adults' Clocks Too Fast or Too Slow?..............267
    10.2.2 Older Adults' Performance on Relative Time Judgment Tasks 272
    10.2.3 Maintaining Attention, Dividing Attention, and Circadian Fluctuations: Distinguishing Aspects of Attention That Influence Older Adults' Perception of Time 276
    10.2.4 Variability Measures 279
    10.2.5 Age Differences in Attention and Interval Timing: Summary and Conclusions 280
10.3 Age Differences in Memory, Processing Speed, and Interval Timing 282
10.4 Interval Timing and Other Areas of Cognitive Aging.................................285
10.5 Summary and Conclusions 288
References..288

10.1 INTRODUCTION

"Time flies when you're having fun," but "a watched pot never boils."

These popular sayings reflect our intuition that the amount of attention we pay to time directly influences our perception of its passing. Many aspects of attention change with advancing age, raising the question: How does time fly for older adults?

This chapter attempts to organize the existing literature on older adults' interval timing performance, that is, judgments about durations in the seconds-to-minutes range. Age differences in the perception of time have both practical and scientific significance. Successful performance of many tasks — from hitting a baseball to cooking the perfect hamburger to changing lanes on a busy freeway — depends critically on the ability to "get the timing right." Age-related disruptions in time perception can make these tasks more difficult for older adults. From a research perspective, there is a great deal of overlap in the areas of interest for investigators in interval timing and cognitive aging. Questions about attention, memory, and frontal-striatal brain circuits are of central importance in both domains.

These shared interests and growth in the fields of both interval timing and cognitive aging have led to a recent upswing in experiments on older adults' perception of time (e.g., Craik and Hay, 1999; Fernandez and Pouthas, 2001; Lustig and Meck, 2001a; Malapani et al., 1998; McCormack et al., 1999; Perbal et al., 2002; Wearden et al., 1997). Some of these studies ask questions about time judgments in tasks that are unique to humans, whereas others use procedures that have already been well studied in the established literature on animal timing. In some cases, neuropsychological or neuroimaging techniques have also been used to explore the brain correlates of age-related behavioral differences in interval timing (e.g., Fernandez and Pouthas, 2001; Malapani et al., 1998). Many of these experiments start from the assumption of age differences in attention and memory functioning and ask how these changes influence older adults' perception of time. Age differences in timing may also influence age differences on other cognitive tasks, particularly those that require speeded performance or the accurate timing and sequencing of multiple operations.

The increase in experiments on older adults' interval timing performance has led to many intriguing results, but in general, there is a lack of an organizing framework within which to understand those results, their relation to each other, and how they might fit into more general theories of interval timing or cognitive aging. This chapter attempts to provide such a framework by considering how the dominant information-processing model of interval timing, scalar expectancy theory (SET) (Gibbon et al., 1984), might be affected by age differences in attention and memory. This model includes components such as attention, memory, and processing speed that are of central importance to cognitive aging researchers, and those components have been heavily investigated using both behavioral and biological techniques. These features make SET an ideal starting point for understanding how interval timing may change with age.

10.1.1 An Information-Processing Model of Interval Timing

Most information-processing models of interval timing, including SET, share a basic structure, as proposed by Treisman (1963). SET is described more fully elsewhere in this volume, including the introduction, but the features important for the present discussion are pictured in Figure 10.1. These models include a pacemaker that emits pulses that then pass through an attention-controlled switch or gate before piling up in an accumulator (e.g., Gibbon et al., 1984; Zakay and Block, 1997; for a discussion of attention's role in different models of interval timing, see Lejeune, 1998). The accumulator values may then be compared to values stored in long-term reference memory to make judgments about current time relative to past durations.

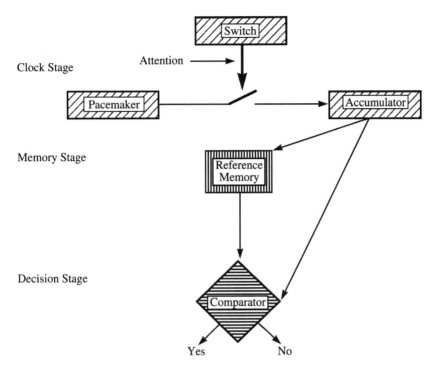

FIGURE 10.1 Basic structure of information-processing models of timing, including scalar expetancy theory (Gibbon et al., 1984). Attention to a to-be-timed stimulus closes the switch, allowing pacemaker pulses to pass into the accumulator. The current accumulator value is compared to a past value representing the target sampled from reference memory to decide if it meets criterion to be judged the same as (yes) or different from (no) the target.

The clock stage of this model has been the target of most questions about older adults' interval timing performance because of its relevance to other investigations of cognitive aging. Extensive animal research, more recently supplemented by human neuropsychological and neuroimaging studies (e.g., Hinton, this volume; Hinton et al., 1996; Malapani and Rakitin, this volume; Pouthas, this volume; Rao et al., 2001), locates this stage in the frontal-striatal circuits of the brain, and its functioning is heavily influenced by dopaminergic and catecholamine neurotransmitter systems (Meck, 1996; Meck and Benson, 2002). Changes in these brain systems are thought to be of critical importance to many of the cognitive changes that occur with age (see, e.g., Cabeza, 2001; Park et al., 2001; Rubin, 1999). In particular, they are thought to be centrally involved in age differences in attentional functioning.

As Figure 10.1 indicates, attention plays an important role in interval timing models. Fluctuations in attention can cause a "flickering" of the switch that allows pulses to pass from the pacemaker to the accumulator, leading to a loss of pulses and thus a lower — and consequently slower — clock reading for a given unit of physical time. These models thus allow for clear predictions about the interval timing performance of individuals and groups that have problems with attentional functioning, including older adults.

Of course, aging affects many aspects of behavior and cognition in addition to attention. For example, depending on the situation, older adults may adopt more or less conservative decision criteria than do young adults, and memory complaints are ubiquitous among older adults (see recent reviews by Craik and Jennings, 1992; Light, 1991; Zacks et al., 2000). Fortunately, research using information-processing models of interval timing has revealed ways of separating out effects on these different functions. In particular, a central tenet of SET is that temporal processes reliant on the clock and memory stages have scalar variability; that is, their variability increases or decreases as a constant proportion of the duration being timed. Manipulations that shift the accuracy of duration judgments but lead to violations of this scalar property are likely to have their effects on nontemporal processes, such as decision criteria or response execution (Gibbon et al., 1984, 1997).

Furthermore, variables that influence attention or other components of the clock stage can often be differentiated from those that influence memory stages by observing the effects of feedback (Meck, 1983). With feedback, participants can learn to adjust to a speeded or slowed clock by increasing or decreasing the number of accumulator counts associated with a response. However, distortions that occur in memory are resistant to feedback, as the correct feedback value will also be stored in a distorted fashion. This distinction is extensively used in animal research on interval timing to differentiate behavioral, pharmacological, and surgical manipulations that have an effect on attention and clock functioning from those that have an effect on temporal memory, and shows similar promise for differentiating these components in older adults (e.g., Malapani et al., 2002b; Rakitin et al., submitted; Wearden et al., 1997).

10.1.2 AGE DIFFERENCES IN ATTENTION RELEVANT TO INTERVAL TIMING

Attention plays an important role in interval timing performance, but as described above, timing can also be affected by influences on memory, the pacemaker, or other components of the information-processing model. Furthermore, the term *attention* is a broad one and has been operationalized in many procedures that have no apparent relation to one another. Not every cognitive function that falls under the rubric of "attention" will have an influence on interval timing (see discussion by Meck and Benson, 2002), and not all of these functions will change with age. Furthermore, many aspects of attention, such as the efficacy of different types of attentional cues, have been studied separately in the contexts of aging (e.g., on visual attention tasks, see Curran et al., 2001; Greenwood et al., 1993) and interval timing (e.g., Chen and O'Neill, 2001; Meck, 1984), but the effects of age differences in those aspects of attention on older adults' interval timing performance have yet to be explored.

The majority of experiments on older adults' interval timing performance, and thus the bulk of this chapter, focus on age differences in the allocation and maintenance of attention to a particular task or stimulus, and the consequences of these differences for older adults' timing accuracy. As described below, there is some controversy in the cognitive aging literature regarding the characterization of and reasons for age differences in attentional function (see reviews by Hartley, 1992; McDowd and Birren, 1990; Rogers and Fisk, 2001).

Despite these controversies, most investigators agree that *automatic* aspects of attention are for the most part spared in normal aging, whereas *controlled* aspects of attention are subject to age-related declines (e.g., Hasher and Zacks, 1979; Jennings and Jacoby, 1993). Automatic aspects of attention are those that occur without intention, do not necessarily give rise to awareness, and do not interfere with other processing (Posner and Snyder, 1975). Examples include the way in which a bright flash of light in a particular location will draw attention to that location, or how we understand the words we are reading without deliberately and effortfully retrieving their meaning. Controlled aspects of attention usually require awareness and intention, and tax the system so that only a limited number can be carried out at one time. One example is directing attention to a peripheral location in response to a centrally located arrow pointing to that location. Controlled attention also plays an important role on the Stroop test, which requires participants to say the ink color of conflicting color names (e.g., the word *red* printed in green ink). In this situation, attention must be effortfully directed away from the word information and toward the color information.

Most cognitive processes exist somewhere on the continuum between controlled and automatic, and their position can change over time. In particular, an initially effortful process (e.g., driving) can become more automatized through repeated practice. In contrast, under adverse conditions, a largely automatized process can require more control. For example, under normal circumstances, most of us do not find it difficult to walk and chew gum at the same time, reflecting that these are both highly automatized processes that do not interfere with each other. In a state of inebriation, however, walking can require a great deal more effort and attempted control. More benignly, when pondering a difficult problem, we may literally stop and think about it — redirecting attention from walking to our thoughts.

There is general agreement that the controlled aspects of attentional performance are the ones that most change with age, but the reasons for those changes are a matter of some debate. One group of theories focuses on the idea that older adults have reduced controlled attention resources. Evidence for these ideas comes from experiments showing that asking young adults to divide their attention between multiple tasks can reduce their performance to the level of older adults who are tested under full attention conditions. This is especially the case when the performance measure is subsequent memory for information studied during the divided attention task (e.g., Craik, 1977; Craik et al., 1996; Anderson et al., 1998). The explanation given for these findings is that the requirements of the additional task reduce the amount of controlled attention that young adults have available to devote to the studied information, and that this experimentally induced reduction in young adults simulates preexisting resource deficits in older adults.

Reduced-resource views are a popular way of explaining age deficits in performance and receive support from a variety of findings. For example, compared to young adults studying under full attention conditions, older adults and participants studying under divided attention conditions may show similar activity reductions in left frontal brain areas associated with later successful memory (Anderson et al., 2000). On behavioral tests, the memory costs of divided attention are often similar for young and older adults, but older adults show larger impairments on the secondary task (Anderson, 1999; Anderson et al., 1998). Data analyzed using the process dissociation procedure (Jacoby, 1991) have also been used to argue that controlled, but not automatic aspects of memory change with divided attention and age (Schmitter-Edgecombe, 1999), consistent with the distinction described above.

Another group of theories focuses on potential age differences in the *allocation* of attentional resources, rather than amount. For example, the inhibitory deficit hypothesis posits that many of older adults' problems in experimental tasks and in daily life stem from a reduced ability to keep attention and working memory free of irrelevant information (Hasher and Zacks, 1988; Hasher et al., 1999; McDowd et al., 1995). This reduced ability to keep irrelevant information out of the focus of attention may result in older adults' functionally being in a divided attention situation much of the time (Hasher et al., 2001).

A third perspective emphasizes the possibility that older adults have different priorities for the allocation of attention, and that declines in perceptual and motor abilities may result in the same task requiring more attention from older adults than from young adults (Baltes, 1997; Baltes and Baltes, 1990; Freund et al., 1999). When driving down unfamiliar roads on a dark, snowy night, most of us will turn off the radio and cut down on conversation because we require more concentration — that is, attention — to perceive what lies ahead of us and control the car. Likewise, impairments in sensory and motor functioning due to age (rather than bad weather) may require older adults to devote more attention to the simple perception and execution of a task than is the case for young adults.

This idea was tested directly in a series of experiments that asked participants to memorize a list of words while walking on a course filled with obstacles (Li, Lindenberger, Freund, and Baltes, 2001). Performance in the dual-task condition was compared to performance on each task performed individually. Over different combinations of task difficulty, older adults showed larger dual-task costs on the memory task than did young adults, whereas the two groups showed equivalent costs on walking performance. Furthermore, older adults used the aid provided for the walking task (a handrail) more often than the aid provided for the memorizing task (more time), but young adults showed the opposite pattern. These results suggest that older adults devoted their attention and efforts to the walking task, whereas young adults focused on memorizing.

In summary, there are different views on how to characterize older adults' difficulties on controlled attention tasks, but all converge on a functional level to suggest that older adults often devote less controlled attention to a particular task than do young adults. As described in the following section, age differences in the allocation and prioritization of controlled attention have important consequences for older adults' perception of time.

10.2 HOW DO AGE DIFFERENCES IN ATTENTION AFFECT OLDER ADULTS' INTERVAL TIMING?

Ideas from the information-processing models of interval timing described above can be combined with characterizations of older adults' attentional functioning to make predictions about older adults' interval timing performance. In particular, any manipulation that decreases the attention available to hold closed the "switch" that allows pulses to pass into the accumulator will cause a flickering of that switch and a loss of pulses, resulting in a slower clock. Older adults will typically be more vulnerable than young adults to such manipulations, which may include external, nontemporal distractions or ongoing tasks that draw attention away from timing, or the need to divide attention among multiple to-be-timed stimuli.

When considering how different attention manipulations may affect older adults' interval timing performance, it is important to consider whether they occur when the critical duration is being learned or during test trials (see Figure 10.2). This is the case because attention-related distortions in time perception occur when there is a difference in attention — and therefore clock speed — between training and test. If controlled attention demands are identical for training and test trials, age differences in interval timing performance may be small or nonexistent.

If controlled attention demands are greater when the duration is learned than when it is tested, older adults will tend to underreproduce that duration during test trials and overestimate the durations presented in test trials relative to the previously learned criterion. Conversely, if controlled attention demands are greater during test trials than during training, older adults will overreproduce the critical duration learned in training and tend to underestimate test durations relative to the critical duration. In many cases, young adults will show patterns that are in the same direction as older adults', but because young adults have better attentional control, the effects for them are much smaller than for older adults.

This section presents a simple framework for understanding older adults' interval timing performance across different situations. This framework is based on predicting how age differences in attention might affect the function of the information-processing models of interval timing that were previously discussed. The following sections summarize the extant literature on older adults' interval timing performance and the degree to which those findings fit into the organization suggested here.

10.2.1 ABSOLUTE TIME JUDGMENTS AND SUBJECTIVE/OBJECTIVE DURATION RATIOS: ARE OLDER ADULTS' CLOCKS TOO FAST OR TOO SLOW?

"This message will self-destruct in 5 seconds."

Humans' use of language allows us to efficiently communicate information about important durations. In most cases, making accurate judgments about these labeled durations (also known as "absolute" judgments of time) does not require that we carefully check our watches. Instead, we rely on our internal clocks.

One very basic question that can be asked is whether older adults tend to perceive time's passing to be faster or slower, compared to young adults. That is, are older adults' clocks too fast or too slow? Older adults' self-reports indicate that they often feel as if time passes more quickly than it did when they were younger (Fraisse, 1984; Schroots and Birren, 1990). This suggests that aging may lead to a slowing of the internal clock, which would make it seem as though external time were passing by quickly. Such slowing would be consistent with general findings that

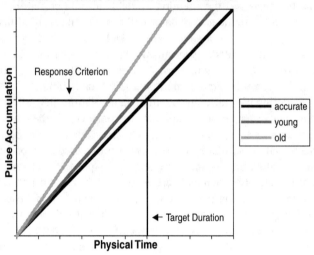

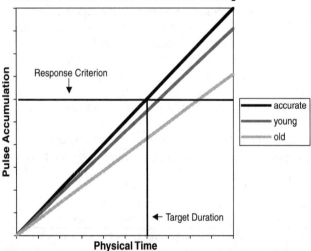

FIGURE 10.2 A schematic for predicting how changes in attentional demands between training and test will affect young and older adults' performance on interval timing tasks. When attentional demands are greater at training than at test (top), pulse accumulation will occur more rapidly during the test trials than during training for the same unit of physical time. As a result, the number of pulses accumulated will reach criterion for response too soon. The opposite pattern occurs when attentional demands are greater during test trials than during training. In both cases, effects will usually be exaggerated for older adults because of their reduced attentional control.

older adults are slower than young adults on many cognitive tasks (e.g., Salthouse, 1996). The idea that older adults have a slower internal clock than do young adults would at first seem consistent with ideas about age differences in controlled attention. That is, older adults' reduced attentional functioning might lead them to accumulate pulses more slowly than do young adults, resulting in a lower — and slower — clock reading.

Surprisingly, a meta-analyis of interval timing experiments using young and old adults found the opposite pattern: older adults had higher ratios of subjective time to objective time than did young adults, a result that at first might seem to suggest that older adults have a faster internal clock (Block et al., 1998).

However, as Block et al. (1998) note, this counterintuitive finding may largely result from the testing procedures used in the studies included in their review. Almost all of these experiments used absolute judgments of empty intervals. That is, participants either assigned a verbal label (e.g., 3 sec) to an experimenter-produced duration or attempted to produce the appropriate duration in response to the experimenter's query (e.g., "Hold this button down until you think 3 seconds has passed"). Within the context of the interval timing models described above, these procedures can be described as asking participants to compare current accumulator values acquired in the experimental setting to reference memory values acquired in daily life.

When asked to time an empty interval in the quiet setting of a laboratory, there is little to pay attention to except for time. Therefore, the attentional switch should remain steadily closed, allowing the accumulation of pulses that measure time. In contrast, daily experience is usually filled with many attentional demands and distractions in addition to time. To the degree that these demands and distractions disrupt attention to time, the function of the attentional switch will be disrupted, and fewer pulses will accumulate than would be the case in the lab. The discrepancy in attentional demands — and therefore pulse accumulation rates — causes the clock to "run faster" in the experimental setting than it does in everyday life.

Absolute time judgment tasks lead to overestimation and underproduction because they ask participants to compare clock readings acquired during the experiment, with a "fast clock," to values associated with labels learned in everyday life, with a "slow clock." Because older adults are more vulnerable to distractions and demands on attention than are young adults, they have a larger discrepancy between experimental-setting clock speed and everyday-life clock speed. This larger discrepancy for older adults leads to their showing exaggerated subjective/objective duration ratios compared to young adults, as shown in the studies reviewed by Block et al. (1998).*

The direction of age differences in subjective/objective duration ratios for absolute time judgments can easily be reversed by simply increasing the attentional burden within the experiment relative to everyday life. One way of doing this is to divide participants' attention within the experiment between timing and some

---

* Young adults had subjective/objective duration ratios slightly over 1.0 in many of the experiments reviewed by Block et al. (1998), indicating that they, too, were affected by the difference in attentional demands between the outside environment and the lab. However, the deviations from 1.0 were typically much smaller than for older adults, and often not statistically significant.

other task. To the degree that the divided attention task presents attentional demands that are greater than those of the external environment, subjective/objective duration ratios should be less than 1.0. Because older adults are more vulnerable to demands that reduce the amount of attention available for timing, they should show a greater shrinking of subjective/objective duration ratios than do young adults under these conditions.

Craik and Hay (1999) tested young and older adults' interval timing performance under divided attention conditions using both estimation and production timing tasks. Rather than making time judgments about empty intervals, as was the case in the studies reviewed by Block et al. (1998), participants in these experiments made judgments about the characteristics (shape, size, brightness, etc.) of visual stimuli during the to-be-timed interval. In the estimation task, participants performed the visual judgment task until interrupted by the experimenter and asked to estimate how much time had passed. In the production task, participants were asked to perform the visual judgment task for a particular duration and to stop when they thought that duration had passed. In both cases, participants had to divide their attention and efforts between timing and the visual judgment task.

The increased attentional demands of the procedures used by Craik and Hay (1999) led to results that were very different from those reviewed by Block et al. (1998). In the Craik and Hay experiments, both age groups underestimated and overproduced time during the experimental trials, leading to subjective/objective duration ratios of less than 1.0. The distortion was relatively mild for young adults (with an overall duration ratio of 0.81), but dramatic for older adults, who had an overall duration ratio of 0.54. That is, older adults reported that about 33 sec had passed when the actual duration was 60 sec. This pattern is the exact opposite of the one found in the studies reviewed by Block et al., where older adults overestimated and underproduced time more than did young adults.

The contrast between the studies reviewed by Block et al. (1998) and those conducted by Craik and Hay (1999) fits nicely with the proposed attention-based framework for understanding older adults' interval timing performance, and demonstrates how divided attention manipulations can be used to test this framework. Both sets of experiments involved absolute judgments of time, where reference memory values obtained during the divided attention conditions of everyday life were compared with current accumulator values obtained in the experiment. The difference between them lies in the attentional demands present during the experiment. When there was little available in the experimental context to distract attention from timing, older adults tended to overestimate and underproduce time in the experiment (Block et al., 1998). In contrast, when the experiment required them to divide attention between the timing task and another task, older adults showed the opposite results, now underestimating and overproducing durations to a greater degree than did young adults (Craik and Hay, 1999).

Despite this general agreement, one aspect of Craik and Hay's (1999) findings might be seen as contradicting the ideas that dividing attention can affect interval timing performance, and that divisions of attention are especially detrimental for older adults. Contrary to their initial predictions and to previous findings (e.g., Brown, 1985; Fortin and Rosseau, 1998; Fortin, this volume), Craik and Hay did not find that timing accuracy decreased as the complexity (and thus attentional demands) of the visual judgment task increased, except at their longest duration (120 sec). They suggest that the lack of task complexity effects is attributable to a problem in their experimental design: all trials of the judgment task were 10 sec long, and participants may have used the consistent duration of the individual trials to help them keep their time judgments consistent across increasing levels of complexity. This explanation might be adequate to explain the general lack of task complexity effects, but is less acceptable as an explanation for the lack of age differences. Even in the 120-sec condition, where there were significant main effects of task complexity, older adults' timing performance was not more affected by task complexity than was that of young adults.

The resolution to this apparent contradiction may lie in the results for the visual judgment task. Although timing performance did not decline as visual judgment complexity (i.e., the number of decisions about shape, color, etc., that had to be made) increased, performance on the visual judgment task itself did. These complexity-related declines in performance were especially pronounced for older adults. This pattern of results parallels that of Li, Lindenberger, Freund, and Baltes (2001), who found that older adults preserved walking performance rather than memory performance when the difficulty of the two tasks was varied. In the Craik and Hay (1999) experiments, the preservation of timing performance in the face of increased visual judgment complexity may have occurred because participants prioritized the timing task and devoted their attentional resources there, rather than to the visual judgment task. This explanation receives some support from the results of a relative time judgment experiment (Vanneste and Pouthas, 1999) in which participants divided their attention between multiple to-be-timed stimuli, rather than between the timing task and multiple nontemporal stimuli, as was the case for the Craik and Hay experiments. Here, timing errors increased with the number of stimuli to be timed, and the increase in errors was more pronounced for older adults.

Are older adults' clocks too fast or too slow? The answer from these absolute judgment tasks seems to be "it depends." More specifically, the answer depends on whether the demands for controlled attention are greater in the outside environment or in the experimental setting. When test trials are conducted under full attention conditions, attentional demands will typically be less than in the outside environment, leading to quicker pulse accumulation and thus a fast clock. When test trials are conducted under divided attention conditions, attentional demands may be greater than in the outside environment, leading to a flickering of the mode switch, slower pulse accumulation, and thus a slow clock. Both of these effects will be exaggerated for older adults because they are more vulnerable than are young adults to demands on controlled attention.

The results for absolute timing judgments therefore fit nicely into the framework described earlier and pictured in Figure 10.2. Importantly, they do not show that older adults simply have slower clocks, as might be predicted from a speed-of-processing perspective; they also do not show a nonspecific increase in errors, as might be predicted from general memory or performance problems. Instead, the direction of older adults' timing errors depends specifically on the attentional demands of the experiment vs. the outside environment.

Furthermore, older adults' interval timing errors meet an important criterion for establishing them as the result of problems with attention or other aspects of clock speed: they are amenable to feedback. As described earlier, sensitivity to feedback is considered a hallmark of clock speed effects, because participants can use feedback to adjust to the new clock speed and relearn the number of pulses associated with the target duration (Meck, 1983). Wearden et al. (1997, experiment 4) asked both young-old (age 60 to 69 years) and older-old (age 70 to 79 years) adults to produce 1-sec intervals. Without feedback, participants overproduced the duration, and this overproduction was more pronounced for the older group. However, after receiving feedback (a display after each trial indicating the actual duration they had produced), participants became highly accurate and maintained this accuracy even after feedback was discontinued. This pattern strongly suggests that the older adults in this experiment were able to use feedback to adjust to the different attentional demands — and thus faster pulse accumulation and clock speed — of the experiment vs. daily life.

Absolute time judgment tasks are easy for participants to understand and have a great deal of face validity. However, they are problematic for investigating attention's role in interval timing because the experimenter does not control the attentional environment in which the target duration is learned and established in reference memory. It is therefore hard to pinpoint *a priori* whether the controlled attention demands are greater inside the experimental context or the everyday environment, and to what degree. Therefore, attention-based explanations of older adults' absolute time judgment performance, including the framework described in this chapter, are necessarily post hoc. Because of this problem, many investigations of older adults' interval timing use relative time judgment tasks, where the experimenter controls the demands for controlled attention at both training and test.

10.2.2 Older Adults' Performance on Relative Time Judgment Tasks

In relative time judgment tasks, participants are exposed to an unnamed critical target duration in a series of training trials, so that a distribution of accumulator values associated with the target duration is built up in reference memory. On test trials, participants make judgments about presented durations relative to that earlier-learned target. Because they do not require the use of language-based duration labels (e.g., 5 sec), these procedures are used for animal experiments on interval timing. Many human experiments build directly off standard procedures from the animal literature (e.g., peak-interval production or temporal bisection), whereas others use procedures specific to a particular study. The critical feature these procedures share in common, and that differentiates them from absolute time judgments, is that both the reference memory values and the current accumulator values are acquired during the experiment.

Under full attention conditions, age differences in interval timing on relative time judgment tasks are typically smaller than those found using absolute time judgment tasks. This is to be expected, as the controlled attention demands should be largely the same during both training (acquisition of reference memory values) and test trials (acquisition of current accumulator values). Because of this, the clock should run at approximately the same speed when the target duration is learned as when it is tested, resulting in very little distortion. However, small age differences are often found on relative time judgment tasks, usually in a direction that suggests that older adults' clocks may be running more slowly during test trials than they were during training.

These differences could occur if older adults' attention to time was greater during the training trials, when they are still "fresh" and under the watchful eye of the experimenter, than it is during the numerous test trials, when participants may be distracted by unrelated thoughts, boredom, or fatigue. While both young and older adults may be more prone to lapses of attention during test trials than during training trials, leading to a slower clock speed, this effect is likely to be greater for older adults because of their reduced attentional control. Consistent with this suggestion, age differences on relative time judgments are often more evident for long durations (when attention would have more opportunity to drift) than for short durations. Furthermore, both the proximity of training trials to the testing trials and the nature of feedback have an effect on age differences on relative time judgment tasks that is consistent with the idea that those differences are due to older adults' decreased control of attention during test trials as opposed to training.

McCormack et al. (1999) tested participants using a temporal generalization procedure in which participants were first trained on a tone of a standard duration (500 msec) and at test were asked to indicate whether presented durations (ranging between 125 and 875 msec) were the same length as the standard. Both young (18 to 22 years) and old (65 to 73 years) participants showed a bias to underestimate the lengths of the long test durations and to say that they were the same length as the standard. Although the differences between these two age groups were not statistically significant, this tendency to underestimate long durations was more pronounced for the older adults than for the young adults. A third group of old-old adults (age 75 to 99 years) did show a significantly greater underestimation of the long durations, replicating previous findings for older adults in a study by Wearden et al. (1997).

Both Wearden et al. (1997) and McCormack et al. (1999) explain their results in terms of memory problems associated with age, but older adults' underestimation of the long durations might also be explained as the result of problems in attention. Wearden et al. (1997) suggest that older adults might have a more variable memory representation of the standard, leading to errors when making decisions about a test duration relative to the standard. However, a memory representation that was simply more variable should lead to more errors for both short and long test durations, rather than a unidirectional bias toward underestimating the long durations. McCormack et al. (1999) modified this idea by suggesting that older adults had a systematic bias in memory such that they remembered the standard as being longer than it really was, whereas young children (who showed the opposite result — that is, overestimation of durations shorter than the standard) had a bias to remember the standard as being shorter than it was. However, there is no *a priori* reason to predict this overremembering on the part of older adults.**

The approach taken in this chapter would locate the source of underestimation in the attentional switch and accumulator, rather than in reference memory. This attention-based account assumes that keeping attention on the timing task will be relatively easy during training, when the participant is under the careful watch of the experimenter, compared to the many test trials, when the participant must begin to deal with interference, boredom, or fatigue. Thus, attention to time is assumed to require more control during testing trials than during training.

If this is the case, then lapses in attention will reduce the number of pacemaker pulses that pass through the attentional switch during test trials, as compared to training trials. Such lapses in attention may be especially likely to occur during long test trials, reducing the accumulator values acquired in those trials. When compared to the reference memory values for the standard acquired during training, the reduced accumulator values will make the long test durations seem more like the standard. Because older adults are more vulnerable to lapses of attention than are young adults, they will be especially sensitive to this effect. The pattern found by McCormack et al. (1999) was in keeping with this prediction: both age groups showed a tendency toward underestimation of the long test trials, but the underestimation effect was larger for older adults because of their problems with attentional control.

Wearden et al. (1997) and McCormack et al. (1999) also tested participants using temporal bisection tasks. On bisection tasks, participants first learn a short (e.g., 200 msec) and a long (e.g., 800 msec) anchor duration. At test, they are presented with intermediate durations and asked to indicate whether the presented duration is closer to the short or the long anchor duration. The primary measure in such tasks is the bisection point, the duration for which participants respond "short" and "long" with equal probability. In these experiments, young and older adults showed equivalent bisection points. However, older adults were slightly less likely than young adults to call the longer test durations (e.g., 600 msec) long. Such a result might occur if older adults were more likely to lose pulses during the long test durations due to lapses in attention, reducing the accumulator values for those durations and thus their perceived similarity to the long anchor.

The results presented by McCormack et al. (1999) and Wearden et al. (1997) for the temporal generalization and temporal bisection tasks do not allow for a clear differentiation between the memory-based explanations proposed by those authors and the attention-based framework used in the current chapter. However, the attention-based framework offers several advantages. First, it can explain why young adults often show effects that are in the same direction as those shown by older adults, but much smaller. Second, it parsimoniously explains the findings for both the relative time judgment experiments discussed here and the absolute time judgment studies discussed in the previous section. Third, it can be tested by examining the effects of feedback and dividing attention.

As described earlier, interval timing effects that are sensitive to feedback are usually considered to be located at the attentional switch or other components of

---

** Some rat studies show an overreproduction with increased age that is not corrected by feedback, indicating a memory effect (Lejeune et al., 1998; Meck et al., 1986; but see Meck, 2001). In humans, older adults' overreproductions are typically corrected by feedback, more consistent with a clock speed effect caused by age differences in attention (e.g., Malapani et al., 1998, 2002b; Wearden et al., 1997). This discrepancy may occur because humans approach the task differently than do animals: humans are more likely to become bored or distracted during test trials, because they are not working for food rewards.

the clock stage, rather than in memory or decision stages of temporal processing (Meck, 1983). Age differences on interval reproduction tasks are influenced by both the timing and nature of feedback. This point can be illustrated by comparing the results of a series of experiments in which participants attempted to reproduce a target duration under various feedback conditions. When there was a significant delay between training and test trials and the test trials occurred without feedback, older adults overreproduced the interval, suggesting reduced attention to time and thus slower clocks during testing than training (Malapani et al., 1998). In contrast, when each test trial occurred immediately after exposure to a target interval, minimizing attentional differences between training and test, no age differences in reproduction were found (Vanneste and Pouthas, 1999).

Feedback that consists of simple reexposure to the target duration does not ameliorate older adults' tendency to overreproduce that duration (Malapani et al., 2002b). This finding seems inconsistent with the idea that older adults' memories are simply more variable (e.g., Wearden et al., 1997) or that older adults overremember the target (McCormack et al., 1999). If either of those possibilities were the case, re-presentation of the target duration would be expected to refresh or correct older adults' memories of the target and improve their performance. Instead, Malapani et al. (2002b) found that older adults' overreproductions were remedied by giving them specific feedback about their performance, potentially alerting them to changes in their clock speed from training to test and the need to pay attention to time.

The impact of age differences in controlled attention on interval timing performance is especially evident in experiments that deliberately manipulate the attentional demands between training and test. By the framework used in this chapter, greater attentional demands at training than at test should lead to underreproduction of the target duration on test trials and overestimation of test-presented durations relative to the target. In contrast, greater attentional demands at test than at training should lead to overreproduction of the target duration on test trials and underestimation of test-presented durations relative to the target. Both of these effects are expected to be greater for older adults than for young adults because of older adults' reduced attentional control.

The existing data are consistent with these predictions. For example, Vanneste and Pouthas (1999) tested young and older adults using a procedure in which three letters were presented on the screen. The presentation time of each letter overlapped with that of the other two letters, but each had a different onset and duration. Participants received both full attention and divided attention trials. For the full attention encoding condition, participants were told which letter's duration would later be tested, and that they should ignore the other two. For the divided attention encoding conditions, participants had to attend to the duration of two of the letters or all three and did not know which one would be tested. Reproduction always occurred under full attention: after all letters were removed, participants were presented with one of the letters and asked to reproduce its just-presented duration.

In this experiment (Vanneste and Pouthas, 1999), attention effects would be revealed by underreproduction of the target interval for those trials on which attention was divided at encoding. In the full attention encoding condition, the amount of pulses accumulated during encoding should be equivalent to the number accumulated during the full attention testing conditions, minimizing distortions in time. Under full attention at encoding, young adults and older adults showed largely equivalent performance. As the number of to-be-timed durations, and thus the need to divide attention, increased, both age groups showed an increasing tendency to underreproduce the target duration. However, older adults were much more affected by the increasing number of to-be-timed durations than were young adults. It is important to note that increasing the number of durations led not only to a general increase in errors but also to an underestimation of encoded time. The directional nature of the errors is highly consistent with the idea that the divided attention manipulation led to a slower clock speed during training than during test trials. In contrast, when the divided attention manipulation occurs at test trials rather than at training, older adults' errors are in a direction that suggests a slower clock speed during test trials than during training (Lustig and Meck, 2002).

Overall, the results for older adults' performance on relative time judgment tasks fit well with the idea that age differences on timing tasks often result from age differences in attentional control. Under full attention conditions, age differences on relative time judgment tasks are usually smaller than on absolute judgment tasks, because current accumulator values are acquired under roughly the same attentional demands as were reference memory values. Small time distortions are often seen, however, usually in the direction that suggests a slower clock speed during test trials than during training. These distortions — and age differences in their degree — are most often seen under conditions (long test trials, long delays between training and test) conducive to lapses of attention during the test. Both young and older adults may be affected by divided attention manipulations that change the attention demands between training and test, but older adults are more vulnerable than young adults to these manipulations.

10.2.3 Maintaining Attention, Dividing Attention, and Circadian Fluctuations: Distinguishing Aspects of Attention That Influence Older Adults' Perception of Time

The previous sections present evidence that strongly suggests that older adults' timing distortions relative to those of young adults are the result of age-related difficulties in maintaining attention to time (Block et al., 1998; Malapani et al., 1998, 2002b; McCormack et al., 1999; Wearden et al., 1997) and in dividing attention between multiple stimuli (Craik and Hay, 1999; Vanneste and Pouthas, 1999). As mentioned previously, automatic attention functions may be largely spared from age-related declines, and not all of the disparate cognitive functions that have been labeled "attention" will have equivalent effects on interval timing performance. Is it possible to distinguish the influence (or lack thereof) of different aspects of attention on age differences on interval timing?

The results of a recent series of experiments that divide participants' attention between to-be-timed stimuli of different modalities suggest that the answer may be "yes." These experiments build on a newly proposed explanation for the classic interval timing finding that sounds are judged longer than lights. That is, an auditory stimulus will be judged to have a longer duration than a visual stimulus of the same physical duration if both modalities are used in the same experimental session and refer to the same standards (e.g., Goldstone and Goldfarb, 1964; Walker and Scott, 1981; Wearden et al., 1998). Penney et al. (1998, 2000; see also Penney, this volume) proposed an explanation for this finding that is grounded in SET and draws upon psychophysical and chronometric studies showing that auditory stimuli tend to capture and hold attention automatically, whereas attending to visual stimuli requires greater control (Posner, 1976; see also Liu, 2001; Spence and Driver, 1997; Schmitt et al., 2000).

By this explanation, auditory stimuli more efficiently and automatically close the attentional switch that allows the accumulation of pulses that mark the passage of time. As a result, more pulses accumulate during an auditory stimulus than during a visual stimulus of the same physical duration (Penney et al., 1998, 2000). The distribution of pulses associated with that duration in reference memory will be mixed, containing both auditory (i.e., relatively large) and visual (i.e., relatively small) values. On test trials, distortions occur when a current accumulator value from one modality is compared to a reference memory value from the other modality. A test duration presented in the auditory modality and compared to a reference memory value formed during a visually presented training trial will seem inappropriately long. In contrast, a test duration presented in the visual modality and compared to a reference memory value formed during an auditorily presented training trial will seem inappropriately short.

This idea was tested in a series of temporal bisection experiments using both visual and auditory stimuli for the anchor durations during training (Penney et al., 1998, 2000). Participants were then presented with intermediate test durations of either modality and asked to judge whether each test duration was closer to the short or the long anchor. Consistent with the differential switch-closing/mixed-memory hypothesis, visual stimuli had to be of a longer duration before they were judged "long" with the same probability as auditory stimuli.

Testing also included divided attention trials that required participants to simultaneously time two stimuli of different modalities, durations, and onsets (e.g., a visual stimulus 3.57 sec in duration and an auditory stimulus that began 1 sec into the visual stimulus and lasted for 4.24 sec). Young adults were equally accurate and sensitive to time on these divided attention trials as on single-stimulus trials. Furthermore, the divided attention trials did not lead to an exaggerated modality effect (Penney et al., 1998, 2000). This pattern suggests that the controlled attention functions responsible for maintaining attention and keeping the switch closed during visual stimuli are different from those required for dividing attention among multiple stimuli.

Interval timing experiments with older adults support the idea that both modality effects and divided effects are related to controlled attention, but that they may tap different functions of attentional control. Older adults show greater modality effects and greater divided attention effects than do young adults, but these effects appear to be independent.

In a mixed-modality bisection experiment using only single-stimulus trials (Lustig and Meck, 2001b), both young and older adults showed a modality effect similar to that shown by Penney et al. (1998, 2000): visual stimuli had to be of a longer duration before they were judged long with the same probability as auditory stimuli. However, this effect was greater for older adults than for young adults, consistent with older adults' greater difficulties with attentional control.

Lustig and Meck (2002) used a peak-interval timing task to examine young and older adults' sensitivity to modality effects under full and divided attention conditions. Participants were first trained to associate a signal of one modality (visual or auditory) with a short duration (8 sec) and the signal of the other modality with a long duration (21 sec). Because one modality was assigned to "short" and the other to "long," there was no mixed modality memory distribution as in the Penney et al. (1998, 2000) experiments. Instead, each duration had a separate reference memory distribution made up of values from only one modality. At test, participants were presented with visual and auditory stimuli and asked to indicate when the signal had been on for the duration associated with it during training. On full attention trials, participants were presented with only a single stimulus of either modality. On divided attention trials, participants were presented with one signal from each modality. On these trials, the long signal always began first, and the short signal onset after a variable delay.

Young adults were highly accurate on all trials, though they were less sensitive to time (had more variable productions) for visual stimuli than for auditory stimuli. Older adults performed very similarly to young adults, with one dramatic exception: older adults overproduced the short stimulus by almost 2 sec on divided attention trials when it was presented in the visual modality. Older adults were also approximately twice as variable in their productions of short visual stimuli on divided attention trials vs. full attention trials. These results suggest that older adults had great difficulty allocating attention to a visual stimulus, which required controlled attention, when they were already timing another stimulus presented in the auditory modality. In contrast, both young and older adults remained accurate for short auditory stimuli on divided attention trials, where they were presented in the context of an ongoing visual long stimulus, though for both age groups variability increased slightly for divided attention trials. This relative preservation of accurate performance for auditory stimuli even in the presence of visual stimuli that already occupy attention is consistent with the idea that auditory stimuli can keep the attentional switch closed relatively automatically.

Another experiment (Lustig and Meck, 2001a) tested young and older adults on a mixed-modality bisection task with both full attention and divided attention trials, and also examined the influence of circadian arousal patterns on these attentional effects. Many physiological variables change over the course of the day, including blood pressure, temperature, and the levels of various hormones and neurotransmitters, and attention and memory variables that affect interval timing may change as well (e.g., Fujiwara et al., 1992; Portman, 2001; for reviews, see Hinton and Meck, 1997; Yoon et al., 1999). In particular, the ability to inhibit or ignore information that is irrelevant to the current task seems to be greatest for older adults in the morning and for young adults in the evening. However, this crossover may be specific to inhibitory aspects of attention; it is not shown by tasks that tap other aspects of attention (May and Hasher, 1998).

This experiment provided additional evidence for the idea that both modality and divided attention manipulations tap functions of controlled attention that are impaired in older adults, but separate from each other. Overall, older adults showed a larger modality effect than did young adults, requiring a visual stimulus of a longer duration before judging it long with the same probability as an auditory stimulus of the same physical duration. Older adults were also more affected by the divided attention manipulation: the two age groups were equally sensitive to time on single-stimulus trials, but older adults were much less sensitive to time on divided attention trials. Although age differences in controlled attention are evident in both modality differences and divided attention effects, these two variables did not interact: for both age groups, the modality effect was the same size on full attention and divided attention trials.

The pattern of circadian arousal effects also suggested a separation between different controlled attention functions. For both young and old adults, modality differences were smaller and sensitivity to time greater in the afternoon than in the morning, but the ability to divide attention among multiple to-be-timed stimuli did not fluctuate with time of day. However, older adults tested in the morning showed a peculiar pattern, displaying very high sensitivity to time for visual trials in the full attention condition. This pattern might have occurred if older adults attempted to compensate for poor controlled attention overall by devoting their efforts to this type of stimulus, which would be effortful but not as difficult as the divided attention trials, and relatively ignoring the other stimulus types. Older adults' ability to focus on one stimulus type and ignore others would be best in the morning, when their inhibitory abilities are relatively high.

Overall, the results of these interval timing experiments strongly suggest that maintaining attention and dividing attention are both controlled attention functions that decline with age, but that these functions are separable and distinct. Older adults' timing performance was more affected than young adults' by modality manipulations of the need to maintain controlled attention on a to-be-timed stimulus and by divided attention manipulations that required the timing of multiple stimuli. However, the effects of these manipulations did not interact and, furthermore, were affected differently by the time of day at which participants were tested. Circadian influences on attention and interval timing may impact the results of many experiments, especially those that include groups whose circadian arousal patterns may differ (e.g., young and older adults). In the Lustig and Meck (2001a) experiment, time of day influenced older adults' choice of one time-related stimulus over another; changes in inhibitory abilities over the course of the day may also impact the ability to prioritize temporal vs. nontemporal tasks in divided attention experiments.

These interval timing experiments illuminate distinct aspects of attention that can affect the accuracy of older adults' interval timing. Understanding the interactions — or lack thereof — between age differences in these different controlled attention functions is likely to be important for understanding older adults' performance on many cognitive tasks. As described in the next section, in addition to accuracy measures, measures of variability in timing performance can also provide insights on older adults' attentional functioning under different conditions.

10.2.4 VARIABILITY MEASURES

In addition to asking about the accuracy of older adults' interval timing performance, one can also ask about its precision or consistency. The question of whether older adults' timing performance is more variable than that of young adults has more often been investigated in studies of rhythmic timing. In these tasks, which typically require the repeated production of durations much shorter than those often used in interval timing (< 2 sec), age differences are very small or nonexistent if participants can synchronize with an external pacer, but older adults are more variable if they have to maintain multiple rhythms simultaneously or self-generate a rhythm (e.g., Krampe et al., 2001).

Variability measures are used much less often in age studies of interval timing, and those papers that do report variability have mixed results. Using very brief intervals (≤ 1 sec), Wearden et al. (1997) found age increases in variability for temporal generalization, but not for temporal bisection or interval production. Rammsayer (2001) did not find age differences in variability for discrimination or reproduction of short intervals (1 sec) or reproduction of a much longer (15 sec) interval. However, Rammsayer also did not replicate standard age differences in reproduction accuracy for the 15-sec reproduction task, suggesting that the experiment may have been insensitive to age differences in general. Vanneste et al. (1999) found that older adults had larger coefficients of variation than did young adults when reproducing intervals ranging in duration from 6 to 8 sec.

In general, age differences in variability seem more pronounced under divided attention conditions. Vanneste et al. (1999) found that while the requirement to divide attention among multiple to-be-timed stimuli increased the coefficient of variation for both young and older adults, this increase in variability was greater for older adults. The sensitivity measure used by Lustig and Meck (2001a) is inversely related to variability; young and older adults were equally sensitive to time on single trials, but older adults were much less sensitive to time on divided attention trials. Perbal et al. (2002) found that dividing participants' attention with a nontemporal task during production and reproduction increased coefficients of variation for both young and older adults, but the increased variability was greater for older adults' reproduction.

Taken together, these results suggest that old age does lead to increased variability on timing tasks, but that age effects on variability are smaller than those on accuracy. Dividing attention seems to increase variability, with older adults being more affected than are young adults. There is also some suggestion that the type of timing task and interval length influence an experiment's sensitivity to age and attention effects on variability, with reproduction tasks and longer intervals most likely to show effects. These are all very tentative conclusions, though, and the picture may change with increased reports of variability in experiments on older adults' timing.

10.2.5 AGE DIFFERENCES IN ATTENTION AND INTERVAL TIMING: SUMMARY AND CONCLUSIONS

Attention plays an important role in our judgments of time. The experiments reviewed in the previous sections use a variety of methods to address the question of how age differences in attention might lead to age differences in interval timing. Each of these methods has its own advantages and disadvantages, but overall the results of these experiments show that age differences in interval timing can be explained by considering how age differences in controlled attention might influence the function of the internal clock central to information-processing models of interval timing, including SET.

Some duration judgment tasks allow a closer examination of attention's role in interval timing than do others. Absolute time judgments, which make use of our verbal labels for time (e.g., 3 sec), have the advantage of being easy to explain to participants and have an obvious connection to the way that we use duration information in everyday life. These features make absolute time judgment tasks very attractive and popular, but because the experimenter does not control the conditions under which durations are learned, they are not ideal for studying attention. Relative time judgments may often be better for asking questions about attention's influence on interval timing performance, because the target duration is both learned and tested within the context of the experiment. However, even on relative time judgment tasks, attentional "drift" between training and test trials can affect performance. This is especially the case for older adults, who may have difficulty maintaining attention at a high level over long periods of time.

Divided attention manipulations, especially on relative time judgment tasks, are among the most powerful ways of examining the effects of attention on interval timing. Older adults are typically more sensitive than young adults to these manipulations on both absolute (e.g., Craik and Hay, 1999) and relative (e.g., Perbal et al., 2002; Vanneste and Pouthas, 1999) time judgment tasks. Experiments with young adults (e.g., Fortin and Massé, 2000; see review by Fortin, this volume) suggest that simply expecting a break or the addition of another task (even if the break or addition is omitted on that trial) can divert attention and influence the perception of time; presumably, these effects would be even greater for older adults. Even with divided attention manipulations, though, care must be taken because participants may prioritize one stimulus over another (e.g., between timing and a nontemporal stimulus, or between multiple to-be-timed stimuli of different modalities), and the order of prioritization may not be the same for young and older adults.

Despite these complexities, the results of most experiments on age differences in interval timing are consistent with the framework described at the beginning of this chapter: older adults' clocks will run slower than young adults' in situations that place strong demands on controlled attention, and age differences in interval timing will appear when those demands are different when a target duration is learned vs. when it is tested.

In most cases, older adults' timing distortions have certain characteristics that mark them as being the result of differences in attention and clock speed, rather than general performance declines or problems with memory or some other function. Attentional manipulations do not just increase older adults' timing errors overall, but instead lead to directional effects that indicate a slower and more variable clock in the high-demand situation. These effects increase with increasing attentional demands (e.g., Vanneste and Pouthas, 1999) and are much reduced when the need for controlled attention is minimized (e.g., Lustig and Meck, 2001a, 2001b, 2002). Feedback about their timing distortions can often improve older adults' performance (e.g., Malapani et al., 2002b; Rakitin et al., submitted; Wearden et al., 1997), and this sensitivity to feedback is considered a hallmark of effects on attention and clock speed (Meck, 1983).

Age differences in controlled attention clearly appear to play an important role in age differences in judgments about time. However, timing and attention are not identical: some aspects of attentional functioning that change with age may not affect timing performance, at least not directly, and other cognitive functions that change with age may also play a role in interval timing performance. The following section examines recent attempts to understand the impact of age differences in these other areas of cognition on older adults' interval timing.

10.3 AGE DIFFERENCES IN MEMORY, PROCESSING SPEED, AND INTERVAL TIMING

Attention is not the only aspect of cognitive function that changes with age; memory problems are a common complaint of older adults (Craik and Jennings, 1992; Light, 1991; Zacks et al., 2000), and an apparent age-related decline in processing speed is often suggested as an underlying reason for age differences on many cognitive tasks (e.g., Salthouse, 1996). As described above, reference memory is a central component of many information-processing models of interval timing. Processing speed plays a special role in mathematical formulations of SET, as substantiated in the K* parameter, the speed at which accumulator values are encoded into and decoded from reference memory (Gibbon et al., 1984; Meck, 1983).

The contributions of these components to age differences in interval timing have not been as heavily researched as the contribution of attention, in part because they are not as amenable to experimental manipulations that can be easily implemented in humans. Furthermore, at least some interval timing experiments using human participants are designed in a way that minimizes the contribution of reference memory as typically defined in information-processing models of interval timing and tested in animal experiments (see Wearden and Bray, 2001). However, recent attempts have been made to examine the relationship between age differences on interval timing tasks and neuropsychological tests designed to target specific cognitive functions (Perbal et al., 2002), and to ask whether older adults may show memory deficits that are specific for time (Rakitin et al., submitted).

Perbal et al. (2002) examined the relations between young and older adults' scores on neuropsychological tests and their performance on different interval timing tasks. The neuropsychological measures included were designed to differentially emphasize processing speed, passive short-term memory storage, working memory (online storage and processing), and long-term, delayed memory storage. The interval timing tasks included a production task under conditions of divided attention (read-aloud digits presented at random intervals) and a reproduction task in which attention was divided during encoding of the standard, but not its reproduction.***

*** Perbal. et al. (2002) also included a control counting condition, in which participants counted aloud while producing the target interval in the production task and while encoding and reproducing the interval in the reproduction task. Participants were extremely accurate in this condition, and there were no age differences, so it will not be discussed further here.

Standard age and divided attention effects were found on the timing tasks, with older adults overproducing and underreproducing the target intervals to a greater extent than did young adults. Performance on the neuropsychological measures of working memory and long-term memory was moderately correlated with accurate performance on the timing tasks, especially reproduction. There was some tendency for these correlations, especially for long-term memory, to be greater for longer (14 and 38 sec) than shorter (5 sec) intervals. Moreover, when age differences in working memory and long-term memory were statistically controlled for, age differences for longer durations in the reproduction task were no longer statistically significant. Much of the age-related variance in the production task was accounted for by processing speed.

Perbal et al. (2002) conclude that working memory and processing speed contributed to the ability to divide attention between the digit-reading and timing tasks. (It should be noted that this "executive function" of working memory is also often described as working attention (e.g., Baddeley, 1993; Engle, 2001).) They explain the correlations between long-term memory and interval timing performance in the longer durations by suggesting that for long durations, pulses collected early in the interval eventually pass out of the accumulator and are stored in long-term memory. If this were the case, older adults' overproduction and underreproduction would result from deficits in two aspects of cognition: First, age deficits in working memory (or attention) would cause them to miss more pulses during initial accumulation. Second, deficits in long-term memory would cause older adults to lose more pulses from storage for the longer durations than would young adults.

The Perbal et al. (2002) experiments represent an important step in relating interval timing performance to performance on standard measures of attention and memory, but several caveats should be kept in mind when interpreting their results. First, an earlier experiment by this group (Vanneste and Pouthas, 1999) using a task that divided attention between multiple to-be-timed stimuli and a different measure of working memory did not find any relation between timing performance and working memory for either young or older adults. Second, the function of long-term memory as described by Perbal et al. is more closely related to accumulator functioning than to reference memory as usually conceived in models of interval timing. Finally, it is important to note that the contributions of both working memory and processing speed as described by Perbal et al. act through attention's mediation of the switch that gates pulses through the accumulator, not the rate of pacemaker pulsation or K^*.

The reference memory component of the interval timing model has been the focus of several recent investigations. McCormack et al. (2002) trained young and older adults on six tones of increasing duration. On test trials, participants were asked to identify which of the six durations had just been presented. Young adults were quite accurate, but older adults misclassified the tones as being shorter than they really were. In contrast, when asked to make judgments about the pitch of tones rather than their duration, older adults made more errors than did young adults but did not show the same systematic underestimation as for duration. McCormack et al. suggested that older adults had a duration-specific distortion in memory such that they remembered the standards as being longer than they really were. This suggestion receives support from previous human and animal studies suggesting an age-related distortion in duration memory (McCormack et al., 1999; Meck, 1986; Wearden et al., 1997), but the same results could also occur if older adults showed a greater drift of attention from training to test trials.

Rakitin et al. (submitted) found age differences on a reproduction task that they attributed to age differences in reference memory for the target durations. They used a peak-interval reproduction procedure in which participants attempted to reproduce a duration learned in an earlier training session. When a single target duration is used, older adults will overreproduce the target more than will young adults (Malapani et al., 1998). As described previously, this result likely occurs because older adults' deficits in controlled attention cause them to miss pulses during test trials, requiring more physical time before current accumulator values match those associated with the target duration in reference memory. Rakitin et al. (submitted) found a very different result when they used two durations in the same session. Under these conditions, older adults overreproduced the shorter of the two intervals, but underreproduced the longer interval. Interestingly, this "migration effect" appeared to be specific to memory for durations; no such effect was found for a line-length reproduction task that also used a short and long standard.

Rakitin et al. (submitted) suggested that older adults' vulnerability to the migration effect for durations may be attributed to age-related declines in dopamine function. A very similar effect is found for Parkinson's patients, whose disease stems from a dopamine depletion (Malapani et al., 1998, 2002a). However, the migration effect in non-Parkinson's older adults was not proportionate: the overestimation of the short duration was greater than the underestimation of the long duration. In fact, underestimation of the long duration was only significant for a subgroup of the older adults. Rakitin et al. (submitted) suggested that the migration effect may be a marker for older adults who have especially pronounced declines in dopamine function and are thus particularly vulnerable to the memory effects.

The idea of a vulnerable subgroup of older adults can explain why only some older adults showed a significant underestimation of the long interval, but it does not account for the uneven effects for the short and long targets. One possibility is that age effects on attention and memory compound each other for the short duration and counteract each other for the long duration. That is, for the short duration, attention problems would lead to a loss of pulses and a tendency to overproduce the interval to match the memory representation of the target, a representation that would be inappropriately large because of migration toward the longer duration. For the long duration, attention problems would likewise lead to a tendency to overproduce the interval to match the memory representation of the target, but in this case, migration would shorten that representation, causing the attention and memory effects to partially cancel each other out.

Patterns of scalar and nonscalar variability for the Parkinson's and non-Parkinson's older adults support the idea of a dopaminergic influence. In addition to its role in cognition, dopamine plays an important role in motor functioning; the motor problems caused by dopamine deficits are the primary reason that Parkinson's patients medicate the disease. As described in the introduction, manipulations that influence clock and memory components of the SET model maintain the scalar property of increasing variability with increasing duration (Gibbon et al., 1984, 1997). Manipulations that lead to nonscalar variability are thought to have their effect through nontemporal avenues, such as response criterion or motor output. In the experiments conducted by Rakitin et al. (submitted; Malapani et al., 1998, 2002a), older adults were accurate in training trials, which were conducted with feedback, and inaccurate on delayed test trials without feedback, but showed nonscalar variability under both conditions. Parkinson's patients tested off the medication that remediates their dopamine deficit were both inaccurate and nonscalar for both training and testing trials, but both the migration effect and nonscalar variability were remediated by medication. The conclusion suggested by this pattern is that nonscalar production results from problems with motor processes, whereas the migration effect is located at attention and memory processes, and that both are influenced by dopamine.

To summarize, most of the research on age differences in interval timing has focused on the role of attention, and much less is known about the influence of memory, processing speed, or other cognitive characteristics that change with age. The experiments described above exemplify methods that may be useful for examining these influences. Correlations between interval timing performance and performance on neuropsychological tests designed to tap specific cognitive abilities can help establish connections between group and individual differences in both domains. In addition, the analysis procedures provided by SET and other information-processing models of timing can help to separate the relative contributions of attention, memory, and nontemporal processes to performance on an interval timing task.

10.4 INTERVAL TIMING AND OTHER AREAS OF COGNITIVE AGING

Most of this chapter has focused on the ways in which age differences in attention affect older adults' interval timing performance relative to that of young adults, but interval timing research can also be useful for gaining a broader understanding of the cognitive changes that occur with advanced age. On a procedural level, the nonverbal and continuous nature of many interval timing tasks minimizes the confounders that can be problematic for aging studies of attention and memory. In addition, age differences in timing performance may contribute to age differences in performance on other tasks, especially speeded tasks or those that require actions to be performed at a certain time or in a particular sequence. Finally, the great deal of research on the neurobiological underpinnings of interval timing performance may provide important insights for the relation between biological and cognitive changes that occur during aging.

For example, many standard memory experiments make use of verbal materials. This often presents a problem for studying age differences in memory, as older adults typically have richer vocabularies than do young adults, and may also suffer from breakdowns in the networks that connect semantic, phonological, and orthographic aspects of language (see reviews by Burke et al., 2000; Kemper and Mitzner, 2001; Wingfield and Stine-Morrow, 2000). Furthermore, older adults are often anxious about their memory performance, and the instructions used in most memory tests (e.g., "Remember the list of words you were shown earlier") can elicit this anxiety and distract them from the memory task, ironically reducing their performance. Older adults may do better in situations where the use of the word *memory* is minimized in the instructions (Rahhal et al., 2001). Timing tasks, which are nonverbal in nature and which typically use instructions that do not emphasize memory (e.g., "Is this duration shorter or longer than the standard?"), may help avoid both of these problems. However, some caution is required here, as older adults may show memory deficits that are specific to duration (McCormack et al., 2002; Rakitin et al., submitted).

Interval timing tasks may also be useful for researchers interested in sustained attention. One problem with many sustained attention or vigilance tasks is that their dependent measure is the detection of a rare, intermittent event, whereas the cognitive construct of interest is the ability to continuously maintain an attentional state. A person's attention may "flicker" or be temporarily diverted from the task, but this lapse will not be detected unless it coincides with the presentation of the rare target event. Thus, these experiments may underestimate age differences in the ability to maintain attentional control. In timing tasks, at least as characterized by pacemaker–accumulator models, the measurement of sustained attention is much more continuous (accumulation of pulses that is measurable by changes in clock speed). Timing procedures may thus be a more sensitive measure of how age and other variables affect the ability to sustain attention and vigilance.

Frequent attempts have been made to relate age differences in timing performance to age-related slowing, a topic of great interest in aging research (e.g., Cerella, 1985; Salthouse, 1996). The metaphor of an internal clock that may run at faster or slower speeds has led some investigators to propose that older adults' clocks may run more slowly because of reductions in processing speed, and others to propose that a slower clock may be a reason for cognitive slowing (e.g., Block et al., 1998; Craik and Hay, 1999; Rabbitt, 2000; Schroots and Birren, 1990; Vanneste et al., 2001).

There probably are strong relations between timing-related concepts such as clock speed or memory storage speed and cognitive slowing as measured by psychometric speed or reaction time tasks. However, this relationship is not likely to be as direct as would be expected from the shared use of the term *speed*. A number of factors complicate interpretation: depending on a particular investigator's background, *processing speed* may mean any of several different variables, including psychomotor speed, decision speed, or perceptual speed (Salthouse, 2000). The degree to which older adults are slower than young adults can vary across these domains, and change depending on response mode, task complexity, and whether verbal or nonverbal processing is involved (e.g., Jenkins et al., 2000; Oberauer and Kliegl, 2001; for a review, see Verhaegen, 2000). It is still unclear how these different aspects of slowing are related to each other and to other cognitive variables, including attention, or what biological changes underlie behavioral changes in processing speed (for discussion, see Park et al., 2001). However, in the context of the issues discussed in this chapter, it is interesting to note that the frontal-striatal brain circuits involved in attention and timing (e.g., Meck and Benson, 2002) have also been suggested as the locus of changes important for age reductions in processing speed (e.g., Backman et al., 2000; Rubin, 1999).

Some attempt to pit clock speed explanations of timing performance against attention and memory explanations, but as has been described here, attention is directly related to the speed of the clock, via its influence on the rate at which pulses accumulate. This problem might be circumvented by restricting the definition of the clock to the pacemaker. However, the idea that older adults have a slower clock than they did in their youth, when they learned the labels associated with particular clock readings (e.g., 3 sec), would predict that older adults should overproduce and underestimate durations on absolute time judgment tasks. The results of the Block et al. (1998) meta-analysis revealed the exact opposite pattern. It is also difficult to discern how a general slowing of the clock could account for the directional distortions older adults show on relative time judgment tasks.

Cognitive slowing could have an effect by reducing the attentional resources available to older adults (Salthouse, 1996). For example, Perbal et al. (2002) found a relation between processing speed as measured by a reaction time task and age differences on an interval production task performed under divided attention conditions. In this case, processing speed may have had its effects by influencing participants' ability to divide attention or switch rapidly between temporal and nontemporal processing. Processing speed's effect on clock speed would thus be indirect, mediated through its effect on attention. In general, age differences in attention functioning appear to provide more proximal and parsimonious explanations of age differences in interval timing than does the idea that older adults generally have a slower clock.

It has also been suggested that age differences in timing may influence performance on reaction time or speed tasks themselves. In such tasks, people tend to use timing information to prepare a response and to locate themselves on a speed–accuracy continuum such that they spend the smallest amount of time on each trial that will allow them to avoid making a mistake (Grosjean et al., 2001). If older adults have inaccurate or more variable representations of the time they take to complete each trial, they may be less able to optimally calibrate their performance on this continuum (Rabbitt, 2000). Age differences in timing may also affect performance on prospective memory tasks that require participants to remember to perform particular actions at particular times, or on executive control tasks that require the timing and sequencing of multiple actions (Krampe et al., 2002; Park et al., 1997).

Besides these relations between interval timing tasks and other measures of behavior, researchers interested in the neurobiological consequences of age may find important clues in both human and animal timing research. A variety of drugs and neurotransmitter systems have been examined in the context of interval timing, and in some cases, relatively clear distinctions have been drawn between their effects on attention (clock speed) and memory (for reviews, see Buhusi, this volume; Cevik, this volume; Mattell and Meck, 2000; Meck, 1996; Meck and Benson, 2002). In particular, dopamine changes have recently received a great deal of attention in human cognitive aging research (e.g., Backman et al., 2000; Braver et al., 2001; Li, Lindenberger, and Sikstrom, 2001; Park et al., 2001; Volkow et al., 1998, 2000). Dopamine functioning has been heavily investigated in interval timing experiments using both animals and human populations (e.g., schizophrenics, Parkinson's patients) with known dopamine dysfunctions. As the Rakitin et al. (submitted) experiments demonstrate, the findings from these populations are likely to be highly relevant for understanding the cognitive effects that occur as the result of neurobiological changes in healthy aging as well.

10.5 SUMMARY AND CONCLUSIONS

The results of many experiments suggest that time often flies less accurately and more variably for older adults than it does for young adults. This chapter has attempted to organize those results in the context of one of the dominant models of interval timing performance, scalar expectancy theory, and to relate them to general questions of age differences in attention and memory. As described throughout the chapter, the evidence regarding some issues (e.g., divided attention effects) is extensive and clear-cut, but in other cases (e.g., variability measures), there have been few or even conflicting reports, and many questions remain completely open and unexplored.

The discussion here has been centered on scalar expectancy theory and other pacemaker–accumulator information-processing models of interval timing. It remains to be seen how other characterizations of interval timing (e.g., the coincidence detection model discussed by Mattell et al., this volume) will account for age effects. The majority of these models consider both cognitive concepts (e.g., attention and memory) and brain systems (especially cortico-striatal circuits) that are of great interest to cognitive aging research. A continuing convergence of research across interval timing and other domains will be important for understanding not only older adults' performance on interval timing tasks, but also the multitude of changes that occur in cognitive functioning as we age.

REFERENCES

Anderson, N.D., The attentional demands of encoding and retrieval in younger and older adults: 2. Evidence from secondary task reaction time distributions, *Psychol. Aging*, 14, 645–655, 1999.

Anderson, N.D., Craik, F.I.M., and Naveh-Benjamin, M., The attentional demands of encoding and retrieval in younger and older adults: 1. Evidence from divided attention costs, *Psychol. Aging*, 13, 405–423, 1998.

Anderson, N.D., Iidaka, T., Cabeza, R., Kapur, S., McIntosh, A.R., and Craik, F.I.M., The effects of divided attention on encoding- and retrieval-related brain activity: a PET study of younger and older adults, *J. Cognit. Neurosci.*, 12, 775–792, 2000.

Backman, L., Ginovart, N., Dixon, R.A., Wahlin, T.B.R., Wahlin, A., Halldin, C., and Farde, L., Age-related cognitive deficits mediated by changes in the striatal dopamine system, *Am. J. Psychiatry*, 157, 635–637, 2000.

Baddeley, A.D., Working memory or working attention? in *Attention: Selection, Awareness, and Control: A Tribute to Donald Broadbent*, Baddeley, A.D. and Weiskrantz, L., Eds., Clarendon Press, New York, 1993, pp. 152–170.

Baltes, P.B., On the incomplete architecture of human ontogeny: selection, optimization, and compensation as foundation of developmental theory, *Am. Psychol.*, 52, 366–380, 1997.

Baltes, P.B. and Baltes, M.M., Psychological perspectives on successful aging: the model of selective optimization with compensation, in *Successful Aging: Perspectives from the Behavioral Sciences*, Baltes, P.B. and Baltes, M.M., Eds., Cambridge University Press, New York, 1990, pp. 1–34.

Block, R.A., Zakay, D., and Hancock, P.A., Human aging and duration judgments: a meta-analytic review, *Psychol. Aging*, 13, 405–423, 1998.
Braver, T.S., Barch, D.M., Keys, B.A., Carter, C.S., Cohen, J.D., Kaye, J.A., Janowsky, J.S., Taylor, S.F., Yesavage, J.A., Mumenthaler, M.S., Jagust, W.J., and Reed, B.R., Context processing in older adults: evidence for a theory relating cognitive control to neurobiology in healthy aging, *J. Exp. Psychol. Gen.*, 130, 746–763, 2001.
Brown, S.W., Time perception and attention: the effects of prospective versus retrospective paradigms and task demands on perceived duration, *Percept. Psychophys.*, 38, 115–124, 1985.
Burke, D.M., Makay, D.G., and James, L.E., Theoretical approaches to language and aging, in *Models of Cognitive Aging*, Perfect, T.J. and Maylor, E.A., Eds., Oxford University Press, New York, 2000, pp. 204–237.
Cabeza, R., Cognitive neuroscience of aging: contributions of functional neuroimaging, *Scand. J. Psychol.*, 42, 277–286, 2001.
Cerella, J., Information processing rates in the elderly, *Psychol. Bull.*, 98, 67–83, 1985.
Chen, Z. and O'Neill, P., Processing demand modulates the effects of spatial attention on the judged duration of a brief stimulus, *Percept. Psychophys.*, 63, 1229–1238, 2001.
Craik, F.I.M., Age differences in human memory, in *Handbook of the Psychology of Aging*, Birren, J.E. and Schaie, K.W., Eds., Van Nostrand Reinhold, New York, 1977, pp. 384–420.
Craik, F.I.M., Govoni, R., Naveh-Benjamin,, M., and Anderson, N.D., The effects of divided attention on encoding and retrieval processes in human memory, *J. Exp. Psychol. Gen.*, 125, 159–180, 1996.
Craik, F.I.M. and Hay, J.F., Aging and judgments of duration: effects of task complexity and method of estimation, *Percept. Psychophys.*, 61, 549–560, 1999.
Craik, F.I.M. and Jennings, J.M., Human memory, in *The Handbook of Aging and Cognition*, Craik, F.I.M. and Salthouse, T.A., Eds., Erlbaum, Hillsdale, NJ, 1992, pp. 51–110.
Curran, T., Hills, A., Patterson, M.B., and Strauss, M.E., Effects of aging on visuospatial attention: an ERP study, *Neuropsychologia*, 39, 288–301, 2001.
Engle, R.W., What is working memory capacity? in *The Nature of Remembering: Essays in Honor of Robert G. Crowder*, Roediger, H.L. and Nairne, J.S., Eds., American Psychological Association, Washington, D.C., 2001, pp. 297–314.
Fernandez, A.M. and Pouthas, V., Does cerebral activity change in middle-aged adults in a visual discrimination task? *Neurobiol. Aging*, 22, 645–657, 2001.
Fortin, C. and Massé, N., Expecting a break in time estimation: attentional time-sharing without concurrent processing, *J. Exp. Psychol. Hum. Percept. Perform.*, 26, 1788–1796, 2000.
Fortin, C. and Rosseau, R., Interference from short-term memory processing on encoding and reproducing brief durations, *Psychol. Res. Psychol. Forsch.*, 61, 269–276, 1998.
Fraisse, P., Perception and estimation of time, *Ann. Rev. Psychol.*, 35, 1–36, 1984.
Freund, A.M., Li, K.Z.H., and Baltes, P.B., Successful development and aging: the role of selection, optimization, and compensation, in *Action and Self-Development: Theory and Research through the Lifespan*, Brandstadter, J. and Lerner, R.M., Eds., Sage, Thousand Oaks, CA, 1999, pp. 401–434.
Fujiwara, S., Shinkai, S., Kurokawa, Y., and Watanabe, T., The acute effects of experimental short-term evening and night shifts on human circadian-rhythm: the oral-temperature, heart-rate, serum cortisol and urinary catecholamines levels, *Int. Arch. Occup. Environ. Health*, 63, 409–418, 1992.
Gibbon, J., Church, R.M., and Meck, W.H., Scalar timing in memory, *Ann. N.Y. Acad. Sci.*, 423, 52–77, 1984.

Gibbon, J., Malapani, C., Dale, C.L., and Gallistel, C.R., Toward a neurobiology of temporal cognition: advances and challenges, *Curr. Opin. Neurobiol.*, 7, 170–184, 1997.

Goldstone, S. and Goldfarb, J.L., Auditory and visual time judgement, *J. Gen. Psychol.*, 70, 369–387, 1964.

Greenwood, P.M., Parasuraman, R., and Haxby, J.V., Changes in visuospatial attention over the adult life-span, *Neuropsychologia*, 31, 471–485, 1993.

Grosjean, M., Rosenbaum, D.A., and Elsinger, C., Timing and reaction time, *J. Exp. Psychol. Gen.*, 130, 256–272, 2001.

Hartley, A.A., Attention, in *The Handbook of Aging and Cognition*, Craik, F.I.M. and Salthouse, T.A., Eds., Lawrence Erlbaum Associates, Hillsdale, NJ, 1992, pp. 3–49.

Hasher, L., Tonev, S.T., Lustig, C., and Zacks, R., Inhibitory control, environmental support, and self-initiated processing in aging, in *Perspectives on Human Memory and Cognitive Aging: Essays in Honor of Fergus Craik*, Naveh-Benjamin, M., Moscovitch, M., and Roediger, H.L., III, Eds., Psychology Press, Philadelphia, 2001, pp. 286–297.

Hasher, L. and Zacks, R.T., Automatic and effortful processes in memory, *J. Exp. Psychol. Gen.*, 108, 356–388, 1979.

Hasher, L. and Zacks, R.T., Working memory, comprehension, and aging: a review and a new view, in *The Psychology of Learning and Motivation*, Vol. 22, Bower, G.H., Ed., Academic Press, San Diego, CA, 1988, pp. 193–225.

Hasher, L., Zacks, R., and May, C.P., Inhibitory control, circadian arousal, and age, in *Attention and Performance*, XVII, Gopher, D. and Koriat, A., Eds., MIT Press, Cambridge, MA, 1999, pp. 653–675.

Hinton, S.C. and Meck, W.H., The 'internal clocks' of circadian and interval timing, *Endeavour*, 2, 82–87, 1997.

Hinton, S.C., Meck, W.H., and MacFall, J.R., Peak-interval timing in humans activates frontal-striatal loops, *Neuroimage*, 3, S224, 1996.

Jacoby, L.L., A process dissociation framework: separating automatic from intentional uses of memory, *J. Mem. Lang.*, 30, 513–541, 1991.

Jenkins, L., Myerson, J., Joerding, J.A., and Hale, S., Converging evidence that visuospatial cognition is more age-sensitive than verbal cognition, *Psychol. Aging*, 15, 157–175, 2000.

Jennings, J.M. and Jacoby, L.L., Automatic versus intentional uses of memory: aging, attention, and control, *Psychol. Aging*, 8, 283–293, 1993.

Kemper, S. and Mitzner, T.L., Language production and comprehension, in *Handbook of the Psychology of Aging*, 5th ed., Birren, J.E. and Schaie, K.W., Eds., Academic Press, San Diego, CA, 2001, pp. 378–398.

Krampe, R.T., Engbert, R., and Kliegl, R., Age-specific problems in rhythmic timing, *Psychol. Aging*, 16, 12–30, 2001.

Krampe, R.T., Engbert, R., and Kliegl, R., The effects of expertise and age on rhythm production: adaptations to timing and sequencing constraints, *Brain Cognit.*, 48, 179–194, 2002.

Lejeune, H., Switching or gating? The attentional challenge in cognitive models of psychological time, *Behav. Process.*, 44, 127–145, 1998.

Lejeune, H., Ferrara, A., Soffie, M., Bronchart, M., and Wearden, J.H., Peak procedure performance in young adult and aged rats: acquisition and adaptation to a changing temporal criterion, *Q. J. Exp. Psychol. Comp. Physiol. Psychol.*, 51B, 193–217, 1998.

Li, K.Z.H., Lindenberger, U., Freund, A.M., and Baltes, P.B., Walking while memorizing: age-related differences in compensatory behavior, *Psychol. Sci.*, 12, 230–237, 2001.

Li, S.C., Lindenberger, U., and Sikstrom, S., Aging cognition: from neuromodulation to representation, *Trends Cognit. Sci.*, 5, 479–486, 2001.

Light, L.L., Memory and aging: four hypotheses in search of data, *Annu. Rev. Psychol.*, 42, 333–376, 1991.

Liu, Y.C., Comparative study of the effects of auditory, visual and multimodality displays on drivers' performance in advanced traveller information systems, *Ergonomics*, 44, 425–442, 2001.

Lustig, C. and Meck, W.H., Paying attention to time as one gets older, *Psychol. Sci.*, 12, 478–484, 2001a.

Lustig, C. and Meck, W.H., Signal Modality Interactions Reveal Developmental/Aging Changes in Interval Timing, paper presented at the 42nd Annual Meeting of the Psychonomic Society, Boca Raton, FL, 2001b.

Lustig, C. and Meck, W.H., Age Deficits and Sparing in Simultaneous Temporal Processing, poster presented at the Cognitive Neuroscience Meeting, San Francisco, CA, 2002.

Malapani, C., Deweer, B., and Gibbon, J., Separating storage from retrieval dysfunction of temporal memory in Parkinson's disease, *J. Cognit. Neurosci.*, 14, 1–12, 2002a.

Malapani, C., Rakitin, B.C., Dube, K., Lobo, S., and Fairhurst, S., Disambiguating the Role of Cognitive Strategy versus Memory Updating in an Age-Related Time Production Deficit, poster presented at the Cognitive Neuroscience Society meeting, San Francisco, CA, April 2002b.

Malapani, C., Rakitin, B., Meck, W.H., Deweer, B., Dubois, B. and Gibbon, J., Coupled temporal memories in Parkinson's disease: a dopamine-related dysfunction, *J. Cognit. Neurosci.*, 10, 316–331, 1998.

Mattell, M.S. and Meck, W.H., Neuropsychological mechanisms of interval timing behaviour, *Bioessays*, 22, 94–103, 2000.

May, C.P. and Hasher, L., Synchrony effects in inhibitory control over thought and action, *J. Exp. Psychol. Hum. Percept. Perform.*, 24, 363–379, 1998.

McCormack, T., Brown, G.D.A., Maylor, E.A., Darby, R.J., and Green, D., Developmental changes in time estimation: comparing childhood and old age, *Dev. Psychol.*, 35, 1143–1155, 1999.

McCormack, T., Brown, G.D.A., Maylor, E.A., Richardson, L.B.N., and Darby, R.J., Effects of aging on absolute indentification of duration, *Psychol. Aging*, 17, 363–378, 2002.

McDowd, J.M. and Birren, J.E., Aging and attentional processes, in *Handbook of the Psychology of Aging*, 3rd ed., Birren, J.E. and Schaie, K.W., Eds., Academic Press, San Diego, CA, 1990, pp. 222–233.

McDowd, J.M., Oseas-Kreger, D.M., and Filion, D.L., Inhibitory processes in cognition and aging, in *Interference and Inhibition in Cognition*, Dempster, F.N. and Brainerd, C.J., Eds., Academic Press, San Diego, CA, 1995, pp. 363–400.

Meck, W.H., Selective adjustment of the speed of the internal clock and memory processes, *J. Exp. Psychol. Anim. Behav. Process.*, 9, 171–201, 1983.

Meck, W.H., Attentional bias between modalities: effect on the internal clock, memory, and decision stages used in animal time discrimination, *Ann. N.Y. Acad. Sci.*, 423, 528–541, 1984.

Meck, W.H., Neuropharmacology of timing and time perception, *Cognit. Brain Res.*, 3, 227–242, 1996.

Meck, W.H., Choline uptake in the frontal cortex is proportional to the absolute error of a temporal memory translation constant in mature and aged rats, *Learn. Motiv.*, 33, 88–104, 2001.

Meck, W.H. and Benson, A.M., Dissecting the brain's internal clock: how frontal-striatal circuitry keeps time and shifts attention, *Brain Cognit.*, 48, 195–211, 2002.

Meck, W.H., Church, R., and Wenk, G.L., Arginine vasopressin inoculates against age-related increases in sodium-dependent high affinity choline uptake and discrepancies in the content of temporal memory, *Eur. J. Pharmacol.*, 130, 327–331, 1986.

Oberauer, K. and Kliegl, R., Beyond resources: formal models of complexity effects and age differences in working memory, *Eur. J. Cognit. Psychol.*, 13, 187–215, 2001.

Park, D.C., Hertzog, C., Kidder, D.P., Morrell, R.W., and Mayhorn, C.B., Effect of age on event-based and time-based prospective memory, *Psychol. Aging*, 12, 314–327, 1997.

Park, D.C., Polk, T.A., Mikels, J.A. Taylor, S.F., and Marshuetz, C., Cerebral aging: integration of brain and behavioral models of cognitive function, *Dialogues Clin. Neurosci.*, 3, 151–165, 2001.

Penney, T.B., Allan, L.G., Meck, W.H., and Gibbon, J., Memory mixing in duration bisection, in *Timing of Behavior: Neural, Psychological, and Computational Perspectives*, Rosenbaum, D.A. and Collyer, C.E., Eds., MIT Press, Cambridge, MA, 1998, pp. 165–193.

Penney, T.B., Gibbon, J., and Meck, W.H., Differential effects of auditory and visual signals on clock speed and temporal memory, *J. Exp. Psychol. Hum. Percept. Perform.*, 26, 1770–1787, 2000.

Perbal, S., Droit-Volet, S., Isinigrini, M., and Pouthas, V., Relationships between age-related changes in time estimation and age-related changes in processing speed, attention, and memory, *Aging Neuropsychol. Cog.*, 9, 201–216, 2002.

Portman, M., Molecular clock mechanisms and circadian rhythms intrinsic to the heart, *Circ. Res.*, 89, 1084–1086, 2001.

Posner, M.I., *Chronometric Explorations of Mind*, Erlbaum, Hillsdale, NJ, 1976.

Posner, M.I. and Snyder, C.R.R., Attention and cognitive control, in *Information Processing and Cognition: The Loyola Symposium*, Solso, R.L., Ed., Lawrence Erlbaum Associates, Hillsdale, NJ, 1975.

Rabbitt, P.M.A., Measurement indices, functional characteristics, and psychometric constructs in cognitive aging, in *Models of Cognitive Aging*, Perfect, T.J. and Maylor, E.A., Eds., Oxford University Press, New York, 2000, pp. 160–187.

Rahhal, T.A., Hasher, L., and Colcombe, S.J., Age differences in memory: now you see them, now you don't, *Psychol. Aging*, 16, 697–706, 2001.

Rakitin, B.C., Stern, Y., and Malapani, C., Age-related time production errors in free-recall are duration dependent, submitted.

Rammsayer, T.H., Aging and temporal processing of durations within the psychological present, *Eur. J. Cognit. Psychol.*, 13, 549–565, 2001.

Rao, S.M., Mayer, A.R., and Harrington, D.L., The evolution of brain activation during temporal processing, *Nat. Neurosci.*, 4, 317–232, 2001.

Rogers, W.A. and Fisk, A.D., Understanding the role of attention in cognitive aging research, in *Handbook of the Psychology of Aging*, 5th ed., Birren, J.E. and Schaie, K.W., Eds., Academic Press, San Diego, CA, 2001, pp. 267–287.

Rubin, D.C., Frontal-striatal circuits in cognitive aging: evidence for caudate involvement, *Aging Neuropsychol. Cognit.*, 6, 241–259, 1999.

Salthouse, T.A., The processing-speed theory of adult age differences in cognition, *Psychol. Rev.*, 103, 403–428, 1996.

Salthouse, T.A., Aging and measures of processing speed, *Biol. Psychol.*, 54, 35–54, 2000.

Schmitt, M., Postma, A., and De Haan, E., Interactions between exogenous auditory and visual spatial attention, *Q. J. Exp. Psychol. Hum. Exp. Psychol.*, 53A, 105–130, 2000.

Schmitter-Edgecombe, M., Effects of divided attention and time course on automatic and controlled components of memory in older adults, *Psychol. Aging*, 14, 331–345, 1999.

Schroots, J.F. and Birren, J.E., Aging and attentional processes, in *Handbook of the Psychology of Aging*, 3rd ed., Birren, J.E. and Schaie, K.W., Eds., Academic Press, San Diego, CA, 1990, pp. 45–64.

Spence, C. and Driver, J., Audiovisual links in exogenous covert spatial orienting, *Percept. Psychophys.*, 59, 1–22, 1997.

Treisman, M., Temporal discrimination and the difference interval: implications for a model of the "internal clock," *Psychol. Monogr.*, 77, 1–31, 1963.

Vanneste, S., Perbal, S., and Pouthas, V., Estimation of duration in young and aged subjects: the role of attentional and memory processes, *Annee Psychol.*, 99, 385–414, 1999 (abstract from PsycINFO).

Vanneste, S. and Pouthas, V., Timing in aging: the role of attention, *Exp. Aging Res.*, 25, 49–67, 1999.

Vanneste, S., Pouthas, V., and Wearden, J.H., Temporal control of rhythmic performance: a comparison between young and old adults, *Exp. Aging Res.*, 27, 83–102, 2001.

Verhaegen, P., The parallels in beauty's brow: time-accuracy functions and their implications for cognitive aging theories, in *Models of Cognitive Aging*, Perfect, T.J. and Maylor, E.A., Eds., Oxford University Press, New York, 2000, pp. 50–86.

Volkow, N.D., Gur, R.C., Wang, G.J., Fowler, J.S., Moberg, P.J., Ding, Y.S., Hitzemann, R., Smith, G., and Logan, J., Association between decline in brain dopamine activity with age and cognitive and motor impairment in healthy individuals, *Am. J. Psychiatry*, 155, 344–349, 1998.

Volkow, N.D., Logan, J., Fowler, J.S., Wang, G.J., Gur, R.C., Wong, C., Felder, C., Gatley, S.J., Ding, Y.S., Hitzemann, R., and Pappas, N., Association between age-related decline in brain dopamine activity and impairment in frontal and cingulated metabolism, *Am. J. Psychiatry*, 157, 75–80, 2000.

Walker, J.T. and Scott, K.J., Perceived duration of lights, tones, and gaps, *J. Exp. Psychol. Hum. Percept. Perform.*, 7, 1327–1339, 1981.

Wearden, J.H. and Bray, S., Scalar timing without reference memory? Episodic temporal generalization and bisection in humans, *Q. J. Exp. Psychol. Comp. Physiol. Psychol.*, 54B, 289–309, 2001.

Wearden, J.H., Edwards, H., Fakhri, M., and Percival, A., Why "sounds are judged longer than lights": application of a model of an internal clock in humans, *Q. J. Exp. Psychol.*, 51B, 97–120, 1998.

Wearden, J.H., Wearden, A.J., and Rabbitt, P.M.A., Age and IQ effects on stimulus and response timing, *J. Exp. Psychol. Hum. Percept. Perform.*, 23, 962–979, 1997.

Wingfield, A. and Stine-Morrow, E.L., Language and speech, in *The Handbook of Aging and Cognition*, 2nd ed., Craik, F.I.M., Fergus, I., and Salthouse, T.A., Eds., Erlbaum, Mahwah, NJ, 2000, pp. 359–416.

Yoon, C., May, C.P., and Hasher, L., Aging, circadian arousal patterns, and cognition, in *Cognition, Aging, and Self Reports*, Schwartz, N., Park, D., Knauper, B., and Sudman, S., Eds., Psychological Press, Washington, D.C., 1999, pp. 117–143.

Zacks, R.T., Hasher, L., and Li, K.Z.H., Human memory, in *The Handbook of Aging and Cognition*, 2nd ed., Craik, F.I.M. and Salthouse, T.A., Eds., Erlbaum, Mahwah, NJ, 2000, pp. 293–357.

Zakay, D. and Block, R.A., Temporal cognition, *Curr. Directions Psychol. Sci.*, 6, 12–16, 1997.

Section II

Neural Mechanisms

11 Neurogenetics of Interval Timing

Münire Özlem Çevik

CONTENTS

11.1 The Forward Genetic (from Phenotype to Gene) Approach to Behavior 298
    11.1.1 Genome-Wide Saturation Mutagenesis Requires Screening Thousands of Phenotypes 298
    11.1.2 Mutant Screens Should Be Both Task Relevant and Efficient 299
        11.1.2.1 Nonassociative Learning Procedures 299
        11.1.2.2 The Temporal Conditioning Procedure 300
        11.1.2.3 Screening Phenotypes en Masse 300
    11.1.3 The Mutated Gene Should Produce a Clear Behavioral Phenotype 301
        11.1.3.1 Loss-of-Timing Mutations 301
        11.1.3.2 Mutations That Change the Speed of Timing 302
        11.1.3.3 Mutations That Affect the Accuracy or Precision of Timing 303
    11.1.4 Why Not Use the Rat? 304
11.2 The Reverse Genetic (from Gene to Phenotype) Approach to Behavior 305
    11.2.1 Producing Gene Knockouts 305
    11.2.2 Strain Differences 306
    11.2.3 Inducible and Tissue-Specific Knockouts 306
    11.2.4 Behavioral Phenotyping of Knockouts 307
    11.2.5 An Example: Interval Timing in Dopamine-Transporter Knockout Mice 308
11.3 What if the Single-Gene Approach Does Not Work? 313
References 313

Research pioneered by Seymour Benzer and his colleagues since the 1960s proved it possible to dissect the genetic basis of complex behavioral processes by analyzing the behavioral and neural modifications caused by single-gene mutations. Neurogenetics uses naturally occurring or created mutations in genes that affect brain structure and function to understand the complex neural pathways between genes and behavior. This chapter presents a brief introduction to the neurogenetic approach and discusses its potential use and feasibility in understanding the neural mechanisms of interval timing.

11.1 THE FORWARD GENETIC (FROM PHENOTYPE TO GENE) APPROACH TO BEHAVIOR

Forward genetics identifies novel genes: it is most useful when one has no prior knowledge or wants to make no assumptions about the genetic basis of the trait being studied. It starts with a random mutagenesis of the genome by exposure to x-ray radiation or chemical mutagens. The mutated genes are transmitted to the next generation when the germline cells are affected during mutagenesis. The progeny is screened for modifications in the behavioral trait of interest, and when such mutant phenotypes are isolated, research proceeds to map and clone the genes underlying the phenotypic defect.

Forward genetic research requires substantial investments of time, space, and funds. The feasibility of forward genetics depends on multiple factors that determine the efficiency of obtaining and isolating mutant phenotypes and the practicality of positional cloning. First of all, the behavioral tests that are used to screen the phenotypes should be appropriate for screening literally thousands of animals over a reasonable period of time. Obviously, the model organism should have a short generation time and low maintenance costs. Further, high-density linkage maps and genomic libraries of the model organism should be available for positional cloning. These factors impose considerable limitations on the choice of the model organism. As unlikely a model of complex behavioral traits as it seems to be, the fruit fly *Drosophila melanogaster* has been the favorite species of behavior geneticists because it is rather unique for fulfilling the criteria for the feasibility of forward genetic analysis. Forward genetic analysis of behavior in the mouse has been taken up only very recently: for example, the circadian *period* gene of *Drosophila* (Konopka and Benzer, 1971) was identified 23 years before the *Clock* gene of the mouse (King et al., 1994; Vitaterna et al., 1994).

11.1.1 GENOME-WIDE SATURATION MUTAGENESIS REQUIRES SCREENING THOUSANDS OF PHENOTYPES

The first step in forward genetic analysis is the random mutagenesis of the genome. Genome-wide saturation mutagenesis refers to mutating every locus in the genome at least once. Theoretically, approaching saturation mutagenesis should allow the detection of a set of genes, each of which produces a phenotypic modification in the behavioral trait of interest when mutated. Conversely, the failure to saturate the genome might result in a failure to mutate and thereby detect the genes that affect the behavioral trait being studied.

Several effective germline mutagens are available for both *Drosophila* and mice, each associated with a different type (e.g., small or large deletions, insertions, point mutations) or rate of mutation. When the purpose is to produce mutant behavioral phenotypes, what is intended in most cases is the point mutation of a gene, resulting in either a gain or loss of function for that gene that would produce a detectable novel phenotype for the behavioral trait being studied. This can be contrasted with small or large deletions in the genome with widespread effects that can considerably reduce viability or fertility, in addition to a modification in the behavioral trait of interest.

How many phenotypes need to be screened to reach saturation mutagenesis? Let me give an example. The alkylating agent ethylnitrosourea (ENU) can produce point mutations at a rate of 0.0015 mutations per locus per gamete in the mouse (Takahashi et al., 1994). In the case of a dominant mutation where each first-generation offspring represents one mutagenized gamete, approximately 2600 gametes are required for an 87.5% chance of mutating any locus. However, because a mutagenized gamete is not always passed onto the progeny, or it might fail to be recognized or recovered after being passed on, even higher numbers of gametes are screened to make sure that a mutation can be isolated when it occurs (Justice, 2000). In practice, approximately 3000 animals are screened to approach saturation mutagenesis in the mouse following ENU treatment (Takahashi et al., 1994). That is a large number. Due to the substantial investment of time, space, and funds it requires, the forward genetic approach to behavior using the mouse as the model organism had not been considered feasible until recently (e.g., Capecchi, 1989).

11.1.2 MUTANT SCREENS SHOULD BE BOTH TASK RELEVANT AND EFFICIENT

Behaviorally screening 3000 animals to isolate mutant phenotypes is a daunting task, and it is only the initial step in identifying genes. Assuming that one animal is screened per day, it would take approximately 10 years of working 6 days a week to reach saturation mutagenesis. Clearly, mutant screens should be extremely rapid and efficient.

Experimental psychologists observe behavior in detail, which takes time. The majority of the procedures developed and used most productively over the years by experimental psychologists take too long and are too inefficient for mutant screening purposes. In the case of interval timing, for example, a variety of temporal discrimination, production, and estimation procedures are widely used to study temporally controlled behavior in rats, pigeons, humans, and other mostly vertebrate species. Most interval timing research is aimed at understanding the quantitative properties of behavior at the steady state when behavior is no longer changing appreciably. Depending on the timing procedure used, it might take 2 to several weeks for behavior to reach the steady state. Assuming it takes 2 weeks to reach the steady state, and subjects are tested in groups of 50, it would take 2.3 years to screen 3000 animals. Obviously, experimental procedures for faster behavioral testing are necessary for using a forward genetic approach to identify genes involved in interval timing.

11.1.2.1 Nonassociative Learning Procedures

Interval timing procedures with extensive associative learning requirements are not useful as mutant screens because they are too long and labor-intensive. Nonassociative learning procedures such as habituation are more appropriate. Habituation refers to the reduction in the amplitude of a response as the result of the repeated presentation of the stimulus that triggers the response. Temporal variables have well-established effects on the rate of habituation: the longer the interstimulus interval, the slower the habituation (e.g., Leaton, 1976; see also Leaton and Tighe, 1976). If the mechanisms of interval timing under associative conditioning procedures overlap with those that control the effects of temporal variables under nonassociative learning procedures, habituation can be used to screen for interval timing mutants (see Sasaki et al., 2001). For example, a mutation that affects the well-established inverse relation between the duration of the interstimulus intervals and the rate of habituation is also expected to affect temporally controlled behavior under a variety of time-based associative conditioning procedures. However, one should be cautious not to select the mutants that fail to habituate since this behavioral phenotype might result from a learning deficit as well as a timing deficit (see Section 11.1.3.1).

Habituation is indeed an appropriate protocol for a mutant screen: First of all, the neural basis of habituation seems to have been conserved across phyla, and it has been subject to extensive research, pioneered by Kandel and associates (e.g., Castelluci et al., 1970; Kandel, 1985; also see Thompson and Spencer, 1966). Second, unlike most interval timing procedures that involve associative learning, habituation does not require extensive training; it can be obtained within minutes using automated apparatus for a variety of behavioral reflexes. Rate of habituation for a given behavioral response depends on multiple factors, including the frequency of stimulus presentation and the intensity of stimulus. The values of these parameters in a habituation protocol can easily be adjusted to ensure rapid testing.

11.1.2.2 The Temporal Conditioning Procedure

The temporal conditioning procedure (also known as fixed-time procedure) might also be useful as a rapid behavioral screen. In this procedure, a reinforcer is presented at fixed temporal intervals regardless of the subject's behavior at the time of reinforcer delivery. The subjects learn to anticipate the reinforcer right before its occurrence under the temporal conditioning procedure. Notice that the experimental protocol is quite similar to that of habituation in that both procedures basically involve periodic presentation of the same stimulus. However, when the stimulus is a biologically important one (e.g., food for a hungry animal), anticipatory responding rather than habituation is observed. Temporal conditioning as a mutant screen can isolate those mutant phenotypes that show changes in the accuracy and precision in the timing of the anticipatory responses (see Section 11.1.3.3). Once again, the selection of the mutants that fail to show anticipatory responses should be avoided, because this behavioral phenotype can possibly be produced by a learning deficit rather than a timing one.

11.1.2.3 Screening Phenotypes en Masse

One strategy to increase the efficiency of mutant screens is testing subjects as a group. For example, Diptera had long been recognized to exhibit associative learning (for reviews, see McGuire, 1984; Tully, 1984), but the early experimental procedures that were used to demonstrate conditioned responding in bees and flies were too long and effortful to serve as mutant screens. It is not a coincidence that *dunce*, the first *Drosophila* memory mutant (Dudai et al., 1976), was isolated shortly after the development of an olfactory avoidance learning procedure that allowed the training and testing of flies en masse (Quinn et al., 1974). A highly efficient and automated Pavlovian olfactory discrimination procedure that allows the testing of approximately 100 flies at a time was later developed (Tully and Quinn, 1985), which led to the expedited discovery of new learning mutants. Group testing of flies has been revolutionary in the genetic dissection of learning and memory (Dubnau and Tully, 1998), in spite of being controversial on the grounds that it fails to show whether individual flies learn (Holliday and Hirsch, 1986).

11.1.3 THE MUTATED GENE SHOULD PRODUCE A CLEAR BEHAVIORAL PHENOTYPE

The choice of the behavioral domain is critical in whether attempts to dissect the genetic basis of the behavioral trait yield a success story. Behavioral mutations can be isolated if they result in a clear, detectable modification in the phenotype. If the behavioral trait of interest is well defined and quantifiable in the wild type, a mutation-induced deviation from the wild-type phenotype will also be clear and detectable. Such is the case for two behavioral domains, namely, circadian rhythms and learning and memory that have been subject to extensive genetic research. For example, mutations that affect circadian rhythms disrupt the phenotypic expression of 24-h rhythms or modify the intrinsic 24-h period by either shortening or lengthening it (Hall, 1998; Low-Zeddies and Takahashi, 2000; Ralph and Menaker, 1989). It is straightforward to measure both the absence of a 24-h rhythm and a change in the period of rhythms. Similarly, in the realm of learning and memory, mutations that affect the acquisition, storage, or retrieval of information can be detected in the form of an inability (or an enhanced ability) to perform a learned task (Dubnau and Tully, 1998; Silva et al., 1997).

The behavioral criteria for isolating interval timing mutants are important for minimizing the false positive and false negative errors in the identification of interval timing mutants. Below is a discussion of phenotypic changes that can be produced by mutations of single genes that are involved in interval timing and whether these behavioral modifications are appropriate for mutant screening purposes.

11.1.3.1 Loss-of-Timing Mutations

Let us first consider the issue of whether we should expect to find genes (or neural mechanisms that involve the products of these genes) that are dedicated to timing such that when mutated, these genes affect solely the timing of stimuli or events, sparing other cognitive and perceptual processes. If learning the changeable temporal relationships in the animal's current environment has a significant effect on its survival, there might have been selection for evolution of genes that control the potential to learn such temporal regularities. But even if such genes exist, should we expect them to be any different than the genes that are involved in learning, or neural plasticity in general? Because everything happens in real time, everything that the animal learns has a temporal dimension that ought to be part of what is learned (e.g., Gallistel and Gibbon, 2000; Miller and Barnet, 1993). So the mechanisms of interval timing should be intricately bound to the real-time dynamics of the learning process, although it has been suggested that the timing of stimuli and events is underlain by an independent mechanism (e.g., Malapani et al., 1998, Matell and Meck, 2000; Matell et al., this volume).

Notice that even if the mechanisms of learning and timing are independent, screening for loss of interval timing mutations can result in misidentification of learning mutants as timing mutants. Unlike the circadian timing mechanism, the interval timing mechanism does not have an intrinsic period. The experimental contingency based on the timed interval is learned under interval timing procedures. That interval timing is reflected in learned behavior makes it very difficult to distinguish a loss-of-timing mutation from a loss-of-learning mutation because both would produce a phenotype that fails to learn a time-based contingency. For example, in a peak-interval procedure, the subjects are reinforced for their first response after a fixed interval (e.g., 30 sec) elapses on some trials. On other trials, the reinforcement is omitted, and the subject's responses are recorded over a period longer than the fixed interval. In these trials, response rate increases smoothly as time elapses, reaches a peak around the time when reinforcement is usually delivered, and decreases fairly symmetrically afterwards. Animals that fail to learn the temporal contingencies in this procedure respond at a constant, undifferentiated rate throughout the trial. This is exactly how mice that express a defective form of the cyclic AMP responsive element binding transcription factor (CREB) perform under the peak-interval procedure (Carvalho et al., 2000). Now, is the failure of the CREB-deficient mice to learn temporal contingency under the peak-interval procedure due to a timing deficit, a learning deficit, or both? CREB plays a well-established role in the formation of long-term memories, and animals that express a deficient form of CREB have impaired performance under other memory tests that do not involve time-based contingencies (Silva et al., 1998). So their impaired performance under the peak-interval procedure might be underlain by a deficit in learning. But because a deficiency in interval timing would produce exactly the same performance profile, it is hard to infer whether these animals have a timing problem as well. Clearly, in screening for interval timing mutants, the better strategy would be to look for mutation-induced modifications in well-established qualitative and quantitative properties of temporally controlled behavior instead of a loss-of-timing mutation.

11.1.3.2 Mutations That Change the Speed of Timing

Let us assume for the sake of argument that timing is controlled by a single polymorphic gene such that individuals that express different allelic forms of the gene fall into discrete phenotypic categories with respect to timing speed. How would the behavior of the phenotypes be different? One would be tempted to say that individuals that have the fast-timing allele would overestimate the duration of intervals under time-based schedules, and vice versa. However, discrete phenotypic categories with respect to timing speed would not produce discrete phenotypic categories of behavior because individuals would learn to estimate durations correctly in relation to their own timing speed. The effects of a change in timing speed will be reflected in behavior only when the subjective speed of time changes for a given individual by an intervention such as the administration of psychoactive drugs (Meck, 1983), which does not provide any information about individual differences in timing speed (see Section 11.2.5). Therefore, speed of timing is not a good behavioral trait to start looking for single-gene effects due to the difficulty of measuring individual differences with respect to this trait.

11.1.3.3 Mutations That Affect the Accuracy or Precision of Timing

What other aspects of temporally controlled behavior are likely to be controlled by single genes with measurable phenotypic effects? Accuracy and precision of interval timing are good candidates. Accuracy of timing refers to the coincidence of the average subjective estimates of duration with the actual duration of timed intervals. Precision of timing refers to the extent of variability in the subjective estimates of duration (Malapani et al., 1998). Both people and animals show individual differences in accuracy and precision of timing, although the heritability of neither measure has yet been reported.* Given that animals are raised and tested under the same conditions show variability with respect to accuracy or precision of timing, individual differences in these measures are likely to have a genetic basis. Whether accuracy or precision would be affected by single-gene mutations is an open experimental question.

Extensive research shows that both accuracy and precision of temporally controlled behavior are remarkably systematic and predictable under both Pavlovian and operant time-based schedules (e.g., Gibbon, 1977), which satisfies the criterion of a well-defined phenotypic trait for mutant screening purposes. A thorough review of the quantitative properties of temporally controlled behavior is beyond the scope of this chapter, but one such property, timescale invariance (Gallistel and Gibbon, 2000), is worth mentioning. Timescale invariance refers to the observation that when a time-based experimental protocol is repeated using different absolute time intervals, data from these experiments superimpose if plotted on relative rather than absolute timescales. In addition to temporal perception being controlled by relative rather than absolute durations, superimposition of curves also requires that the variability in the estimates of time increase in proportion with the length of the temporal intervals being estimated. Timescale invariance is the manifestation of a strong form of Weber's law, which states that a difference of equal ratio (but not of equal absolute amount) in a continuous variable (e.g., brightness, temperature, weight, etc.) yields equal perceptual discrimination. That temporally controlled behavior obeys Weber's law shows that both accuracy and precision of interval timing change systematically under interval timing schedules.

Accuracy and precision of timing are appropriate behavioral criteria to be used in mutant screens to isolate interval timing mutants. First of all, it is possible to detect individual differences in both accuracy and precision of timing. Second,

---

* This experiment is currently in progress in my lab.

although there are exceptions, Weber's law applies for temporally controlled behavior under most interval timing schedules (e.g., Gibbon, 1977; Platt, 1979; Stubbs, 1979). Mutation-induced deviations from Weber's law would be straightforward to detect and measure as long as one uses a time-based schedule under which the behavior of the wild-type controls is shown to obey Weber's law. Further, Weber's law in interval timing holds true across species (Green et al., 1999), suggesting that the mechanisms of accuracy and precision in interval timing might be evolutionarily conserved. In fact, because it applies to the perception of all continuous variables, Weber's law might reflect a constraint common to all sensory-perceptual processing.

11.1.4 Why Not Use the Rat?

So far, the feasibility of the forward genetic approach has been discussed in relation to the availability of rapid behavioral screens and existence of detectable phenotypes. But when research moves from the phenotype to the gene, even the existence of a clear, detectable phenotype, and the availability of efficient and valid behavioral tests to screen for that phenotype, does not ensure that the underlying molecular defect can be found easily.

How can novel genes be discovered if the mutant phenotype is not associated with a known molecular defect? The answer lies in positional cloning for which no information about gene function is necessary. Positional cloning starts with linkage analysis to map the location of the novel gene on a chromosome. When the gene is located on a narrow enough region on the chromosome through iterative linkage analysis, the DNA from that region is cloned, and candidate genes on the clone are identified.

Very briefly, here is how it works: The lower the distance between a pair of genes on a chromosome, the higher the probability that they will be inherited together. Linkage maps that depict the proximity and order of genes on a chromosome can be constructed by arranging crosses between parental strains that are polymorphic for a set of genes and then analyzing the offspring for the pattern of segregation between genes. As a result of the widespread use of molecular techniques that allow genotypic detection, not only the genes with visible phenotypic effects, but also other types of molecular markers in noncoding regions of DNA that exhibit polymorphisms can now be used to construct high-resolution whole-genome maps. These maps are used in positional cloning to locate the novel gene by determining its proximity to known markers using linkage analysis.

Note that the marker alleles used in the construction of high-density genome maps should be homozygous for a given inbred strain and polymorphic between strains for the experimental crosses between inbred strains to be informative. This requires that genotypic information on several inbred strains of a species be available. One of the early goals of the Human Genome Project was to construct high-density linkage maps of increasing resolution for model organisms, including the fruit fly *Drosophila melanogaster* and the mouse *Mus musculus*. The higher the resolution of a high-density linkage map, the easier and quicker it is to map and clone the novel genes using positional cloning. Unfortunately, the resolution of the linkage map for the rat, the favorite species of experimental psychologists, is not nearly as high as that of the mouse. One can obtain and isolate behavioral mutants in the rat, but given the low resolution of the linkage maps, positional cloning of the genes that produce the mutant phenotype would take several years. This is the main reason why behavior geneticists do not use the rat as a model organism in spite of the abundance of data on its behavioral profile. Mouse and *Drosophila* linkage maps have been developed over the years with the cumulative effort of several labs, and such an extensive effort is necessary for the rat before it becomes a popular model organism for genetic analysis.

11.2 THE REVERSE GENETIC (FROM GENE TO PHENOTYPE) APPROACH TO BEHAVIOR

The reverse genetic approach (also known as from gene to phenotype, or gene targeting approach) is used to study the function of genes that have already been identified and sequenced. This approach is based on producing targeted mutations in genes to alter their function. Several techniques are available to change the level of expression of a gene, each appropriate for a different type of experimental question. A knockout animal is produced if the targeted mutation prevents the expression of the gene, eliminating its function. A knockdown animal is produced if the targeted mutation reduces the gene function without completely eliminating it. Animals that have multiple copies of a particular gene are used to observe the effects of overexpression of that gene. A transgenic animal expresses a gene of another organism; these animals are often used as animal models of human diseases that have been linked to single genes (e.g., Huntington's disease).

The rationale behind using knockout animals is to understand the function of the targeted gene by observing the phenotypic defects that are produced in its absence. It should be emphasized that knockout animals do not provide direct information about gene function. Therefore, one should be rather conservative and cautious in making inferences about gene function based on the phenotypic changes produced by the absence of genes observed in the knockout animals.

11.2.1 PRODUCING GENE KNOCKOUTS

Although gene targeting is possible in other species as well, the mouse is the favorite animal model for a reverse genetic approach to behavior. A knockout mouse strain can be produced as follows: once a gene has been identified and sequenced, its fragments can be introduced into mouse embryonic stem (ES) cells where they would insert themselves into the chromosome by undergoing homologous recombination with the endogenous DNA, causing a mutation at the locus they replace. In order to select the ES cells that have taken up the mutant gene, a selective marker such as the neomycin resistance gene is inserted into the targeting vector along with the homologous sequence. Then the ES cells are plated on neomycin, and only the ones that express the resistance gene, which are also those that express the mutant form of the targeted gene, survive. The transfected ES cells are then implanted into normal blastocysts, which in turn are implanted into pseudopregnant females. The resulting mouse pups will be chimeric, i.e., some of their cells will have originated from transfected ES cells, whereas others will have originated from the normal ES cells of the recipient blastula. If the transfected ES cells develop into germline cells, the mutation will be passed on to the offspring, which are bred to produce a mutant strain (Hasty et al., 2000).

11.2.2 Strain Differences

The genetic background is anything but trivial in assessing the effects of a targeted gene mutation because a targeted gene might have different phenotypic effects, depending on the genetic background (Choi, 1997; Dubnau and Tully, 1998; Silver, 1995). This is because genes do not always affect the phenotype on their own. Rather, interactions between genes (i.e., epistasis) or between the two copies of the same gene (i.e., dominance) play important roles in determining phenotypic expression. For example, it is possible for a targeted mutation to produce a behavioral effect in one strain, but no phenotypic difference in another due to epistatic interactions (Crusio, 2002). Not surprisingly, different strains of mouse show considerable differences with respect to several behavioral traits, including learning and memory (Wehner and Silva, 1996). Therefore, it is advisable to test the effects of a knockout on at least two different backgrounds.

In most cases, the ES cells that are used to produce knockouts are derived from the 129 inbred mouse strain. However, 129 is not the best genetic background to study for the behavioral effects of a modified gene because certain substrains of 129, such as the 129/J, fail to develop a normal corpus callosum (Wahlsten, 1982), and they have impaired memory (e.g., Montkowski et al., 1997). Behavioral phenotyping of knockout strains is usually carried out after the 129-derived ES line is expressed in either a C57BL/6 or BALB/c background by repeatedly backcrossing the mice that carry the knockout gene to a C57BL/6 or BALB/c background (for breeding strategies, see Wolfer et al., 2002).

Behavioral testing should be carried out with the homozygous$^{-/-}$, heterozygous$^{+/-}$, and wild-type littermates obtained from breeding heterozygous$^{+/-}$ parents. Note that the genetic background might still be segregating for the three genotypes after an insufficient number of backcrosses, which itself can produce behavioral differences, or prevent the expression of the potential behavioral difference produced by the targeted gene. Similarly, one should avoid breeding the homozygous$^{-/-}$ and the wild-type animals amongst themselves for the sake of efficiency of obtaining each genotype because this breeding strategy too can result in the segregation of the backgrounds, such that the homozygous$^{-/-}$ genotype will be expressed in a 129 background from which the ES cells were derived, and the wild-type alleles will be expressed in another (e.g., C57BL/6) background.

11.2.3 Inducible and Tissue-Specific Knockouts

Conventional knockout animals lack a specific gene, and hence the protein product of that gene. They are not good models if one would like to understand how a gene is involved in the manifestation of behavior as it occurs in the adult organism. This is because genes that are involved in controlling the manifestation of a particular behavior during adulthood might also serve other functions in the central nervous system or other parts of the body during development. The absence of such genes might affect the viability, development, and general health of the organism to various degrees. Conversely, the behavioral phenotype of the knockout animals might appear to be normal due to the compensatory effects of other genes that have taken over during development (Crusio, 2002).

Inducible and tissue-specific mutations provide a good solution to the above problems by limiting the genetic lesions in time and space. In inducible mutations, the transcription of the mutant gene is controlled by the administration of a specific substance (e.g., the antibiotic tetracycline) or exposure to a specific event (e.g., heat shock). Tissue-specific mutations are obtained by putting the expression of the mutant gene under the transcriptional control of another gene that is expressed exclusively in the tissue of interest. Mutations can be both inducible and tissue specific, which limits the effects of the mutation in both time and space as required by the experimenter. In essence, then, genetic lesions produced by inducible and tissue-specific mutations are the equivalent of extremely selective drugs that temporarily affect the tissue of interest only.

11.2.4 BEHAVIORAL PHENOTYPING OF KNOCKOUTS

The targeted mutation of a gene can affect a behavioral trait if the gene is involved in either the development of the neural substrates for behavior or the actual manifestation of behavior. In both instances of behavioral control by single genes, it is reasonable to expect that the gene be expressed in the tissues that serve the behavioral function (see Baker et al., 2001). On the other hand, mutations of genes that are involved in sensory or motor processes can also produce a modified phenotype with respect to the behavioral trait, and one should be cautious not to ascribe specific behavioral functions to such genes. Detailed descriptions of the test batteries that are designed to assess the general health and the motor and sensory-perceptual abilities of the knockout mice, as well as specific behavioral traits such as learning, memory, social behaviors, and emotion, are available (e.g., Crawley, 2000).

If the targeted gene plays a critical role in the normal development of the embryo, the homozygous$^{-/-}$ or the heterozygous$^{+/-}$ animals might not be as healthy as the wild-type littermates, making it difficult to obtain behavioral measures from all three phenotypes. In some cases, the differential viability of the phenotypes requires that the homozygous$^{-/-}$ or the heterozygous$^{+/-}$ animals be maintained under special conditions. For example, mice that lack dopamine transporters (DAT-KO) have an unusually high metabolism, so they are healthier when fed a high-fat diet. However, the high-fat diet causes the wild-type littermates to gain extra weight. Under these conditions, either the type or the amount of food has to be different for the homozygous and the wild-type littermates, which is a concern when food reinforcement is used during behavioral testing (see Section 11.2.5). Because maintenance conditions in the home cage environment might also affect behavior, caution should be addressed in treating the littermates of all genotypes as equally as possible.

Finally, the availability of valid, task-relevant behavioral tests can be a major problem, depending on the behavioral domain. The majority of the behavioral tests currently used were originally designed for rats and later adapted for mice. It seems to be an underappreciated fact that rats and mice are not as similar in behavior as they are in sight. A behavioral procedure that has not been appropriately adjusted for the mice might lead to erroneous or biased conclusions about this organism.

11.2.5 An Example: Interval Timing in Dopamine-Transporter Knockout Mice

A pilot interval timing experiment carried out with dopamine transporter knockout (DAT-KO) mice is presented here to provide an example of the reverse genetic approach to interval timing.**

The neurotransmitter dopamine is suggested in the timing of temporal intervals in the seconds-to-minutes range because a disruption or imbalance of dopaminergic activity alters the perception of temporal intervals. For example, following self-administration of amphetamines, humans report that "the world looks like it is in slow motion." Similarly, rats overestimate short temporal intervals following systemic injections of methamphetamine (e.g., Meck, 1983), suggesting that they too perceive time as being stretched out under amphetamine.

Amphetamine reverses the action of the dopamine transporter by causing it to release dopamine from the presynaptic terminals instead of clearing dopamine from the synapse. The net result of this process is increased and prolonged availability of dopamine in the synaptic cleft, causing a nonselective activation of different classes of dopamine receptors (Jones et al., 1998). Although all monoamine transporters in the central nervous system are substrates for amphetamine and its derivatives, amphetamine has higher affinity to the dopamine transporter relative to the serotonin transporter, and it does not release norepinephrine as much as dopamine. Therefore, at low to moderate doses, most of the behavioral and perceptual effects of amphetamine are due to its action on the dopamine system (Feldman et al., 1997).

A targeted mutation of the dopamine transporter (DAT), much like an acute amphetamine administration, prolongs the availability of dopamine in the synaptic cleft in mice that are homozygous ($DAT^{-/-}$) or heterozygous ($DAT^{+/-}$) for the mutation because the excess dopamine cannot be cleared from the synaptic cleft in the absence of functional dopamine transporters. On the other hand, the absence of the dopamine transporter also causes compensatory changes during development: the amount of dopamine released per electrical pulse is lower, dopamine synthesis is reduced (Jones et al., 1998), and the sensitivity of both pre- and postsynaptic receptors is altered in complex ways (Jones et al., 1999). Nevertheless, in spite of the compensatory changes, the net effect of a dopamine transporter knockout is functional hyperdopaminergia in both $DAT^{-/-}$ and $DAT^{+/-}$ mice (Gainetdinov et. al., 1999a).

The behavioral effects of the absence of the dopamine transporter are also quite similar to those of moderate doses of amphetamine. $DAT^{-/-}$ mice exhibit marked

---

** These data were collected at Duke University when the author was a post-doctoral fellow working with Warren Meck.

hyperactivity along with an impairment in learning, which has led to the suggestion of the phenotype as a model for human attention deficit hyperactivity disorder (ADHD) (Gainetdinov et al., 1999b). In support of this view, an association between ADHD and polymorphisms in the noncoding regions of the human DAT gene has been suggested (Cook et al., 1995; Gill et al., 1997). Interestingly, ADHD patients have been reported to have altered temporal perception. In fact, it has been suggested that the perception of time being stretched out, or passing slowly, might contribute to the frequently reported boredom of ADHD patients, and that the impulsivity of these individuals can possibly be controlled by preventing this boredom (although see Goddard, 2000).

The purpose of the present experiment was to study temporal discrimination in dopamine transporter knockout mice. Male C57BL/129SvJ mice homozygous ($DAT^{-/-}$) or heterozygous ($DAT^{+/-}$) for DAT deletion and their wild-type (WT) littermates were used as subjects. They were generated by crossing $DAT^{+/-}$ parents and 3-month-old mice at the beginning of the experiment. In previous studies that reported the behavioral and physiological profile of the DAT knockouts, the mice had been backcrossed to a C57BL/6 background for seven to nine generations over an approximately 2-year period (e.g., Gainetdinov et al., 1999a). In the present experiment, knockout strains had not been backcrossed to a C57BL/6 background for more than two generations, so it is possible that the genetic background was still segregating for $DAT^{-/-}$, $DAT^{+/-}$, and WT mice. Although the physical appearance and the behavioral profile of the $DAT^{-/-}$ or $DAT^{+/-}$ mice used in this experiment were similar to those reported previously (e.g., Gainetdinov et al., 1999b; Spielowoy et al., 2000), a segregation in the genetic background might possibly have produced behavioral differences between the three genotypes in addition to the DAT deletion.

Although all three genotypes had originally been planned to be included in the experiment, $DAT^{-/-}$ animals were later dropped. The present experiment involved administration of approximately 80 standard pellets (of which about 65 were collected) daily as food reinforcement. This required that the subjects be kept at their 85% body weight. $DAT^{-/-}$ mice could not tolerate this food deprivation regimen because they are dwarf with very low-fat reserves, and they have an unusually high basal metabolism. In previous behavioral studies that involved food reinforcement, food was removed from the home cages 5 h before the beginning of daily experimental sessions, which seems to be the most that the $DAT^{-/-}$ mice can tolerate (e.g., Gainetdinov et al., 1999b). When we tried the same 5-h food deprivation schedule, neither the $DAT^{-/-}$ mice nor the others were hungry enough to eat 50 pellets during the experiment. Therefore, we decided to continue the experiment with $DAT^{+/-}$ and WT mice only. Notice that because the $DAT^{+/-}$ animals have one functional copy of the DAT gene, a comparison of $DAT^{+/-}$ and WT mice reveals the effects of a 50% reduction in the dosage rather than the absence of the gene. Also, using $DAT^{+/-}$ mice avoids confounders introduced by the absence of the gene during development. Although the $DAT^{+/-}$ mice too have functional hyperdopaminergia, they are indistinguishable in general appearance and health from their WT littermates. There were six subjects in each group at the beginning of the experiment, but one of the $DAT^{+/-}$

mice died due to a brain tumor, and a WT mouse failed to learn the task. So the experiment continued with five subjects per group.***

The experiment started with training the subjects to press levers for food. Once a subject learned to press the levers for food, intertrial intervals (ITIs) of progressively longer duration were inserted between lever presentations until the ITIs were 45 sec long on average. Lever press training lasted until the subjects learned to press both the left and right levers reliably. During the two-signal training phase, each trial started with the onset of a light located approximately an inch above the food hopper. The light remained on for either 2.52 (short) or 12.7 (long) sec. The levers were presented immediately after the light was off, and the first response on either lever caused both levers to be retracted. A food pellet was delivered immediately if the left lever was pressed following the short signal, or the right lever following the long signal. If the mice failed to press either lever, the levers were retracted after 5 sec, and a null response was recorded. Trials were separated by ITIs that were 45 sec long on average. All lights, including the houselight, were off during the ITIs. Each group was run in two short daily sessions because the mice got satiated and stopped responding if they were run in a single long session. Each session consisted of 40 trials, so the animals were exposed to 80 trials daily. Half of the trials involved the presentation of the short signal, and the other half involved the presentation of the long signal.

A correction procedure was used to avoid preferences for either lever. When the subjects made an incorrect response, the same trial was repeated until a correct response was made. If the subject failed to make the correct response over two consecutive trial presentations, only the correct lever was presented during the third presentation of the same trial. Animals that were performing poorly were exposed to a cued training procedure where only the correct lever was presented to them following signal presentation. However, a cued training procedure, if used extensively, can retard rather than enhance learning because it prevents the animal from making a discrimination by presenting it with the correct response alternative only. So cued training was used only for those mice who had stopped responding during two-signal correction training, and it was terminated as soon as the animal started responding reliably. Finally, if a subject continued to have a position bias regardless of the correction training, it was exposed to intermittent sessions that involved the presentation of either the short or the long signal only.

Eight-signal training started after 40 days of two-signal training (including the cued and correction training). In order to prevent position biases, the first daily session was eight-signal training, and the second daily session involved two-signal correction training for each group. Eight-signal training involved the presentation of six additional signal durations (3.2, 4.0, 5.04, 6.4, 8.0, and 10.8 sec), spaced at equal geometric intervals between the short (2.52 sec) and the long (12.7 sec) signals. Each signal was presented five times within a session. A left lever press was reinforced following the presentation of the first four signals (2.52, 3.2, 4.0, and 5.04 sec), and a right lever press was reinforced following the presentation of the

---

*** In general, it is advisable to use at least 10 subjects per group (Crawley, 2000), but because this was a pilot study, we used smaller groups.

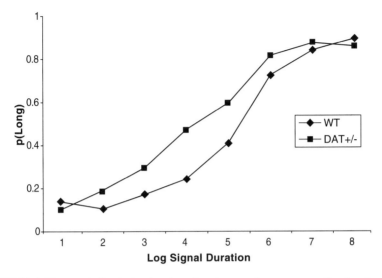

FIGURE 11.1 Temporal discrimination in mice that are heterozygous for the dopamine transporter deletion (DAT$^{+/-}$) and their wild-type littermates (WT).

last four signals (6.4, 8.0, 10.8, and 12.7 sec). Eight-signal training lasted 13 days, and it was followed by four test days during which each group was tested in a single daily session, each consisting of 40 trials. There were no correction trials during the test days.

Figure 11.1 shows that P(Long) increased as a sigmoidal function of signal duration for both DAT$^{+/-}$ and WT mice ($F(7, 56) = 3.62$, $P = .003$). Data points represent the proportion of long responses for each signal, averaged first across sessions and then across subjects. Interestingly, the psychometric function of the DAT$^{+/-}$ group is shifted to the left relative to that of the WT group, although this effect was not statistically significant ($P = .310$).

Why do the DAT$^{+/-}$ mice have a higher tendency than their WT littermates to make a "long" response for any given signal duration? Does this behavioral effect reflect a perceptual difference whereby the DAT$^{+/-}$ mice perceive time as more stretched out than do the WT mice? In rats, systemic injections of methamphetamine, a hyperdopaminergia-causing agent, produce a similar leftward shift in psychometric functions. At first glance, one is tempted to argue that because DAT$^{+/-}$ mice also have functional hyperdopaminergia, the leftward shift of their temporal discrimination function relative to that of the WTs should also reflect a difference in temporal perception between the two groups. But does it?

An altered state of duration perception cannot be observed directly; rather, it is inferred as the perceptual change leads to a behavioral change. Perceptual differences can be detected if the performance of the same individuals is measured under different perceptual states using the same experimental protocol. For example, if a group of rats is trained to perform a temporal discrimination procedure under saline, and then tested under methamphetamine, a methamphetamine-induced leftward shift is observable in full extent during the first session of drug administration (Meck,

1983). Further, under certain experimental conditions, the leftward shift is perfectly parallel; i.e., neither the slope nor the asymptote of the function is affected, indicating that other performance factors such as discriminability are spared (Cevik, 2000). Under these conditions, the leftward shift of the psychometric function is highly likely to have resulted from a change in temporal perception, which causes the signals to be perceived as longer than they were in the absence of methamphetamine. However, when the performance in a temporal discrimination procedure of two groups of rats — one trained under saline and the other under methamphetamine — is compared, no difference in behavior can be observed. The psychometric functions overlap in spite of the difference in the perceptual states because each group of subjects has learned to perform the task in relation to its own perceptual state (Meck, 1983). In general, when performance can be calibrated to the current state of perception, a between-subjects comparison is not appropriate to detect perceptual differences.

Why, then, did the psychometric functions of the $DAT^{+/-}$ and WT groups not overlap in the present experiment? If discrimination of durations can be calibrated to one's speed of timing to yield optimal performance, the psychometric functions of the $DAT^{+/-}$ and WT groups should have overlapped with their temporal bisection point at the geometric mean (Church and Deluty, 1977). Notice that the current procedure involved the reinforcement of intermediate as well as the shortest and longest signals, which forced the bisection point to be at the geometric mean. Nevertheless, the $DAT^{+/-}$ group slightly underestimated the geometric mean, whereas the WT group overestimated it to a somewhat larger extent. So what, if not a perceptual difference, accounts for the deviation of the $DAT^{+/-}$ and WT functions from each other and from the geometric mean?

The psychometric functions were parallel with similar slope and asymptote values, indicating that discriminability and overall performance accuracy were similar for $DAT^{+/-}$ and WT mice. The proportion of correct responses (which is equal to the obtained probability of reinforcement) was 0.76 and 0.78 for $DAT^{+/-}$ and WT groups, respectively, and both groups responded during more than 90% of the trials. Further, the learning rate was not different for the two groups; the number of training sessions to steady-state performance was similar for $DAT^{+/-}$ and WT mice. Taken together, these results suggest that neither overall learning ability nor performance factors that are not related to timing were likely to have affected the position of the psychometric functions.

What else, if not a difference in perception, learning, or motivation, can account for the observed divergence in temporal discrimination functions of the $DAT^{+/-}$ and WT mice? A difference in attentional processes is a likely candidate. Although attention is too elusive a concept to be offered as a satisfactory explanation for behavioral differences, we are all familiar with the feeling that time passes too slowly as we wait, counting the seconds, for something to happen. In other words, time seems to be stretched out as we pay attention to it (e.g., waiting for the stats class to be over), but it flies by if we are too happily involved with something else to pay attention to time (e.g., being on vacation). Notice, that although homozygous $DAT^{-/-}$ have been suggested as a model for human ADHD, the behavioral profile of the heterozygous $DAT^{+/-}$ is suggestive of an enhanced, rather than impaired, attention.

DAT$^{+/-}$ mice might have enhanced attention relative to both DAT$^{-/-}$ and WT mice because attention is an inverted-U-shaped function of the effective levels of dopamine transmission. For example, it is a well-known effect that attention and vigilance change as inverted-U-shaped functions of amphetamine dose (e.g., Grilly et al., 1998). Consistent with this argument is the fact that reaction times were lower for the DAT$^{+/-}$ group, indicating that they were indeed more vigilant than the WTs.

In sum, hyperdopaminergia of DAT$^{+/-}$ seems to have caused these mice to become more likely to respond "long" under a temporal discrimination procedure, which might reflect higher attentiveness of these animals. This hypothesis can be tested by exposing these animals to more detailed tests of attention and vigilance. Alternatively, the divergence in the temporal discrimination functions of the DAT$^{-/-}$ and WT mice might reflect a difference in temporal perception; however, whether the chronic state of hyperdopaminergia of the DAT$^{+/-}$ produces the same perceptual state as an acute methamphetamine administration is disputable.

11.3 WHAT IF THE SINGLE-GENE APPROACH DOES NOT WORK?

Single-gene mutations do not always produce major effects on the phenotypic expression of a complex behavioral trait that is under polygenic control. In several instances (e.g., schizophrenia), single genes with major phenotypic effects elude identification (for a review, see Plomin et al., 1994). In such cases, the phenotypic expression of the behavioral trait seems to be controlled in a continuous, quantitative manner by a set of genes called the quantitative trait loci (QTL). Although individual QTL are inherited in a Mendelian manner, they do not result in discrete, discontinuous phenotypic classes since each exerts an effect of relatively small size. The mutation of any one of these loci might therefore fail to affect the behavioral phenotype in any clear, detectable manner in spite of the actual involvement of that locus in the control of the behavioral trait of interest (Moore and Nagle, 2000).

Linkage techniques that are used to identify single genes can be extended to detect QTL. Alternatively, one can carry out allelic association studies that instead of linking the expression of a phenotypic trait to a chromosomal locus, attempt to associate variations on a behavioral phenotype with specific allelic forms of genes. Both approaches are based on exploiting the naturally occurring variations in a behavioral trait. Therefore, the feasibility of research aimed at QTL identification, much like the forward genetic approach, depends critically on the generation time and maintenance costs of the model organism, efficient and task-relevant behavioral essays that can detect differences in the behavioral phenotype, availability of inbred strains to be used in linkage analysis, and the presence of a high-density linkage map to be used in positional cloning (Taylor, 2000).

REFERENCES

Baker, B.S., Taylor, B.J., and Hall, J.C., Are complex behaviors specified by dedicated regulatory genes? Reasoning from *Drosophila*, *Cell*, 105, 13–24, 2001.

Capecchi, M.R., The new mouse genetics: altering the genome by gene targeting, *Trends Genet.*, 5, 70–76, 1989.

Carvalho, O.M., Silva, A.J., and Balleine, B.W., Temporal perception in mutant mice, *Soc. Neurosci. Abstr.*, 26, 1747, 2000.

Castelluci, V., Pinsker, H., Kupfermann, I., and Kandel, E.R., Neuronal mechanisms of habituation and dishabituation of the gill withdrawal reflex in *Aplysia*, *Science*, 167, 1745–1748, 1970.

Cevik, M.O., Effects of changes in signal durations and methamphetamine on temporal discrimination, *Diss. Abstr. Int.*, 61, 6181, 2000 (Doctoral dissertation, Indiana University, 2000).

Choi, D.W., Background genes: out of sight but not out of brain, *Trends Neurosci.*, 20, 499–500, 1997.

Church, R.M. and Deluty, M.Z., Bisection of temporal intervals, *J. Exp. Psychol. Anim. Behav. Process.*, 3, 216–228, 1977.

Cook, E.H., Stein, M.A., Krasowski, M.D., Cox, N.J., Olkon, D.M., Kieffer, J.E., and Leventhal, B.L., Association of the attention deficit disorder and the dopamine transporter gene, *Am. J. Hum. Genet.*, 56, 993–998, 1995.

Crawley, J.N., *What's Wrong with My Mouse?* Wiley-Liss, New York, 2000.

Crusio, W.E., Genetic dissection of mouse exploratory behaviour, *Behav. Brain Res.*, 125, 127–132, 2002.

Dubnau, J. and Tully, T., Gene discovery in *Drosophila*: new insights for learning and memory, *Ann. Rev. Neurosci.*, 21, 407–444, 1998.

Dudai, Y., Jan Y.N., Byers, D., Quinn, W., and Benzer, S., *Dunce*, a mutant of *Drosophila melanogaster* deficient in learning, *Proc. Natl. Acad. Sci. U.S.A.*, 73, 1684–1688, 1976.

Feldman, R.S., Meyer, J.S., and Quenzer, L.F., *Principles of Neuropsychopharmacology*, Sinauer Associates, Boston, 1997.

Gainetdinov, R.R., Jones, S.R., and Caron, M.G., Functional hyperdopaminergia in dopamine transporter knockout mice, *Biol. Psychiatry*, 46, 303–311, 1999a.

Gainetdinov, R.R., Wetsel, W.C., Jones, S.R., Levin, E.D., Jaber, M., and Caron, M.G., Role of serotonin in the paradoxical calming effect of psychostimulants on hyperactivity, *Science*, 283, 397–401, 1999b.

Gallistel, C.R. and Gibbon, J., Time, rate, and conditioning, *Psychol. Rev.*, 107, 289–344, 2000.

Gibbon, J., Scalar expectancy theory and Weber's law in animal timing, *Psychol. Rev.*, 84, 279–325, 1977.

Gill, M., Daly, G., Heron, S., Hawi, Z., and Fitzgerald, M., Confirmation of association between attention deficit hyperactivity disorder and a dopamine transporter polymorphism, *Mol. Psychiatry*, 2, 311–313, 1997.

Goddard, J., Perceived passage of time: its possible relationship to attention deficit hyperactivity disorder, *Med. Hypotheses*, 55, 351–352, 2000.

Green, J.T., Ivry, R.B., and Woodruff-Pak, D.S., Timing in eye-blink classical conditioning and timed-interval tapping, *Psychol. Sci.*, 10, 19–23, 1999.

Grilly, D.M., Pistell, P.J., and Simon, B.B., Facilitation of stimulus detection performance of rats with d-amphetamine: a function of dose and level of training, *Psychopharmacology*, 140, 272–278, 1998.

Hall, J.C., Molecular neurogenetics of biological rhythms, *J. Neurogenet.*, 12, 115–181, 1998.

Hasty, P., Abuin, A., and Bradley, A., Gene targeting, principles, and practice in mammalian cells, in *Gene Targeting: A Practical Approach*, Joyner, A.L., Ed., Oxford University Press, New York, 2000, pp. 1–35.

Holliday, M. and Hirsch, J., A comment on the evidence for learning in Diptera, *Behav. Genet.*, 16, 439–447, 1986.

Jones, S.R., Gainetdinov, R.R., Hu X-T, Cooper, D.C., Wightman, R.M., White, F.J, and Caron, M.G., Loss of autoreceptor functions in mice lacking the dopamine transporter, *Nat. Neurosci.*, 2, 649–655, 1999.

Jones, S.R., Gainetdinov, R.R., Wightman, R.M., and Caron, M.G., Mechanisms of amphetamine action revealed in mice lacking the dopamine transporter, *J. Neurosci.*, 18, 1979–1986, 1998.

Justice, M.J., Mutagenesis of the mouse germline, in *Mouse Genetics and Transgenics: A Practical Approach*, Jackson, I.J. and Abbott, C.M., Eds., Oxford University Press, New York, 2000, pp. 186–215.

Kandel, E.R., Cellular mechanisms of learning and the biological bases of individuality, in *Principles of Neural Science*, 2nd ed., Kandel, E.R. and Schwartz, J.H., Eds., Elsevier, New York, 1985.

King, D.P., Zhao, Y., Sangoram, A.M., Wilsbacher, L.D., Tanaka, M., Antoch, M.P., Steeves, T.D., Vitaterna, M.H., Kornhauser, J.M., Lowrey, P.L., Turek, F.W., and Takahashi, J.S., Positional cloning of the mouse circadian *Clock* gene, *Cell*, 89, 641–653, 1994.

Konopka, R.J. and Benzer, S., Clock mutants of *Drosophila melanogaster*, *Proc. Natl. Acad. Sci. U.S.A.*, 68, 2112–2116, 1971.

Leaton, R.N., Long term retention of the habituation of lick suppression and startle response produced by a single auditory stimulus, *J. Exp. Psychol. Anim. Learn. Process.*, 2, 248–259, 1976.

Leaton, R.N. and Tighe, T., *Habituation: Perspectives from Child Development, Animal Behavior and Neurophysiology*, Erlbaum, Hillsdale, NJ, 1976.

Low-Zeddies, S.S. and Takahashi, J.R., Genetic influences on circadian rhythms in mammals, in *Genetic Influences on Neural and Behavioral Functions*, Pfaff, D.W., Berrettini, W.H., Joh, T.H., and Maxson, S.C., Eds., CRC Press, Boca Raton, FL, 2000, pp. 293–305.

Malapani, C., Rakitin, B., Levy, R., Meck, W.H., Deweer, B., Dubois, B., and Gibbon, J., Coupled temporal memories in Parkinson's disease: a dopamine related dysfunction, *J. Cognit. Neurosci.*, 10, 316–331, 1998.

Mattell, M.S. and Meck, W.H., Neuropsychological mechanisms of interval timing behavior, *Bioessays*, 22, 94–103, 2000.

McGuire, T.R., Learning in three species of Diptera: the blow fly *Phormina regina*, the fruit fly *Drosophila melanogaster*, and the house fly *Musca domestica*, *Behav. Genet.*, 14, 479–526, 1984.

Meck, W.H., Selective adjustment of the speed of internal clock and memory processes, *J. Exp. Psychol. Anim. Behav. Process.*, 9, 171–201, 1983.

Miller, R.R. and Barnet, R.C., The role of time in elementary associations, *Curr. Directions Psychol. Sci.*, 2, 106–111, 1993.

Montokowski, A., Poettig, M., Mederer, A., and Holsboer, F., Behavioral performance in three substrains of mouse strain 129, *Brain Res.*, 762, 12–18, 1997.

Moore, K.J. and Nagle, D.L., Complex trait analysis in the mouse: the strengths, the limitations, and the promise yet to come, *Annu. Rev. Genet.*, 34, 653–686, 2000.

Platt, J.R., Temporal differentiation and psychophysics of time, in *Advances in Analysis of Behavior, Vol. 1: Reinforcement and Organization of Behavior*, Zeiler, M.D. and Harzem, P., Eds., Wiley, New York, 1979, pp. 1–29.

Plomin, R., Owen, M.J., and McGuffin, P., The genetic basis of complex human behaviors, *Science*, 264, 1733–1739, 1994.

Quinn, W.G., Harris, W.A., and Benzer, S., Conditioned behavior in *Drosophila melanogaster*, *Proc. Natl. Acad. Sci. U.S.A.*, 71, 707–712, 1974.

Ralph, M.R. and Menaker, M., A mutation of the circadian system in golden hamsters, *Science*, 241, 1225–1227, 1989.

Sasaki, A., Wetsel, W.C., Rodriguiz, R.M., and Meck, W.H., Timing of the acoustic startle response in mice: habituation and dishabituation as a function of the interstimulus interval, *Int. J. Comp. Psychol.*, 14, 258–268, 2001.

Silva, A.J., Kogan, J.H., Frankland, P.W., and Kida, S., CREB and memory, *Annu. Rev. Neurosci.*, 21, 127–148, 1998.

Silva, A.J., Smith, A.M., and Giese K.P., Gene targeting and the biology of learning and memory, *Annu. Rev. Genet.*, 31, 527–546, 1997.

Silver, L.M., *Mouse Genetics: Concepts and Applications*, Oxford University Press, New York, 1995.

Spielowoy, C., Roubert, C., Hamon, M., Nosten, M., Betancur, C., and Giros, B., Behavioral disturbances associated with hyperdopaminergia in dopamine transporter knockout mice, *Behav. Pharmacol.*, 11, 279–290, 2000.

Stubbs, A., Temporal discrimination and psychophysics, in *Advances in Analysis of Behavior: Reinforcement and Organization of Behavior*, Zeiler, M.D. and Harzem, P., Eds., Wiley, New York, 1979, pp. 341–369.

Takahashi J.S., Pinto, L.H., and Vitaterna, M.H., Forward and reverse genetic approaches to behavior in the mouse, *Science*, 264, 1724–1733, 1994.

Taylor, B.A., Mapping phenotypic trait loci, in *Mouse Genetics and Transgenics: A Practical Approach*, Jackson, I.J. and Abbott, C.M., Eds., Oxford University Press, New York, 2000, pp. 87–120.

Thompson, R.F. and Spencer, W.A., Habituation: a model phenomenon for the study of neuronal substrates of behavior, *Psychol. Rev.*, 73, 16–43, 1966.

Tully, T., *Drosophila* learning: behavior and biochemistry, *Behav. Genet.*, 14, 527–557, 1984.

Tully, T. and Quinn, W., Classical conditioning and retention in normal and mutant *Drosophila melanogaster*, *J. Comp. Physiol. Sensory Neural Behav. Physiol.*, 157A, 263–277, 1985.

Vitaterna, M.H., King, D.P., Chang, A.-M., Kornhauser, J.M., Lowrey, P.L., McDonald, J.D., Dove, W.F., Pinto, L.H., Turek, F.W., and Takahashi, J.S., Mutagenesis and mapping gene, Clock, essential for circadian behavior, *Science*, 264, 719–725, 1994.

Wahlsten, D., Deficiency of corpus callosum varies with strain and supplier of mice, *Brain Res.*, 239, 329–347, 1982.

Wehner, J.M. and Silva, A., Importance of strain differences in evaluations of learning and memory null mutants, *Ment. Retard. Dev. Disabil. Res. Rev.*, 2, 243–248, 1996.

Wolfer, D.P., Crusio, W.E., and Lipp H.P., Knockout mice: simple solutions to the problems of genetic background and flanking genes, *Trends Neurosci.*, 25, 336–340, 2002.

12 Dopaminergic Mechanisms of Interval Timing and Attention

Catalin V. Buhusi

CONTENTS

12.1 Introduction 318
12.2 Dopamine and the Internal Clock..319
    12.2.1 An Information-Processing Model of Interval Timing....................319
    12.2.2 The Peak-Interval Procedure..319
    12.2.3 Dopamine and the Clock Pattern...321
12.3 Dopamine and Attention Sharing 321
    12.3.1 The Switch Hypothesis 322
    12.3.2 The Decay Hypothesis 323
    12.3.3 Dopamine and the Attention-Sharing Pattern..................................323
12.4 Behavioral and Pharmacological Evidence for Attention Sharing in the Gap Procedure 325
    12.4.1 Comparative Studies in Pigeons and Rats.......................................326
    12.4.2 Behavioral Manipulations of Attention Sharing in Rats 328
    12.4.3 Evidence for Dopaminergic Involvement in Attention Sharing......329
12.5 Conclusions 331
    12.5.1 Behavioral Dissociation of Attention Sharing and Time Accumulation 331
    12.5.2 Dissociation of Clock and Attentional Effects of Methamphetamine and Haloperidol...332
    12.5.3 Implications for Theories of Interval Timing in Humans and Other Animals 334
Acknowledgments...335
References...335

12.1 INTRODUCTION

Humans and other animals process temporal information as if they use an internal stopwatch that can be stopped and reset, and whose speed of ticking is adjustable (Church, 1978). Support for this idea comes from data in bees, fish, turtles, birds, rodents, monkeys, and human infants and adults (e.g., Bateson and Kacelnik, 1997, 1998; Brannon et al., 2002; Brodbeck et al., 1998; Gallistel, 1990; Lejeune and Wearden, 1991; Matell and Meck, 2000; Paule et al., 1999; Richelle and Lejeune, 1980; Talton et al., 1999). The internal clock concept was first introduced in the seminal work of Treisman (1963). According to Treisman (1963), pulses emitted at regular intervals by a pacemaker are stored in an accumulator whose content represents the current subjective time. The rate of the pacemaker was proposed to be related to participant's arousal, and the response of the participant was proposed to be controlled by a comparator mechanism. Although the structure of Treisman's (1963) internal clock can still be recognized in current models of interval timing as applied to a variety of species, these models differ in regard to attentional processing. This is not surprising, considering the differences in the behavioral protocols used in the fields of human cognition and animal behavior. For example, data obtained from human participants performing in dual-task paradigms show that the use of a second task performed in parallel with a temporal task results in temporal underestimation. This effect has been observed with various nontemporal tasks, such as perceptual, mental arithmetic, or motor tracking (e.g., Brown, 1985, 1995, 1997; Hicks et al., 1976, 1977; Macar et al., 1994; Zakay, 1989; Zakay et al., 1983). It is inferred from these results that increasing the demand in processing results in fewer temporal pulses to be accumulated, possibly due to the sharing of attention between the internal clock and other processes (see Fortin, this volume; Lustig, this volume; Penney, this volume). Cognitive models of interval timing developed to explain the effects of nontemporal parameters assume that (a) temporal information is obtained from both a timer (temporal processor) and other sources (general processor), and (b) attention is shared between these processors, a mechanism that we will refer to as attention sharing (e.g., Block and Zakay, 1996; Fortin and Masse, 2000; Thomas and Weaver, 1975; Zakay, 1989, 2000). This view predicts that manipulations that would increase the resources allocated to general processing would result in the time processor losing resources, and consequently losing its ability to accumulate pulses and time accurately.

In contrast to human studies of interval timing, studies of interval timing in rats and pigeons are typically conducted in single-task paradigms designed to explore the specifics of the time processor, but less apt to address the possible relation with general processing. The use of such paradigms promotes psychological theories that attempt to explain the effect of all manipulations as changes in the time processor, rather than possible interactions with the other processes. Here we review behavioral and pharmacological data supporting the dissociation of the pacemaker of an internal clock from attention sharing between the timer and other processes. The dissociation of these processes at the behavioral level is critical, because both processes depend on intact dopaminergic function (Buhusi and Meck, 2002; Malapani and Rakitin, this volume).

12.2 DOPAMINE AND THE INTERNAL CLOCK

12.2.1 AN INFORMATION-PROCESSING MODEL OF INTERVAL TIMING

A representation of the mapping between the stopwatch metaphor (Church, 1978) and an information-processing model of interval timing (Gibbon et al., 1984) is given in Figure 12.1A. The information-processing model of interval timing continues the tradition of Treisman's (1963) internal clock in that pulses emitted by a pacemaker are temporarily stored in an accumulator, and that upon delivery of reinforcement, the number of pulses from the accumulator is stored in reference memory (Gibbon et al., 1984). The model implements the scalar expectancy theory (Gibbon, 1977) in that the response of the subjects is controlled by the ratio comparison between the current time (stored in the accumulator) and the criterion interval (stored in the reference memory). Figure 12.1B shows that according to the model, subjective time, stored in the accumulator, is linearly related to objective time, and that the error in time estimation increases with the to-be-timed interval. Therefore, two groups of subjects trained to time 20- and 40-sec criterion intervals will have similar response functions in terms of shape (shown in the left-hand side of Figure 12.1B), but the width of the response function will be twice as large in the 40-sec group as in the 20-sec group. However, the information-processing model (Gibbon et al., 1984) ignores the possible interactions between interval timing and other cognitive processes (see next section). The effects of manipulations of nontemporal aspects of the paradigms are attributed to changes in the speed of the clock or in delays in the activation of the various components of the model (see e.g., Church, 1978, 1984). For example, the model proposes that changes in clock speed account for data showing that bright lights are judged to be longer than dim lights by pigeons (Kraemer et al., 1997a) and rats (Kraemer et al., 1995), and filled intervals are judged to be longer than empty intervals by pigeons (Mantanus, 1981). Accordingly, Figure 12.1C shows that changes in clock speed are proposed to affect the slope of the linear relationship between subjective and objective time. The behavioral effects of such changes are predicted to be proportional to the criterion interval, a feature usually referred to as the clock pattern: a manipulation that supposedly changes the clock speed is predicted to shift the response function twice as much in a group of subjects trained to time a 40-sec interval as in the group trained with a 20-sec interval. Conversely, should the effects of a drug on time perception be proportional to the criterion interval, this would suggest that the drug affects interval timing by changing the speed of the clock. The important point here is that the model has specific predictions for manipulations of each of its components.

12.2.2 THE PEAK-INTERVAL PROCEDURE

How would one go about testing the above predictions experimentally? Data discussed in this chapter were collected using the peak-interval (PI) procedure. In the PI procedure subjects are exposed to two types of trials: fixed-interval (FI) trials and PI trials. In FI trials, subjects are presented with a signal for a fixed duration. The first response after the fixed duration terminates the to-be-timed signal and

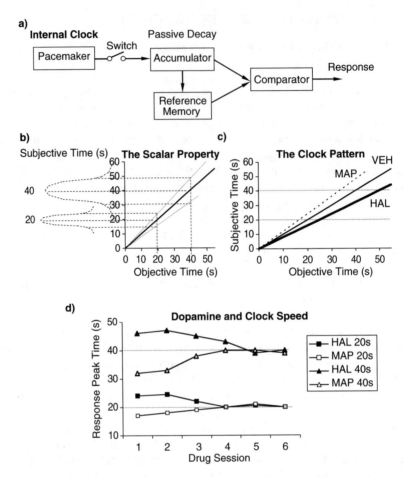

FIGURE 12.1 Dopamine and clock speed. (A) An information-processing model of interval timing. (Adapted from Gibbon, J. et al., *Annals of the New York Academy of Sciences: Timing and Time Perception*, New York Academy of Sciences, New York, 1984, pp. 52–77.) (B) The scalar property as implemented in the information-processing model: the peak time response function is twice as wide when subjects time a 40-sec interval as when they time a 20-sec interval. (C) Theoretical increase in speed of accumulation of temporal units by indirect dopamine agonist methamphetamine and theoretical decrease in speed of accumulation of temporal units by dopamine antagonist haloperidol. Effects on the peak time of responding are predicted to be proportional to criterion interval. (D) Effect of systemic administration of MAP and HAL on the peak time of responding. (Adapted from Meck, W.H., *Cognit. Brain Res.*, 3, 227–242, 1996.) MAP produces a proportional leftward shift of the response function, and HAL produces a proportional rightward shift of the response function.

triggers the delivery of reinforcement. In PI trials, the to-be-timed signal is presented for about three times the FI, and subjects' responses are not reinforced. In these trials the mean response rate increases after the onset of the to-be-timed signal, reaches a peak about the time when subjects are (sometimes) reinforced, and gradually declines afterwards (Catania, 1970; Church, 1978). According to the information-processing model of interval timing (Gibbon et al., 1984), in a PI procedure the current value in the accumulator is compared to the reference memory and the response rate reaches a peak at the time when subjects' responses are (sometimes) reinforced. Therefore, according to the model, the experimentally estimated time at which the response rate reaches a peak in PI trials can be taken as an estimation of the criterion stored in reference memory. Should a manipulation change the clock speed after acquisition of a temporal criterion, the response function would shift leftward (if the speed of the clock increases) or rightward (if the speed of the clock decreases). Conversely, should a drug shift the peak function in a manner proportional to the intervals being timed, this would give a strong indication that the drug acts at the pacemaker or accumulator level of an internal clock.

12.2.3 DOPAMINE AND THE CLOCK PATTERN

Psychopharmacological studies (Maricq and Church, 1983; Maricq et al., 1981; Meck, 1983, 1986) have provided considerable support for the neural basis of some of the parameters of the information-processing model of time perception (Gibbon et al., 1984), proposing that dopaminergic drugs selectively affect the subjective speed of an internal clock, while cholinergic drugs alter memory storage (e.g., Meck, 1983, 1996). An acceleration of the subjective clock speed is suggested by a variety of timing studies in rats and pigeons using indirect dopaminergic agonists such as methamphetamine (MAP) (e.g., Kraemer et al., 1997b; Maricq and Church, 1983; Maricq et al., 1981; Meck, 1983) and cocaine (e.g., Lau et al., 1999; Matell et al., 2002). On the other hand, dopaminergic antagonists such as haloperidol (HAL) have been shown to produce a deceleration of the subjective clock speed in proportion to their affinity to dopamine D_2 receptor (Meck, 1986; see also Maricq and Church, 1983). The effects of MAP and HAL on the peak time of responding in a PI procedure (Meck, 1983, 1996) are shown in Figure 12.1D: MAP administration immediately shifts the response function leftward (decreases the mean peak time), and HAL administration immediately shifts the response function rightward (increases the mean peak time). Both effects are proportional to the criterion interval. Indeed, as shown in Figure 12.1D, the drugs shift the response peak time twice as much in rats trained to time a 40-sec interval as in rats trained to time a 20-sec interval. These results suggest that dopaminergic drugs selectively affect the speed of an internal clock (Gibbon et al., 1984).

12.3 DOPAMINE AND ATTENTION SHARING

In contrast to the above neuropharmacological mapping, some reports suggest that dopaminergic drugs affect attention to temporal signals without selectively altering the speed of an internal clock (e.g., Santi et al., 1995; Stanford and Santi, 1998). These results are congruent with a separate line of evidence that tends to suggest a different role for the dopaminergic synapse: an attention-getting device used to track significant events (see Gray et al., 1997; Lee et al., 1998; Schultz et al., 1997). These data suggest that dopamine involvement in the speed of an internal clock needs to be dissociated from the possible confounder with attentional effects.

To this end, one could examine the effect of dopaminergic drugs in a behavioral task in which both the timing and the attentional components are well understood. For example, behavioral data from dual-task performance in humans support the cognitive implementation of the time processor that acknowledges attentional processing (more specifically, attention sharing). Indeed, a behavioral study by Lejeune et al. (1999), using a paradigm in which pigeons were exposed to an analog of the dual-task procedure used to test attention sharing in humans, found data similar to those collected in humans, thus giving support to attention sharing between the time processor and the general processor (Block and Zakay, 1996; Lejeune, 1998; Zakay, 2000) in pigeons. Therefore, one could examine the effect of dopaminergic drugs on dual-task performance in animals. Examination of the effect of dopaminergic drugs in such procedures might allow the differentiation of the effect of the drug on clock speed and attention.

In our work, we pursued a different approach involving a single-task procedure previously used to study timing and memory for time in animals (Catania, 1970; Meck et al., 1984; Roberts, 1981; Roberts and Church, 1978). We examined a variation of the PI procedure in which subjects (rats or pigeons) have to filter out the gaps that (sometimes) interrupt timing (Catania, 1970; Roberts, 1981). In the gap procedure, the subjects are exposed to three types of trials: FI trials, PI trials, and gap trials. Gap trials are similar to PI trials, but the signal is interrupted for a brief duration called a gap. Typically, in gap trials the mean response rate increases in the pregap interval, declines during the gap, and then increases again after the gap and reaches a peak that is delayed relative to the peak time during PI trials. Evidence discussed below suggests that in this paradigm the delay in peak time is controlled by an attention-sharing process that can be differentiated from the pacemaker stage of the internal clock both behaviorally (Buhusi and Meck, 2000; Buhusi et al., 2002) and pharmacologically (Buhusi and Meck, 2002). This differentiation is critical, because both processes are at least partly dependent on dopaminergic function (Buhusi and Meck, 2002). Before we discuss the differentiation of pacemaker and attention-sharing components, it is important to discuss alternative interpretations of the data in the gap procedure.

12.3.1 THE SWITCH HYPOTHESIS

In rats, the peak time in gap trials was found to be delayed relative to PI trials for approximately the duration of the gap (Meck et al., 1984; Roberts, 1981; Roberts and Church, 1978). These data were taken to suggest that rats "stop" their timing process during the gap and resume it where they left off after the gap. To address these data, Gibbon et al. (1984) proposed an on–off switch mechanism controlled by the presence of the to-be-timed stimulus. In PI trials, the switch is closed, pulses from the pacemaker reach the accumulator, and the response rate reaches a peak near the time of reinforcement. During the gap, the switch is open, so that pulses from the pacemaker fail to reach the accumulator. The pulses accumulated during the pregap interval are not lost, however, due to their proposed maintenance in working memory. Therefore, in accord with experimental data in rats (Meck et al., 1984; Roberts, 1981; Roberts and Church, 1978), the peak time in gap trials is delayed by the duration of the gap (stop rule).

12.3.2 THE DECAY HYPOTHESIS

On the other hand, in gap procedures that use gap durations similar to those used with rats, pigeons' mean response rate after the gap is delayed relative to the time of reinforcement with approximately the duration of the gap plus the duration of the fixed interval (e.g., Roberts et al., 1989). These data suggest that during the gap pigeons "reset" the interval timing process and restart timing after the gap from the beginning of the interval (reset rule). Importantly, a parametric study in pigeons suggests that the delay in peak time during gap trials increases with the duration of the gap (Cabeza de Vaca et al., 1994). Based on this observation, Cabeza de Vaca et al. (1994) proposed a rather different mechanism by which the gap might affect interval timing, namely, that the subjective time — stored in the accumulator — "decays" passively during the gap. This passive decay process is presumably minimal for short gaps, but rather large for longer gaps, thus accounting for both the stop and reset rules. In summary, previous studies in rats and pigeons interpret the results obtained in a gap procedure in terms of two mechanisms, a switch and a passive decay process, which are an integral part of the time processor and are exclusively controlled by the temporal parameters of the procedure.

12.3.3 DOPAMINE AND THE ATTENTION-SHARING PATTERN

We recently examined the involvement of dopamine in putative processes like the pacemaker, the switch, and the passive decay of accumulated time in the gap procedure (Buhusi and Meck, 2002). The predicted results of acute administration of the dopaminergic agonist MAP in the gap procedure, given the assumption that dopaminergic drugs affect solely the pacemaker level of the internal clock, are shown in Figure 12.2. The upper panels of Figure 12.2 show the predicted effect of the gap under vehicle: according to the switch hypothesis, the presentation of the gap should delay the peak time for approximately the duration of the gap (left upper panel). In contrast, according to the decay hypothesis, the presentation of the gap should delay the peak time more than the duration of the gap (right upper panel). The lower panels of Figure 12.2 show the predicted effect of the acute administration of MAP in PI and gap trials. Under the assumption that dopaminergic drugs affect solely the pacemaker level of the internal clock, the lower panel of Figure 12.2 shows that during PI trials, as well as before and after a gap in gap trials, the rate of accumulation is predicted to increase under MAP. Indeed, while the accumulation reaches the 30-sec criterion after 30 sec under vehicle (upper panels), accumulation is predicted to be faster under MAP and to reach the temporal criterion after, for example, 25 sec. Critically for the logic of the experiment, irrespective of the operational hypothesis (switch or decay) the peak time is predicted to be delayed in gap trials relative to PI trials with about the same duration under MAP (lower panels) than under vehicle (upper panels).

To evaluate the latter prediction, we estimated the shift in peak time, computed as follows:

$$\text{Shift} = PT_{Gap} - PT_{PI} - \text{Gap}$$

where PT_{Gap} is the estimated peak time in gap trials, PT_{PI} is the estimated peak time in PI trials, and Gap is the duration of the gap. A null shift would suggest that rats use a stop rule, while a shift equal to the pregap interval (15 sec in this experiment) would suggest that rats use a reset rule. The results of this experiment reported by Buhusi and Meck (2002) are shown in Figure 12.3. In PI trials (left panel), acute administration of MAP decreased the peak time, and acute administration of HAL increased the peak time, supporting the proposal that dopaminergic drugs affect the rate of the pacemaker of an internal clock. However, in contrast to the predictions shown in Figure 12.2, acute administration of both MAP and HAL was found to affect the shift in response peak time in gap trials relative to PI trials. Predictions shown in Figure 12.2 suggest that the difference in peak time between PI and gap trials should be about the same, irrespective of drug manipulation. In contrast, we found that MAP determined a leftward displacement in peak time in PI trials (left panel of Figure 12.3), but it also increased the shift in peak time in gap trials relative to PI trials, toward the reset (right panel of Figure 12.3). In similarity to the predictions shown in Figure 12.2, the slower accumulation of time under HAL should

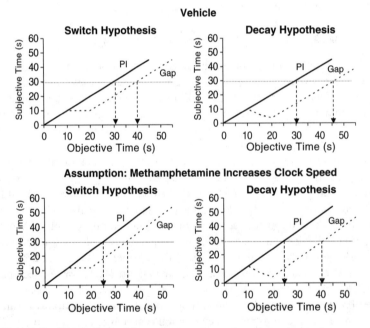

FIGURE 12.2 Predicted effect of dopamine agonist methamphetamine in the gap procedure. Upper panels: Predicted accumulation of temporal units in peak-interval and gap trials under vehicle. The switch hypothesis (left) predicts a rightward shift in peak time equal to the duration of the gap; the decay hypothesis predicts that the rightward shift of the peak time increases with the duration of the gap. Lower panels: Predicted faster accumulation of temporal units under indirect dopamine agonist MAP. In both PI and gap trials the peak time is predicted to shift leftward due to faster accumulation. However, the difference in peak time between PI and gap trials under MAP (lower panels) is predicted to be similar to that in the vehicle condition (upper panels).

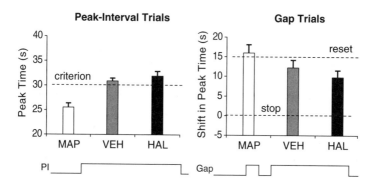

FIGURE 12.3 Dopaminergic effects in the gap procedure. Rats were trained in a 30-sec PI procedure and subsequently tested under acute administration of vehicle (VEH), dopamine agonist methamphetamine, and dopamine antagonist haloperidol. Left panel: MAP produces a leftward shift of the response function, and HAL produces a rightward shift of the response function in PI trials. Right panel: Contrary to prediction, the shift in peak time between PI and gap trials is affected by dopaminergic drugs: MAP increases the shift, and HAL decreases the shift relative to the vehicle condition. (Adapted from Buhusi, C.V. and Meck, W.H., *Behav. Neurosci.*, 116, 291–297, 2002.)

not affect the shift in peak time between PI and gap trials. In contrast, in our experiment, besides determining a rightward displacement of the response functions in PI trials (left panel of Figure 12.3), HAL also decreased the shift in peak time in gap trials relative to PI trials (right panel of Figure 12.3). The different pattern of effects of MAP and HAL in trials with and without gaps suggests that, besides affecting the clock speed, dopaminergic drugs are involved in a separate process in the gap procedure.

To explore the nature of this second process, we pursued a series of behavioral and pharmacological manipulations of the gap procedure. Most of these studies, discussed below, involved two groups of subjects or two conditions that do not differ in terms of temporal parameters, but only in terms of nontemporal aspects (stimulus intensity, stimulus content, stimulus similarity, etc.). Yet we found that these manipulations affect the response function after the gap. Taken together, the studies reviewed below suggest that the gap procedure engages attention sharing, and that this is the process underlying the effect of dopaminergic drugs in the gap procedure.

12.4 BEHAVIORAL AND PHARMACOLOGICAL EVIDENCE FOR ATTENTION SHARING IN THE GAP PROCEDURE

A possible interpretation of the findings discussed above in the gap procedure is that during the gap attentional resources are reallocated between the timer and the general processor. According to this view, the greater the extent that attention is paid to extraneous events — such as the gap — the more processing resources are diverted away from the timing component of the task, and the more pulses are lost, resulting in a larger delay of the peak time (peak time shift). We tested this hypothesis both at the behavioral and pharmacological levels.

At the behavioral level, the basic switch hypothesis predicts that the delay in peak time is always equal to the duration of the gap, while the basic decay hypothesis predicts that the effect of the gap depends solely on its duration. In contrast, attention sharing predicts that manipulations of nontemporal aspects of the gap procedure would affect the timing of the response. In particular, attention sharing might depend on the content of the gap, on the degree to which the gap can be discriminated from the intertrial interval (ITI), and on the intensity of stimulus events.

At the pharmacological level, assuming that MAP and HAL affect solely the speed of an internal clock, one would expect that, relative to vehicle administration, the peak of the response rate would occur proportionally sooner under MAP and proportionally later under HAL in PI trials (Maricq et al., 1981; Meck, 1983, 1986), but that the shift in peak time between PI and gap trials would not be affected by these manipulations. On the other hand, assuming that dopaminergic agonists increase the perceived salience of events (Gray et al., 1997), one might expect the gap to shift the peak time more than the PI trials under MAP, so that the response rate would peak later (reset rule). Similarly, HAL might reduce the effect of the gap on allocation of attentional resources and might cause the response rate to peak sooner after the gap (stop rule). Critically, should the effects of these behavioral and pharmacological manipulations be expected to act on the same chain of processes, related to attention sharing, they should counteract one another.

12.4.1 COMPARATIVE STUDIES IN PIGEONS AND RATS

Previous data suggest that in the gap procedure rats use a stop strategy (Meck et al., 1984; Roberts, 1981; Roberts and Church, 1978) while pigeons are resetters (Cabeza de Vaca et al., 1994; Roberts et al., 1989). The upper left panel of Figure 12.4 shows that a 5-sec gap determines a rightward shift of the response function with about 5 sec in rats, suggesting that rats stop timing during the gap (Roberts, 1981). In contrast, the upper right panel shows that a 6-sec gap delays the response function with about 15 sec in pigeons, supporting the proposal that after the gap pigeons restart timing from the beginning of the interval (Cabeza de Vaca et al., 1994).

In contrast to the species-specific strategy interpretation of the results, recent findings from our laboratory suggest that nontemporal parameters of the to-be-timed event influence the response rule adopted by both rats (Buhusi and Meck, 2000) and pigeons (Buhusi et al., 2002) in the gap procedure. For example, Buhusi and Meck (2000) examined time perception and memory for timing the presence and absence of a visual stimulus. In accord with previous results, Buhusi and Meck (2000) found that when timing for the presence of a visual stimulus, a 5-sec standard (dark) gap prompts rats to stop timing, i.e., to delay their response function with about 5 sec in gap trials relative to PI trials (middle left panel of Figure 12.4). However, when timing for the absence of the stimulus, a 5-sec reversed gap during which the stimulus was turned on, prompted rats to reset the entire timing process after the gap (middle right panel of Figure 12.4). This reset result occurred at gap durations as short as 1 sec (Buhusi and Meck, 2000).

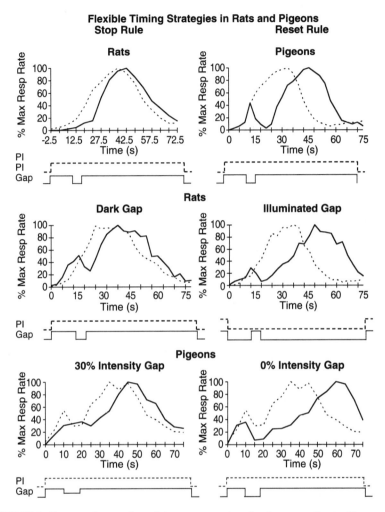

FIGURE 12.4 Comparative studies of the gap procedure in pigeons and rats. Upper panels: Initial accounts of species differences in the gap procedure; rats are "stoppers" (left) (adapted from Roberts, S., *J. Exp. Psychol. Anim. Behav. Process.*, 7, 242–268, 1981) and pigeons are "resetters" (right) (adapted from Cabeza de Vaca, S. et al., *J. Exp. Psychol. Anim. Behav. Process.*, 20, 184–198, 1994). Middle panels: Rats stop timing during a dark gap (left) and reset timing after an illuminated gap (right). (Adapted from Buhusi, C.V. and Meck, W.H., *J. Exp. Psychol. Anim. Behav. Process.*, 26, 305–322, 2000.) Lower panels: Pigeons stop timing during a gap of 30% light intensity (left) and reset timing after a gap of 0% intensity (right). (Adapted from Buhusi, C.V. et al., *J. Comp. Psychol.*, 2002.)

Moreover, Buhusi et al. (2002) found that pigeons flexibly use the stop and reset rules depending on the intensity of the interrupting event in the specific trial (see also Wilkie et al., 1994). In accord with previous results, Buhusi et al. (2002) found that when timing for the presence of a visual stimulus (100% intensity), a 10-sec dark (0% intensity) gap prompted pigeons to restart timing after the gap (lower right panel of Figure 12.4). However, a 10-sec gray (30% intensity) gap prompted pigeons to stop timing during the gap and to delay their response with about the duration of the gap (lower left panel of Figure 12.4). These findings support the proposal that in both rats and pigeons the effect of the gap depends to a large extent on nontemporal components. The next step in our exploration was to experimentally differentiate the switch, the passive decay, and the attention-sharing hypotheses.

12.4.2 Behavioral Manipulations of Attention Sharing in Rats

To differentiate the switch, decay, and attention-sharing hypotheses, we manipulated the content (illuminated vs. dark) of the to-be-timed visual signal, the similarity in auditory content between the gap and the ITI, and the intensity of the stimuli in rats. While attention sharing might depend on these features, the switch and decay hypotheses predict that only temporal aspects of the gap procedure are relevant. Therefore, to differentiate these hypotheses, in our experiments we maintained the same temporal parameters and manipulated only nontemporal aspects of the gap procedure. As shown in Figure 12.5, we found that all these nontemporal variables affect the peak time of responding in the gap procedure. For example, standard (dark) gaps prompt rats to stop timing, and reversed (illuminated) gaps prompt rats to reset the entire timing process after the gap (left panel of Figure 12.5) (Buhusi and Meck, 2000), possibly due to allocation of more resources to the general processing of filled gaps than of empty gaps.

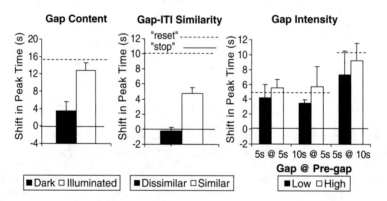

FIGURE 12.5 Behavioral manipulations of nontemporal parameters of the gap procedure in rats. In all experiments the temporal parameters were kept constant in all groups and conditions. Left panel: Rats stop timing during a dark gap (filled bar) and reset timing after an illuminated gap (empty bar). (Adapted from Buhusi, C.V. and Meck, W.H., *J. Exp. Psychol. Anim. Behav. Process.*, 26, 305–322, 2000.) Center panel: Rats stop timing during a silent gap when the ITI is filled with a white noise (dissimilar condition, filled bar) and tend to reset when both the gap and the ITI are silent (similar condition, empty bar). (Adapted from Buhusi, C.V. and Meck, W.H., *Behav. Neurosci.*, 116, 291–297, 2002.) Right panel: Rats reset timing after a high-intensity gap (high, empty bars), but not after a low-intensity gap (low, filled bars). (Adapted from Buhusi, C.V. et al., *J. Comp. Psychol.*, 2002.)

In a different set of experiments, a group of rats were trained to time a 20-sec visual signal, and the ITI was dark and silent (similar condition) (Buhusi and Meck, 2002). The empty bar in the middle panel of Figure 12.5 shows the shift in peak time determined by a 5-sec dark, silent gap. Afterwards, the same rats were trained to time the same 20-sec visual signal, but the ITI was signaled by a white noise (dissimilar condition) (Buhusi and Meck, 2002). The middle panel of Figure 12.5 shows that when the ITI was noisy, a 5-sec dark, silent gap determined a significantly smaller shift in peak time (filled bar in the middle panel of Figure 12.5). In fact, in the dissimilar condition the rats used a perfect stop strategy. Presumably, rats allocate few resources to timing in the ITI, and during a gap similar to the ITI, and therefore tend to reset in the similar condition.

We also evaluated whether rats' response rule in the gap procedure is influenced by the intensity of the gap. Rats were trained to time the absence of a 20-sec visual signal, and the ITI was illuminated by a 2500-lux light. In the test phase, reversed (illuminated) gaps were presented for either 5 sec at a 5-sec pregap interval, 5 sec at a 10-sec pregap interval, or 10 sec at a 5-sec pregap interval. Most importantly, in each test session the intensities of both the gap and ITI were manipulated. The shift in the peak time of responding in test sessions with baseline intensity gaps, 2500 lux, relative to test sessions with gaps of higher intensity, 10,000 lux, is shown in the right panel of Figure 12.5. When gaps of high intensity, 10,000 lux, were presented halfway into the to-be-timed dark interval, rats reset their timing mechanism (open bars). In contrast, when gaps of low intensity, 2500 lux, were presented halfway into the to-be-timed dark interval, rats shifted their peak time significantly less (Buhusi et al., 2002). These results support the notion that the stop–reset mechanism is controlled by the intensity of the gap. Presumably, gaps of high intensity determine a reallocation of attentional resources away from the timing component of the task. Left with fewer resources, the timer is unable to correctly keep timing, and subjects restart timing after the gap. Conversely, signals of low intensity allow rats to use more resources, maintain in memory the pregap interval, and resume timing after the gap where they left off. These results support the hypothesis that interval timing is sharing attentional resources with other cognitive processes, and that, in addition to temporal parameters, the nontemporal features of events control the stop–reset functions of an internal clock (Buhusi et al., 2002).

12.4.3 Evidence for Dopaminergic Involvement in Attention Sharing

Dopaminergic neurons respond equally to reinforcers and to novel, salient stimuli that elicit orienting reactions (Lee et al., 1998; Schultz et al., 1993) regardless of whether the salience derives from reward properties or from physical characteristics of the stimulus (Horvitz et al., 1997). Moreover, genetic manipulations of the dopamine reuptake transporter in mice suggest that an increase in the levels of striatal dopamine correlates with a state of hyperactivity, suggesting a deficit in selective attention (Gainetdinov et al., 1998). These data suggest that dopamine involvement in the speed of an internal clock needs to be dissociated from the possible confounder with attentional effects.

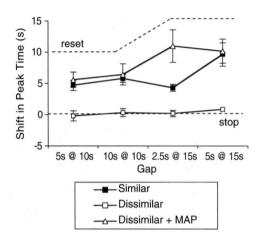

FIGURE 12.6 Behavioral and pharmacological manipulations act on the same chain of processes. Rats were first trained with a silent ITI and tested with silent gaps (similar condition.) When trained with a noisy ITI and tested with silent gaps (dissimilar condition), rats substantially decreased their shift after a gap. Acute administration of methamphetamine in the dissimilar condition (dissimilar + MAP) counteracted the effect of filling the ITI with white noise. (Adapted from Buhusi, C.V. and Meck, W.H., *Behav. Neurosci.*, 116, 291–297, 2002.)

Assuming that dopaminergic agonists increase the perceived salience of interrupting events (see Gray et al., 1997), one might expect the gap to affect timing more under the influence of MAP (reset rule) and less under the influence of HAL (stop rule). Indeed, the results reviewed in Section 12.3.3 suggest that the stop–reset mechanism is affected by acute administration of dopaminergic drugs: MAP tends to increase the shift in peak time toward resetting timing, and HAL tends to decrease the shift in peak time toward stopping timing.

In a separate set of experiments we addressed the question of whether the behavioral and dopaminergic manipulations discussed above affect the same chain of processes, possibly related to attention sharing. Specifically, if these manipulations affect the same chain of processes, one would expect that the stopping effect of filling the ITI discussed in Section 12.4.2 would counteract the resetting effect of acute administration of MAP discussed in Section 12.3.3 when used simultaneously. Figure 12.6 shows the results of a set of experiments directed at testing this prediction.

A group of rats was trained to time a 20-sec visual stimulus, and the ITI was dark and silent. Rats were subsequently tested in four types of gap trials with dark, silent gaps (similar condition) of various durations, positioned at various pregap intervals. Figure 12.6 shows the shift in peak time produced by each of the four types of gaps. Subsequently, the same rats were retrained to time the same 20-sec visual signal, but the ITI was filled with white noise, and tested with the same silent gaps (dissimilar condition). In line with results discussed in Section 12.4.2, we found that dark, silent gaps produce a smaller shift when the ITI was noisy (dissimilar condition) than when the ITI was silent (similar condition). In the dissimilar condition, rats also received acute administration of indirect dopamine agonist methamphetamine. In line with data discussed in Section 12.3.3, acute administration of

MAP shifted the response functions rightward (dissimilar + MAP condition), to about the same delay as in the similar condition. Importantly, in this experiment MAP administration determined a shift in peak time of about 10 sec, i.e., 50% of criterion interval, which is about three times more than previously reported clock speed effects (about 10 to 15% of criterion time) (Meck, 1983, 1986). This result strongly suggests that besides its clock effect, MAP affected another, perhaps attentional, mechanism. The effect of dopaminergic drugs during gap trials might reflect an alteration in the perceived salience of the interrupting event (e.g., Gray et al., 1997). This interpretation is supported by the fact that the shift in peak time in gap trials critically depends on nontemporal aspects of the paradigm, as shown by the differences between the similar and dissimilar conditions. Because MAP prevented the stopping effect of filling the ITI with white noise — a manipulation that affects the discriminability of the gap from the ITI — results suggest that MAP reset timing due to its attentional (stimulus controlled) effect rather than its clock effect. Most importantly, these behavioral and pharmacological manipulations seem to affect the same chain of processes because they antagonize each other when used simultaneously, as shown in this experiment.

12.5 CONCLUSIONS

12.5.1 BEHAVIORAL DISSOCIATION OF ATTENTION SHARING AND TIME ACCUMULATION

The experiments reviewed here were designed to dissociate the switch, decay, and attention-sharing hypotheses by manipulating temporal and nontemporal parameters of the gap procedure. In accord with the attention-sharing hypothesis, data reviewed here show that increasing the salience of the interrupting event by (a) manipulating the content (illuminated vs. dark) of the event, (b) increasing the similarity between the gap and the ITI, or (c) increasing the intensity of the stimuli prompted rats to further delay their response functions after the gap. These results are in accord with the interpretation that more processing resources are allocated to the (temporal or general) processing of salient stimuli. On one hand, bright lights are judged to be longer than dim lights by pigeons (Kraemer et al., 1997a) and rats (Kraemer et al., 1995), possibly due to allocation of more resources to the timer. Conversely, data reviewed here show that high-intensity gaps displace the response function more than low-intensity gaps in both rats (Figure 12.5) (Buhusi et al., 2002) and pigeons (Figure 12.4) (Buhusi et al., 2002), possibly due to allocation of more resources to the general processor. Similarly, filled intervals are judged to be longer than empty intervals by pigeons (Mantanus, 1981), possibly due to allocation of more resources to timing of filled intervals than of empty intervals (Grondin et al., 1998; Thomas and Weaver, 1975). Conversely, filled gaps were found to displace the response function more than empty gaps (Figure 12.4) (Buhusi and Meck, 2000), possibly due to allocation of more resources to the general processing of filled gaps than of empty gaps. Moreover, few resources might be dedicated to timing during the ITI, thus, during a gap similar to the ITI, resulting in a larger effect of the gap in the similar condition than in the dissimilar condition (Figure 12.5) (Buhusi and Meck,

2002). These results complement previous findings suggesting that automatic or controlled variations in the allocation of attentional resources influence interval timing (e.g., Casini and Macar, 1997, 1999; Macar et al., 1994; Penney, this volume; Zakay, 2000). The present findings suggest that besides temporal parameters of the procedure, nontemporal features of the events influence the timing of the response in the gap procedure in both rats (Buhusi and Meck, 2000, 2002; Buhusi et al., 2002) and pigeons (Buhusi et al., 2002). Because in each experiment the temporal parameters were identical among groups and conditions, these results cannot be easily accounted for by either the switch or decay hypothesis. These findings suggest that the timer shares attentional resources with other concurrent processes, and that this sharing of resources controls the response of rats and pigeons in the gap procedure (Figure 12.7A).

Taken together, the data presented here support the proposal that there might be a continuum of outcomes in the gap procedure, dependent on the salience of the stimuli involved. Both the decay and attention-sharing hypotheses allow for a continuum of possibilities in between the stop and reset extremes. However, while the decay hypothesis predicts that only variations in temporal parameters would yield such a continuum, the attention-sharing hypothesis suggests that the continuum is also related to nontemporal characteristics of the interrupting event. Therefore, in relation to temporal manipulations of the gap procedure, the present data reject the basic form of the switch hypothesis, but cannot fully differentiate the decay and attention-sharing hypotheses. Importantly, such a differentiation might not be required. Attention sharing is a mechanism concurrent with interval timing, and the decay of accumulated time might represent the effect of reallocation of attention on interval timing. As such, rather than decaying passively, the representation of accumulated time might decay at a variable rate, dependent on the attention paid to the gap (left lower panel of Figure 12.7). Consequently, within the framework of the information-processing model of interval timing (Gibbon et al., 1984), an active decay mechanism whose rate of decay is controlled by an attention-sharing mechanism seems most adept at addressing the manipulations of both temporal and nontemporal characteristics of the gap.

12.5.2 Dissociation of Clock and Attentional Effects of Methamphetamine and Haloperidol

Another way to differentiate the attention-sharing process from the pulse generator of the timer is to evaluate the effect of pharmacological manipulations on the gap procedure. Buhusi and Meck (2002) showed that MAP, a drug known to increase distractibility (Crider et al., 1982; Gray et al., 1992; Iwanami et al., 1995; McKetin and Solowij, 1999; McKetin et al., 1999), resets timing, while HAL, a neuroleptic drug known to increase selective attention (reviewed by Gray et al., 1997), stops timing (see Figure 12.3). The attentional resetting effect of MAP was eliminated in rats trained with a filled ITI and tested with empty gaps (Buhusi and Meck, 2002) (see Figure 12.6), suggesting that the behavioral and pharmacological manipulations affect the stop–reset mechanism of interval timing by acting on the same processes. Most importantly, Buhusi and Meck (2002) found that MAP and HAL shift the peak

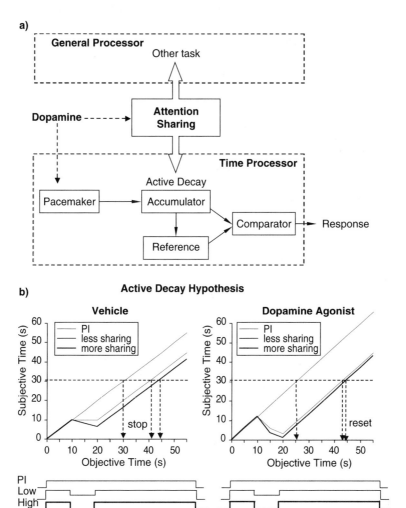

FIGURE 12.7 Dopamine, clock speed, and attention sharing. (A) Relation between time processing, attention sharing, and general processing. Data reviewed here suggest that dopaminergic drugs affect the accumulation of temporal units by two independent processes: changes in the speed of the pacemaker and changes in allocation of attentional resources. (B) A possible implementation of an active decay mechanism controlled by attention sharing. Lower left panel: The subjective time stored in the accumulator decays during the presentation of the gap at a rate proportional to the attention paid to these events. The salience, content, and similarity with the ITI affect the rate of decay. Lower right panel: Dopaminergic agonists increase the rate of the pacemaker (increase the slope of the accumulation line) and increase the rate of decay during the gap by promoting a reduction in the attentional resources allocated to the timing component of the task. Horizontal broken line, criterion interval; broken arrows, peak times. The diagrams under the graphs depict the to-be-timed stimulus in PI trials and in trials with low-intensity and high-intensity gaps.

time in opposite directions in PI trials and gap trials (right panel of Figure 12.3), suggesting that the attentional effects of these drugs are not due to their possible effects on the speed of the internal clock (Maricq et al., 1981; but see Santi et al., 1995). For example, dopaminergic drugs might affect active decay of accumulated time. For example, the right lower panel of Figure 12.7 shows that MAP might (a) increase the rate of the pacemaker, and (b) facilitate the decay of accumulated time during interrupting events by facilitating reallocation of attentional resources away from the timing component of the task. In accord with data discussed here, MAP would (a) produce a leftward shift in PI trials (left panel of Figure 12.3), and (b) promote a resetting of timing after the gap (right panel of Figure 12.3).

The clock and attention-sharing patterns of dopaminergic drugs might be related to the affinity of dopaminergic drugs to the D_1 and D_2 dopamine receptors, respectively (see Stanford and Santi, 1998; Meck, 1986). In this scenario, drugs that activate both receptors, such as amphetamine, are expected to determine both effects, while drugs that selectively affect the D_2 dopamine receptor, such as quinpirole, are expected to determine only the attentional effect (Frederick and Allen, 1996; Stanford and Santi, 1998; but see Castner et al., 2000; Müller et al., 1998). Indeed, in our experiments MAP shifted the peak time in opposite directions in probe trials with or without gaps, indicative of both clock and attentional effects, while HAL shifted the peak time significantly in trials with gaps, but only marginally in trials without gaps, indicative of a predominantly attentional effect rather than a clock speed effect (see Figure 12.3). On the other hand, results might be due to the activation of the noradrenergic (Al-Zahrani et al., 1998) or serotonergic (Chiang et al., 1999) systems. Further pharmacological, neurophysiological, and genetic manipulations using the behavioral paradigms described here will help elucidate the biological substrates of the attentional mechanisms of interval timing.

12.5.3 IMPLICATIONS FOR THEORIES OF INTERVAL TIMING IN HUMANS AND OTHER ANIMALS

The possibility that the gap procedure engages an attentional mechanism was also recently examined in the human literature (Fortin, this volume; Fortin and Masse, 2000). For example, Fortin and Masse (2000) examined whether single-task procedures, like the gap procedure, engage attention-sharing processes in humans. When human participants were extensively trained with both PI and gap trials, after the gap they delayed their response more at longer pregap durations. These data were taken to suggest that participants developed an expectation of the gap, and that the longer the pregap duration, the more this expectation reduced the attentional resources allocated to timing. Therefore, Fortin and Masse's (2000) data suggest that results from the gap procedure can be interpreted in terms of attention sharing between the timer and other cognitive processes (Thomas and Weaver, 1975; Block and Zakay, 1996; Zakay, 1998, 2000). Together with the results from the experiments described in this chapter these findings support the proposal that the attention-sharing model (Block and Zakay, 1996; Thomas and Weaver, 1975; Zakay, 1989, 2000) can be fruitfully applied to the results from the gap procedure in both humans and other animals. In contrast to possible species differences in interval timing strategies (e.g.,

Bateson, 1998; Brodbeck et al., 1998; Roberts et al., 1989), these findings support an integrative approach to the data patterns obtained in interval timing procedures from rodents, birds, and humans.

ACKNOWLEDGMENTS

This work was supported by grants DA13344 from the National Institute on Drug Abuse and MH65561 from the National Institute for Mental Heath to C.V.B. The author thanks Ioana Scripa for help with editing.

REFERENCES

Al-Zahrani, S.S.A., Al-Ruwaitea, A.S.A., Ho, M.-Y., Bradshaw, C.M., and Szabadi, E., Effect of destruction of noradrenergic neurons with DSP4 on performance on a free-operant timing schedule, *Psychopharmacology*, 136, 235–242, 1998.

Bateson, M., Species Differences in Timing Intervals with Gaps: What Do They Mean? paper presented at the International Conference on Comparative Cognition, Melbourne, FL, March 12–14, 1998.

Bateson, M. and Kacelnik, A., Starling's preferences for predictable and unpredictable delays to food, *Anim. Behav.*, 53, 1129–1142, 1997.

Bateson, M. and Kacelnik, A., Risk-sensitive foraging: decision making in variable environments, in *Cognitive Ecology: The Evolutionary Ecology of Information Processing and Decision Making*, Dukas, R., Ed., Chicago University Press, Chicago, 1998.

Block, R.A. and Zakay, D., Models of psychological time revisited, in *Time and Mind*, Helfrich, H., Ed., Kirkland, WA, Hogrefe und Huber, 1996, pp. 171–195.

Brannon, E., Wolfe, L., Meck, W., and Woldorff, M., Electrophysiological correlates of timing in human infants, *Soc. Neurosci. Abstr.*, 28, 589–11, 2002.

Brodbeck, D.R., Hampton, R.R., and Cheng, K., Timing behaviour of blackcapped chickadees (*Parus atricapillus*), *Behav. Process.*, 44, 183–195, 1998.

Brown, S.W., Time perception and attention: the effects of prospective versus retrospective paradigms and task demands on perceived duration, *Percept. Psychophys.*, 38, 115–124, 1985.

Brown, S.W., Time, change, and motion: the effects of stimulus movement on temporal perception, *Percept. Psychophys.*, 57, 105–116, 1995.

Brown, S.W., Attentional resources in timing: interference effects in concurrent temporal and non-temporal working memory tasks, *Percept. Psychophys.*, 59, 1118–1140, 1997.

Buhusi, C.V. and Meck, W.H., Timing for the absence of a stimulus: the gap paradigm reversed, *J. Exp. Psychol. Anim. Behav. Process.*, 26, 305–322, 2000.

Buhusi, C.V. and Meck, W.H., Differential effects of methamphetamine and haloperidol on the control of an internal clock, *Behav. Neurosci.*, 116, 291–297, 2002.

Buhusi, C.V., Sasaki, A., and Meck, W.H., Temporal integration as a function of signal/gap intensity in rats (*Rattus norvegicus*) and pigeons (*Columba livia*), *J. Comp. Psychol.*, 116, 381–390, 2002.

Cabeza de Vaca, S., Brown, B.L., and Hemmes, N.S., Internal clock and memory processes in animal timing, *J. Exp. Psychol. Anim. Behav. Process.*, 20, 184–198, 1994.

Casini, L. and Macar, F., Effects of attention manipulation on perceived duration and intensity in the visual modality, *Mem. Cognit.*, 25, 812–818, 1997.

Casini, L. and Macar, F., Multiple approaches to investigate the existence of an internal clock using attentional resources, *Behav. Process.*, 45, 73–85, 1999.

Castner, S.A., Williams, G.V., and Goldman-Rakic, P.S., Reversal of antipsychotic-induced working memory deficits by short-term dopamine D_1 receptor stimulation, *Science*, 287, 2020–2022, 2000.

Catania, A.C., Reinforcement schedules and psychophysical judgments: a study of some temporal properties of behavior, in *The Theory of Reinforcement Schedules*, Schoenfeld, W.N., Ed., Appleton-Century-Crofts, New York, 1970, pp. 1–42.

Chiang, T.-J., Al-Ruwaitea, A.S.A., Ho, M.-Y., Bradshaw, C.M., and Szabadi, E., Effect of central 5-hydroxytryptamine depletion on performance in the free-operant psychophysical procedure: facilitation of switching, but no effect on temporal differentiation of responding, *Psychopharmacology*, 143, 166–173, 1999.

Church, R.M., The internal clock, in *Cognitive Processes in Animal Behavior*, Hulse, S.H., Fowler, H., and Honig, W.K., Eds., Erlbaum, Hillsdale, NJ, 1978, pp. 277–310.

Church, R.W., Properties of the internal clock, in *Annals of the New York Academy of Sciences: Timing and Time Perception*, Vol. 423, Gibbon, J. and Allan, L.G., Eds., New York Academy of Sciences, New York, 1984, pp. 566–582.

Crider, A., Solomon, P.R., and McMahon, M.A., Disruption of selective attention in the rat following chronic d-amphetamine administration: relationship to schizophrenic attention disorder, *Biol. Psychiatry*, 17, 351–361, 1982.

Fortin, C. and Masse, N., Expecting a break in time estimation: attentional time-sharing without concurrent processing, *J. Exp. Psychol. Hum. Percept. Perform.*, 26, 1788–1796, 2000.

Frederick, D.L. and Allen, J.D., Effect of selective dopamine D_1- and D_2-agonists and antagonists on timing performance in rats, *Pharmacol. Biochem. Behav.*, 53, 759–764, 1996.

Gainetdinov, R.R., Wetsel, W.C., Jones, S.R., Levin, E.D., Jaber, M., and Caron, M.G., Role of serotonin in the paradoxical calming effect of psychostimulants on hyperactivity, *Science*, 283, 397–401, 1998.

Gallistel, C.R., *The Organization of Learning*, MIT Press, Cambridge, MA, 1990.

Gibbon, J., Scalar expectancy theory and Weber's law in animal timing, *Psychol. Rev.*, 84, 279–325, 1977.

Gibbon, J., Church, R.M., and Meck, W.H., Scalar timing in memory, in *Annals of the New York Academy of Sciences: Timing and Time Perception*, Vol. 423, Gibbon, J. and Allan, L.G., Eds., New York Academy of Sciences, New York, 1984, pp. 52–77.

Gray, J.A., Buhusi, C.V., and Schmajuk, N.A., The transition from automatic to controlled processing, *Neural Networks*, 10, 1257–1268, 1997.

Gray, N.S., Pickering, A.D., Hemsley, D.R., and Dawling, S., Abolition of latent inhibition by a single 5 mg dose of d-amphetamine in man, *Psychopharmacology*, 107, 425–430, 1992.

Grondin, S., Meilleur-Wells, G., Oullette, C., and Macar, F., Sensory effects on judgments of short time-intervals, *Psychol. Res.*, 61, 261–268, 1998.

Hicks, R.E., Miller, G.W., Gaes, G., and Bierman, K., Concurrent processing demands and the experience of time-in-passing, *Am. J. Psychol.*, 90, 431–446, 1997.

Hicks, R.E., Miller, G.W., and Kinsbourne, M., Prospective and retrospective judgments of time as a function of amount of information processed, *Am. J. Psychol.*, 89, 719–730, 1976.

Horvitz, J.C., Stewart, T., and Jacobs, B.L., Burst activity of ventral tegmental dopamine neurons is elicited by sensory stimuli in the awake cat, *Brain Res.*, 759, 251–259, 1997.

Iwanami, A., Kanamori, R., Suga, I., and Kaneko, T., Reduced attention-related negative potentials in methamphetamine psychosis, *J. Nerv. Ment. Dis.*, 183, 693–697, 1995.

Kraemer, P.J., Brown, R.W., and Randall, C.K., Signal intensity and duration estimation in rats, *Behav. Process.*, 44, 265–268, 1995.

Kraemer, P.J., Randall, C.K., and Brown, R.W., The influence of stimulus attributes on duration matching-to-sample in pigeons, *Anim. Learn. Behav.*, 25, 148–157, 1997a.

Kraemer, P.J., Randall, C.K., Dose, J.M., and Brown, R.W., Impact of d-amphetamine on temporal estimation in pigeons tested with a production procedure, *Pharmacol. Biochem. Behav.*, 58, 323–327, 1997b.

Lau, C.E., Ma, F., Foster, D.M., and Falk, J.L., Pharmacokinetic-pharmacodynamic modeling of the psychomotor stimulant effect of cocaine after intravenous administration: timing performance deficits, *J. Pharmacol. Exp. Ther.*, 288, 535–543, 1999.

Lee, R.S., Koob, G.F., and Henriksen, S.J., Electrophysiological responses of nucleus accumbens neurons to novel stimuli and exploratory behavior in the awake, unrestrained rat, *Brain Res.*, 799, 317–322, 1998.

Lejeune, H., Switching or gating? The attentional challenge in cognitive models of psychological time, *Behav. Process.*, 44, 127–145, 1998.

Lejeune, H., Macar, F., and Zakay, D., Attention and timing: dual-task performance in pigeons, *Behav. Process.*, 45, 141–157, 1999.

Lejeune, H. and Wearden, J.H., The comparative psychology of fixed-interval responding: some quantitative analyses, *Learn. Motiv.*, 22, 84–111, 1991.

Macar, F., Grondin, S., and Casini, L., Controlled attention sharing influences time estimation, *Mem. Cognit.*, 22, 673–686, 1994.

Mantanus, H., Empty and filled interval discrimination by pigeons, *Behav. Anal. Lett.*, 1, 217–224, 1981.

Maricq, A.V. and Church, R.M., The differential effects of haloperidol and methamphetamine on time estimation in the rat, *Psychopharmacology*, 79, 10–15, 1983.

Maricq, A.V., Roberts, S., and Church, R.M., Methamphetamine and time estimation, *J. Exp. Psychol. Anim. Behav. Process.*, 7, 18–30, 1981.

Matell, M.S., King, G.R., and Meck, W.H., Differential adjustment of interval timing by the chronic administration of intermittent or continuous cocaine, submitted.

Matell, M.S. and Meck, W.H., Neuropsychological mechanisms of interval timing behaviour, *Bioessays*, 22, 94–103, 2000.

McKetin, R. and Solowij, N., Event-related potential indices of auditory selective attention in dependent amphetamine users, *Biol. Psychiatry*, 45, 1488–1497, 1999.

McKetin, R., Ward, P.B., Catts, S.V., Mattick, R.P., and Bell, J.R., Changes in auditory selective attention and event-related potentials following oral administration of D-amphetamine in humans, *Neuropsychopharmacology*, 21, 380–390, 1999.

Meck, W.H., Selective adjustments of the speed of the internal clock and memory processes, *J. Exp. Psychol. Anim. Behav. Process.*, 9, 171–201, 1983.

Meck, W.H., Affinity for the dopamine D_2 receptor predicts neuroleptic potency in decreasing the speed of an internal clock, *Pharmacol. Biochem. Behav.*, 25, 1185–1189, 1986.

Meck, W.H., Neuropharmacology of timing and time perception, *Cognit. Brain Res.*, 3, 227–242, 1996.

Meck, W.H., Church, R.M., and Olton, D.S., Hippocampus, time, and memory, *Behav. Neurosci.*, 98, 3–22, 1984.

Müller, U., von Cramon, D.Y., and Pollmann, S., D_1- versus D_2-receptor modulation of visuospatial working memory in humans, *J. Neurosci.*, 18, 2720–2728, 1998.

Paule, M.G., Meck, W.H., McMillan, D.E., Bateson, M., Popke, E.J., Chelonis, J.J., and Hinton, S.C., The use of timing behaviors in animals and humans to detect drug and/or toxicant effects, *Neurotoxicol. Teratol.*, 21, 491–502, 1999.

Richelle, M. and Lejeune, H., *Time in Animal Behaviour*, Pergamon, Oxford, U.K., 1980.

Roberts, S., Isolation of an internal clock, *J. Exp. Psychol. Anim. Behav. Process.*, 7, 242–268, 1981.

Roberts, W.A., Cheng, K., and Cohen, J.S., Timing light and tone signals in pigeons, *J. Exp. Psychol. Anim. Behav. Process.*, 15, 23–35, 1989.

Roberts, S. and Church, R.M., Control of an internal clock, *J. Exp. Psychol. Anim. Behav. Process.*, 4, 318–337, 1978.

Santi, A., Weise, L., and Kuiper, D., Amphetamine and memory for the event duration in rats in pigeons: disruption of attention to temporal samples rather than changes in the speed of the internal clock, *Psychobiology*, 23, 224–232, 1995.

Schultz, W., Apicella, P., and Ljungberg, T., Responses of monkey dopamine neurons to reward and conditioned stimuli during successive steps of learning a delayed response task, *J. Neurosci.*, 13, 900–913, 1993.

Schultz, W., Dayan, P., and Montague, P.R., A neural substrate of prediction and reward, *Science*, 275, 1593–1599, 1997.

Stanford, L. and Santi, A., The dopamine D_2 agonist quinpirole disrupts attention to temporal signals without selectively altering the speed of the internal clock, *Psychobiology*, 26, 259–266, 1998.

Talton, L.E., Higa, J.J., and Staddon, J.E.R., Interval schedule performance in the goldfish *Carassius auratus*, *Behav. Process.*, 45, 193–206, 1999.

Thomas, E.A.C. and Weaver, W.B., Cognitive processing and time perception, *Percept. Psychophys.*, 17, 363–367, 1975.

Treisman, M., Temporal discrimination and the indifference interval: implications for a model of the "internal clock," *Psychol. Monogr. Gen. Appl.*, 77, 1–31, 1963.

Wilkie, D.M., Saksida, L.M., Samson, P., and Lee, A., Properties of time-place learning by pigeons, *Columbia livia*, *Behav. Process.*, 31, 39–56, 1994.

Zakay, D., Subjective time and attentional resource allocation: an integrated model of time estimation, in *Time and Human Cognition: A Life-Span Perspective*, Levin, I. and Zakay, D., Eds., Elesevier/North-Holland, Amsterdam, 1989, pp. 365–397.

Zakay, D., Gating or switching? Gating is a better model of prospective timing (a response to 'switching or gating?' by Lejeune), *Behav. Process.*, 50, 1–7, 2000.

Zakay, D., Nitzan, D., and Glickson, J., The influence of task difficulty and external tempo on subjective time estimation, *Percept. Psychophys.*, 34, 451–456, 1983.

13 Electrophysiological Correlates of Interval Timing

Shogo Sakata and Keiichi Onoda

CONTENTS

13.1 Introduction 339
13.2 Electroencephalogram and Interval Timing Behavior...............................340
    13.2.1 How to Measure the EEG during Interval Timing Tasks in Rats..341
    13.2.2 Correlation between EEG and Learning 343
    13.2.3 Implications of Electrophysiological Activity 343
13.3 Hippocampal Theta Wave 344
    13.3.1 Analysis of Hippocampal Theta 344
    13.3.2 Hippocampal Theta and Interval Timing...344
    13.3.3 Function of the Hippocampus in Models of Interval Timing 345
13.4 Discussion 346
Acknowledgments..347
References..348

13.1 INTRODUCTION

There are many interesting aspects of timing and time perception in living animals. It goes without saying that timing is one of the most important properties of behavior. The behavioral output from the brain always includes timing mechanisms. The brains of animals process many sensory inputs from the environment through the eyes, ears, nose, mouth, and skin and control the timing of these sensory processes. Where are the primary regions in the brain for processing this sensory input and making decisions about timing? The purpose of this study is to evaluate the roles of various brain regions in timing and time perception using electrophysiological field potentials. The foci of interest in these experiments are the frontal cortex, hippocampus, and cerebellum. These brain areas correspond to the major components of the information-processing model of scalar timing theory proposed by Gibbon et al. (1984) as well as the neuropsychological findings of Ivry and his colleagues (e.g., Diedrichsen et al., this volume). A version of the information-processing model of

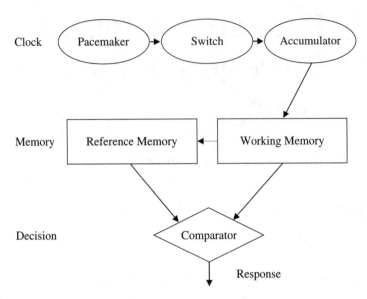

FIGURE 13.1 A summary of the information-processing model of timing adapted from Gibbon, J. et al., in *Annals of the New York Academy of Sciences: Timing and Time Perception,* Vol. 423, Gibbon, J. and Allan, L., Eds., New York Academy of Sciences, New York, 1984, 52–77.

scalar timing theory is illustrated in Figure 13.1. More recently, it has been proposed that the timing of durations in the seconds-to-minutes range involves frontal-striatal circuits (Malapani and Rakitin, this volume; Matell et al., this volume; MacDonald and Meck, this volume; Meck, 1996; Meck and Benson, 2002). In this chapter we will discuss the different stages of temporal processing that may involve separate brain regions in the frontal cortex, hippocampus, and cerebellum.

13.2 ELECTROENCEPHALOGRAM AND INTERVAL TIMING BEHAVIOR

Many studies have been conducted in the past decade to measure the electroencephalogram (EEG) during discrimination task performance in rats (e.g., Ehlers et al., 1999; Givens and Olton, 1994; Sato and Sakata, 1999). Virtually all of these studies have involved simple auditory or visual memory tasks or discriminative learning tasks not involving temporal discrimination or timing. Only a few studies have investigated interval timing behavior and EEG correlations in animals (e.g., Onoda and Sakata, submitted). The event-related potentials (ERPs) are voltage fluctuations that are associated in time with some physiological or mental occurrence (e.g., Picton et al., 2000). In human studies, the timing of stimuli is a very important factor affecting ERPs, even more so for animal studies. Temporal discrimination involves a variety of complex mental operations in animals. Consequently, it is useful to conduct ERP analysis for studying the brain mechanisms involved in timing and time perception.

13.2.1 How to Measure the EEG during Interval Timing Tasks in Rats

There are three major methods to study time perception in animals: temporal discrimination such as the temporal bisection procedure, temporal differentiation procedures, and peak-interval reproduction procedures (Catania, 1970). The temporal bisection procedure deals with the concept of time using short and long anchor durations. The bisection point refers to the point of subjective equality in timescale, which indicates the durations that subjects classify as equally short or long. The temporal differentiation procedure deals with the decision to judge the appropriate point in time to make a response. It requires that a subject suppress its response until a criterion time has been reached. This procedure has both strengths and weaknesses for the study of interval timing behavior; one strength is that it is easy to train, but one weakness is that it is difficult to evaluate the cause of premature responses. The peak-interval procedure is perhaps the best method for studying interval timing behavior in animals. The behavioral responses (e.g., lever presses) occur more naturally and reflect the temporal expectancy of the availability of reinforcement.

EEG is reflected by the brain activity associated with certain decision processes. Typically, it is used in conjunction with the study of stimulus discrimination. The stimulus comes to have meaning for the animal with repeated experience. Changes in brain responsiveness to an external stimulus can be investigated by the measurement of evoked potentials. The evoked potential will increase its amplitude by presenting the stimulus in combination with certain experiences (e.g., Takeuchi et al., 2000). In the same way, an internal stimulus such as the judgment of a temporal criterion will evoke an ERP. Unfortunately, because of the variety of response artifacts, our method for recording ERPs is more easily adapted to a simple temporal discrimination procedure than to the peak-interval procedure. Consequently, we trained rats to make a left response following a 2-sec duration and a right response following an 8-sec duration. We then used this temporal discrimination method to obtain the correlation with ERPs.

In order to record ERPs, one needs to surgically implant chronically indwelling microelectrodes that are targeted as specific regions in the brain. We selected the target regions based on the proposal that the frontal cortex, hippocampus, and the cerebellum are involved in timing (see Meck, 1996, this volume). After sufficient training of the temporal discrimination task, we implanted the electrodes (0.2 mm in diameter) in rats. The stimuli were 2- or 8-sec 2000-Hz, 80-dB tones. The EEG sampling phases were the onset and offset of the stimulus, as shown in Figure 13.2.

In this study we trained nine male rats. All rats were trained on the same temporal discrimination task (T-task) in the first stage of training. After the rats maintained over 85% of correct responses in three consecutive sessions, rats were implanted with electrodes. The EEG was recorded during the T-task in two consecutive sessions, and then rats were trained a simple reaction time control task (C-task). In the C-task, only a 2-sec tone was presented and only one lever was inserted into the box. This simple reaction time task was used as a control for both timing and motor response. The mean ERP for the nine rats is presented in Figure 13.3.

342　　　　　　　　　　　　　　　Functional and Neural Mechanisms of Interval Timing

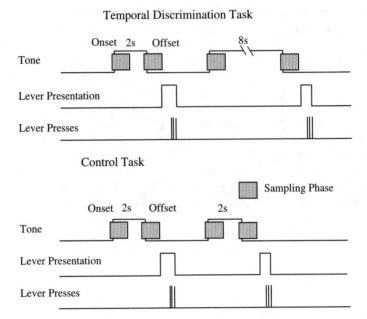

FIGURE 13.2 The experimental procedure involved a temporal discrimination task (2 vs. 8 sec) and a simple reaction time task. ERPs were corrected in a sampling phase.

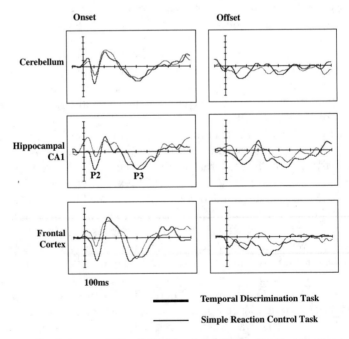

FIGURE 13.3 Grand average ERPs elicited by the auditory stimuli at the onset and offset (n = 9). Temporal discrimination task corrected by 2- and 8-sec auditory stimuli. Simple reaction time task corrected by only 2-sec auditory stimulus. Negative is up.

The ERPs consisted of N1, P2, and P3 components. In the onset ERPs, P3 components were consistently observed for the frontal cortex, hippocampal CA1, and cerebellar cortex in the T-task. The P2 component showed a significant main effect of amplitude that was larger for the T-task than for the C-task. For all statistical tests, the significance level was set at .05. A significant main effect of region for the P2 component latency was obtained with analysis of variance (ANOVA) in three brain regions across the two tasks. Post hoc tests revealed that the P2 component latency was longer for the frontal cortex than for the hippocampus and the cerebellum.

13.2.2 CORRELATION BETWEEN EEG AND LEARNING

It is well known that there is a strong correlation between EEG frequency and various sleep–wake states. EEG frequency is one of the indices used to measure states of arousal. The pattern of electrical activity in a fully awake animal is a mixture of many frequencies dominated by waves of relatively fast frequencies and low amplitude. The EEG will change to low frequencies and high amplitude depending on the stage of sleep (Datta and Hobson, 2000). Visual evoked potentials (VEPs) in the rat showed the same pattern of state-dependent changes in amplitude (Meeren et al., 1998). On the other hand, there have been relatively few studies of cognitive activity based on the analysis of the ongoing EEG. Sato and Sakata (1999) recorded EEG from the rat during the performance of a delayed nonmatching-to-sample task. Between the hippocampal CA1 and the entorhinal cortex, and the CA1 and the cingulated cortex, EEG synchronization was higher in three task-related phases (sample, delay, and comparison) than in the control phase. There are few data showing a clear correlation between the EEG and learning processes in spite of the fact that many investigators believe this correlation to be present. It is also unclear what the relationship is between neural firing and EEG as a field potential. In the analysis of neural firing associated with learning, it is important to take into account the ongoing background EEG as a reflection of the current state.

13.2.3 IMPLICATIONS OF ELECTROPHYSIOLOGICAL ACTIVITY

The temporal and spatial information provided by EEG may be used to understand how the brain implements behavioral change, feeling, and memory. Electrophysiological activity is the origin of behavior or processing by the brain. We can measure the EEG when animals are performing specific tasks. What does the EEG reflect, input, or output? Many investigators have tended to assume the origin of EEG activity is the mind, the so-called "source of controlled, observable variability" (e.g., Picton et al., 2000). How about a timing event? "Time flies when you're having fun," is an often-quipped adage that demonstrates our sensitivity to the time course of events in our everyday lives (Matell and Meck, 2000). How could we explain interval timing by neural mechanisms? Mattel and Meck (2000) proposed a real-time neuropsychological model, the striatal beat frequency (SBF) model, which consists of oscillating neurons. The SBF model serves as a bridge between the firing neurons and field potentials. The point of the SBF model is to identify the clock process of the information-processing model of interval timing. Nevertheless, it is still difficult to envision to what extent timing plays a role in behavior without additional neural indices.

13.3 HIPPOCAMPAL THETA WAVE

Hippocampal theta cells can be distinguished from theta interneurons in freely moving rats on the basis of their electrophysiological properties (Wiebe and Staubli, 2001). The hippocampal theta wave is called theta rhythm or rhythmic slow activity (RSA). Theta rhythm is a local field potential known as a sinusoidal-like EEG signal occurring at frequencies within the bandwidth from 6 to 12 Hz in rats. There are many studies showing a strong correlation between theta rhythm and voluntary motor behaviors such as running, rearing, and jumping (e.g., Bland, 1986; Whishaw and Vanderwolf, 1973). The correlation between the temporal dynamics of theta frequency and nonsteady wheel-running speed within a single trial tends to be positive (Shin and Talnov, 2001). A few studies have reported that learning-related theta cell firing correlates well with performance in simple perceptual discrimination tasks (e.g., Deadwyler et al., 1996), as well as task-related theta frequency changes associated with performance in an auditory discrimination (e.g., Brankack et al., 1996).

13.3.1 ANALYSIS OF HIPPOCAMPAL THETA

The hippocampal theta has long been considered to be a reflection of the neural processing occurring in the structures of the hippocampus. In the analysis of the ongoing hippocampal theta, there are two ways to conduct a time series analysis; one is to remove random variation by averaging and to derive the ERPs. These potentials can be recorded in front and back of the stimulus presentation point from specific brain regions during task performance in rats. ERPs are extracted from the ongoing EEG by means of filtering and signal averaging. ERPs can be evaluated in both frequency and time domains at the point of stimulus presentation. The other method is to evaluate the time series by a power frequency analysis. In this analysis, it is customary to subdivide the EEG signals into quasi-stationary epochs and to characterize them by a number of statistical parameters, such as frequency spectra. The power spectrum gives the distribution of the squared amplitude of different frequency components. These spectral distributions are calculated using the Fast Fourier transform.

13.3.2 HIPPOCAMPAL THETA AND INTERVAL TIMING

It has been proposed that the hippocampus serves as a working memory buffer for many types of information, including temporal information (e.g., Meck et al., 1984). We recorded hippocampal theta in both the T-task and C-task. The rat had to evaluate the duration of auditory stimuli in the former task, either 2- or 8-sec durations, and choose to respond to one of the two levers. On the other hand, the rat does not need to attend to the duration of the stimulus in the latter task; only a 2-sec tone is presented, and the rat is only required to respond on one lever. No significant differences were observed between reaction times in the T-task and the C-task (724 ± 70 and 757 ± 122 msec, respectively; $F(1, 8) = 0.13$, not significant (n.s.)). Those

Electrophysiological Correlates of Interval Timing   345

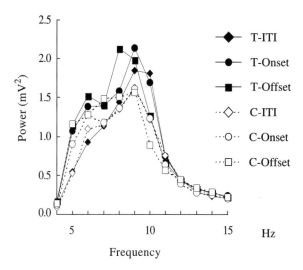

FIGURE 13.4 Mean power spectra of the hippocampal CA1. T refers to the temporal discrimination task, and C indicates the simple reaction time task used as a control. Onset, 1-sec duration after tone onset; offset, 1-sec duration after tone offset.

spectral distributions of hippocampal theta were calculated by fast Fourier transforms (FFT) to compare the involvement of working memory in both tasks. We removed the data for one animal because of contamination by artifacts. We also calculated hippocampal power spectra in three conditions: onset of the auditory stimuli, offset of tones, and intertrial intervals (ITIs). These results are shown in Figure 13.4.

The data show a clear peak in power at a frequency range between 6 and 12 Hz. In the statistical analysis, there were two main effects. First, the hippocampal theta power increases more in the T-task than in the C-task at the onset of the auditory stimuli ($F(1, 7) = 17.8$, $P < .01$). Second, the peak frequency shifted for higher ranges in the T-task than in the C-task ($F(2, 14) = 5.90$, $P < .05$). There were no significant differences among onset, offset, and ITIs in the C-task ($F(2, 14) = 1.89$, n.s.).

13.3.3 FUNCTION OF THE HIPPOCAMPUS IN MODELS OF INTERVAL TIMING

There have been many theories of hippocampal function, timing, and theta rhythm presented in the literature (e.g., Grossberg and Schmajuk, 1989; Kesner, 2002; Meck, 2002a, 2002b; Schmajuk, 1990). The hippocampal system is related to the formation of episodic memory of space, time, objects, etc. The main function of the hippocampus appears to be some form of memory storage or consolidation. In the information-processing model of interval timing, the hippocampus is thought to be primarily involved in working memory (e.g., Meck et al., 1984). As indicated above, Gibbon et al. (1984) proposed three main processing stages in their model: clock, memory, and decision stages. The clock stage is hypothesized to consist of a pacemaker that emits pulses that are transferred to an integrator through a switch.

The brain areas proposed to be involved in this clock stage have included the striatum (Mattel and Meck, 2000; Matell et al., this volume) and the cerebellum (Diedrichsen et al., this volume; Gibbon et al., 1997; Ivry, 1996; Ivry and Richardson, 2002). The memory stage is as a proposed mechanism for clock input to be transferred to reference memory. In the transfer of temporal information to memory, the encoding process first needs to reset the working memory at the beginning of the trial. The final stage is the comparison stage, which makes the decision to respond. In the T-task of this experiment, the decision occurs at a point in time after the offset of auditory stimuli. Rats must make a decision to choose either the left or the right response lever. After the 2-sec stimulus presentation, rats will make a judgment at the offset of the signal. However, in the case of the 8-sec stimulus, rats will make a judgment sometime during the stimulus presentation after the 2-sec criterion has been passed. Consequently, we should compare ERPs following the 2-sec stimulus in both the T-task and C-task. We can evaluate the differences between these two tasks by using the offset ERP.

13.4 DISCUSSION

The purpose of this study was to evaluate the roles of specific brain regions that may be involved in a temporal discrimination task (e.g., 2 vs. 8 sec) by using ERPs and hippocampal theta spectra. The data from the temporal discrimination task were compared to a simple reaction time control task that did not require animals to make a temporal discrimination. We discuss the brain areas involved in temporal processing with attention to the frontal cortex, hippocampus, and cerebellum in order to evaluate the information-processing model proposed by Gibbon et al. (1984). The ERPs recorded from the three regions following stimulus onset consisted of N1, P2, and P3 components, as shown in Figure 13.3. There were no significant differences in N1 components. There was a main effect of task in the P2 component amplitude that was larger for the T-task than for the C-task. Interestingly, the hippocampus and the cerebellum seem to show almost the same amplitude and latency. These results suggest that the cerebellum is involved in temporal processing of durations in the seconds range. Meck et al. (Mattel and Meck, 2000; Matell et al., this volume; Meck, 1996; Meck and Benson, 2002) have proposed that timing for durations in the seconds-to-minutes range is subserved by a different system, involving frontal-striatum circuits. Little attention has been given to the relation between the cerebellar cortex and these frontal-striatal circuits, although there are projections from the cerebellum to the basal ganglia (e.g., Middleton and Strick, 1994). It is debatable whether the cerebellum and frontal-striatal circuits are independent timing systems or work in conjunction with each other (Diedrichsen et al., this volume; Ivry, 1996; Ivry and Richardson, 2002).

The P2 components in the hippocampus were prominent during the T-task, and the P2 latency was almost the same time for the hippocampus and the cerebellum. This result suggested that the information processing related to working memory might be carried out simultaneously in the course of the task and be observed concurrently with the activation of the clock mechanism. The idea here is that certain aspects of memory processing may be required for the initialization of working memory at the beginning of a trial. The P2 latency was significantly longer for the recordings in the frontal cortex than for the hippocampus and the cerebellum. This result showed that the information processing in the frontal cortex was delayed and might involve the integration of temporal processing as represented by the comparison stage (see Meck, this volume; Meck et al., 1987).

The P3 components were observed in all three brain regions at the time of stimulus onset. We used a time window from 300 to 600 msec in positive peak to define the P3 component. The P3 components are relatively long-latency ERPs and have commonly been employed to investigate cognitive processing in a variety of tasks. This component could be elicited by task-relevant orienting stimuli. The tone onset indicates the opportunity for reward in both tasks. There were no significant differences in the P3 components at the onset of the signal for both tasks. In contrast, the ERPs following stimulus offset consisted of the observed positive potential for the frontal cortex and the hippocampus. The mean amplitude of the positive potential had a tendency to be larger for the T-task than for the C-task. This result suggests that the frontal cortex mediates the decision/comparison stage of temporal processing. This component could not be thought of as the same as the onset P3 component. The P3-like component is typically elicited by an "oddball task" in rats (e.g., Jodo et al., 1995). The onset P3 most likely reflects the detection of the reward value of the signal, whereas the offset P3 likely reflects decision making about the duration of the signal. The onset P3 amplitudes were larger than the offset P3 amplitudes. It is important to investigate these offset ERPs in spite of their lower amplitude. In future studies, we propose that it would be better to design experiments in which the period defined by the onset starting stimulus to the onset ending stimulus is used to evaluate the processing of stimulus duration.

The hippocampal theta power increased more in the T-task than in the C-task at the onset of stimuli. The peak frequency shifted toward a higher range in the T-task than in the C-task, as shown in Figure 13.4. These results suggest that the hippocampus was initialized for temporal processing by the resetting of working memory at stimulus onset. According to the information-processing model outlined above, the pacemaker will generate pulses that go through a switch into an accumulator. On a new trial, the timing process may need to initialize or reset working memory before transferring pulses from the accumulator to the hippocampus. At this point, the hippocampus continues working to process the current temporal memories that may reflect hippocampal theta activity (e.g., Meck, 2000b). It is clear that frontal activation related to temporal processing at the beginning of the interval is also observed. This is the first study to investigate the correlations between EEG and proposed interval timing mechanisms in animals (for a description of related work in humans, see Pouthas, this volume). It will be necessary to gather more EEG data to obtain a better understanding of the relations between the neuronal activity in various brain structures and temporal discrimination processes.

ACKNOWLEDGMENTS

We thank Erika Takahashi for her assistance in the research reported here. This research was supported by Grant-in-Aid for Scientific Research (C) 13610090 from the Japan Society for Promotion of Science to Shogo Sakata. We are grateful to Warren Meck for his helpful comments on the manuscript. All correspondence should be addressed to Dr. Shogo Sakata, Department of Behavioral Sciences, Hiroshima University, Higashi-Hiroshima, 739-8521, Japan. E-mail: ssakata@hiroshima-u.ac.jp

REFERENCES

Bland, B.H., The physiology and pharmacology of hippocampal formation theta rhythms, *Prog. Neurobiol.*, 26, 1–54, 1986.

Brankack, J., Seidenbecher, T., and Muller, G.H.W., Task-relevant late positive component in rats: is it related to hippocampal theta rhythm? *Hippocampus*, 6, 475–482, 1996.

Catania, A.C., Reinforcement schedules and psychophysical judgments: a study of some temporal properties of behavior, in *The Theory of Reinforcement Schedules*, Schoenfeld, W.N., Ed., Appleton-Century-Crofts, New York, 1970, pp. 1–42.

Datta, S. and Hobson, J.A., The rat as an experimental model for sleep neurophysiology, *Behav. Neurosci.*, 114, 1239–1244, 2000.

Deadwyler, S.A., Bunn, T., and Hampson, R.E., Hippocampal ensemble activity during spatial delayed-nonmatch-to-sample performance in rats, *J. Neurosci.*, 16, 354–372, 1996.

Ehlers, C.L., Somes, C., Lumeng, L., and Li, T.K., Electrophysiological response to neuropeptide Y (NPY) in alcohol-naive preferring and non-preferring rats, *Pharmacol. Biochem. Behav.*, 63, 291–299, 1999.

Gibbon, J., Church R.M., and Meck, W.H., Scalar timing in memory, in *Annals of the New York Academy of Sciences: Timing and Time Perception*, Vol. 423, Gibbon, J. and Allan, L., Eds., New York Academy of Sciences, New York, 1984, pp. 52–77.

Gibbon, J., Malapani, C., Dale, C.L., and Gallistel, C.R., Toward a neurobiology of temporal cognition: advances and challenges, *Curr. Opin. Neurobiol.*, 7, 170–184, 1997.

Givens, B. and Olton, D.S., Local modulation of basal forebrain: effects on working and reference memory, *J. Neurosci.*, 14, 3578–3587, 1994.

Grossberg, S. and Schmajuk, N.A., Neural dynamics of adaptive timing and temporal discrimination during associative learning, *Neural Networks*, 2, 79–102, 1989.

Ivry, R.B., The representation of temporal information in perception and motor control, *Curr. Opin. Neurobiol.*, 6, 851–857, 1996.

Ivry, R.B. and Richardson, T.C., Temporal control and coordination: the multiple timer model, *Brain Cognit.*, 48, 117–132, 2002.

Jodo, E., Takeuchi, S., and Kayama, Y., P3b-like potential of rats recorded in an active discrimination task, *Electroencephalogr. Clin. Neurophysiol.*, 96, 555–560, 1995.

Kesner, R.P., Neural mediation of memory for time: role of the hippocampus and medial prefrontal cortex, in *Animal Cognition and Sequential Behavior: Behavioral, Biological, and Computational Perspectives*, Fountain S.B., Bunsey, M.D., Danks, J.H., and McBeath, M.K., Eds., Kluwer Academic Press, Boston, 2002, pp. 201–226.

Matell, M.S. and Meck, W.H., Neuropsychological mechanisms of interval timing behavior, *Bioessays*, 22, 94–103, 2000.

Meck, W.H., Neuropharmacology of timing and time perception, *Cognit. Brain Res.*, 3, 227–242, 1996.

Meck, W.H., Choline uptake in the frontal cortex is proportional to the absolute error of a temporal memory translation constant in mature and aged rats, *Learn. Motiv.*, 33, 88–104, 2002a.

Meck, W.H., Distortions in the content of temporal memory: neurobiological correlates, in *Animal Cognition and Sequential Behavior: Behavioral, Biological, and Computational Perspectives*, Fountain S.B., Bunsey, M.D., Danks, J.H., and McBeath, M.K., Eds., Kluwer Academic Press, Boston, 2002b, pp. 175–200.

Meck, W.H. and Benson, A.M., Dissecting the brain's internal clock: how frontal-striatal circuitry keeps time and shifts attention, *Brain Cognit.*, 48, 195–211, 2002.

Meck, W.H., Church, R.M., and Olton, D.S., Hippocampus, time, and memory, *Behav. Neurosci.*, 98, 3–22, 1984.

Meck, W.H., Church, R.M., Wenk, G.L., and Olton, D.S., Nucleus basalis magnocellularis and medial septal area lesions differentially impair temporal memory, *J. Neurosci.*, 7, 3505–3511, 1987.

Meeren, H.K., Van Luijtelaar, E.L., and Coenen, A.M., Cortical and thalamic visual evoked potentials during sleep-wake states and spike-wave discharges in the rat, *Electroencephalogr. Clin. Neurophysiol.*, 108, 306–319, 1998.

Middleton, F.A. and Strick, P.L., Anatomical evidence for cerebellar and basal ganglia involvement in higher cognitive function, *Science*, 266, 458–461, 1994.

Onoda, K. and Sakata, S., Event-related potentials in the frontal cortex, hippocampus, and cerebellum during temporal discrimination task in rats, submitted.

Picton, T.W., Bentin, S., Berg, P., Donchin, E., Hillyard, S.A., Johnson, J.R., Miller G.A., Ritter, W., Ruchkin, D.S., Rugg, M.D., and Taylor, M.J., Guidelines for using human event-related potentials to study cognition: recording standards and publication criteria, *Psychophysiology*, 37, 127–152, 2000.

Sato, N. and Sakata, S., Hippocampal theta activity during delayed nonmatching-to-sample performance in rats, *Psychobiology*, 27, 331–340, 1999.

Schmajuk, N.A., Role of the hippocampus in temporal and spatial navigation: an adaptive neural network, *Behav. Brain Res.*, 39, 205–229, 1990.

Shin, J. and Talnov, A., A single trial analysis of hippocampal theta frequency during nonsteady wheel running in rats, *Brain Res.*, 897, 217–221, 2001.

Takeuchi, S., Jodo, E., Suzuki, Y., Matsuki, T., Hoshino, K.Y., Niwa, S.I., and Kayama, Y., ERP development in the rat in the course of learning two-tone discrimination task, *Neuroreport*, 11, 333–336, 2000.

Whishaw, I.Q. and Vanderwolf, C.H., Hippocampal EEG and behavior: changes in amplitude and frequency of RSA (theta rhythm) associated with spontaneous and learned movement patterns in rats and cats, *Behav. Biol.*, 8, 461–484, 1973.

Wiebe, S.P. and Staubli, U.V., Recognition memory correlates of hippocampal theta cells, *J. Neurosci.*, 21, 3955–3967, 2001.

14 Importance of Frontal Motor Cortex in Divided Attention and Simultaneous Temporal Processing

Kevin C.H. Pang and J. Devin McAuley

CONTENTS

14.1 Introduction 351
14.2 Short-Interval Timing..352
    14.2.1 Scalar Expectancy Theory 353
    14.2.2 Peak-Interval and Simultaneous Temporal Processing Tasks 354
    14.2.3 Basic Findings...355
    14.2.4 Effects of Neurobiological Manipulations 355
    14.2.5 The Attentional Switch Hypothesis 357
    14.2.6 Summary 358
14.3 Divided Attention and the Frontal Cortex 358
    14.3.1 Functional Imaging Studies 358
    14.3.2 Effects of Brain Damage 359
    14.3.3 Summary 360
14.4 Recent Electrophysiological Recordings in the Frontal Cortex..................360
    14.4.1 Evidence for Divided Attention Neurons 360
    14.4.2 Alternative Explanations 362
14.5 Conclusions 365
Acknowledgments..365
References..366

14.1 INTRODUCTION

Research on attention in the context of interval timing has a well-established history (Brown, 1985; Hicks et al., 1977; Macar et al., 1994; Thomas and Weaver, 1975; Zakay and Block, 1996). Despite this history, attention is not a well-defined concept.

General definitions of attention usually include reference to the speed, efficiency, or depth of processing, with the prediction that attended stimuli are processed more quickly, efficiently, and deeply than unattended stimuli (Sternberg, 1999). Within the field of attention research, one important distinction is between *selective attention* and *divided attention*. Selective attention requires a subject to selectively process (focus attention on) one of several possible stimuli, whereas divided attention requires a subject to coordinate the processing of multiple sources of information.

To assess divided attention, many researchers have used what are typically referred to as *dual-task paradigms* (Johnston et al., 1995; Pashler, 1993). Dual-task paradigms involve the concurrent performance of two tasks that vary in complexity. The extent to which the tasks interfere is taken as a measure of the extent to which the two tasks involve common processes (or shared attentional resources). Studies of divided attention using dual-task paradigms have addressed a range of questions, including the types of stimuli or tasks that interfere with one another, the likely site of dual-task interference (in terms of stage of information processing), and under what conditions (e.g., sleep deprivation, drug influence, general fatigue) attention is most impaired (Pashler and Johnston, 1998). The focus of this chapter is on the role of frontal cortical areas in divided attention and simultaneous temporal processing (STP).

Simultaneous temporal processing refers to the concurrent performance of two or more timing tasks. For example, subjects in a typical STP task might be asked to simultaneously reproduce two different durations, each associated with a different cue (e.g., light or tone). Interval timing tasks have often been incorporated into studies of divided attention because timing performance is sensitive to different types of behavioral and neurobiological manipulations of attention (Macar et al., 1994; Meck, 1996; Meck and Williams, 1997; Olton et al., 1988). The primary goal of this chapter is to explore some ideas about the neurobiology of timing and divided attention within the context of the STP paradigm. Toward this end, we provide evidence for neural correlates of divided attention obtained from single-cell recordings of the frontal motor cortex of the rat.

The chapter is divided into three sections. First, we review relevant theory and data on short-interval timing. Next, we consider the role of frontal cortical areas in divided attention within the context of information-processing theories of timing. Finally, we describe a recent electrophysiological study from our lab that supports the idea that neurons in the frontal motor cortex are directly involved in divided attention. We then consider some alternative interpretations of these data and suggest some directions for future research.

14.2 SHORT-INTERVAL TIMING

Timing is a guiding force in behavior. Accurate and flexible perception of timing affords predictions about when events in the environment are likely to occur and how long they are expected to last. In the literature, short-interval timing refers specifically to event durations in the seconds-to-minutes range. Examples of human behaviors that involve short-interval timing range from having expectations about when a traffic light will turn from red to green to dancing to the beat of music (Matell and Meck, 2000).

A wide variety of theories of short-interval timing have been proposed (Church and Broadbent, 1991; Gibbon, 1977; Jones, 1976; Killeen and Fetterman, 1988; Large and Jones, 1999; McAuley, 1996; McAuley and Kidd, 1998; Miall, 1989; Treisman, 1963). These theories generally fall into two classes: approaches from a dynamical system perspective that involve coupled oscillators (Large and Jones, 1998; McAuley and Kidd, 1998) and approaches from an information-processing perspective (Church and Broadbent, 1991; Gibbon, 1977; Treisman, 1963).

An emerging distinction between models concerns the perception of durations shorter than a couple of seconds vs. the perception of durations longer than a couple of seconds. One proposal is that different systems are involved in timing events on these two scales (Ivry, 1996; Ivry and Hazeltine, 1995). For event durations shorter than a couple of seconds (e.g., the time intervals defined by successive beats in a musical performance), it has been suggested that temporal regularity plays more of a role in timing processes than explicit memory because patterns of event durations on this scale form directly perceivable rhythms (e.g., music and speech); hence, timing in this range is predictable by virtue of the exogenous timing cues provided by stimulus markers (Fraisse, 1963; Jones, 1976; Jones and Boltz, 1989; Port, 1995). This issue and related ones pertaining to effects of rhythmic context on time perception have recently generated interest from dynamical systems and information-processing theorists (Barnes and Jones, 2000; Ivry and Hazeltine, 1995; McAuley and Jones, 2002; McAuley and Kidd, 1998; Pashler, 2001). The emphasis of this chapter, however, is on the perception of events that occur in isolation (in the absence of a prevailing temporal context) and that last longer than a couple seconds. On this timescale, perceived rhythm tends to break down, and it is more clear-cut that behaviors rely on explicit memory for duration (e.g., knowing how long the red light lasts at a particular intersection).

14.2.1 SCALAR EXPECTANCY THEORY

Most theories that incorporate explicit memory for time are information-processing models, which involve three independent components: an internal clock used to estimate duration, a reference memory used to store information about duration, and a comparison mechanism used to make judgments about how much time has elapsed relative to a remembered (expected) standard duration (Church and Broadbent, 1991). Within this framework, scalar expectancy theory (SET) has been particularly influential because it has been successfully applied to both human and animal data (Gibbon, 1977; Gibbon et al., 1984; Rakitin et al., 1998). SET posits a neural pacemaker that emits a continuous stream of pulses. Stimulus events marking the beginning and ending of event durations trigger the closing and opening of a switch that gate pulses into an accumulator. The count of the pulses accumulated over the target event duration represents a subjective duration code that is stored in reference memory. Successive time intervals are estimated independently, with relative duration judgments about time intervals involving a comparison between a working memory representation of the accumulator and a criterion time sampled from reference memory. A schematic of the various components of SET is shown in Figure 14.1.

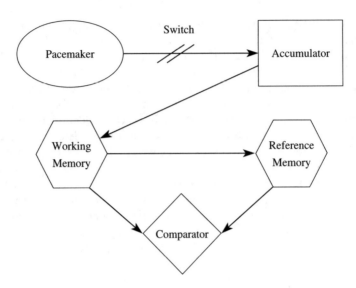

FIGURE 14.1 Schematic diagram of scalar expectancy theory depicting the relation between important components of the model.

14.2.2 PEAK-INTERVAL AND SIMULTANEOUS TEMPORAL PROCESSING TASKS

SET has been used to model human and animal timing in a variety of behavioral tasks. One task, in particular, that is frequently used to assess animal timing is the peak-interval (PI) procedure (Roberts, 1981). The PI procedure is a variant of a fixed-interval (FI) reinforcement schedule. Participants are trained to associate the onset of a light or tone stimulus with the expectation of reinforcement after a fixed time interval. On PI (probe) trials, the stimulus stays on for twice the duration of the FI and no reinforcement is given. The data of interest are the lever presses that the animal makes in the absence of feedback about duration (i.e., probe trials). Human versions of the PI procedure involve essentially the same task, but differ primarily in the method of feedback (Rakitin et al., 1998). Because human participants can be directly instructed about the task, feedback about the duration on FI trials is simply the offset of the stimulus. Similar to probe trials for animals, the stimulus remains on for at least twice the duration of the fixed interval, and the data of interest are the responses that the participant makes in the absence of feedback about the target duration.

During a typical simultaneous temporal processing task, participants are asked to perform two concurrent PI procedures involving pairs of fixed intervals (Meck, 1987; Meck and Church, 1984; Olton et al., 1988). The shorter of the pair of fixed intervals is referred to as the short stimulus, and the longer of the pair as the long stimulus. As in the standard PI procedure, half of the trials provide the opportunity for reinforcement, and the other half are unreinforced probe trials. Data are analyzed only for probe trials, which can be simple or compound. On simple trials, participants are presented with a single stimulus and can focus attention on reproducing the expected duration of the stimulus. On compound trials, both stimuli are presented together with the short stimulus generally embedded within the long stimulus, and participants divide their attention between the two stimulus durations.

14.2.3 BASIC FINDINGS

In both the standard and STP versions of the PI procedure, *peak functions* are obtained by pooling probe trial data across trials and constructing composite relative frequency histograms for different target durations (i.e., fixed intervals). The precise location of the peak on probe trials is referred to as peak time and is taken as the expected time of the target. In both human and animal studies using the PI procedure, three robust behavioral findings concern the mean, variability, and shape of peak functions. For normal participants, peak functions in both focused and divided attention conditions (simple and compound trials) are approximately (1) centered on the target duration, (2) scalar in variability, and (3) normal in shape (Church et al., 1994; Meck and Williams, 1997; Olton et al., 1988; Rakitin et al., 1998). An especially surprising aspect of STP performance is that in some cases, there is little decrement in timing performance when participants must divide attention between timing two durations compared with timing one duration (Meck and Williams, 1997; Olton et al., 1988). For example, in previous animal studies, normal rats have been shown to accurately time up to three durations, as evidenced by the good correspondence among peak times on compound trials and single PI trials (Meck, 1987).

Results obtained from human participants in our lab during focused attention conditions illustrate the three basic features of the PI procedure (Figure 14.2). Participants were trained to estimate the duration of two time intervals (5 and 8 sec), which were associated with two tones differentiated in pitch (low vs. high). Peak functions were obtained by pooling probe trial data across participants and sessions and constructing composite relative frequency histograms for the 5- and 8-sec target durations. Peak functions on an absolute timescale (Figure 14.2A) and on a relative timescale (Figure 14.2B) were constructed.

As can be seen in Figure 14.2, the distributions of produced time intervals are approximately centered on the target duration. For the 5-sec target, the peak time was 4.91 sec, whereas for the 8-sec target, the peak time was 7.78 sec. In addition, the mean coefficients of variability for the 5- and 8-sec targets were approximately equal (0.143 vs. 0.138, respectively), demonstrating the scalar property. Finally, plotting response times on a relative timescale yields near-perfect superimposition of the two response distributions (Figure 14.2B).

14.2.4 EFFECTS OF NEUROBIOLOGICAL MANIPULATIONS

Previous research has shown that both transient and permanent shifts in peak time (corresponding to under- or overestimation of the criterion duration) occur following various behavioral and neurobiological manipulations. Transient shifts refer to changes in peak time that disappear following continued training with the criterion duration, whereas permanent shifts refer to changes in peak time that remain following continued training with the criterion duration.

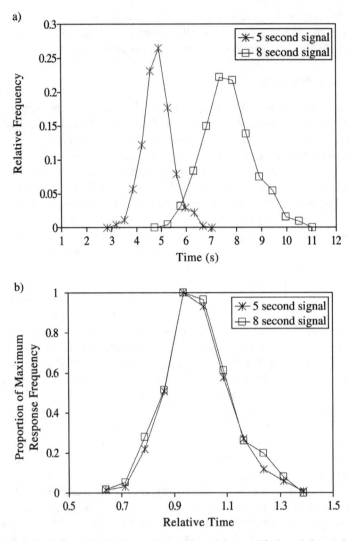

FIGURE 14.2 Performance of adult human subjects in a modified peak-interval procedure during focused attention.

Transient shifts in peak time are observed following manipulations of the dopaminergic system, presumably of the nigrostriatal pathway (Meck, 1986, 1996). Drugs that block dopaminergic receptors, specifically the D2 subtype, produce a rightward shift in peak time (corresponding to overestimates of duration), whereas drugs that stimulate D2 dopaminergic receptors produce a leftward shift (corresponding to underestimates of duration).

One explanation of transient shifts is that they reflect changes in clock speed. According to SET, a change in pacemaker rate under the influence of drugs will alter the rate that clock pulses collect in the accumulator. The result is that it takes more or less time to accumulate the number of pulses that correspond to the previously reinforced count. However, with continued training under the influence of the drug, the number of pulses corresponding to the old criterion duration is gradually updated with a new count corresponding to the expected time of reinforcement based on a clock with the altered speed. These findings have led researchers to posit that dopaminergic neurons in the basal ganglia function as the clock (Meck, 1996). However, see Matell and Meck (2000) and Meck and Benson (2002) for an alternative interpretation.

Permanent shifts in peak time occur following damage of the hippocampus and frontal cortex (Olton et al., 1988). An interesting, yet unexplained, finding by Olton et al. (1988) is that shifts in peak time following hippocampal and frontal cortex damage are in opposite directions. Hippocampal lesions produce leftward shifts in peak time, whereas frontal cortex lesions produce rightward shifts in peak time. These effects of brain lesions remained despite extended training, in contrast to the effects of dopaminergic drugs. Thus, it has been suggested that different components of the timing system were affected by dopaminergic drugs and lesions of the hippocampal and frontal cortex (Meck, 1996). One proposal is that the hippocampus and frontal cortex are involved in maintaining reference memory. SET captures these effects by scaling the output of the accumulator by a proportional factor, referred to as a memory constant (Meck et al., 1986; Meck and Williams, 1997). This account, however, does not identify the specific role that the hippocampus and frontal cortex have in the reference memory system, and it does not explain why peak time should shift in opposite directions when these brain structures are damaged.

14.2.5 THE ATTENTIONAL SWITCH HYPOTHESIS

There have been several proposals about how attention might influence timing performance within the scalar expectancy framework. One proposal is that attention influences the probability that participants' behavior is controlled by stimulus timing on a given trial (Church and Gibbon, 1982; Meck and Williams, 1997). In this view, divided attention decreases the probability that attention is focused on any single stimulus, resulting in an increased asymmetry of the response distributions for each stimulus (Roberts, 1981). An alternative proposal is the attentional switch hypothesis. We have chosen to focus on this second proposal because, in our view, it provides the most consistent explanation of STP data from both human and animal participants (Buhusi, this volume; Fortin, this volume).

The attentional switch hypothesis proposes that attention operates as a switch at the clock stage. The attentional switch influences timing by altering the efficiency with which pulses from the pacemaker are transferred to the accumulator (Allan, 1992; Lejeune, 1998; Macar et al., 1994; Meck and Benson, 2002). Under focused attention conditions, pulses accumulate as a function of time, and the subjective experience of duration is directly proportional to the count of the number of pulses that occur over the temporal extent of the stimulus. However, when attention is divided between two tasks (e.g., a temporal and a nontemporal task or two timing tasks, as in the STP procedure), the assumption is that some pulses may be "lost," with the proportion of lost pulses inversely related to the amount of attention allocated to the timing task.

The implication for behavior is that if attention is directed away from the timing task or is divided between two or more timing tasks, then the effective rate at which pulses collect in the accumulator is reduced due to lost pulses. If participants respond when a given number of pulses have collected in the accumulator (i.e., the count reaches a criterion value), then systematic errors in time estimation should occur. Indeed, many previous studies with human participants have reported data consistent with this view (Brown et al., 1992; Brown and West, 1990; Fortin and Masse, 1999, 2000; Hicks et al., 1977; Macar et al., 1994; Zakay, 1989).

In multiple timing tasks, similar to the STP procedure, Brown et al. (1992; Brown and West, 1990) have found that time judgments during divided attention conditions are less accurate and more variable than during focused attention. Moreover, increasing the number of timed intervals produces a progressive deterioration in time judgment performance (Brown et al., 1992; Brown and West, 1990). When participants are asked to provide time estimates while concurrently performing a nontemporal task (e.g., mental arithmetic), the accuracy of verbal estimations and reproductions decreases monotonically with increasing processing demands of the concurrent nontemporal task (Brown, 1985; Brown and West, 1990; Hicks et al., 1977; Macar et al., 1999; Zakay, 1989). Similarly, Fortin and Masse (1999, 2000) have shown that dividing attention during a time production task results in longer productions (overestimates of the target duration) than in focused attention conditions.

14.2.6 SUMMARY

In summary, SET has been used to explain timing behavior in rats and humans for intervals greater than a couple of seconds. The three major components of SET are the clock, reference memory, and comparator. One way that attention has been incorporated in SET is as an attentional switch at the clock stage. This view is consistent with both human and animal data. Neurobiological studies have suggested that the dopaminergic nigrostriatal pathway may be part of or interact directly with the clock. Other evidence supports the view that the hippocampus and frontal cortex have integral roles in maintaining reference memory. A number of recent functional imaging and electrophysiological studies in animals and humans also link various areas of the frontal cortex (including the prefrontal, premotor, anterior cingulate, and supplementary motor areas) to aspects of temporal processing other than reference memory (Harrington et al., 1998; Macar et al., 1999; Mangels et al., 1998; Macquet et al., 1996; Monfort et al., 2000; Rao et al., 1997; Rubia et al., 1998; Tracy et al., 2000). The remainder of this chapter focuses on the involvement of the frontal cortex in attention and timing (Olton and Pang, 1992).

14.3 DIVIDED ATTENTION AND THE FRONTAL CORTEX

14.3.1 FUNCTIONAL IMAGING STUDIES

Brain imaging studies involving human participants consistently show differential activation of various frontal cortical areas in tasks involving behavioral manipulations of attention. Pardo et al. (1990) observed that the anterior cingulate, left premotor cortex, and supplementary motor cortex show differential activity on congruent and incongruent trials of the Stroop task (Pardo et al., 1990). Using a more direct assessment of divided attention (Corbetta et al. 1997; Vandenberghe et al., 1997) have reported differential activation of the anterior cingulate and premotor cortex, as well as the dorsolateral prefrontal cortex, when participants attended to a single object dimension, compared to when they attended to multiple object dimensions.

14.3.2 Effects of Brain Damage

Consistent with these imaging results, damage of the frontal lobe impairs the ability to divide attention between two stimuli or two stimulus dimensions in nontiming tasks (Godefroy et al., 1996; Godefroy and Rousseaux, 1996) or to divide attention between a temporal and a nontemporal task (Casini and Ivry, 1999). Overall, there is an accumulating body of evidence that supports a general role for frontal cortical areas in attentional processes, including divided attention. Based on these studies, the implication for timing is that frontal cortical areas should be differentially activated on simple and compound trials of the STP task.

No previous imaging or brain damage studies with human participants have directly investigated divided attention during STP for durations in the seconds-to-minutes range. However, several variants of the STP task have been used with rats. Although normal rats can accurately time up to three stimuli presented concurrently, as evidenced by similar peak times on simple and compound trials (Meck, 1987; Meck and Williams, 1997; Olton et al., 1988), rats with damage to the frontal cortex show a pattern of results that support the involvement of the frontal cortex in divided attention (Olton et al., 1988). In the STP procedure, rats with frontal cortex damage produce peak times for the compound long stimulus that overestimate the actual duration of the long fixed interval by an amount approximately equal to the duration of the short fixed interval (Olton et al., 1988). This contrasts with the relatively accurate timing by the same rats for short and long fixed intervals on simple trials. Thus, during divided attention conditions, it appears that rats with frontal cortex damage exhibit a complete breakdown of divided attention, timing the two stimulus durations in a serial fashion, rather than in parallel.

The results from rats with frontal cortex damage support the attentional switch hypothesis in the following way. If frontal cortex lesions impair divided, but not focused attention, then timing will be accurate on simple trials of the STP task, but not on compound trials. On compound trials, rats will initially attend to and time the long stimulus. At the onset of the short stimulus, normal animals are able to divide attention between short and long stimuli. However, rats with damage of the frontal cortex are unable to divide attention between stimuli. Instead, these rats stop attending to the long stimulus, which stops the timing of the long stimulus, and start attending to and timing the short stimulus. At the end of the fixed interval for the short stimulus, rats with lesions stop attending to (and timing) the short stimulus and switch attention back to the long stimulus. The switch of attention back to the long stimulus resumes timing of the long stimulus. Thus, lesions of the frontal cortex limit attention to only one stimulus at any instant, and pacemaker pulses can only be accumulated for the attended stimulus, resulting in two concurrent stimuli being timed serially. This interpretation is consistent with the view that attention is necessary for transfer of pacemaker pulses to an accumulator, and that the frontal cortex allows animals to attend to multiple stimuli and accumulate pacemaker pulses simultaneously for both short and long stimuli.

14.3.3 Summary

Overall, functional imaging of normal human participants and behavioral assessment of humans and animals with brain damage have provided evidence that the frontal cortex is a critical component of divided attention. Information-processing models of timing that incorporate an attentional switch help to explain how attention and the frontal cortex might influence short-interval timing. However, the exact role of the various areas of the frontal cortex in divided attention is not well understood. Single-cell recording can help to identify functions that various areas of the frontal cortex contribute to divided attention and timing.

14.4 RECENT ELECTROPHYSIOLOGICAL RECORDINGS IN THE FRONTAL CORTEX

In an attempt to better understand how frontal cortex neurons contribute to attention and timing, we recorded from neurons in the lateral agranular region of the frontal cortex during the STP procedure (Pang et al., 2001). The lateral agranular region of the frontal cortex was selected for recordings because rats with damage to this area have impaired performance on compound, but not simple trials of the STP procedure, suggesting a deficit in divided attention (Olton et al., 1988). Rats were trained on the STP procedure using 10 sec as the fixed interval for one stimulus (short stimulus) and 20 or 40 sec as the fixed interval for the second stimulus (long stimulus). Electrophysiological data were acquired and analyzed from probe trials exclusively. Peri-event time histograms were constructed for the onset of each stimulus. Confidence intervals were determined on the basis of neuronal activity during the 10 sec prior to stimulus onset.

In behavioral analyses, peak time for the short stimulus was similar on simple and compound trials, and peak time for the long stimulus was similar on simple and compound trials (Figure 14.3). Similar peak times on simple and compound trials for each stimulus demonstrate that timing two stimuli simultaneously was just as accurate as timing a single stimulus, and support the idea that normal rats can divide attention and accurately time two stimuli simultaneously.

14.4.1 Evidence for Divided Attention Neurons

Task-sensitive neurons were those cells having firing rates that exceeded the 95% confidence intervals for at least 4 consecutive seconds. A total of 125 neurons were recorded in the frontal cortex, and 65 of these were task-sensitive neurons by our definition. Of the task-sensitive neurons, four general firing patterns were observed. Type 1 cells (60%) responded to simultaneous presentation of both stimuli on compound trials, but not to stimuli on simple trials (Figure 14.4). Type 2 cells (10%)

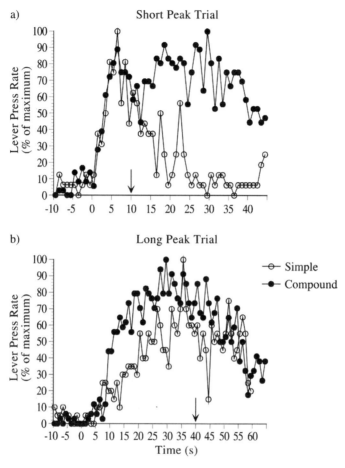

FIGURE 14.3 Example of behavioral data from a rat trained in a simultaneous temporal processing procedure. This rat was trained to associate a tone with a fixed interval of 10 sec (A, arrow) and a light with a fixed interval of 40 sec (B, arrow). Simple trials are plotted with open circles, and compound trials are plotted with closed circles. Peak times in simple and compound trials were similar for each stimulus. (Reprinted from Pang, K.C., Yoder, R.M., and Olton, D.S., *Neuroscience*, 103, 615–628, 2001. With permission.)

responded to both stimuli regardless of the trial type. Type 3 cells (27%) responded to a single stimulus regardless of trial type. Type 4 neurons (3%) altered their activity in response to one stimulus during simple trials, but not compound trials.

The most interesting firing pattern with respect to the focus of this chapter was that demonstrated by type 1 cells. These cells responded during compound, but not simple trials and were easily the most common type of firing pattern recorded. Importantly, the onset of the neuronal response occurred not at the onset of the first compound stimulus, but at the onset of the second stimulus. This is the exact time that divided attention is needed. Although divided attention is only necessary for the duration of the short fixed interval, neuronal responses generally outlasted the short

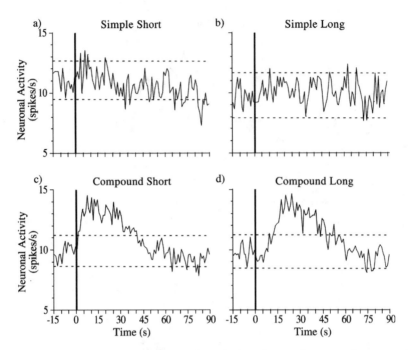

FIGURE 14.4 Example of a type 1 neuron from the lateral agranular cortex that was excited on compound, but not simple stimuli. For this rat, the short fixed interval was 10 sec and the long fixed interval was 40 sec. The time on the abscissa represents time from stimulus onset. The two horizontal lines in each graph represent the 95% confidence limits of the baseline firing rate. Bin width = 1 sec. (Reprinted from Pang, K.C., Yoder, R.M., and Olton, D.S., *Neuroscience*, 103, 615–628, 2001. With permission.)

fixed interval, possibly due to the fact that recordings were obtained on probe trials and participants may be somewhat uncertain of the end of the short fixed interval.

In summary, recordings of lateral agranular neurons in the frontal cortex support the view that this area is important in divided attention. The most common response pattern was one in which cells responded at the time that divided attention is most necessary (i.e., simultaneous presentation of both stimuli). These neurons may allow attention to be switched rapidly between stimuli so that clock pulses can be efficiently collected in the respective accumulators for each stimulus. When these cells are damaged, as in frontal cortex lesions, the participant may be unable to switch attention between tasks and becomes focused on only one of the stimuli, subsequently timing only the attended stimulus, even though multiple stimuli are presented. Although this interpretation is consistent with the results from rats with lesions of the frontal cortex, alternative interpretations exist, and some of these are explored in the next section.

14.4.2 ALTERNATIVE EXPLANATIONS

In this section, we entertain some alternative explanations for the response pattern of type 1 neurons and suggest some directions for future research. The two obvious

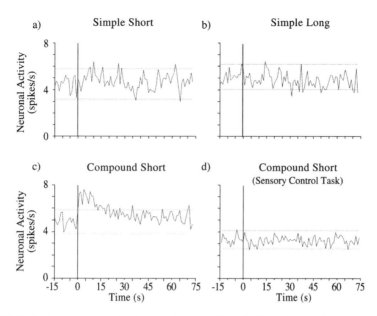

FIGURE 14.5 Another example of a type 1 neuron from the lateral agranular cortex. For this rat, the short fixed interval was 10 sec and the long fixed interval was 40 sec. This cell was recorded during the simultaneous temporal processing procedure (A to C) and during a sensory control task (D). Although the cell responded on compound trials during the STP task, the cell did not respond during the same stimulus presentation (compound short stimulus) in the sensory control task. The overall lower firing rate in the sensory control task (D) than in the STP task (C) suggests that motor responses contribute partly to the overall activity of these cells. The time on the abscissa represents time from stimulus onset. The two horizontal lines in each graph represent the 95% confidence limits of the baseline firing rate. Bin width = 1 sec.

alternatives are sensory and motor interpretations. A sensory explanation proposes that type 1 cells are sensory neurons that fire to the combination of visual and auditory stimulation (the two modalities of stimuli used in the study). Two results make the sensory interpretation unlikely. First, as mentioned above, neuronal recordings were obtained from probe trials. If type 1 cells behaved as conjunctive sensory neurons, we would expect these cells to respond to the simultaneous presentation of both stimuli, which on probe trials lasted the duration of the trial. Instead, type 1 neurons had responses that had transient rises or falls that returned to baseline prior to the termination of the trial. Second, on the day following recording in the STP procedure, cells were recorded in a sensory control task (Figure 14.5D). During this task, response levers were unavailable to the rats and the sensory stimuli were presented as in the STP procedure. During the sensory control task, type 1 neurons did not respond to the simultaneous presentation of both stimuli on "compound" trials, as they did in the STP procedure. Thus, type 1 neurons do not appear to have purely sensory correlates for the conjunction of short and long stimuli.

A motor interpretation may be more likely. The lateral agranular cortex is thought to be the rat analog of the primary motor cortex. This conclusion is based on microstimulation studies and anatomical connections with thalamic motor areas

(Donoghue et al., 1979; Donoghue and Wise, 1982; Krettek and Price, 1977; Neafsey et al., 1986; Wang and Kurata, 1998). It is therefore possible that type 1 neurons might be responding to some motor aspect of the task. Although the patterns of neuronal activity and of lever responses were similar on compound trials, similar response patterns were not observed for simple trials. The rate of lever pressing increased during simple trials, with a peak near the fixed interval associated with each of the stimuli (Figure 14.3). In contrast, neuronal activity of type 1 cells was fairly constant throughout the simple trials (Figure 14.4A and B). In our study, motor activity may be reflected in the basal firing rate of the neuron. This can be seen as a lower overall firing rate in the sensory control task (Figure 14.5D) than in the STP task (Figure 14.5C). As mentioned earlier, response levers were inaccessible to the rats during the sensory control task, although all other aspects of the STP procedure were present. Future studies will have to explore the motor interpretation more carefully, possibly by using nontemporal tasks that require lever responses.

One interpretation that was not addressed in our study and will need to be examined is that the firing of type 1 cells reflects a general increase in cognitive demand, rather than a specific role in divided attention. Compound trials require more cognitive effort than simple trials, possibly leading to the differential firing pattern that we observed. One way that divided attention and cognitive demand might be separated is to increase the difficulty in performing a single-task procedure, such as making stimuli less distinguishable in a discrimination task or increasing the retention interval in a task with a memory component. Although it may be difficult to distinguish between cognitive demand and divided attention, this distinction is an important one for models that attempt to explain the interaction of timing and attention.

As mentioned earlier, proposed timing functions of the frontal cortex include involvement in the representation of duration and its storage in memory (Meck, 1996; Niki and Watanabe, 1979; Olton et al., 1988). Consistent with this view, type 2 cells responded on all types of trials in the STP procedure, supporting their general involvement in timing processes. Type 2 cells, however, constituted only 10% of the cells that responded in the task, suggesting that only a small proportion of the cells in this brain region may serve a direct clock or memory function. The most commonly observed cells were type 1 cells, which responded only on compound trials. A somewhat provocative interpretation of this finding is that type 1 cells are divided attention neurons with the primary function of coordinating the processing of multiple sources of information.

We would argue, however, that it is more natural to propose that a cognitive function such as divided attention resides in secondary motor areas than the primary motor cortex. This view is consistent with electrophysiological and imaging studies of normal participants and with studies in brain-damaged individuals, which support the involvement of the premotor and supplementary motor cortex in divided attention and timing (Casini and Ivry, 1999; Corbetta et al., 1991; Godefroy et al., 1996; Godefrey and Rousseaux, 1996; Pardo et al., 1990; Vanderberghe et al., 1997). One possibility is that the divided attention neuronal correlates we observed in the primary motor cortex are merely reflecting a response that is originating from a region that projects to the primary motor cortex. To resolve this issue, future studies are needed that record from the premotor and supplementary motor cortex to determine if, indeed, response patterns similar to those observed in the primary motor cortex are found.

Finally, it is interesting that many of the critical components of short-interval timing have been proposed to lie within the motor system (Ivry, 1993, 1996; Meck, 1996). Key motor areas that are directly involved in short-interval timing are the basal ganglia, cerebellum, and frontal motor cortex. Frontal motor areas may also serve a coordinating role because of their direct reciprocal connections with both the basal ganglia and cerebellum (Diedrichsen et al., this volume). One possibility is that simple oscillatory neural networks between the frontal cortex, basal ganglia, and cerebellum are responsible for the perception and production of durations under a second that comprise directly perceivable rhythms. These oscillatory networks may form the basis of a system that is capable of encoding and remembering longer durations. A recent approach along these lines is the striatal beat frequency model (Matell and Meck, 2000; Matell et al., this volume; Miall, 1989).

14.5 CONCLUSIONS

One of the future challenges in this field is to determine the neural mechanisms of the various components of timing. It is through the integration of information from a variety of approaches like lesions, brain imaging, and electrophysiology that we strive to understand the neural basis of timing. In this chapter, we have attempted to bring together results from the various approaches. Initial attempts at mapping brain areas to the components of SET have been made, which suggest that the basal ganglia and cerebellum have clock functions, while the temporal cortex and frontal cortex support reference memory (see Malapani and Rakitin, this volume; Meck, 1996). This type of mapping of brain areas to functional components in the timing system is only the first step. Future studies will need to define how cell assemblies in different brain areas contribute to timing. In this vein, we have presented our first step to understand the manner in which neurons in the frontal motor cortex may be important in short-interval timing (i.e., by influencing divided attention). We are just at the beginning of understanding timing at the neuronal level. Only by actively integrating results from human and animal studies, using multiple research approaches, can we continue to make progress in understanding the functional and neural basis of timing.

ACKNOWLEDGMENTS

We recognize the contribution and influence of David Olton in our studies of the frontal cortex and divided attention. We also thank J.P. Miller, Nathaniel Miller, Jacquelyn Toft, Ryan Yoder, and the other members of the Timing Research Group at Bowling Green State University, Bowling Green, Ohio, for their contributions to the preparation of this chapter.

REFERENCES

Allan, L.G., The internal clock revisited, in *Time, Action and Cognition: Toward Bridging the Gap*, Macar, F., Pouthas, V., and Friedman, W.J., Eds., Kluwer Academic Publishers, Dordrecht, Netherlands, 1992, pp. 191–202.

Barnes, R. and Jones, M.R., Expectancy, attention, and time, *Cognit. Psychol.*, 41, 254–311, 2000.

Brown, S.W., Time perception and attention: the effects of prospective versus retrospective paradigms and task demands on perceived duration, *Percept. Psychophys.*, 38, 115–124, 1985.

Brown, S.W., Stubbs, D.A., and West, A.N., Attention, multiple timing, and psychophysical scaling of temporal judgments, in *Time, Action and Cognition: Toward Bridging the Gap*, Macar, F., Pouthas, V., and Friedman, W.J., Eds., Kluwer Academic Publishers, Dordrecht, Netherlands, 1992, pp. 191–202.

Brown, S.W. and West, A.N., Multiple timing and the allocation of attention, *Acta Psychol.*, 75, 103–121, 1990.

Casini, L. and Ivry, R.B., Effects of divided attention on temporal processing in patients with lesions of the cerebellum or frontal lobe, *Neuropsychology*, 13, 10–21, 1999.

Church, R.M. and Broadbent, H.A., A connectionist model of timing, in *Models of Behavior: Neural Network Models of Conditioning*, Commons, M.L., Grossberg, S., and Staddon, J.E.R., Eds., Lawrence Erlbaum Associates, Hillsdale, NJ, 1991, pp. 225–240.

Church, R.M. and Gibbon, J., Temporal generalization, *J. Exp. Psychol. Anim. Behav. Process.*, 8, 165–186, 1982.

Church, R.M., Meck, W.H., and Gibbon, J., Application of scalar timing theory to individual trials, *J. Exp. Psychol. Anim. Behav. Process.*, 20, 135–155, 1994.

Corbetta, M., Miezin, F.M., Dobmeyer, S., Schulman, G.L., and Petersen, S.E., Selective and divided attention during visual discriminations of shape, color and speed: functional anatomy by positron emission tomography, *J. Neurosci.*, 11, 2383–2402, 1991.

Donoghue, J.P., Kerman, K.L., and Ebner, F.F., Evidence for two organizational plans within the somatic sensory-motor cortex of the rat, *J. Comp. Neurol.*, 183, 647–663, 1979.

Donoghue, J.P. and Wise, S.P., The motor cortex of the rat: cytoarchitecture and microstimulation mapping, *J. Comp. Neurol.*, 212, 76–88, 1982.

Fortin, C. and Masse, N., Order information in short-term memory and time estimation, *Mem. Cognit.*, 27, 54–62, 1999.

Fortin, C. and Masse, N., Expecting a break in time estimation: attentional time-sharing without concurrent processing, *J. Exp. Psychol. Hum. Percept. Perform.*, 26, 1788–1796, 2000.

Fraisse, P., *The Psychology of Time*, Harper & Row, New York, 1963.

Gibbon, J., Scalar expectancy theory and Weber's law in animal timing, *Psychol. Rev.*, 84, 279–325, 1977.

Gibbon, J., Church, R.M., and Meck, W.H., Scalar timing in memory, in *Annals of the New York Academy of Sciences*, Vol. 423, *Timing and Time Perception*, Gibbon, J. and Allan, L.G., Eds., New York Academy Press, New York, 1984, pp. 52–77.

Godefroy, O., Lhullier, C., and Rousseaux, M., Non-spatial attention disorders in patients with frontal or posterior brain damage, *Brain*, 119, 191–202, 1996.

Godefroy, O. and Rousseaux, M., Divided and focused attention in patients with lesion of the prefrontal cortex, *Brain Cognit.*, 30, 155–174, 1996.

Harrington, D.L., Haaland, K.Y., and Knight, R.T., Cortical networks underlying mechanisms of time perception, *J. Neurosci.*, 18, 1085–1095, 1998.

Hicks, R.E., Miller, G.W., Gaes, G., and Bierman, K., Concurrent processing demands and the experience of time-in-passing, *Am. J. Psychol.*, 90, 431–446, 1977.

Ivry, R., Cerebellar involvement in the explicit representation of temporal information, *Ann. N.Y. Acad. Sci.*, 682, 214–230, 1993.

Ivry, R.B., The representation of temporal information in perception and motor control, *Curr. Opin. Neurobiol.*, 6, 851–857, 1996.

Ivry, R.B. and Hazeltine, R.E., Perception and production of temporal intervals across a range of durations: evidence for a common timing mechanism, *J. Exp. Psychol. Hum. Percept. Perform.*, 21, 3–18, 1995.

Johnston, J.C., McCann, R.S., and Remington, R.W., Chronometric evidence for two types of attention, *Psychol. Sci.*, 6, 365–369, 1995.

Jones, M.R., Time, our lost dimension: toward a new theory of perception, attention, and memory, *Psychol. Rev.*, 83, 323–355, 1976.

Jones, M.R. and Boltz, M., Dynamic attending and responses to time, *Psychol. Rev.*, 96, 459–491, 1989.

Killeen, P.R. and Fetterman, J.G., A behavioral theory of timing, *Psychol. Rev.*, 95, 274–295, 1988.

Krettek, J.E. and Price, J.L., The cortical projections of the mediodorsal nucleus and adjacent thalamic nuclei in the rat, *J. Comp. Neurol.*, 171, 157–191, 1977.

Large, E.W. and Jones, M.R., The dynamics of attending: how we track time varying events, *Psychol. Rev.*, 106, 119–159, 1999.

Lejeune, H., Switching or gating? The attentional challenge in cognitive models of psychological time, *Behav. Process.*, 44, 127–145, 1998.

Macar, F., Grondin, S., and Casini, L., Controlled attention sharing influences time estimation, *Mem. Cognit.*, 22, 673–686, 1994.

Macar, F., Vidal, F., and Casini, L., The supplementary motor area in motor and sensory timing: evidence from slow brain potential changes, *Exp. Brain Res.*, 125, 271–280, 1999.

Mangels, J.A., Ivry, R.B., and Shimizu, N., Dissociable contributions of the prefrontal and neocerebellar cortex to time perception, *Brain Res. Cognit. Brain Res.*, 7, 15–39, 1998.

Maquet, P., Lejeune, H., Pouthas, V., Bonnet, M., Casini, L., Macar, F., Timsit-Berthier, M., Vidal, F., Ferrara, A., Degueldre, C., Quaglia, L., Delfiore, G., Luxen, A., Woods, R., Mazziotta, J.C., and Comar, D., Brain activation induced by estimation of duration: a PET study, *Neuroimage*, 3, 119–126, 1996.

Matell, M.S. and Meck, W.H., Neuropsychological mechanisms of interval timing behavior, *Bioessays*, 22, 94–103, 2000.

McAuley, J.D., Perception of time as phase: toward an adaptive-oscillator model of rhythmic pattern processing, *Diss. Abstr. Int.*, 56, 4988, 1996 (Doctoral dissertation, Indiana University, 1996).

McAuley, J.D. and Jones, M.R., Rhythmic expectations in time judgment behavior: implications for entrainment and interval-based models of time perception, submitted.

McAuley, J.D. and Kidd, G.R., Effect of deviations from temporal expectations on tempo discrimination of isochronous tone sequences, *J. Exp. Psychol. Hum. Percept. Perform.*, 24, 1786–1800, 1998.

Meck, W.H., Affinity for the dopamine D_2 receptor predicts neuroleptic potency in decreasing the speed of an internal clock, *Pharmacol. Biochem. Behav.*, 25, 1185–1189, 1986.

Meck, W.H., Vasopressin metabolite neuropeptide facilitates simultaneous temporal processing, *Behav. Brain Res.*, 23, 147–157, 1987.

Meck, W.H., Neuropharmacology of timing and time perception, *Cognit. Brain Res.*, 3, 227–242, 1996.

Meck, W.H. and Benson, A.M., Dissecting the brain's internal clock: how frontal-striatal circuitry keeps time and shifts attention, *Brain Cognit.*, 48, 195–211, 2002.

Meck, W.H. and Church, R.M., Simultaneous temporal processing, *J. Exp. Psychol. Anim. Behav. Process.*, 10, 1–29, 1984.

Meck, W.H., Church, R.M., and Wenk, G.L., Arginine vasopressin innoculates against age-related increases in sodium-dependent high affinity choline uptake and discrepancies in the content of temporal memory, *Eur. J. Pharmacol.*, 130, 327–331, 1986.

Meck, W.H. and Williams, C.L., Simultaneous temporal processing is sensitive to prenatal choline availability in mature and aged rats, *Neuroreport*, 8, 3045–3051, 1997.

Miall, C., The storage of time intervals using oscillating neurons, *Neural Comput.*, 1, 359–371, 1989.

Monfort, V., Pouthas, V., and Ragot, R., Role of frontal cortex in memory for duration: an event-related potential study in humans, *Neurosci. Lett.*, 286, 91–94, 2000.

Neafsey, E.J., Bold, E.L., Haas, G., Hurley-Gius, K.M., Quirk, G., Sievert, C.F., and Terreberry, R.R., The organization of the rat motor cortex: a microstimulation mapping study, *Brain Res.*, 396, 77–96, 1986.

Niki, H. and Watanabe, M., Prefrontal and cingulate unit activity during timing behavior in the monkey, *Brain Res.*, 171, 213–224, 1979.

Olton, D.S. and Pang, K., Interactions of neurotransmitters and neuroanatomy: it's not what you do, it's the place that you do it, in *Neurotransmitter Interactions and Cognitive Function*, Levin, E.D., Decker, M.W., and Butcher, L.L., Eds., Birkhauser, Boston, 1992, pp. 275–286.

Olton, D.S., Wenk, G.L., Church, R.M., and Meck, W.H., Attention and the frontal cortex as examined by simultaneous temporal processing, *Neuropsychologia*, 26, 307–318, 1988.

Pang, K.C., Yoder, R.M., and Olton, D.S., Neurons in the lateral agranular frontal cortex have divided attention correlates in a simultaneous temporal processing task, *Neuroscience*, 103, 615–628, 2001.

Pardo, J.V., Pardo, P.J., Janer, K.W., and Raichle, M.E., The anterior cingulate cortex mediates processing selection in the Stroop attentional conflict paradigm, *Proc. Natl. Acad. Sci. U.S.A.*, 87, 256–259, 1990.

Pashler, H., Dual-task interference and elementary mental mechanisms, in *Attention and Performance*, XIV, Meyer, D.E. and Kornblum, S., Eds., MIT Press, Cambridge, MA, 1993, pp. 245–264.

Pashler, H., Perception and production of brief durations: beat-based versus interval-based timing, *J. Exp. Psychol. Hum. Percept. Perform.*, 27, 485–493, 2001.

Pashler, H. and Johnston, J.C., Attentional limitations in dual-task performance, in *Attention*, Pashler, H., Ed., Psychology Press, Ltd., East Sussex, U.K., 1998, pp. 155–189.

Port, R.F., *Mind as Motion: Explorations in the Dynamics of Cognition*, MIT Press, Cambridge, MA, 1995.

Rakitin, B.C., Gibbon, J., Penney, T.B., Malapani, C., Hinton, S.C., and Meck, W.H., Scalar expectancy theory and peak-interval timing in humans, *J. Exp. Psychol. Anim. Behav. Process.*, 24, 15–33, 1998.

Rao, S.M., Harrington, D.L., Haaland, K.Y., Bobholz, J.A., Cox, R.W., and Binder, J.R., Distributed neural systems underlying the timing of movements, *J. Neurosci.*, 17, 5528–5535, 1997.

Roberts, S., Isolation of an internal clock, *J. Exp. Psychol. Anim. Behav. Process.*, 7, 242–268, 1981.

Rubia, K., Overmeyer S., Taylor, E., Brammer, M., Williams, S., Simmons, A., Andrew, C., and Bullmore, E., Prefrontal involvement in "temporal bridging" and timing movement, *Neuropsychologia*, 36, 1283–1293, 1998.

Sternberg, R.J., *Cognitive Psychology*, 2nd ed., Harcourt Brace & Company, New York, 1999.

Thomas, E.A.C. and Weaver, W.B., Cognitive processing and time perception, *Percept. Psychophys.*, 17, 363–367, 1975.

Tracy, J.I., Faro, S.H., Mohamed, F.B., Pinsk, M., and Pinus, A., Functional localization of a "time keeper" function separate from attentional resources and task strategy, *Neuroimage*, 11, 228–242, 2000.

Treisman, M., Temporal discrimination on the indifference interval: implications for the model of the "internal clock," *Psychol. Monogr. Gen. Appl.*, 77, 1–31, 1963.

Vandenberghe, R., Duncan, J., Dupont, P., Ward, R., Poline, J.B., Bormans, G., Michiels, J., Mortelmans, L., and Orban, G.A., Attention to one or two features in left or right visual field: a positron emission tomograph study, *J. Neurosci.*, 17, 3739–3750, 1997.

Wang, Y. and Kurata, K., Quantitative analyses of thalamic and cortical origins of neurons projecting to the rostral and caudal forelimb motor areas in the cerebral cortex of rats, *Brain Res.*, 781, 135–147, 1998.

Zakay, D., Subjective time and attentional resource allocation: an integrated model of time estimation, in *Advances in Psychology*, Vol. 59, *Time and Human Cognition: A Life-Span Perspective*, Levin I. and Zakay, D., Eds., Elsevier, Amsterdam, 1989, pp. 365–397.

Zakay, D. and Block, R.A., The role of attention in time estimation processes, in *Time, Internal Clocks and Movement*, Pastor, M.A. and Artieda, J., Eds., Elsevier, Amsterdam, 1996, pp. 143–164.

15 Integration of Behavior and Timing: Anatomically Separate Systems or Distributed Processing?

Matthew S. Matell, Warren H. Meck, and Miguel A.L. Nicolelis

CONTENTS

15.1 Introduction 371
15.2 Components of an Interval Timer..372
    15.2.1 Generalized Timing Model 372
    15.2.2 Interdependence of Processes 374
15.3 Neural Timing Models..374
15.4 Electrophysiological Data of Striatum and Cortex during Timing.............376
    15.4.1 Methods..376
    15.4.2 Neural Firing Patterns...376
    15.4.3 Discussion 378
15.5 Localized or Distributed Striatal Timing?...381
    15.5.1 Multiple Processes 381
    15.5.2 Localized Proposal..382
    15.5.3 Distributed Proposal..383
    15.5.4 Neural Evidence 383
    15.5.5 Competitive Interactions 384
15.6 Temporal and Sequential Control Interactions 387
15.7 Future Directions..388
15.8 A Feedback Loop to the Beginning 388
References...389

15.1 INTRODUCTION

With the recent development of powerful methods to study brain–behavior relations, the study of interval timing has rapidly shifted from primarily behavioral analyses elucidating the psychological constructs of timing to investigations aimed at identifying the anatomical and physiological underpinnings of the interval timing system. This transition to the study of the biological substrates of interval timing is well timed to stimulate further model development. Because the various interval timing models are already extremely accurate at predicting the behavioral data (Church and Broadbent, 1991; Gibbon, 1977; Killeen and Fetterman, 1988; Staddon and Higa, 1999), much of their attractiveness is associated with their philosophical approach (i.e., behaviorism vs. cognitivism), rather than their predictive accuracy. Although these models fare quite well at explaining behavioral data, because of their fundamental differences, they do not provide us with an unbiased framework from which to search for the neural mechanisms of interval timing. As such, we believe that a theory-free model of interval timing would be valuable. Such a general timing model is a much needed "place to hang our hats" when searching for the neural processes associated with timing and time perception.

In this chapter, we will discuss the basic components of an interval timer and describe how these components may be realized in the brain. To this end, we will present electrophysiological data showing duration-specific activity in the striatum and discuss the implications these data carry for current and future models of interval timing.

15.2 COMPONENTS OF AN INTERVAL TIMER

15.2.1 GENERALIZED TIMING MODEL

It has previously been proposed that information-processing models of interval timing are composed of clock, memory, and decision stages (Church, 1997) — a framework we will refer to as a generalized timing model (GTM). The *clock stage* encompasses the *production of a temporal percept*, with processes ranging from the generation of a utilizable temporal signal to the *integration of this signal* into a meaningful output. The temporal signal is usually characterized as single or multiple, fast or slow oscillators (Church et al., 1991; Gibbon, 1977; Miall, 1989). However, the temporal signal could conceivably be any type of patterned signal, so long as it can be reliably repeated, albeit with some error, for each opportunity to time. In contrast, the integration of the temporal signal is often characterized as a monotonically changing function, e.g., linear (Gibbon and Church, 1981) or log (Staddon and Higa, 1999), so that the percept of time varies in a systematic manner. The clock stage may also incorporate *processes that allow temporal integration to start anew* upon a biologically relevant signal onset (e.g., starting the pacemaker or resetting the accumulation process). Upon occurrence of an important event, the *output of the clock stage is stored in long-term memory*, thereby composing the *memory stage*. Given subsequent opportunities to time a similar event, the component processes associated with the clock stage are initiated, and the *current clock reading is compared to previously stored values in long-term memory* to produce a similarity function, thereby composing the *decision stage*. The temporally predictive behaviors of the organism are, by definition, based on the output of this decision stage. Each of these stages is conceived as being relatively orthogonal and serially organized from an information-processing perspective, in that information is passed in a single

Integration of Behavior and Timing

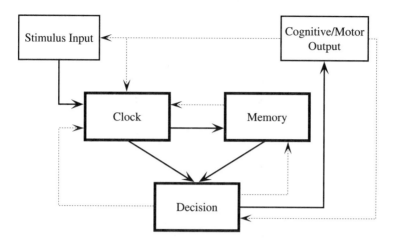

FIGURE 15.1 Information-processing diagram for a generalized timing model. The core components specific to the processing of time are the clock, memory, and decision processes, indicated in dark boxes. The arrows indicate direction of information flow. Addition of stimulus input and cognitive and motor output extends this GTM to the processes necessary to produce a properly behaving organism. Although not dictated in terms of information flow, the dotted lines indicate the extensive array of feedback loops that may significantly modulate the processing of information.

direction (e.g., from clock to decision stage). The organization of the GTM is shown in Figure 15.1.

In the timing literature, reference to these processing stages has become common, and while we fully agree with this idea within an information-processing capacity, we are concerned that there may be an underlying tendency to parse these stages into anatomically separate brain regions, with each region processing information in a serial manner, akin to the GTM. The temptation may also exist to separate these timing components from other cognitive and motor systems in which such temporal information is utilized. We argue that assuming anatomical and computational independence for the component processes of interval timing is not valid. Similarly, assuming that the cognitive and motor systems which utilize temporal information are distinct from the processes composing the timing system is also unreasonable. Instead, we propose that all of these systems are highly interrelated.

Unlike the timing signal and the temporal integration function, which cannot be determined *a priori*, the output of the decision stage must be the basis of the temporally controlled behavior of the organism, and will therefore be highly similar in temporal patterning to these behaviors (i.e., the neural discharge of the decision process will be difficult to dissociate from the motor output in terms of their temporal structure). Temporal generalization procedures, such as the commonly used peak-interval procedure (Church et al., 1991; Roberts, 1981), require the subject to indicate when it expects the occurrence of a temporally predictable reward. The distribution of these behavioral indices (e.g., lever pressing) over time in the trial is generally described by a Gaussian-shaped peak centered very close to the normal time of reinforcement, with a width that is proportional to the peak time. This proportionality in spread has been termed the scalar property (Gibbon, 1977) and is a defining property of interval timing in the seconds-to-minutes range. We suggest that this response peak is isomorphic with the decision stage output function occurring in the organism while it is performing these production procedures. The output function may also form the basis for other interval timing judgments by serving as the basic substrate onto which other computations may be performed (e.g., from a rat's perspective, the duration bisection procedure might be a two-duration temporal generalization procedure, rather than a discrimination procedure).

15.2.2 INTERDEPENDENCE OF PROCESSES

In terms of a temporal percept (i.e., the current clock reading or a similarity comparison of now vs. not now), the GTM describes all of the processes necessary for timing. However, to understand how the behavior of an organism unfolds in time, two additional information-processing components are required (see Figure 15.1). These components are the stimulus input and cognitive and motor output functions that are needed to detect the signal to be timed and then to respond in a temporally meaningful manner. Although incorporation of these components may seem like unnecessary baggage for a GTM, their importance becomes clear upon the realization that such external processes are neither independent of the functioning of the timing processes nor in the determination of the actual behavior of the animal. The lack of independence between the timing process itself and the processing of the stimulus to be timed is demonstrated by experiments showing that auditory stimuli tend to drive the clock stage at a faster rate than visual stimuli (Penney, this volume; Penney et al., 2000). In terms of the actual motor output of the organism, it is perhaps unnecessary to state that even in our best experimental designs the behavior of an individual subject has many internal and external influences other than its temporal percept. Although behavioral data are frequently averaged across trials, sessions, and subjects with the hopes of characterizing the underlying decision function, averaged data of this sort cannot be used in real-time analyses such as electrophysiological or brain imaging studies, in which the data of interest must be interpreted with respect to an individual subject's behavior. Furthermore, assuming independence between the timing processes and subsequent motor output excludes the investigation of feedback loops, which may play an important role in interval timing (see below). Due to these additional sources of influence, a GTM that incorporates these input–output components would be a more useful framework for investigating the neural components of the interval timing system.

15.3 NEURAL TIMING MODELS

We have recently proposed that the interval timing system is highly dependent upon functioning cortico-striatal circuits (Matell and Meck, 2000, 2002). A thorough evaluation of the anatomical and electrophysiological characteristics of these areas led us to propose that these neural circuits can provide the majority of the necessary computations for the clock, memory, and decision stages of the GTM. Briefly, each striatal spiny neuron receives 10,000 to 30,000 unique cortical inputs (Wilson et al., 1983). The synaptic strength of these inputs can be modulated by a dopamine reward signal, and through such modulation, the striatal neuron can be "trained" to respond to specific patterns of cortical activity (Beiser and Houk, 1998; Houk, 1995). Given the similarity between this recognition process and the one utilized by the beat frequency model of timing (Miall, 1989), we proposed that striatal spiny neurons detect beat frequencies of oscillating cortical neurons and, in so doing, are able to "time" intervals. Simulations of this striatal beat frequency (SBF) model, which involved adding variance to both the cortical oscillatory frequencies and striatal firing threshold and implementing a physiologically realistic two-state membrane potential in the striatal neuron, produced peak-shaped firing patterns similar to the behavioral data obtained in the peak-interval procedure (Matell and Meck, 2000, in press).

In contrast to an anatomically distinct interval timing system, which might be surmised from the GTM, almost all of the information-processing components of SBF lie within single striatal neurons. Although the SBF model proposes that the clock signal comes from cortical neurons, the clock stage integration process occurs through an alteration of the membrane potential of a striatal neuron. Similarly, temporal memories are stored as specific combinations of synaptic weights of the cortico-striatal synapses. When the proper combination of cortical neurons fires synchronously, the striatal neuron fires. This firing thereby indicates that the current clock value (cortical firing pattern) matches the temporal memory (synaptic strength template), thereby serving as a similarity function. As such, the firing of these spiny neurons over time can be viewed as the decision stage output.

Irrespective of whether this model is an accurate account of interval timing, some interesting processes occur within this model that we would like to elucidate. First, this model shows that it is feasible to build physiologically realistic interval timing models in which the information-processing components have very little anatomical separation. Secondly, close inspection of the SBF model shows that the processing of time in this model does not progress in a serial manner from clock stage through decision stage. Rather, the clock signal is filtered, prior to integration, by those synapses that represent the temporal memories. In other words, the clock stage is not upstream from the memory stage. Similarly, the anatomical circuitry of the cortico-striatal system suggests that the clock stage is not solely upstream from the decision stage: output from the striatum passes through the basal ganglia output nuclei (e.g., both segments of the globus pallidus, the substantia nigra pars reticulata, and the subthalamic nucleus), through the thalamus, and back to the cortex, thereby making up cortico-striato-thalamo-cortical loops. The output pathways of the striatum are believed to return to the same cortical regions that innervated the striatal neuron in question (Strick et al., 1995) and imply that the functioning of these regions is dynamic. In terms of the SBF model, decision stage activity in the striatal neuron produces changes in the cortical clock signal, which thereby alters subsequent striatal activity. The dynamic nature of these circuits or loops implies that unlike the GTM described above, the neural implementation of an interval timer is anything but independent and serial.

15.4 ELECTROPHYSIOLOGICAL DATA OF STRIATUM AND CORTEX DURING TIMING

We have recently begun testing the hypotheses generated by the SBF model by recording from ensembles of neurons in both the striatum and cingulate cortex of rats while they perform an interval timing task (Matell et al., in press; for general methods on ensemble electrophysiology, see Nicolelis et al., 1997). As with all electrophysiological investigations in behaving animals, it is necessary to design the behavioral task so that the neural activity related to the cognitive variables of interest (in this case, the perception of time) can be dissociated from the motor activity resulting from task performance. While never an easy feat in freely behaving animals, this dissociation is all the more difficult in the study of timing, as the behaviors of the rats evolve as a function of time (Fetterman et al., 1998).

15.4.1 Methods

To overcome this difficulty, we utilized a matched behavior analysis, in which we compared the same behavior across different periods in time, so that time itself was the primary variable that differed. To this end, we trained the rats on a multiple-duration, fixed-interval procedure. Specifically, rats were trained that food reinforcement could be earned for the first lever press either 10 or 40 sec after stimulus onset. Responses prior to the criterion time had no consequence. Only one response lever was available, and a barrier was constructed around this lever so that the rat could only respond with its right forepaw. In this manner, all responses on the lever would be largely identical in terms of motor activity. The trial ended after the delivery of reinforcement at either the early (10 sec) or late (40 sec) duration. No information was provided to the rat as to which duration would be reinforced on each trial.

This reinforcement schedule produced two bursts of lever pressing on each trial, one press burst occurring around the short duration and, if reinforcement was not delivered (i.e., if the late duration was primed), a second burst of pressing occurred around the late duration. Over trials, the distribution of lever pressing on late trials had a peak around 10 sec and a scallop up to 40 sec. By choosing reinforcement probabilities for the two criterion durations that were inversely proportional to the reinforcement densities, the rats' peak rates of lever pressing at these two durations were nearly identical. The distribution of lever pressing on late trials from an individual rat in this task is shown in Figure 15.2.

15.4.2 Neural Firing Patterns

Given the similarity of response rates for pressing at the two durations, the moment-by-moment behavior of the rat was approximately the same when the rat was pressing for the early-duration reward and when it was pressing for the late-duration reward. Therefore, the neural activity corresponding to these periods of pressing should be identical in terms of overt motor aspects, and as such, differences in neural activity can be interpreted as resulting from the processing of different durations. The firing pattern of a striatal neuron while the rat behaved on the late trials is shown in

Integration of Behavior and Timing

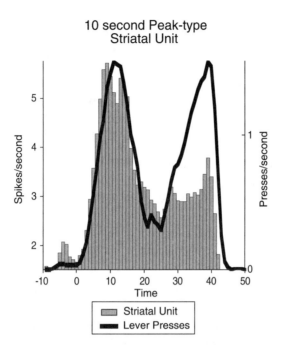

FIGURE 15.2 Distribution of lever presses and striatal activity as a function of trial time. Rats were trained that they could earn food at either 10 or 40 sec from stimulus onset (0 sec). Data are from trials in which food was available at 40 sec. Rates of lever pressing peaked at both 10 and 40 sec, demonstrating the rat's temporally specific expectation of food reward. In contrast, the striatal neuron fired maximally at 10 sec, but showed significantly less activity at 40 sec, thereby demonstrating its temporal specificity. The shape of the neural activity is classified as a peak because the magnitude of firing rate change from the baseline at 10 sec is greater than twice the change at 40 sec. Data have been smoothed by a 3-sec running mean, and the y-axis is partially shown to emphasize the match between striatal activity and behavior.

Figure 15.2. As can be seen, there was a phasic increase in neural activity around 10 sec without a robust increase as time approached 40 sec.

In order to quantitatively evaluate whether this neural activity at 10 sec was significantly different from that at 40 sec, we compared the number of spikes that occurred during an early vs. late press period. The lengths of these press periods were matched so that the number of presses, width of the press window, and therefore mean rate of pressing were all identical across the two durations. We found significantly different firing rates in 28% of the striatal neurons recorded and 20% of the cortical neurons recorded. Of these differentially active striatal neurons, 93% had activity levels during one or both of the press periods that crossed a 99% confidence interval, indicating that they were firing at near-maximal levels. Similarly, 91% of the differentially active cortical neurons crossed this confidence interval. Having maximal (or minimal) activity at the criterion durations and having differences in activity across durations suggest that these neurons were primarily involved in the encoding of specific signal durations.

Of those striatal neurons showing both differences across duration and maximal activity at these signal durations, 71% showed activity patterns similar to that illustrated in Figure 15.2, which we classified as peak shaped because the magnitude of change in firing rate during the early-duration press period was more than twice the magnitude of change during the late-duration press period. In addition to this peak firing pattern, 14% of the primarily timing striatal neurons showed a modulatory firing pattern, which we defined as showing peaks in firing associated with one or both of the behavioral peaks, but where the magnitude of change across durations was less than twofold. Also, 14% of the primarily timing striatal neurons showed ramp-like activity patterns, in which the firing rate showed monotonic changes as a function of time. In contrast, 40% of the primarily timing cortical neurons had peak-type changes, 50% were classified as modulatory, and 20% as ramp (see Figure 15.3).

Further inspection of the rats' behavior during these press periods using a fine-grained video analysis showed that their behavior alternated between bursts of pressing and checking the food magazine for reward. Because the levels of magazine checking often varied as a function of time in the trial, and could potentially account for the neural activity differences found across signal durations in the mean functions, the data were reanalyzed by comparing the neural firing rates during only those periods of time in which the rat was engaged in a press burst. Any differences in press topography (e.g., press duration, interpress interval, release-press interval) that may have changed as a function of signal duration were controlled for by using these values as covariates in the analysis.

The results of this motor-controlled analysis showed that a total of 12 of 54 (22%) striatal and 8 of 54 (15%) cortical neurons had differences in neural activity as a function of signal duration that could not be explained by any measured motor variables. Of these neurons, eight striatal and four cortical neurons were retained from the earlier mean function analysis, and four striatal and four cortical neurons were added (i.e., they did not show significant difference when testing without the press covariates). Of those striatal neurons retained, 75% were peak, 12% were modulatory, and 12% were ramp. Of the cortical neurons retained, 0% were peak, 50% were modulatory, and 50% were ramp. Those neurons that were added were not given shape classifications because their mean functions were not indicative of the duration-related differences found in the latter analysis.

15.4.3 Discussion

Finding a different magnitude of neural activity at one vs. the other criterion duration provides compelling evidence that striatal and cortical neurons are intimately involved in interval timing. However, the shapes of their neural activity patterns tells us considerably more than general involvement in timing. As can be seen in Figure 15.2, the firing rate modulation of this striatal neuron was peak shaped, firing maximally around the first criterion time of 10 sec, which matches the decision stage output function hypothesized by the SBF model (Matell and Meck, 2000, in press) and other interval timing models (Gibbon and Church, 1984; Staddon and Higa, 1999). Because the majority of striatal neurons showing a difference in firing rate

Integration of Behavior and Timing

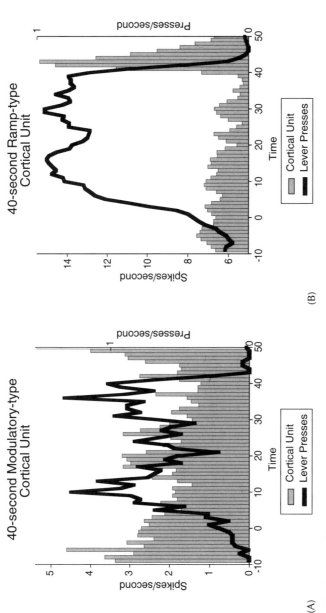

FIGURE 15.3 Distribution of lever presses and cortical activity as a function of trial time. (A) Cortical neuron showing modulatory activity in relation to time in the trial. This classification was given to neurons showing significant differences in their spike rate between 10 and 40 sec, but with the magnitude of change from the baseline to the peak at 40 sec less than twice that from the baseline to the peak at 10 sec. (B) Cortical neuron showing ramp activity in relation to time in the trial. This classification was given to those neurons that had significant differences in firing rate between 10 and 40 sec, and visual inspection showed monotonic changes across the trial. Figure notation and data smoothing are identical to that in Figure 15.2.

as a function of time had similar peak-shaped activity patterns, these data suggest that a subset of striatal neurons either are the decision stage output of an interval timer or are downstream from it.

In addition to being representative of a decision stage output function, the striatal activity pattern is also quite similar to the lever press distribution occurring around the short criterion duration. Such a match between the decision stage and motor output is prescribed in the GTM and would, under many experimental circumstances, prevent one from being able to discern in which information-processing component these striatal neurons should be placed in (i.e., decision stage or output stage). However, as can also be seen in Figure 15.2, this behavioral-neural firing match becomes dissociated as a function of time in the trial, as there is a much smaller secondary increase in firing during the behavioral scallop at 40 sec. Given this clear dissociation between the neural activity pattern and the overt behavioral expression of the rat as time approaches the second reinforcement duration, we can conclude that striatal activity does not directly encode the motor programs involved in pressing, and its role in the timing circuit must therefore be upstream from the neural circuitry producing the behavioral output. Thus, we are led to conclude that this striatal firing encodes the decision stage output of the internal clock.

In contrast to the clear peak observed in Figure 15.2, the cortical neurons showed only modulatory or ramp-like firing patterns (Figure 15.3). As can be seen in Figure 15.3A, this modulatory firing pattern generally matched the lever press distribution across the entire trial, with only a slight, although significant, difference in firing rate across the two criterion times. This sort of firing pattern does not match any prescribed functions of the GTM and, given its similarity to the entire lever press distribution, suggests that this type of cortical activity falls on the behavioral output end of the process.

The cortical ramp pattern is, on the other hand, similar to functions that have been proposed to serve as the integration function of the clock signal, notably the linear signal of scalar expectancy theory (Gibbon, 1977) and the logarithmic decay signal of the multiple timescales model (Staddon and Higa, 1999). Despite this similarity, we are not yet comfortable making any inferences regarding the precise role of these cells, for several reasons. First, we had only a small number of cells that showed ramp-like activity during the trial (two cortical cells and one striatal cell). Second, of these ramp cells, both the striatal cell and one of the two cortical cells showed the ramp pattern beginning at delivery of food reward without any alteration in slope at signal onset, suggesting that these cells were not under stimulus control. Because we utilized variable-length intertrial intervals, the failure to show stimulus control suggests that these cells were not serving as an integration function. However, because all of these ramp patterns provide systematically changing levels as a function of time in the trial (i.e., they are isochronic), they may be a potential clock signal and warrant future investigation.

Given that we found peak firing patterns in a subset of striatal cells, but failed to find these same peak patterns in the cortex, these data suggest that the striatum generates, rather than relays, the decision stage output function for an internal clock. Although we acknowledge the possibility that a peak-shaped input may be passed to the striatum by way of other cortical neurons or areas than those from which we recorded, we chose to investigate the cingulate cortex, as it is primarily afferent to the striatal area from which these peak data are derived (Sesack et al., 1989). Therefore, we currently favor the proposal that these striatal neurons are performing the necessary computations, either individually or as an ensemble, to generate a decision stage output.

15.5 LOCALIZED OR DISTRIBUTED STRIATAL TIMING?

The generation of a decision stage output in the firing of striatal neurons is consistent with the predictions of the SBF model. Although we do not yet have conclusive data regarding the anatomical location of the clock and memory processes, the SBF model proposes that clock integration and memory stages also occur within these same peak-generating neurons. Given the possibility that single striatal neurons can function as the majority of an interval timing system, we are suddenly confronted with two possible scenarios for implementation of an internal clock in the brain. One notion is that interval timing striatal neurons may be localized to a specific striatal area. In this scenario, interval timing would be carried out by an anatomically localizable group of striatal neurons, all performing roughly the same operations, with the output of these neurons sent to all motor and cognitive cortical areas requiring temporal information. The alternative idea is that single neurons functioning as interval timers may be distributed throughout the striatum, such that they temporally modulate the output of the functional area in which they are localized. As such, interval timing may not be a separate function occurring in one region of the striatum, but an important component operating within this behavioral and cognitive control system. An evaluation of both the anatomical arrangement of the cortico-striato-thalamo-cortical circuit and the physiological data from behaving primates suggests that the distributed interval timing hypothesis is a more likely scenario for temporal modulation of behavioral and cognitive processes.

15.5.1 MULTIPLE PROCESSES

An example of the many separate pieces of temporal information utilized by an organism while interacting with its environment may be useful in arguing this point. One common experimental task utilized in primates is the delayed match-to-sample procedure. This procedure progresses through the following phases:

1. The subject rests its hand on a central button.
2. An instructional cue is briefly presented after some delay (e.g., sample A or sample B).
3. An action cue is presented following another delay.
4. The subject performs the behavior designated by the instruction cue (e.g., press button associated with sample A).
5. The reward is delivered following a third delay.

It is clear that many different temporal relations exist within this task, and that the subject must modulate its cognitions and behaviors to conform to these temporal relations. First, attention shifts in a temporally dependent manner to the onset of the instruction cue, action cue, and reward output. Along with these preparatory attentional shifts are changes in gaze that will also have a nonuniform temporal distribution. Second, the subject needs to prepare to make a response upon onset of the action cue. Third, processing related to the upcoming delivery of reward is undoubtedly occurring, such as preparatory increases in salivation. At the same time, inhibition of other competing behaviors and attentions must be heightened in a temporally modulated manner in order to prevent behavioral inefficiency.

Because all of these behaviors and cognitions utilize different sources of temporal information (e.g., the preparation of a motor response occurs at a different time than the expectation of reward), there will be a variety of different temporal decisions produced by those striatal neurons performing temporal computations. Electrophysiological data from these types of studies demonstrate that such varied temporal processing, or at least preparatory- and expectation-related processing, is occurring. Specifically, separate striatal neurons have been found to fire in a preparatory manner for the instructional cues, the action cues, and reward (Schultz and Romo, 1992). Differently timed preparatory motor activity for the different behavioral requirements has also been well documented in the striatum (Jaeger et al., 1993, 1995). Thus, in order to modulate the numerous behavioral and cognitive processes required for successful completion of the task, the appropriate temporal information will need to be utilized by the appropriate control structure (presumably anatomically and functionally separate cortical areas). In other words, this temporal information needs to be sent to the appropriate output region of the GTM.

15.5.2 Localized Proposal

In the localized hypothesis, a variety of temporal criteria would be output by a single striatal area that is specialized for timing. Although this localization hypothesis is consistent with the functional topography that has been shown to exist in the striatum (Alexander et al., 1990; Gerfen and Keefe, 1994; McGeorge and Faull, 1989; Selemon and Goldman-Rakic, 1985; Webster, 1961), in order for these timing neurons to impact a variety of cognitive and motor processes, the temporally informative output would eventually need to project to the variety of cortical areas that will utilize this information (e.g., motor cortex for behavioral modulation, cingulate cortex for attentional modulation). However, the anatomical separation of function found in the striatum is preserved as it moves through the basal ganglia output channels on its way back to the cortex (Hoover and Strick, 1993). Therefore, these anatomical data would force the processes of selection and transmission of the appropriate temporal signal to occur within the cortex (i.e., different temporal signals would need to be sent from the cortical area receiving the timing information to the cortical areas requiring the timing information). Although such a cortical selection–transmission requirement is not unfeasible, we feel that it is a less efficient system and that the process is more vulnerable to disruption than that utilized by a distributed timing system.

15.5.3 DISTRIBUTED PROPOSAL

In a distributed timing system, the criterion time computed by a particular striatal region is restricted to those times that are utilized by the functional region of the cortex to which the striatal region projects. In other words, because temporal information regarding the expected time of the reward is not output from the forearm motor region of the striatum, time-of-reward information does not reach the region of the cortex involved in controlling forearm movement. Instead, only the temporal information specifying when a forearm movement is to be made is passed to this cortical motor region. The anatomical specificity that is required for selecting and transmitting which durations to associate with which behavioral and cognitive processes is already in place, and thereby eliminates the need for the cortex to perform these processes. As such, processing time in a distributed manner seems more likely in terms of the efficiency of information transfer.

Distributed striatal processing of time is also consistent with the cortico-striatal topography and predicts that anatomically focused striatal and cortical regions would be activated during timing tasks. However, the specific cortical and striatal regions that are activated would be highly dependent on the behavioral and cognitive demands of the task. Moreover, the electrophysiological data from the delayed match-to-sample studies described above came from various dorsal and ventral striatal areas, which suggests that such preparatory activity is indeed distributed, rather than localized to a single striatal region. For these reasons, *we propose that the interval timing system provides its information in the form of temporally specific activation of striatal neurons that are primarily involved in processing motor or cognitive information through the basic mechanisms proposed in the SBF model.*

15.5.4 NEURAL EVIDENCE

Our ensemble recording data also support the notion that the processing of time is distributed in the striatum. One consequence of a distributed timer is that those striatal neurons showing decision stage output functions are embedded within functional striatal areas that perform processes other than timing. Indeed, we found that the large majority of striatal neurons had small, secondary activity peaks at the other criterion time (e.g., a large neural peak at 10 sec and a smaller neural peak at 40 sec). These different-sized peaks may be the resultant sum of both lever pressing-related activity and temporal expectation-related activity. Specifically, the smaller neural peak at 40 sec would reflect activity solely related to lever pressing, whereas the larger peak at 10 sec would reflect both lever pressing-related activity and the temporally informative decision stage activity.

In order to evaluate this possibility, we compared the neural activity associated with lever pressing during the 10-sec press period with the neural activity associated with lever pressing during the 40-sec press period. As can be seen in Figure 15.4, there is a phasic increase in the firing rate of this neuron immediately before and after a lever press. This modulation in firing rate occurs during both the early and late presses. The behaviorally locked decrease in firing seen in this figure indicates that this neuron encodes some motor or cognitive process associated with lever

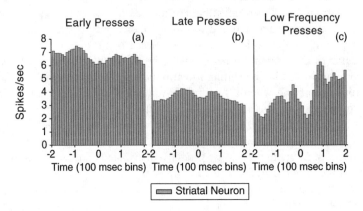

FIGURE 15.4 Striatal neuron activity in relation to lever presses occurring at 10 sec (a) or 40 sec (b). Increases in spike rate immediately before and after a lever press indicate that neural activity is time-locked to this behavior. Note the large difference in the mean spike rate across early and late presses, indicating the temporal specificity of this neuron. The extent of lever press-induced variation in firing rate is masked by the high frequency of pressing during the early and late press periods. Panel (c) shows a more extensive firing rate variation by plotting the spikes resulting from low-frequency presses (1-sec interpress interval). This neuron is the same as that shown in Figure 15.2. Lever presses occur at 0 sec, and data have been smoothed by a 3-bin running mean for presentation.

pressing. Therefore, these data suggest that this neuron does not fall within a pure interval timing region of the striatum. Additional evidence suggesting that this temporally modulated striatal neuron is embedded within a striatal area that is involved in lever pressing is demonstrated by plotting the average lever press-related firing rate on each trial as a function of the average interpress interval on each trial (Figure 15.5). As can be seen, the firing rate of this neuron increased as a function of the interpress interval, indicating that this neuron is encoding detailed features of the behavioral response. Although the correlation was only significant for the late presses (early, $r^2 = .05$, $P > .05$; late, $r^2 = .25$, $P < .01$), the slopes of the regression line are identical in both cases (the differing correlation could be due to the extensive duration modulation seen during the early press period). In addition to this motor modulation, this neuron is clearly providing temporally specific information in its firing rate as well. As seen in Figure 15.4, the background rate is much higher during early presses than late presses. This effect is also evident in Figure 15.5, in which the average early press-related firing rate is generally greater than the average late press-related firing rate.

15.5.5 Competitive Interactions

If timing in the striatum is a distributed process, what predictions does it make regarding the other computations that occur in an interval timing system? We

Integration of Behavior and Timing

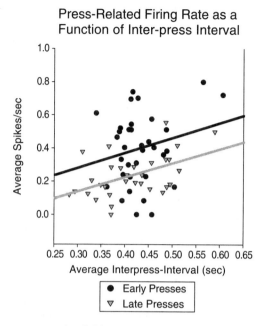

FIGURE 15.5 Average press-related firing rate in a striatal neuron as a function of average interpress interval. The spike rate and interpress interval were computed during press bursts occurring during the 10- or 40-sec peak. The positive slopes for the regression lines indicate that this neuron encodes aspects of the press topography. Only the regression for the late presses was significant ($r^2 = .25$, $P < .01$). This is the same neuron that is shown in Figure 15.2.

discussed earlier the likelihood that the striatal decision stage output influences its subsequent input, thereby creating a dynamic loop. The cortico-striato-thalamo-cortical circuit is thought to be composed of anatomically separate loops (e.g., sensory loop, motor loop, association loop, limbic loop, etc.) (Alexander et al., 1990). However, there is undoubtedly some cross talk occurring at many junctions of the circuit. One dramatic source of cross talk occurs at the input from the cortex to the striatum. Although the predominant source of cortical input to any particular striatal region comes from the matched functional cortical region (and eventually returns to that region), input from multiple cortical areas impinge on single striatal neurons (Finch, 1996). This cross talk leads to the ability of different components of the interval timing systems to modulate not only their own input, but that of the other systems as well, at least to some degree. In other words, *a cortico-striatal interval timing system, although distributed in terms of its immediate timing processes, may be integrated in terms of the structuring of behavior and cognition.* Such integration across systems might thereby induce the formation of behavioral and cognitive chains — chains that are so extensively a part of an organism's temporal behavior that they have been hypothesized to serve as the foundation of timing in the behavioral theory of timing (BET) (Killeen and Fetterman, 1988).

In contrast to BET, which utilizes a single pacemaker for transition from behavioral state to behavioral state and thereby allows only a single state to occur at any one time, a distributed striatal timing process would allow multiple temporally modulated processes to be activated simultaneously (e.g., the delayed match-to-sample processes of attention, gaze, motor preparation, reward expectation, etc.). Although the temporal processing for each of these behavioral and cognitive processes will dynamically influence each other through the mechanisms described above, each circuit would continue to support timing even if one of the other processes is prevented or disrupted. However, if such disruption occurs, the decision stage function for the remaining processes would likely be shifted or altered. The extent of this decision stage alteration would depend on the extent of cross talk between the systems in question. Presumably, preventing the expression of an early motor behavior (e.g., gaze direction toward the instruction cue) will have a greater effect on a subsequent motor behavior (e.g., gaze toward action cue) than on the behavioral expression of reward expectation (e.g., salivation). This proposed influence, rather than dependence, of one set of processes would predict that a subject's temporally informative behavior (i.e., its terminal behavior) would be altered, but not eliminated, by blocking its expression of an early component of its behavior chain. Indeed, Frank and Staddon (1974) have shown that restraining pigeons from producing adjunctive behaviors causes a change in the form of its fixed-interval scallop, but does not dramatically disrupt this behavior in the steady state. Interestingly, their interpretation of this result was that behavioral chains were not necessary for timing, and that adjunctive and terminal behaviors competitively influenced each other.

This competitive transition from one behavioral state to another might evolve naturally through a winner-takes-all process in the striatum. In a winner-takes-all system, a single output signal is selected from a variety of output signals based on which one has the greatest signal strength. Such a network property is thought to arise in systems with lateral inhibitory networks, as the fastest firing neuron (or region) produces the greatest inhibition of its neighbors. As the neighboring neurons are inhibited, their degree of lateral inhibition is also dampened, thereby further increasing the signal of the loudest neuron. Such a mechanism has been proposed to operate in the striatum (Kropotov and Etlinger, 1999; Wickens et al., 1996), through its inhibitory interneurons (Kawaguchi et al., 1995), or through axon collaterals of the GABAergic striatal spiny neurons, which innervate neighboring dendritic fields and may inhibit their likelihood of firing (Kawaguchi et al., 1990).

Interestingly, a winner-takes-all mechanism might explain the slight rightward skew frequently found in peak procedure data (Gibbon and Church, 1984), as well as the finding of the duration bisection at a point at or to the right of the geometric mean of the anchor points (Church and Deluty, 1977; Wearden, 1991), by extending the period in which the strongest signal produces the winning output.* To illustrate this point, imagine an experiment in which a subject is required to time two different

---

* In a bisection procedure the rat is asked to respond on the left lever after presentation of a short anchor duration (e.g., 2 s) and respond on the right lever after presentation of a long anchor duration (e.g., 8 s). Once this discrimination is well learned, intermediate length stimuli (e.g., 3, 4, 5, 6, or 7 s) are presented as probes (i.e., no reinforcement is given), and the rat's classification of the stimuli (i.e., short or long) is determined. The point of subjective equality (i.e., the duration the rat classifies as short 50% of the time and long 50% of the time) is frequently found to occur at the geometric mean of the anchor durations (e.g., 4 s).

durations and respond on two different manipulanda, e.g., bisection procedure (Meck, 1983) or tripeak procedure (Matell and Meck, 1999). One possible algorithm for performing these procedures would entail timing each of the two durations in an independent manner and responding on the appropriate manipulanda in direct proportion to the decision stage output of each timer. If these curves were simply Gaussian in shape, a function that has been used as an approximation of peak data, they would cross at the harmonic mean, a point that occurs to the left of the geometric mean, thereby indicating that additional computations are required or that a Gaussian approximation is incorrect. One possible solution to this problem is to introduce rightward skew into the Gaussian approximation of the decision stage output function. The effect of adding rightward skew to the curves plotted in our bisection example would be to induce a rightward shift in the point at which these curves meet. The precise point at which these curves would cross depends on the amount of skew introduced. Although a rightward skew can be seen in the majority of peak-interval data, and suggests that it is a common feature of timing data, Church et al. (1991) have shown that this skew can be greatly minimized, or eliminated, by varying the length of the probe trials or the intertrial interval. These data therefore indicate that the extent of skew in the temporally informative behavior of an organism is not fixed, but depends on the task parameters. In other words, rightward skew might not be an inherent property of the decision stage output, but may arise through the interactions across multiple timed processes. Such interactions would be expected to occur in a distributed striatal timing model. Specifically, the winner-takes-all mechanism proposed to operate in the striatum might allow such rightward skew to develop naturally simply through the inhibition of competing processes. To reiterate, the combination of a dynamic timing system (i.e., feedback loops from decision stage to clock stage) and a distributed timing system embedded within functional regions of the striatum may lead to the development of competitive interactions between functionally separate striatal outputs, thereby producing the behavioral chains frequently found in timing tasks.

15.6 TEMPORAL AND SEQUENTIAL CONTROL INTERACTIONS

The proposition that interval timing is an important function of the striatum is compatible with its role in other behavioral structuring functions. Striatal involvement in sequence processing has been demonstrated in studies with rats (Aldridge and Berridge, 1998; Berridge and Fentress, 1987), monkeys (Kermadi and Joseph, 1995; Miyachi et al., 1997; Shidara et al., 1998), and humans (Rauch et al., 1997; Toni et al., 1998). Like the control of behavior proposed for interval timing, efficient sequence processing involves directing behavior and attention to sequential elements of the target. Similar to our proposal for distributed striatal processing of time, sequence-related processing has been found across multiple striatal locations and appears to depend on the type of sequence processed and the degree to which the sequence execution has become automatic (Miyachi et al., 1997).

Given the role of the striatum in both sequencing and interval timing, it is not surprising that these two processes are closely related. This phenomenon can be seen anecdotally when humans are asked to time an interval, as they immediately begin counting (i.e., they utilize a sequential behavior strategy). The interrelation of these two processes can be further appreciated given the formulation of models such as BET (Killeen et al., 1988), where the temporal control of behavior does not result from a temporal percept, but instead from the sequential behavior produced by the organism. In contrast to relying on such sequential behavior for timing, we have proposed that sequential-temporal behavioral interactions may develop naturally due to the interacting processes of timing. The converse sequential-temporal interactions are also seen in models of sequence processing that rely on temporal information (Dominey and Boussaoud, 1997), and such an interaction is supported by evidence showing that disrupting the temporal relationship of sequential elements impairs sequence recognition (Dominey, 1998). Extrapolating from these interacting timing and sequencing examples, we conclude that the striatum may be involved in composing or directing behavior by utilizing numerous sources of information so that behavior is executed in an efficient manner. In other words, the striatum may act to guide behavior by controlling the when, which, and how much of motor and cognitive activity.

15.7 FUTURE DIRECTIONS

Although the SBF model successfully predicted the peak-type striatal data described above, we found no evidence of oscillating cortical cells, which were proposed to serve as the clock signal in the model (Matell and Meck, 2000, in press). Furthermore, although there is frequent mention of oscillatory processes in the recording literature (Ahissar and Vaadia, 1990; Gray et al., 1989; Steriade, 1997), cortical oscillations in an appropriate range are most frequently found during inactivity or rest, rather than when the animal is actively behaving — which would be necessary given the existence of the behavioral chains described above. As such, we feel that although the global framework of the SBF model quite likely describes timing in the brain, the reliance on cortical oscillations is a potential weakness of the model. We have not yet fully explored the effects of feedback loops on the generation of a peak output in our simulations. However, given that our model produced peak output without feedback, it is likely that adding feedback would disrupt its output. Similar disruptive effects may also be seen in other timing models following addition of feedback (although in some respects, BET is built on feedback). As such, we have begun to explore whether a cortico-striatal coincidence detection circuit coupled with a feedback loop can produce peak output without the reliance on a variety of oscillatory clock signals. Although we have not yet achieved an acceptable degree of similarity to the decision stage peak output, our results have been encouraging, and we intend to pursue these mechanisms further. Given the known circuitry of the cortico-striatal system, we believe that the investigation of feedback loops in other timing models will be a fruitful source of exploration.

15.8 A FEEDBACK LOOP TO THE BEGINNING

Like our main thesis, we return to our initial statements in which we have outlined the logical possibility that interval timing in the brain might not be composed of independent, serial mechanisms, but rather operate via feedback loops in which the output of an internal clock serves to dynamically modify its own input. We argued that reference to a generalized timing model incorporating both input and output processes could provide guidance in the search for the neural mechanisms of timing because such processes likely play an important part in the production of a temporal percept. We have shown that striatal neurons can indicate times of expected reward through fluctuations in their firing rates, thereby suggesting a prominent role of the striatum in timing and time perception. Thus, it may be possible that the output of individual striatal neurons serves as a decision stage function of an interval timer. Furthermore, based on both our own recording data and anatomical evidence, it appears that the striatal timers are distributed throughout all functional areas of the striatum and are thereby embedded within regions organized for other behavioral and cognitive functions. Such an embedded interval timing system would be an efficient mechanism of organization for a process that is so heavily involved in every facet of behavior. Further, the distributed nature of the timing system may be an important contributor to the behavioral sequences that are engrained in the behavioral expression of temporal perception. Finally, we believe that a thorough investigation of the effects of feedback processes within interval timing models may steer the field of timing and time perception in promising new directions.

REFERENCES

Ahissar, E. and Vaadia, E., Oscillatory activity of single units in a somatosensory cortex of an awake monkey and their possible role in texture analysis, *Proc. Natl. Acad. Sci. U.S.A.*, 87, 8935–8939, 1990.

Aldridge, J.W. and Berridge, K.C., Coding of serial order by neostriatal neurons: a "natural action" approach to movement sequence, *J. Neurosci.*, 18, 2777–2787, 1998.

Alexander, G.E., Crutcher, M.D., and DeLong, M.R., Basal ganglia-thalamocortical circuits: parallel substrates for motor, oculomotor, "prefrontal" and "limbic" functions, *Prog. Brain Res.*, 85, 119–146, 1990.

Beiser, D.G. and Houk, J.C., Model of cortical-basal ganglionic processing: encoding the serial order of sensory events, *J. Neurophysiol.*, 79, 3168–3188, 1998.

Berridge, K.C. and Fentress, J.C., Disruption of natural grooming chains after striatopallidal lesions, *Psychobiology*, 15, 336–342, 1987.

Church, R.M., Timing and temporal search, in *Time and Behaviour: Psychological and Neurobehavioural Analyses*, Bradshaw, C.M. and Szabadi, E., Eds., Elsevier, Amsterdam, 1997, pp. 41–78.

Church, R.M. and Broadbent, H.A., A connectionist model of timing, in *Neural Network Models of Conditioning and Action: Quantitative Analyses of Behavior Series*, Michael, S.G.J.E.R.S. and Commons, L., Eds., Erlbaum, Hillsdale, NJ, 1991, pp. 225–240.

Church, R.M. and Deluty, H.Z., The bisection of temporal intervals, *J. Exp. Psychol. Anim. Behav. Process.*, 3, 216–228, 1977.

Church, R.M., Miller, K.D., Meck, W.H., and Gibbon, J., Symmetrical and asymmetrical sources of variance in temporal generalization, *Anim. Learn. Behav.*, 19, 207–214, 1991.

Dominey, P.F., A shared system for learning serial and temporal structure of sensori-motor sequences? Evidence from simulation and human experiments, *Cognit. Brain Res.*, 6, 163–172, 1998.

Dominey, P.F. and Boussaoud, D., Encoding behavioral context in recurrent networks of the fronto-striatal system: a simulation study, *Cognit. Brain Res.*, 6, 53–65, 1997.

Fetterman, J.G., Killeen, P.R., and Hall, S., Watching the clock, *Behav. Process.*, 44, 211–224, 1998.

Finch, D.M., Neurophysiology of converging synaptic inputs from the rat prefrontal cortex, amygdala, midline thalamus, and hippocampal formation onto single neurons of the caudate/putamen and nucleus accumbens, *Hippocampus*, 6, 495–512, 1996.

Frank, J. and Staddon, J.E.R., Effects of restraint on temporal discrimination behavior, *Psychol. Rec.*, 24, 123–130, 1974.

Gerfen, C.R. and Keefe, K.A., Neostriatal dopamine receptors, *Trends Neurosci.*, 17, 2–3, 1994.

Gibbon, J., Scalar expectancy theory and Weber's law in animal timing, *Psychol. Rev.*, 84, 279–325, 1977.

Gibbon, J. and Church, R.M., Time left: linear versus logarithmic subjective time, *J. Exp. Psychol. Anim. Behav. Process.*, 7, 87–108, 1981.

Gibbon, J. and Church, R.M., Sources of variance in an information processing theory of timing, in *Animal Cognition*, Roitblat, H.L., Bever, T.G., and Terrace, H.S., Eds., Erlbaum, Hillsdale, NJ, 1984, pp. 465–488.

Gray, C.M., Konig, P., Engel, A.K., and Singer, W., Oscillatory responses in cat visual cortex exhibit inter-columnar synchronization which reflects global stimulus properties, *Nature*, 338, 334–337, 1989.

Hoover, J.E. and Strick, P.L., Multiple output channels in the basal ganglia, *Science*, 259, 819–821, 1993.

Houk, J.C., Information processing in modular circuits linking basal ganglia and cerebral cortex, in *Models of Information Processing in the Basal Ganglia*, Houk, J.C., Davis, J.L., and Beiser, D.G., Eds., MIT Press, Cambridge, MA, 1995, pp. 3–10.

Jaeger, D., Gilman, S., and Aldridge, J.W., Primate basal ganglia activity in a precued reaching task: preparation for movement, *Exp. Brain Res.*, 95, 51–64, 1993.

Jaeger, D., Gilman, S., and Aldridge, J.W., Neuronal activity in the striatum and pallidum of primates related to the execution of externally cued reaching movements, *Brain Res.*, 694, 111–127, 1995.

Kawaguchi, Y., Wilson, C.J., Augood, S.J., and Emson, P.C., Striatal interneurones: chemical, physiological and morphological characterization, *Trends Neurosci.*, 18, 527–535, 1995.

Kawaguchi, Y., Wilson, C.J., and Emson, P.C., Projection subtypes of rat neostriatal matrix cells revealed by intracellular injection of biocytin, *J. Neurosci.*, 10, 3421–3438, 1990.

Kermadi, I. and Joseph, J.P., Activity in the caudate nucleus of monkey during spatial sequencing, *J. Neurophysiol.*, 74, 911–933, 1995.

Killeen, P.R. and Fetterman, J.G., A behavioral theory of timing, *Psychol. Rev.*, 95, 274–295, 1988.

Kropotov, J.D. and Etlinger, S.C., Selection of actions in the basal ganglia-thalamocortical circuits: review and model, *Int. J. Psychophysiol.*, 31, 197–217, 1999.

Matell, M.S. and Meck, W.H., Reinforcement-induced within-trial resetting of an internal clock, *Behav. Process.*, 45, 159–172, 1999.

Matell, M.S. and Meck, W.H., Neuropsychological mechanisms of interval timing behavior, *Bioessays*, 22, 94–103, 2000.

Matell, M.S. and Meck, W.H., Cortico-striatal circuits and interval timing: coincidence-detection of oscillatory processes, *Cog. Brain Res.*, in press.

Matell, M.S., Meck, W.H., and Nicolelis, M.A.L., Interval timing and the encoding of stimulus duration by striatal and cortical neurons, *Behav. Neurosci.*, in press.

McGeorge, A.J. and Faull, R.L.M., The organization of the projection from the cerebral cortex to the striatum of the rat, *Neuroscience*, 29, 509–537, 1989.

Meck, W.H., Selective adjustment of the speed of internal clock and memory processes, *J. Exp. Psychol. Anim. Behav. Process.*, 9, 171–201, 1983.

Miall, R.C., The storage of time intervals using oscillating neurons, *Neural Comput.*, 1, 359–371, 1989.

Miyachi, S., Hikosaka, O., Miyashita, K., Karadi, Z., and Rand, M.K., Differential roles of monkey striatum in learning of sequential hand movement, *Exp. Brain Res.*, 115, 1–5, 1997.

Nicolelis, M.A., Ghazanfar, A.A., Faggin, B.M., Votaw, S., and Oliveira, L.M., Reconstructing the engram: simultaneous, multisite, many single neuron recordings, *Neuron*, 18, 529–537, 1997.

Penney, T.B., Gibbon, J., and Meck, W.H., Differential effects of auditory and visual signals on clock speed and temporal memory, *J. Exp. Psychol. Hum. Percept. Perform.*, 26, 1770–1787, 2000.

Rauch, S.L., Whalen, P.J., Savage, C.R., Curran, T., Kendrick, A., Brown, H.D., Bush, G., Breiter, H.C., and Rosen, B.R., Striatal recruitment during an implicit sequence learning task as measured by functional magnetic resonance imaging, *Hum. Brain Mapping*, 5, 124–132, 1997.

Roberts, S., Isolation of an internal clock, *J. Exp. Psychol. Anim. Behav. Process.*, 7, 242–268, 1981.

Schultz, W. and Romo, R., Role of primate basal ganglia and frontal cortex in the internal generation of movements: I. Preparatory activity in the anterior striatum, *Exp. Brain Res.*, 91, 363–384, 1992.

Selemon, L.D. and Goldman-Rakic, P.S., Longitudinal topography and interdigitation of corticostriatal projections in the rhesus monkey, *J. Neurosci.*, 5, 776–794, 1985.

Sesack, S.R., Deutch, A.Y., Roth, R.H., and Bunney, B.S., Topographical organization of the efferent projections of the medial prefrontal cortex in the rat: an anterograde tract-tracing study with Phaseolus vulgaris leucoagglutinin, *J. Comp. Neurol.*, 290, 213–242, 1989.

Shidara, M., Aigner, T.G., and Richmond, B.J., Neuronal signals in the monkey ventral striatum related to progress through a predictable series of trials, *J. Neurosci.*, 18, 2613–2625, 1998.

Staddon, J.E.R. and Higa, J.J., Time and memory: towards a pacemaker-free theory of interval timing, *J. Exp. Anal. Behav.*, 71, 215–251, 1999.

Steriade, M., Synchronized activities of coupled oscillators in the cerebral cortex and thalamus at different levels of vigilance, *Cereb. Cortex*, 7, 583–604, 1997.

Strick, P.L., Dum, R.P., and Picard, N., Macro-organization of the circuits connecting the basal ganglia with the cortical motor areas, in *Models of Information Processing in the Basal Ganglia*, Houk, J.C., Davis, J.L., and Beiser, D.G., Eds., MIT Press, Cambridge, MA, 1995, pp. 117–130.

Toni, I., Krams, M., Turner, R., and Passingham, R.E., The time course of changes during motor sequence learning: a whole-brain fMRI study, *Neuroimage*, 8, 50–61, 1998.

Wearden, J.H., Human-performance on an analog of an interval bisection task, *Q. J. Exp. Psychol. Comp. Physiol. Psychol.*, 43B, 59–81, 1991.

Webster, K.E., Cortico-striate interrelations in the albino rat, *Proc. Natl. Acad. Sci. U.S.A.*, 87, 7050–7054, 1961.

Wickens, J.R., Begg, A.J., and Arbuthnott, G.W., Dopamine reverses the depression of rat corticostriatal synapses which normally follows high-frequency stimulation of cortex in vitro, *Neuroscience*, 70, 1–5, 1996.

Wilson, C.J., Groves, P.M., Kitai, S.T., and Linder, J.C., Three-dimensional structure of dendritic spines in the rat neostriatum, *J. Neurosci.*, 3, 383–398, 1983.

16 Time Flies and May Also Sing: Cortico-Striatal Mechanisms of Interval Timing and Birdsong

Christopher J. MacDonald and Warren H. Meck

CONTENTS

16.1 Introduction 393
16.2 Birdsong: A Temporal Hierarchy..394
    16.2.1 The Development of Avian Vocal Behavior 395
    16.2.2 Behavioral Parallels between Interval Timing and Birdsong..........395
16.3 Parallels between the Mammalian and Avian Telencephalon.....................397
16.4 Areas in the Avian Telencephalon Important for Vocal Learning 398
    16.4.1 Posterior Motor Pathway 400
    16.4.2 Is the Auditory Template Contained within the Anterior Forebrain Pathway? 402
    16.4.3 An Alternative Site for the Auditory Template 403
    16.4.4 What Type of Biological Substrate Is Required for Song Learning?...404
16.5 Can Parallel Cortico-Striatal Modules Mediate Song Learning? 405
    16.5.1 The Cortico-Striatal Module as a Pattern Detector.........................406
    16.5.2 The Auditory Processing Module 407
    16.5.3 Role of the Comparison Signal's Transmission to the AFP in Relation to the Posterior Motor Pathway 408
16.6 Reinterpretation of Some Experimental Results 410
16.7 Conclusion and Future Perspectives..411
Acknowledgments..412
References...412

16.1 INTRODUCTION

There is ample evidence that humans and other animals can time multiple events embedded within a hierarchical structure when presented with complex schedules of reinforcement or stimulus sequences (e.g., Leak and Gibbon, 1995; Meck and Church, 1982, 1984; Pang and McAuley, this volume; Rousseau and Rousseau, 1996). It is unclear, however, whether there are specialized systems for the perception and production of serial-ordered behavior that rely on the same interval timing mechanisms that are engaged by the types of stimuli that are typically presented in laboratory studies of timing and time perception. One such system might be the vocal learning circuit of songbirds, which is highly structured and specific.

To date, the study of interval timing and birdsong acquisition has remained largely detached (but see Hills, this volume; Weisman et al., 1999). Laboratory research on interval timing has traditionally used rats, pigeons, and humans. In contrast, the songbird may be an ideal organism for studying certain aspects of interval timing because the neural pathways that are involved in song learning and production are well described from both a molecular and electrophysiological perspective. Although this description is by no means complete, the interaction between song learning and production coupled with the neurobiological findings makes the songbird amenable to bio-behavioral study. Consequently, this chapter will begin with a brief introduction to the temporal patterns of behavior occurring in birdsong and a hypothesis of a connection to interval timing. The latter part of this chapter will explore the neural systems underlying birdsong in the context of what is known about the neurophysiology of interval timing in mammals. This is followed by the proposal of an alternative solution for song learning that relies on the same interval timing mechanisms studied in other species.

16.2 BIRDSONG: A TEMPORAL HIERARCHY

Songbirds comprise nearly half of the known living avian species and thus make up the largest suborder of birds. Ethologists divide songbird vocalizations into songs and calls. In general, songs are longer and more syntactically complex sounds relative to calls, although this is not an absolute distinction. Songs are organized by a temporal hierarchical structure. The simplest individual sounds that a bird can make are termed *notes*. A bird can string together several notes without a silent interval, in which case it is called a *syllable*. These syllables are further arranged into sequences to form song *types* or *motifs*. The durations of song types can also vary among species from the 10-sec utterances of the male winter wren to the 1-sec hiccup of the winter Henslow. Characteristically, an individual songbird will produce several song types, called its *repertoire*. However, the repertoire number can vary greatly from species to species.

Typically, a songbird will produce *bouts* of singing by arranging discrete song types into a sequence. The song types are separated by variable intervals of time that may be as short as 5 sec during an intense bout of singing. The bouts themselves also vary among species. A bout will typically contain many vocalizations whereby a bird will systematically cycle through its song types (e.g., A, A, A, B, B, B, B, C, C, C, ...). The number of songs in a bird's repertoire, and how they are presented, is thought to be an important functional aspect of communication (Hartshorne, 1956; Krebs, 1977; Kroodsma, 1980). Accordingly, birdsong is a complex motor sequence that brings to light a dilemma that commonly arises in the study of serial-ordered behavior (Lashley, 1951). How do songbirds learn and express motor programs that are marked by rigid ordinality and variable tempo while maintaining a consistent hierarchical structure?

16.2.1 THE DEVELOPMENT OF AVIAN VOCAL BEHAVIOR

Up to this point, all studies conducted on the oscine suborder of songbird species have revealed that song is learned (Ball and Hulse, 1998; Kroodsma and Baylis, 1982). Although the specifics in the development of birdsong can vary among species, some useful generalizations have emerged. Vocal learning in the songbird can be divided into two stages, a sensory phase and a sensorimotor integration phase (Konishi and Nottebohm, 1969; Marler, 1970).

In the *sensory phase*, a young songbird is exposed to conspecific utterances from either its father or other males that are within hearing distance. The juvenile songbird is believed to form a sensory template that represents the memorized vocalizations from the tutor so that it may be used to guide song development. Before the sensorimotor integration phase begins, the bird goes through a period called subsong. Subsong is characterized by amorphous and unstructured vocalizations akin to a soft babbling and is considered nonimitated vocal motor practice. This leads to the *sensorimotor integration phase*, where subsong progresses to plastic song and many of the elements that are common to the memorized template begin to appear; however, the final organization is incomplete (Marler and Peters, 1982). Recent findings suggest that juvenile zebra finches that are not exposed to song will begin vocalizing by instinctively producing a string of back-to-back syllable prototypes. The acquisition of an auditory representation permits different syllables to emerge from these prototypes (Tchernichovski et al., 2001). Lastly, the final form of the birdsong develops, which represents the adult *crystallized* song type.

16.2.2 BEHAVIORAL PARALLELS BETWEEN INTERVAL TIMING AND BIRDSONG

The presiding thought many years ago was that the auditory template guiding singing was tantamount to a motor tape so that each vocal production of a song type was considered invariant (e.g., Konishi and Nottebohm, 1969; Marler, 1970). However, a song type may be better thought of as a theme that underscores a number of vocalizations, some of these being improvisations that are scarcely repeated (Podos et al., 1992). As a result, in song sparrows it has been said that the "motor representation of song appears to be probabilistic, with each song type stored as an abstract average that carries with it probabilities describing an allowable range of within-type variation" (Podos et al., 1992, p. 104). Another way of interpreting this definition is that a song type may be defined by the probability of there being systematic behavioral regularities as a function of time.

Defining birdsong in this manner parallels some of the properties of interval timing (see Church, 1984, this volume). Scalar timing theory accounts for the decision of when to begin responding during an interval with a ratio comparison (e.g., Church and Gibbon, 1982; Gibbon, 1977; Gibbon et al., 1984). The comparison is made between two values. The first value is the current estimate of elapsed time since signal onset. The second value is the representation of the expected time of signal termination, selected from a normal distribution of expected times in memory. Once the current estimate of elapsed time exceeds a fixed fraction of the expected time to signal termination, the animal begins to respond in that trial. With regard to the development of birdsong, interval timing could mediate the juvenile songbird's perception of the temporal relationship among the syllables within the song during the sensory phase. Once an auditory representation of the adult song is formed, it may serve to impose temporal structure onto a motor sequence. This modulation of the vocal motor sequence would typically occur during the sensorimotor phase of development.

How would the auditory representation be formed in the first place? Each syllable in the song is equivalent to a subdividing stimulus that provides information about the hierarchical structure of the song. Within this structure, song completion serves as a common reference point, as there is evidence that it could serve as salient feedback (e.g., Adret, 1993; Stevenson, 1967, 1969). This arrangement is amenable to a system that mediates behavior where multiple timing mechanisms would operate from syllable onsets until song completion. The ability to independently time each perceived syllable within the song in relation to song completion may facilitate the emergence of specific patterns of motor production, as observed in other examples of simultaneous temporal processing (e.g., Meck and Church, 1984).

Because a syllable's duration can vary, there is a potential problem in determining what point in time within the syllable is credited with stimulus onset. This problem is avoided when one considers that categorical perception underlies a songbird's discrimination between syllables (Marler, 1989). By imposing a boundary on the syllable's continuum, discriminations among syllables within a given category are made ineffectual, while discriminations among syllables that straddle the boundary are made with relative ease. Therefore, in order for the syllable to be a reliable *time marker* a salient feature must be categorically perceived at some consistent time point within the syllable. The perception of this time marker within the syllable is a function of its overall dynamic spectra. How the songbird selects the conspecific syllable from the hodgepodge of sounds that pervade its auditory space is not known. There is evidence, however, that conspecific and heterospecific syllables are not equipotential as song learning stimuli (Marler and Peters, 1988). Moreover, some species of songbird learn songs by perceiving salient, introductory whistles that appear to draw attention to the song that follows (Soha and Marler, 2000).

There are indications that the syllable can be used as a reliable time marker in order to predict song completion. Canaries can learn to use perceived syllables as time markers during song playback if the song's offset is consistently paired with a noxious stimulus, like in a shock avoidance paradigm (Jarvis et al., 1995). The canaries first initiated an avoidance response at the onset of the song. However, with continued training, they gradually transferred the avoidance response to the final syllable in the song, right before electric shock was given. It was as if once the specific time of the shock's occurrence in relation to song onset was determined, it was used to guide the canary's decision of when to produce a fear response (cf. Gibbon, 1977).

Is it possible that the songbird's neural system evolved in such a way that it was able to exploit an ability to attend to brief durations of numerous auditory stimuli that simultaneously elapse (e.g., Meck and Church, 1984; Rousseau and Rousseau, 1996)? Would this allow the songbird to process the various auditory signals in parallel for the guidance of vocal development? In order to substantiate this hypothesis, a logical place to begin would be to first describe the similarities between the mammalian and avian brain. This would allow us to compare the properties of the various neural circuits implicated in interval timing in mammals with the avian neural circuits that have been found to contribute to birdsong.

16.3 PARALLELS BETWEEN THE MAMMALIAN AND AVIAN TELENCEPHALON

An extensive number of structural, functional, embryological, and genetic studies support the hypothesis that the telencephalon is composed of three parts: the pallium, striatum, and pallidum regions (e.g., Puelles et al., 2000; Smith-Fernendez et al., 1998; Striedter, 1999; Swanson, 2000). The *pallium* can be further separated into four subdivisions. The dorsal pallium is the most conspicuous part of the mammalian telencepalon. The greater part of the mammalian telencephalon is dorsal pallium, similar to an external mantle encasing the brain that can be segregated into different areas depending on lamination patterns. Broadly speaking, these areas are functionally differentiated. However, the avian brain has significantly less dorsal pallium that makes up the telencephalon. The majority of the avian telencephalon is derived from the ventral and lateral pallium. In mammals, the ventral and lateral pallium each contribute parts of the amygdala and give rise to the claustrum and pyriform cortex. A fourth subdivision exists in the avian and mammalian telecephalon — the medial pallium. In the mammal, the medial pallium develops into the hippocampal formation.

The *striatum* is separated into two parts. The dorsal striatum consists of the caudate and putamen, which together are often referred to as the striatum in the literature, though this obfuscates the so-called ventral striatum. There is a diffuse projection from the whole of the striatum to the midbrain dopaminergic neurons (MDNs) in mammals (Gerfen, 1985; Parent and Cicchetti, 1998; Swanson, 2000), and this is paralleled in the avian brain (Brauth et al., 1978; Karten and Dubbeldam, 1973; Kitt and Brauth, 1981; Reiner, 2002). The dorsal striatum also sends axons to the dorsal pallidum, or the globus pallidus, while the ventral striatum sends projections to the ventral pallidum. In some cases, the striatal efferents to the pallidum are collaterals that branch from those striatal efferents that project to the MDNs.

The striatum receives excitatory input from diffuse topographical projections that originate throughout the entirety of the pallium (McGeorge and Faull, 1989; Webster, 1961). In fact, the principal cell type in the striatum, the spiny cell, has been estimated to have between 10,000 and 30,000 dendritic spines, all of which are hypothesized to be contacted by different isocortical or thalamic neurons in mammals (e.g., Groves et al., 1995; Wilson, 1995). The avian striatum is densely innervated by pallium afferents as well (Brauth et al., 1978; Reiner, 2002; Veenman et al., 1995). The cellular constituents of the avian striatum are electrophysiologically and morphologically comparable to those found in the mammalian striatum. Furthermore, there is a large degree of convergence on avian spiny striatal cells, much like the mammalian striatum (Reiner, 2002; Reiner et al., 1998).

In both mammals and songbirds, the striatum serves as a nexus for the functions that different pallial areas subserve (e.g., vision, audition, motor production, and so forth). Therefore, the striatum is considered an ideal candidate for a locus of sensory and motor integration. In primates, there is evidence to suggest that reciprocal striato-nigral-striatal connections allow an interface among the different areas of the striatum and, as a result, functionally distinct areas of the pallium (Haber et al., 2000). In the mammalian (Swanson, 2000) and avian (Reiner, 1998, 2002) brain, there are diffuse reciprocal projections from the MDNs back to the entire striatum. The evidence suggests that reciprocal connections between the striatum and MDNs encourage a distributed interplay between these two systems (Joel and Weiner, 2000). One consequence of this interaction may be sensorimotor learning.

Another form of organization is superimposed on this system within the mammal. The cortical efferent to the striatum is followed by a striatal projection to the *pallidum* that represents another degree of convergence because the number of spiny striatal cells in relation to its primary target cells (both the globus pallidus and MDNs) reflects a 30:1 ratio in the rat and an 80:1 ratio in the monkey (e.g., Oorschot, 1996). The globus pallidus sends collaterals to the thalamus. The thalamus, in turn, sends an excitatory projection back to the same primary cortical area to form topographical, segregated loops (see Alexander et al., 1986; Matell et al., this volume). The loop is closed in the sense that the projections throughout the loop remain in register (Hoover and Strick, 1999; Kelley and Strick, 1999). This structural arrangement lends itself well to information-processing models based on coincidence detection mechanisms (e.g., Beiser et al., 1997). The presence of cortico-striatal modules gives further support to the emerging belief that the mammalian striatum contributes to cognition in addition to motor processes (e.g., Brown et al., 1997; Malapani and Rakitin, this volume; Middleton and Strick, 2000; White, 1998).

16.4 AREAS IN THE AVIAN TELENCEPHALON IMPORTANT FOR VOCAL LEARNING

Vocal learning is a rare trait that enables an organism to imitate sounds that it hears. To date, vocal learning is known to exist in only three avian orders, which are thought to have evolved this trait independently (e.g., Brenowitz, 1997; Nottebohm, 1970). There is some evidence to suggest that the phylogenesis of vocal learning arises in parallel with a systematic constraint in the telencephalon (Jarvis et al., 2000). All three orders of vocal learners have vocal learning nuclei in the same relative positions within their respective telencephalons. Furthermore, the connectivity between them is conserved so that there are two distinct pathways: a posterior motor/production pathway and an anterior forebrain pathway (AFP). The posterior motor pathway is directly involved with singing throughout the bird's life. The anterior forebrain pathway is necessary for song learning, but under normal circumstances is not required after the final song is established.

Cortico-Striatal Mechanisms of Interval Timing and Bird Song

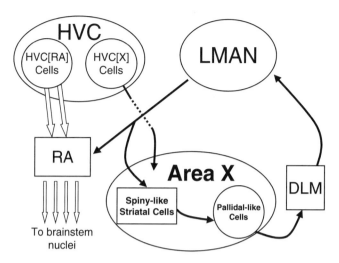

FIGURE 16.1 The connections of the posterior motor pathway and anterior forebrain pathway. The posterior motor pathway (white arrows) begins in the high vocal center and extends to the robust nucleus of the archistriatum via HVC[RA] projection neurons. RA neurons extend to lower motor neurons/respiratory neurons in the songbird throat (see text). For the AFP (black arrows), Area X receives input from the HVC from HVC[X] projection neurons. The spiny striatal cells in Area X extend to pallidal-like cells, also within Area X. The pallidal-like cells leave Area X and input to neurons in the medial nucleus of the dorsolateral thalamus that proceed to send efferents to the lateral magnocellular nucleus of the anterior neostriatum. Bifurcating neurons within LMAN project to RA and back to Area X. The Area X to DLM to LMAN pathway is closed and remains in register (see text).

The *posterior motor pathway* encompasses a nucleus in the avian pallium called the high vocal center (HVC) and one of its projections to the nucleus, called robustus archistriatalis (RA). The HVC also contributes to the AFP by sending a projection to Area X, as illustrated in Figure 16.1. Area X is found within the avian equivalent of the striatum. Area X sends an efferent to the dorsolateral anterior thalamic nucleus (DLM) that in turn projects back up to another ventral pallial area called the lateral magnocellular nucleus of the anterior neostriatum (LMAN). The LMAN projection to RA is topographic so that the projections are horizontally layered across RA. This organization is compatible with a myotopic organization of the RA efferents that extend to the throat musculature (Vicario, 1991). LMAN projection neurons bifurcate so that one axon extends to RA, and a collateral extends to Area X. The LMAN projection to Area X forms a feedback loop to give rise to distinct modular domains.

The AFP is a pathway that parts from HVC only to append with the posterior motor pathway at the level of RA. What is more, an additional cell type is present in Area X that resembles those found in the mammalian dorsal pallidum, or globus pallidus (Farries and Perkel, 2002). Therefore, the AFP is recursive and forms a closed cortico-striatal loop: the parts of the loop are Area X, DLM, and LMAN (Luo et al., 2001). Current thinking is that the spiny striatal cell in Area X projects to the pallidal-like cells, which then project to the thalamic DLM, which sends efferents to LMAN. The pallidal-like cells in Area X are far outnumbered by the spiny cells, suggestive of a high degree of convergence. Thus, all of the elements of the mammalian cortico-striatal module are embraced within the avian cortico-striatal counterpart. The loop is topographically organized in a manner that accords with the myotopic architecture of RA.

The avian cortico-striatal-like module may not be precisely analogous to its mammalian counterpart. The mammalian module receives projections from the dorsal pallium, while the avian module includes projections from the ventral and lateral pallium — what will loosely be considered the avian cortex. Nevertheless, from an interval timing perspective, these findings imply that the evolution of integrated sensorimotor behavior may entail the development of a common unit of processing. The modular organization in Area X suggests that these are functional domains that process information via a closed loop in a manner that preserves myotopic organization (Vates and Nottebohm, 1995). The main question is whether this closed topographical loop specialized for the learning and production of birdsong also plays a role in the more general process of interval timing? In order to address how the vocal pathways are involved in sensorimotor learning, we will first begin at the level of motor production. From there, we will ask how the hierarchical structure of a learned song template can be used to shape the song's production.

The zebra finch (*Taenopyggia guttata*) is the preferred choice of species for those who look at birdsong from a neurobiological perspective. This is partly due to the fact that the zebra finch has one motif (song type) that is syntactically simple and deterministic. The motif is composed of several syllables, each syllable made up of notes. The zebra finch will engage in singing bouts where it will repeat its motif. The variation in the motif over a bout is very small, although this is influenced by different social contexts (Sossinka and Bohner, 1980).

16.4.1 POSTERIOR MOTOR PATHWAY

The HVC is thought to be an avian homologue of the mammalian primary motor cortex, layers II and III, and the RA nucleus is thought to be a homologue of layer V, which projects to lower motor neurons (Karten, 1969). Specifically, RA projects to a set of brain stem nuclei that coordinate respiration and vocalization; the most prominent are the tracheosyringeal part of the hypoglossal nerve motor nucleus (nXIIts).

It was established early on that both HVC and RA are required for production of song, as lesions of either HVC or RA result in silent song (Nottebohm et al., 1976). Silent song was distinguished by the observation that the songbird engaged in all other mannerisms that were characteristically associated with singing, but did not produce song itself. These results suggested that both RA and HVC are necessary to produce movement in the songbird's throat musculature. This may not be surprising given that lesions of the mammalian motor isocortex lead to severe deficits in motor performance. It is difficult to find parallels in the mammalian interval timing literature regarding primary motor isocortical lesions. If the behavioral output is severely disrupted, one cannot properly assess deficits in interval timing (see Sasaki et al., 2001).

Single-unit recordings in HVC cells identified as projection neurons to RA (HVC[RA]) exhibit bursting only at a specific time in a motif. That is to say, each HVC[RA] cell that was recorded bursted reliably at exactly one point in the sequence (Hahnloser et al., 2002). In addition, simultaneous recordings in RA revealed that a single RA projection neuron to the throat musculature will burst multiple times during a single motif, but each occasion appears to be tightly time-locked to the single firing of a specific HVC[RA] neuron. Moreover, there is strong evidence supporting a causative relationship between the burst of an HVC[RA] neuron and an ensuing burst in an RA neuron that occurs with a minimal latency. Thus, it would seem that a single HVC[RA] neuron fires dependably at one time in the motif and immediately yields a burst in an RA cell. A single RA cell may burst multiple times during a motif, but presumably each burst is driven by a different HVC[RA] cell that is time-locked to fire at that point in time. It has been proposed that there is a direct temporal map imposed on the various HVC[RA] neurons (Hahnloser et al., 2002). Different subpopulations of HVC[RA] neurons are firing at a predetermined time once per motif, and this drives the firing of different RA cells projecting onto the songbird's throat musculature.

If a syllable is repeated twice within the same motif, will the subpopulation of HVC[RA] cells used to code the syllable's first occurrence be used to induce the same RA cell activity during its second occurrence? In other words, how specific is the activity of HVC[RA] cells during a motif with respect to time? A purely temporal code would imply that a unique subpopulation of neurons would burst fire only once during a given motif's duration. Although this has not been directly addressed, there is sufficient evidence to offer a hypothesis.

Each RA neuron generates a mean of approximately ten bursts for each motif so that 15% of the total RA population is active at any one time during a motif (Hahnloser et al., 2002). Identical patterns generated by a group of RA neurons lead to identical syllables, whereas similar, but not identical syllables that share many notes in common can have drastically different RA patterns. Therefore, it seems likely that if a syllable is repeated twice within the same motif, the same population of HVC[RA] neurons codes the sequence of each syllable (M. Fee, personal communication, June, 2002). The different subpopulations of HVC[RA] neurons that are time-locked for an RA sequence serve as indicators that "point" to a group of RA neurons corresponding to a sound that is intended to be produced at that point in time. This is entirely different from the earlier song code mechanisms that were proposed (e.g., Yu and Margoliash, 1996). Previously, HVC and RA were functionally dichotomized. HVC was thought to code for structure over an interval that corresponded to a syllable, and RA for much shorter durations at the level of a note. In the most recent proposal, the sequence of activity that unfolds over time is inherent to the individual subpopulations of cells in HVC. This sequence of activity is projected onto a "muscle map" in RA to create a song so that HVC and RA are functionally connected; i.e., a temporal code operates on a spatial code. The results presume that the time at which these HVC[RA] neurons are firing in a motif is already ingrained within their respective subpopulations. These conclusions do not focus on *how* the neurons in HVC[RA] come to be tuned to fire at an appropriate point in time in order to select the sequence of RA neurons corresponding to a particular syllable.

How is the auditory template in the songbird used to guide the formation of connections between the HCV[RA] and RA? Before answering this question, the location of the auditory template in the avian brain must be uncovered.

16.4.2 Is the Auditory Template Contained within the Anterior Forebrain Pathway?

The finding that the AFP is crucial for the development of complex vocalizations within the songbird, but not for maintenance under normal conditions in adulthood, was a significant discovery for those searching for the auditory template in the songbird (Bottjer et al., 1984; Scharff and Nottebohm, 1991). Auditory properties were first described in the HVC of anesthetized zebra finch; however, they were soon found throughout the AFP as well. Additionally, auditory responses have been recorded within RA of the zebra finch. The responses within RA are thought to depend on HVC input due to the fact that lesions of LMAN (also afferent to RA) do not abolish auditory responses (Doupe and Konishi, 1991).

An auditory property is established within a cell if it is observed to preferably respond to one sound stimulus over another. Many neurons in HVC fire in preference to an autogenous song (what will henceforth be referred to as the bird's own song (BOS)) over a conspecific or heterospecific song. With this in mind, the auditory properties in HVC have been considerably elaborated on by exposing the adult bird to different types of sounds such as the BOS. Multiunit recordings of HVC neurons fired vigorously during the BOS, more than what was observed when the bird was presented with its tutor's song or other conspecific song (Margoliash, 1986; Margoliash and Konishi, 1985). What's more, often these neurons will exhibit temporal combination sensitivity (Lewicki and Konishi, 1995; Margoliash, 1983; Margoliash and Fortune, 1992). This property illustrates a neuron's sensitivity to two attributes in a complex acoustical stimulus like birdsong: order selectivity and combination sensitivity. *Order selectivity* is demonstrated when the neuron fires at a higher rate to the BOS than when the BOS syntactic organization is manipulated. This includes playing back the song in reverse (a global song manipulation), reversing the temporal order within the syllables, but maintaining the same syllable sequence (a local song manipulation) or presenting the syllables in reverse order so that the local temporal order within each syllable remains the same (a more specific global manipulation). *Combination sensitivity* describes an observation that a neuron will fire in response to a combination of certain syllables at a greater rate than what would be predicted by summing the responses to each individual syllable component.

Consequently, the AFPs necessity for song learning, the existence of auditory properties within the AFP that are specific to the juvenile BOS, and association with a posterior pathway that is necessary for motor production called to mind the possibility that the AFP contained the auditory template (e.g., Doupe, 1993; Doupe and Solis, 1997). However, the vast majority of experiments that confirm auditory responses in various song learning nuclei are performed on the anesthetized songbird. It is very difficult to obtain auditory responses from the HVC and other vocal nuclei in the awake bird. The auditory responses recorded in HVC and other vocal nuclei are a product of the behavioral state that is generated by the anesthetic (Capsius and Leppelsack, 1996; Dave et al., 1998; Margoliash, 1997; Schmidt and Konishi, 1998). The auditory properties obtained from HVC[RA] cells in the anesthetized bird compare favorably with recent results displaying the feature of one burst per motif, reminiscent of the "temporal code" (Hahnloser et al., 2002). Thus, the auditory responses may be a reflection of the motor response one would expect in the awake, behaving bird rather than true acoustic-based properties. For that reason, it is equally likely that the AFP represents a motor-like conduit that can use an auditory-related signal generated elsewhere. That is, the AFP may not be the genuine locus of auditory memories, but may serve as a specialized motor pathway that conducts a signal used to facilitate the formation of accurate HVC[RA] and RA synapses in the posterior motor pathway. Indeed, it came as a surprise to some avian researchers that the AFP was active during song production, as revealed by robust gene expression within the AFP nuclei (Jarvis and Nottebohm, 1997).

Another point to consider about the AFP in the context of song development is the importance of auditory feedback. The final song is made to be abnormal after depriving the songbird of auditory feedback by deafening after the memorization phase but preceding the sensorimotor phase (Konishi and Nottebohm, 1969). What is more, the nature of crystallized adult song is not so straightforward. The maintenance of the adult final song requires real-time auditory feedback (Leonardo and Konishi, 1999; Nordeen and Nordeen, 1992; Okanoya and Yamaguchi, 1997). For example, deafening the adult songbird or perturbing the auditory feedback can result in decrystallization of the adult song. On the other hand, lesioning parts of the AFP that disrupt this circuit in a deafened adult prevent song deterioration (Brainard and Doupe, 2000). Therefore, the AFP is also thought to play a role in evaluating auditory feedback.

16.4.3 AN ALTERNATIVE SITE FOR THE AUDITORY TEMPLATE

We propose that the source of the auditory template resides in areas that are involved in basic auditory processing in songbirds and nonsongbirds. Field L is the primary auditory region in the songbird and is the avian homologue of mammalian auditory cortex. It is a large and laminated structure that is subdivided and composed of multiple nuclei that abut with indistinct borders (Fortune and Margoliash, 1992; Vates et al., 1996). Field L is functionally complicated, abounding with reciprocal connections among its subdivisions, and is organized in a tonotopic manner (Muller and Leppelsack, 1985). One of its subdivisions sends a dense projection to a nearby area known as the caudal nucleus of the neostriatum (NCM). The NCM is thought to be an important component of the song system due to its apparent involvement in song perception. There is a vigorous increase of electrophysiological activity at the onset of the sensory phase in the NCM following conspecific song presentation relative to other songs or simple acoustic stimuli (e.g., tone bursts or white noise) in the awake, behaving zebra finch (Stripling et al., 1997, 2001). More recently, it was found that the presentation of individual syllables to an adult canary gives rise to unique patterns of early gene expression in the NCM (Ribeiro et al., 1998). A complex stimulus like the syllable appears to be represented by distinct subsets of the NCM cells that are highly integrated and are recruited in a unique manner to encode the stimulus. All of these findings intimate that different syllable-related activity within the NCM may interact in a way that makes the formation of auditory representations possible.

Moreover, subpallial regions appear to be involved with auditory processing. Gene expression studies in both the zebra finch (Mello and Clayton, 1996) and the hummingbird (Jarvis et al., 2000) indicate that the caudaldorsal paleostriatum is active upon hearing birdsong. The position of the caudal paleostriatum (PC) situated under the auditory pallial regions is consistent with it being an auditory area of the striatum. There is evidence that the NCM is reciprocally connected with the PC (Mello and Clayton, 1996), and anterograde tracers indicate innervation from fields L1, L2b, and L3 (Vates et al., 1996). Furthermore, the entire avian striatum, comprising the PC, projects to the dorsal pallidum, and the neurochemical attributes of the projections mimic those striatal efferents in mammals. The pallidal outputs project to the MDNs and thalamus (Medina and Reiner, 1997). Functionally, there is very little else known about the PC; however, its close association with pallial auditory regions and input from the auditory thalamic nucleus (Durand et al., 1992) intimate a fundamental role in audition and may contribute along with the NCM in an interface between song perception and production. Certainly, the connections that include the PC are amenable to a closed cortico-striatal loop, though this has not been confirmed. It should be noted that unlike Area X, the pallidal component is not part of the PC, so the striatal projection is not as local, though the possibility of a pallidal-like cell component still remains.

16.4.4 WHAT TYPE OF BIOLOGICAL SUBSTRATE IS REQUIRED FOR SONG LEARNING?

The faithful reproduction of a learned song requires first discriminating the temporal relationships among the sequence of perceived elements in the song. This information should be encoded in a manner that permits the construction of motor sequences that match the auditory template. The comparison between what is produced (real-time auditory feedback) and what ought to be produced (the auditory template) need not be made within the AFP, though the AFP is expected to play a role in transmitting the results of this comparison. One potential source of this *comparison signal* is in the MDNs, which project to Area X in the avian striatum. The source of this signal is significant because dopamine (DA) is thought to play an important role in song learning, although its specific role is far from understood. During the sensorimotor phase of the zebra finch, DA levels rise intensely and peak throughout the AFP (Harding et al., 1998). Furthermore, DA modulates the excitability of spiny neurons within Area X (Ding and Perkel, 2002).

In mammals, there is substantial evidence to suggest that the MDNs encode both the occurrence and the time of reward as assessed during Pavlovian conditioning. This is indicative of a role in expectation or prediction (Hollerman and Schultz, 1998). During Pavlovian conditioning, the MDNs respond vigorously to the unconditional stimulus (US). However, as learning progresses, there appears to be a transfer of excitatory activity from the US to the onset of the conditional stimulus (CS). This transfer of activity is suggestive of a process where the neuronal activity is coactive with the detection of the earliest stimulus that best predicts a reward. However, if the delivery of reward is delayed following training, there is a depression in activity at the time reward is normally expected, followed by a burst of activity when the reward is delivered. The nature of this inhibitory activity implies that it does "not constitute a simple neuronal response, but reflect[s] an expectation process based on an internal clock tracking the time of predicted reward" (Schultz, 1998, p. 4).

The MDNs are also involved in situations that involve learning of sequential CS. The activity in a population of dopaminergic neurons in response to a US following a series of CS-US pairings progressively decreases and is transferred to the onset of each CS (Schultz et al., 1993). This result occurs as long as each CS maintains the same temporal relationship with the US. In the case of vocal learning, if the comparison signal that is transmitted to Area X in the songbird is a function of a discrepancy in the temporal relationship between the real-time auditory feedback and the auditory representation, it could be used to shape the spiny cell output of Area X. In the mammal, most of the MDN input to the caudate nuclei is neuromodulatory, this input shapes the spiny neuron's response to cortical inputs, as opposed to providing direct excitatory or inhibitory inputs. If the avian spiny neurons in Area X of the AFP could be shaped so that they fire at a tempo corresponding to a copy of the auditory template, this process could facilitate the formation of appropriate connections in the posterior motor pathway given the two pathways' connective relationship to one another. However, the major question is whether the physiology of the avian telencephalon is conducive to such a system. At this point, one may question the significance of the cortico-striatal loop motif underlying both the auditory template and the AFP.

16.5 CAN PARALLEL CORTICO-STRIATAL MODULES MEDIATE SONG LEARNING?

The existence of parallel cortico-striatal loops in the songbird that interface through the midbrain dopaminergic system could provide a substrate to oversee the production and transmission of a ratio comparison signal. We suggest that there is a cortico-striatal loop in the auditory center of the songbird, henceforth referred to as the auditory processing module. It is involved with generating a comparison signal through real-time auditory feedback that is transmitted to the AFP (i.e., the second cortico-striatal module) via the MDNs. The major components of the auditory-based cortico-striatal loop are recapitulated in Figure 16.2. In this way, a sequence of stimuli (i.e., syllables) that change in an orderly manner as a function of time can cause the auditory processing module to pass a comparison signal to the AFP, which in turn will influence the appropriate connections made in the posterior motor pathway. We will begin our discussion with the role of the cortico-striatal module in interval timing. This will be followed by a discussion of the auditory processing module, and then the AFP and the outcome on the posterior motor pathway.

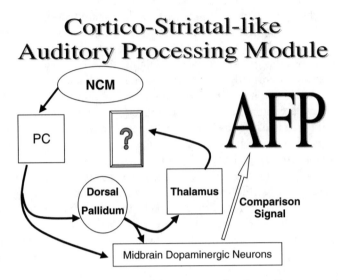

FIGURE 16.2 A proposed cortico-striatal loop representing the auditory processing module. A cortico-striatal module comprising auditory regions of the avian telencepahlon is proposed to make up the auditory processing module. The caudal nucleus of the neostriatum sends a projection to the underlying striatal area, caudal paleostriatum. PC projects to dorsal pallidum, and dorsal pallidal neurons extend to the thalamus. The thalamic neurons project back to the pallium. The operational aspect of the auditory processing module is represented by two axonal collaterals to the midbrain dopaminergic neurons. The axonal collaterals we allude to originate from two sources: those that bifurcate from PC efferents to the dorsal pallidum, and those that diverge from the dorsal pallidum projection to the DLM. This anatomical feature provides an interface with the anterior forebrain pathway.

16.5.1 The Cortico-Striatal Module as a Pattern Detector

The spiny striatal neuron is well suited to detecting patterned isocortical activity to the extent that the cortical firing pattern reflects a unique, behaviorally significant context (Houk, 1995). In both the mammalian and avian telencephalon, the spiny striatal cell's membrane potential oscillates between a low state, when it is hyperpolarized, and a high state, when it is in a slightly depolarized state. The downstate does not allow for action potentials to occur, whereas the upstate is particularly favorable to neural output (Wilson, 1993). In particular, the vast convergence onto single spiny striatal cells requires patterns of temporally coherent activation in order to drive these cells into an upstate. A specific detection of cortical input may emerge through coincidence detection mechanisms so that a cortical pattern comes to be favored by long-term potentiation (LTP)–mediated plasticity (e.g., Charpier and Deniau, 1997).

In light of the observed properties of these neural circuits, the striatal beat frequency (SBF) model was proposed to account for interval timing within the mammalian brain (see Matell and Meck, 2000; Matell et al., this volume). The SBF model builds on the conceptual framework embodied by contemporary views of cortico-striatal architecture and integrates it with an earlier neural network model of interval timing (Miall, 1989) by hypothesizing that event durations may be detected through the spiny striatal neuron's capacity to detect patterns of cortical input. Given the number of different cortical afferents that project to a single spiny neuron within the striatum, many different permutations of cortical afferents will fire simultaneously onto a spiny neuron throughout a timed interval. The idea of the SBF model is that persistent feedback is responsible for selectively weighting a pattern of simultaneous cortical inputs onto a spiny neuron that happens to be active at the time of occurrence of a biologically pre-potent event. Additionally, the SBF model assumes that ensembles of spiny striatal neurons are recruited to time an interval rather than individual neurons so that ensemble activity would progressively increase and reach a maximum average around the criterion time. Furthermore, members of the ensemble must have their established pattern of cortical inputs start firing together at the beginning of the interval so that the temporally effective pattern of coincident activity will be maximal at the criterion time. The MDNs could serve this function as their activation is gradually transferred from the end of the interval to the beginning with repeated training. This excitatory response to predictive stimulus onsets is assumed to synchronize the cortical inputs and innervate the ensembles of spiny neurons in order to initialize them for the timing of another interval (Matell and Meck, 2000; Matell et al., this volume).

16.5.2 THE AUDITORY PROCESSING MODULE

In this manner, the architecture of an auditory template in the auditory processing module permits the development of a temporal representation of the tutor song in order to guide vocal learning. The perception of a time marker (see Section 16.2.2) within each syllable engages a population of NCM neurons, and these are afferent to spiny cells in the PC. The consistent temporal relationship among the time markers that compose the song and the feedback provided by the completion of the song permits the ensembles of spiny neurons in the PC to be trained so that their mean firing rate is expected to gradually rise and peak at the time of the expected song completion. However, due to secondary conditioning (see Rescorla, 1980) and the song's consistent temporal organization, the various syllables would be expected to accrue associative strength so that they too could serve as feedback for temporally antecedent syllables. This brings up the potential for recruiting additional PC spiny neurons in order to coordinate their activity in relation to temporal criteria that fall *between* syllables, rather than only adjusting their firing rate in relation to song completion. Encoding the time of song completion is paralleled by the MDNs transferring an excitatory signal associated with this event to the onset of each syllable's time markers, as they are temporally predictive of feedback.

This system embodies the crux of simultaneous temporal processing as described by Meck and Church (1984). Several signals (syllables in our case), each one indicating the onset of a different temporal criteria, can be combined in such a way that they provide temporal information about the occurrence of a single biologically pre-potent stimulus. In this fashion, the songbird could come to behave as if it is independently timing each syllable without interference (see Pang and McAuley, this volume).

For the moment, a foundation has been established in order to explain the generation of a comparison signal due to real-time auditory feedback. One could suppose that the extent to which a perceived sequence of syllables matches the auditory template, in terms of both the spectral characteristics of each syllable and the temporal relationships among them, would influence the degree of overlap in the PC spiny neurons activated with relation to those originally recruited to encode the auditory template. Therefore, this overlap establishes the degree of the excitatory signal generated within the MDNs that is associated with predictive stimuli. Due to the reciprocal connections within the striatum via the MDNs (see Section 16.3), this excitatory signal is transmitted to the spiny cells within Area X of the AFP, modulating its activity (see Section 16.4.3).

16.5.3 Role of the Comparison Signal's Transmission to the AFP in Relation to the Posterior Motor Pathway

Recall that the songbird learns its song by producing prototypes and using the auditory template to shape the primitive song into the final song (Tchernichovski et al., 2001) (see Section 16.2.1). The activity in HVC that occurs with the onset of song production can precede the song by up to 2 sec (McCasland, 1987). The premotor drive for each prototype comes to transmit two motor signals: a *primary signal* toward RA via HVC[RA] and a *secondary signal* to Area X (see Figure 16.2). The secondary signal for each syllable prototype will activate a population of HVC[X] neurons. Keep in mind that with each vocalized syllable, there is ongoing real-time auditory feedback into the auditory processing module that contains a representation of the memorized tutor's song. The objective for the juvenile songbird is to modulate the prototype syllables so that vocal feedback engages these same PC ensembles that are tuned to the memorized song. In doing this continuously, a sequence of dopaminergic bursts is transmitted to the spiny striatal neurons in Area X in a consistent temporal pattern. This provides a substrate with which the Area X spiny neurons may be temporally entrained using the same SBF-like process that was used to construct the auditory representation. The only difference is that populations of HVC[X], rather than NCM neurons, are recruited due to the secondary signal. Individual HVC[X] neurons typically burst reliably a few times in the song motif. The bursts are not as robust or reproducible as the HVC[RA] neurons. Thus, at the level of individual HVC[X] neurons, there is no visible relationship between their bursts and song elements (M. Fee, personal communication, June, 2002). From our standpoint, however, individual HVC[X] neurons need not have a relationship with specific elements of the song because the SBF model stipulates that populations of regularly firing HVC[X] neurons are recruited as a unique pattern to project onto Area X spiny neurons.

Various ensembles of Area X spiny neurons would modulate their activity such that the net output from its target neurons, the pallidal-like cells in Area X, would come to reflect a pattern of activity corresponding to the learned durations in the auditory representation. There is a paucity of published experimental data characterizing spiny-neuronal activity in Area X of an awake, behaving songbird. Some preliminary single-unit recordings from Area X demonstrate time-locked activity for each syllable in the song, though other units had a modulatory shape in their activity profile marked by inhibition (Margoliash, 1997). The latest findings of single-unit activity in Area X found an increase in firing rate during a vocalized motif relative to no singing, while no particular relationship between the firing rate increase and elements in the song was found. On the other hand, the properties of the single cells were similar to the pallidal-like cell types found in Area X (Hessler and Doupe, 1999).

Interestingly, the increase in firing rate observed in these pallidal-like cells of Area X was accompanied by an increase in momentary pauses that interrupted activity; these pauses may be indicative of spiny-like striatal inputs, which are inhibitory (Hessler and Doupe, 1999). From the published data alone, the extensive convergence onto pallidal cells makes it difficult to draw any strong conclusions regarding the temporal pattern of the spiny striatal input. These pauses are important for current computational models involving cortico-striatal loops. The pallial location of LMAN and connectivity relative to the other telencephalic structures suggest a homology with the mammalian area of the brain equivalent to the frontal cortex (Bottjer and Johnson, 1997). This area includes the dorsolateral prefrontal cortex (dlPFC), an area most often used in computational models for cortico-striatal modules. In applying these models to the AFP, pauses in pallidal cell firing would ultimately lead to disinhibition of LMAN neurons that would permit them to go into a persistently active state. Faithfully matching the auditory template would result in a consistent temporal pattern of activity innervating LMAN, permitting a reorganization in its circuitry that would reflect this temporal-structured sequence. This process may be guided through presynaptic LTP mechanisms (Bottjer and Johnson, 1997).

In fact, during development, stabilized spatial patterns of activity appear to emerge in LMAN that reflect an accurate rendition of the bird's learned song (Boettiger and Doupe, 2001). Blocking N-methyl-D-aspartate (NMDA) receptors in LMAN prior to a juvenile bird's exposure to the tutor song impairs its ability to acquire song (Basham et al., 1996). The biophysical properties of the NMDA receptor are particularly amenable to LTP mechanisms and to the capability for sustaining activations (Wang, 1999). Moreover, depriving the young songbird of auditory information before the sensory phase prevents the development of the topographical organization from LMAN to RA even though all other connections to and from LMAN remain topographical (Iyengar and Bottjer, 2002). These results suggest that the proper connectivity between LMAN and RA requires the existence of an auditory representation, the necessity for vocal feedback, or both. The intrinsic circuitry, or perhaps the recurring pathway back to the same topographical point in Area X, might help stabilize the activation of LMAN while the interval is being timed. Similar types of reasoning have been used in computational models of the cortico-striatal modules.

How does this process facilitate encoding of the primary signal passing through the posterior motor pathway? The circuitous root of the AFP imparts a delay on the secondary signal to RA relative to the primary signal's arrival in RA. The delay on the secondary signal is sufficiently great, such that the activity from the secondary signal evoked by LMAN cells may coincide with the arrival of the primary signal for the *subsequent* syllable in the sequence. The significance of this is that the vocal learning system may be able to construct a song by associating the response to a syllable's secondary signal via the AFP with the pattern of activity evoked by the subsequent syllable's primary signal (Dave and Margoliash, 2000).

16.6 REINTERPRETATION OF SOME EXPERIMENTAL RESULTS

An explicit separation of the auditory template from the vocal learning nuclei is agreeable with some of the findings in the birdsong literature. Electrolytic lesions of Area X or LMAN do not affect the zebra finch's capacity to learn discriminations between conspecific songs signaled by a motor behavior in an operant go/no-go conditioning paradigm. The lesioned zebra finch does take longer than controls to acquire the BOS discrimination (Scharff et al., 1998). Within the framework of our proposal, the auditory template would still be intact after an AFP lesion. However, a disruption in learning may be expected if the motor system makes use of some information residing within the AFP, or through cells that traverse the AFP, to benefit from song-related, perceptual information. If this is the case, the results suggest that the songbird could discriminate between songs, but was hindered in its ability to convey this to the observer because of difficulty in integrating the template's perceptual information with the operant response. This would imply that the AFP is not entirely devoted to vocal production of birdsong.

Another result worth bringing up again is the observation that lesioning the AFP circuit in the adult songbird prevents song deterioration that is induced by deafening. From our perspective, a song decrystallizes in a deafened bird due to "drift" introduced into the system from the interaction between the auditory template and the AFP. The entry point of this drift is unknown, though it is tempting to think that it is a memory effect such that the removal of auditory feedback deprives the auditory template of a consistent "update." With this in mind, one might expect different types of song deterioration depending on the nature of the perturbation introduced to the songbird. Perturbations that actively distort the auditory feedback, presumably effecting a change to the auditory template, in an otherwise normal songbird might be expected to induce a more profound change in song relative to those songbirds that are deafened. Indeed, the distortion of temporal cues by providing delayed feedback (Leonardo and Konishi, 1999) to the adult songbird induces a more profound degradation than deafening the adult songbird, which leads to a gradual loss of song stability (Nordeen and Nordeen, 1992).

Alternatively, it may be useful to interpret results from interval timing experiments using our proposal as an outline. It was revealed that lesions in LMAN early in vocal development disrupted song significantly (Bottjer et al., 1984; Scharff and Nottebohm, 1991), though these same lesions in the adult songbird have no effect (Scharff and Nottebohm, 1991). Indeed, these lesion effects revealed that songbirds seemed to prematurely crystallize songs to the extent that the song was rendered a sequence of simplistic, often-recurring syllables. Also, the intervals between syllables became more variable. In light of our current proposal, lesions of this sort would have eradicated the capacity to form stabilized activations of spatial patterns within LMAN, thus preventing their association with descending input to RA via HVC[RA] projection neurons. In this case, the song system was forced to stabilize the rudimentary pattern of activity in RA that was developed up until that point.

A similar distributed, interactive cortico-striatal motif may be present in mammals insofar that a sensory module may influence the acquisition of behavioral output through a motor module. Thus, one would presume that frontal cortical lesions in rats that were previously trained in an interval timing task such as the peak-interval (PI) procedure would not show a significant deficit in the temporal control of behavior. Indeed, after training the animal in the PI procedure, the sensory module would have had ample time to impose its temporal content on the motor units constituting the behavior. Interestingly, lesions within the dlPFC homologue of the rat after it has reached steady-state timing performance do not drastically affect its ability to discriminate time, although slight rightward shifts in the timing functions are sometimes observed (Dietrich et al., 1997; Meck et al., 1987).

However, given our interpretation of the effects of LMAN lesions in the juvenile songbird, the largest effects of frontal cortical lesions in rats would best be observed when *learning* a behavioral output that required temporal reorganization. That is, frontal cortical lesions may not bring about substantial effects on a temporally organized behavior that previously had been established, but would disrupt the acquisition of the same behavior. To be sure, frontal lesions in rats drastically impair their ability to learn a temporal discrimination, indicated by significantly more trials until acquisition. Moreover, the behavioral output as a function of time is significantly flattened in the rats that did acquire the behavior, although 33% of the rats never learned the discrimination (Dietrich and Allen, 1998).

16.7 CONCLUSION AND FUTURE PERSPECTIVES

This proposal likens the process of song learning to trial and error. The system progressively transforms the prototype syllables into a sequence that matches an auditory representation of song that originally entailed hedonic feedback. Indeed, because the MDNs are sensitive to temporal discrepancy, they may contribute to acquiring a spatial representation of serially arranged events (e.g., the syllable) that also encode the correct temporal relationships among the constituent elements in the sequence (J. Houk, personal communication, August, 2002). Our proposal explains the nature of this sensitivity to temporal discrepancy as a product of SBF mechanisms. One goal of this chapter was to integrate the fields of timing and time perception and songbird neuroethology in a manner that could benefit the experimental approaches of both fields. To this end, our proposal of integrated, parallel cortico-striatal modules has, at the very least, introduced some potential avenues of study for both sides. In particular, the role of the MDNs in the songbird needs to be elucidated along with the role of the subpallial region PC.

The hierarchical structure found in songbird behavior need not end at the level of the song. Many experiments have observed the songbird for temporal patterning in song delivery during a bout. Indeed, there is a hierarchical organization in the song types that nightingales copy that, remarkably, may exceed 200. This hierarchical organization is reflected in the song bout so that some song types in the bout are consistently found in given positions (for a review, see Todt and Hultsch, 1998). The nature of song type switching during a bout (i.e., A, A, A, B, B, B, C, C, ...) may reflect an interval timing process. In the chaffinch, bouts with many repetitions are consistently delivered at a high rate, while bouts with few repetitions are delivered at a low rate. This is suggestive of a temporal window that is allocated and for which the bird delivers a song type sequence (Riebel and Slater, 1999). Given that there is evidence that the songbird discriminates song types by means of categorical perception (Searcy et al., 1999), a startling possibility begins to unfold. Could interval timing mediate birdsong to the extent that it contributes to a hierarchical structure ranging from seconds to minutes?

Our knowledge of the neurobiological underpinnings of interval timing is in its infancy, though enormous strides have been taken within the last few years (see chapters in this volume by Diedrichsen et al.; Malapani and Rakitin; Matell et al.; Pang and McAuley; Pouthas; Sakata and Onoda). One uncertainty in the field concerns the extent to which the neurophysiological substrate for interval timing is confined to a specified area in the brain. The proposal in this chapter clearly implicates a distributed system (see Matell et al., this volume) such that sensorimotor behavior is the product of an interaction between temporal processing occurring in the auditory system and the motor system. Both systems share the utility of a cortico-striatal module, and the interaction is permitted by an interface through striato–nigral–striatal pathways. The implication is that there may be separate pallial-striatal modules for different stimulus modalities and that "clocks" may be both centralized (e.g., amodal) and distributed (e.g., modality specific), depending on the need for global synchronization and reduction in variability (see chapters in this volume by Diedrichsen et al.; Hopson; Malapani and Rakitin; Matell et al.; Penney).

ACKNOWLEDGMENTS

We would especially like to thank Erich Jarvis for his valuable contributions that shaped the path this chapter undertook. In addition, the author would like to express gratitude to Steve Nowicki for his comments on earlier drafts and Michale Fee for his helpful correspondence.

REFERENCES

Adret, P., Operant conditioning, song learning and imprinting to taped song in zebra finch, *Anim. Behav.*, 46, 149–159, 1993.

Alexander, G.E., DeLong, M.R., and Strick, P.L., Parallel organization of functionally segregated circuits linking basal ganglia and cortex, *Annu. Rev. Neurosci.*, 9, 357–391, 1986.

Ball, G.F. and Hulse, S.H., Birdsong, *Am. Psychol.*, 53, 37–58, 1998.

Basham, M.E., Nordeen, E.J., and Nordeen, K.W., Blockade of NMDA receptors in the anterior forebrain impairs sensory acquisition in the zebra finch (*Taenopygia guttata*), *Neurobiol. Learn. Mem.*, 66, 295–304, 1996.

Beiser, D.G., Hua, S.E., and Houk, J.C., Network models of the basal ganglia, *Curr. Opin. Neurobiol.*, 7, 185–190, 1997.

Boettiger, C.A. and Doupe, A.J., Developmentally restricted synaptic plasticity in a songbird nucleus required for song learning, *Neuron*, 31, 809–818, 2001.
Bottjer, S.W. and Johnson, F., Circuits, hormones and learning: vocal behavior in songbirds, *J. Neurobiol.*, 33, 602–618, 1997.
Bottjer, S.W., Miesner, E.A., and Arnold, A.A., Forebrain lesions disrupt development but not maintenance of song in passerine birds, *Science*, 224, 901–903, 1984.
Brainard, M.S. and Doupe, A.J., Auditory feedback in learning and maintenance of vocal behaviour, *Nat. Neurosci.*, 1, 31–40, 2000.
Brauth, S.E., Fergusson, J.L., and Kitt, C.A., Prosencephalic pathways related to the paleostriatum of the pigeon, *Brain Res.*, 147, 205–221, 1978.
Brenowitz, E.A., Comparative approaches to the avian song system, *J. Neurobiol.*, 33, 517–531, 1997.
Brown, L.L., Schneider, J.S., and Lidsky, T.I., Sensory and cognitive functions of the basal ganglia, *Curr. Opin. Neurobiol.*, 7, 157–163, 1997.
Capsius, B. and Leppelsack, H.J., Influence of urethane anesthesia neural processing in the auditory cortex analogue of a songbird, *Hear. Res.*, 96, 59–70, 1996.
Charpier, S. and Deniau, J.M., In vivo activity-dependent plasticity at cortico-striatal connections: evidence for physiological long-term potentiation, *Proc. Natl. Acad. Sci. U.S.A.*, 94, 7036–7040, 1997.
Church, R.M., Properties of the internal clock, in *Annals of the New York Academy of Sciences*, Vol. 423, *Timing and Time Perception*, Gibbon, J. and Allan, L.G., Eds., New York Academy of Sciences, New York, 1984, pp. 566–582.
Church, R.M. and Gibbon, J., Temporal generalization, *J. Exp. Psychol. Anim. Behav. Process.*, 8, 165–168, 1982.
Dave, A.S. and Margoliash, D., Song replay during sleep and computational rules for sensorimotor vocal learning, *Science*, 290, 812–816, 2000.
Dave, A.S., Yu, A.C., and Margoliash, D., Behavioral state modulation of auditory selectivity in a vocal motor system, *Science*, 282, 2250–2254, 1998.
Dietrich, A. and Allen, J.D., Functional dissociation of the prefrontal cortex and hippocampus in timing behavior, *Behav. Neurosci.*, 112, 1043–1047, 1998.
Dietrich, A., Frederick, D.L., and Allen, J.D., The effects of total and subtotal prefrontal cortical lesions on the timing ability of the rat, *Psychobiology*, 25, 191–201, 1997.
Ding, L. and Perkel, D.J., Dopamine modulates excitability of spiny neurons in the avian basal ganglia, *J. Neurosci.*, 22, 5210–5218, 2002.
Doupe, A.J., A neural circuit specialized for vocal learning, *Curr. Opin. Neurobiol.*, 116, 104–111, 1993.
Doupe, A.S. and Konishi, M., Song-selective auditory circuits in the vocal control system of the zebra finch, *Proc. Natl. Acad. Sci. U.S.A.*, 88, 11339–11343, 1991.
Doupe, A.S. and Solis, M.M., Song and order-selective neurons develop in the songbird anterior forebrain during vocal learning, *J. Neurobiol.*, 33, 694–709, 1997.
Durand, S.E., Tepper, J.M., and Cheng, M.E., The shell region of the nucleus ovoidalis: a subdivision of the avian auditory thalamus, *J. Comp. Neurol.*, 323, 495–518, 1992.
Farries, M.A. and Perkel, D.J., A telencephalic nucleus essential for song learning contains neurons with physiological characteristics of both striatum and globus pallidus, *J. Neurosci.*, 22, 3776–3787, 2002.
Fortune, E.S. and Margoliash, D., Cytoarchitectonic organization and morphology of cells of the field L complex in male zebra finches (*Taenopygia guttata*), *J. Comp. Neurol.*, 325, 388–404, 1992.
Gerfen, C.R., The neostriatal mosaic: I. Compartmental organization of projections from the striatum to the substantia nigra in the rat, *J. Comp. Neurol.*, 236, 454–476, 1985.

Gibbon, J., Scalar expectancy theory and Weber's law in animal timing, *Psychol. Rev.*, 84, 279–325, 1977.

Gibbon, J., Church, R.M., and Meck, W.H., Scalar timing in memory, *Ann. N.Y. Acad. Sci.*, 423, 52–77, 1984.

Groves, P.M., Garcia-Munoz, M., Linder, J.C., Manley, M.S., Martone, M.E., and Young, S.J., Elements of the intrinsic organization and information processing in the neostriatum, in *Models of Information Processing in the Basal Ganglia*, Houk, J.C., Davis, J.L., and Beiser, D.G., Eds., MIT Press, Cambridge, MA, 1995, pp. 51–96.

Haber, S.H., Fudge, J.L., and McFarlan, N.R., Striatonigrostriatal pathways in primate form an ascending spiral from the shell to the dorsolateral striatum, *J. Neurosci.*, 20, 2369–2382, 2000.

Hahnloser, R.H.R., Kozhevnikov, A.A., and Fee, M.S., An explicit representation of time underlies the generation of neural sequences in a songbird, *Nature*, 419, 65–70, 2002.

Harding, C.F., Barclay, S.R., and Waterman, S.A., Changes in catecholamine levels and turnover rates in hypothalamic, vocal control, and auditory nuclei in male zebra finches during development, *J. Neurobiol.*, 34, 329–346, 1998.

Hartshorne, C., The monotony threshold in singing birds, *Auk*, 83, 176–192, 1956.

Hessler, N.A. and Doupe, A.J., Singing related activity in a dorsal forebrain–basal ganglia circuit of adult zebra finches, *J. Neurosci.*, 19, 10461–10481, 1999.

Hollerman, J.R. and Schultz, W., Dopamine neurons report an error in the temporal prediction of reward during learning, *Nat. Neurosci.*, 1, 304–309, 1998.

Hoover, J.E. and Strick, P.L., The organization of cerebellar and basal ganglia outputs to primary motor cortex as revealed by retrograde, transneuronal transport of herpes simplex virus type 1, *J. Neurosci.*, 19, 1446–1463, 1999.

Houk, J.C., Information processing in modular circuits linking basal ganglia and cerebral cortex, in *Models of Information Processing in the Basal Ganglia*, Houk, J.C., Davis, J.L., and Beiser, D.G., Eds., MIT Press, Cambridge, MA, 1995, pp. 3–10.

Iyengar, S. and Bottjer, S.W., The role of auditory experience in the formation of neural circuits underlying vocal learning in zebra finches, *J. Neurosci.*, 22, 946–958, 2002.

Jarvis, E.D., Mello, C.V., and Nottebohm, F., Associative learning and stimulus novelty influence the song-induced expression of an intermediate early gene in the canary forebrain, *Learn. Mem.*, 1995, 62–80, 1995.

Jarvis, E.D. and Nottebohm, F., Motor-driven gene expression, *Proc. Natl. Acad. Sci. U.S.A.*, 94, 4097–4102, 1997.

Jarvis, E.D., Ribeiro, S., deSilva, M.L., Ventura, D., Viellard, J., and Mello, C., Behaviorally driven gene expression reveals song nuclei in hummingbird brain, *Nature*, 406, 628–632, 2000.

Joel, D. and Weiner, I., The connections of the dopaminergic system with the striatum in rats and primates: an analysis with respect to the functional and compartmental organization of the striatum, *Neuroscience*, 96, 451–474, 2000.

Karten, H.J., The organization of the avian telencephalon and some speculations on the phylogeny of the amniote telencephalon, *Ann. N.Y. Acad. Sci.*, 167, 164–179, 1969.

Karten, H.J. and Dubbeldam, J.L., The organization and projections of the paleostriatal complex in the pigeon (*Columbia liva*), *J. Comp. Neurol.*, 148, 61–90, 1973.

Kelley, R.M. and Strick, P.L., Retrograde transneuronal transport of rabies virus through basal-ganglia corticothalamic circuits of primates, *Soc. Neurosci. Abstr.*, 25, 1925, 1999.

Kitt, C.A. and Brauth, S.E., Projections of the paleostriatum upon the midbrain tegmentum in the pigeon, *Neuroscience*, 6, 1551–1566, 1981.

Konishi, M. and Nottebohm, F., Experimental studies in the ontogeny of avian vocalizations, in *Bird Vocalizations*, Hinde, R.A., Ed., Cambridge University Press, London, 1969, pp. 29–48.

Krebs, J.R., The significance of song repertoires: the Beau-Geste hypothesis, *Anim. Behav.*, 37, 266–292, 1977.

Kroodsma, D.E., Continuity and versatility in birdsong: support for the monotony threshold hypothesis, *Nature*, 274, 681–683, 1980.

Kroodsma, D.E. and Baylis, J.R., Appendix: a world survey of evidence for vocal learning in birds, in *Acoustic Communication in Birds*, Vol. 2, Kroodsma, D.E. and Miller, E.H., Eds., Academic Press, New York, 1982, pp. 311–337.

Lashley, K., The problem of serial order in behavior, in *Cerebral Mechanisms of Behavior*, Jeffres, L.A., Ed., John Wiley & Sons, New York, 1951, pp. 112–136.

Leak, T.M. and Gibbon, J., Simultaneous timing of multiple intervals: implications of the scalar property, *J. Exp. Psychol. Anim. Behav. Process.*, 21, 3–19, 1995.

Leonardo, A. and Konishi, M., Decrystallization of adult birdsong by perturbation of auditory feedback, *Nature*, 399, 466–470, 1999.

Lewicki, M.S. and Konishi, M., Mechanisms underlying the temporal sensitivity of songbird forebrain neurons to temporal order, *Proc. Natl. Acad. Sci. U.S.A.*, 92, 5582–5586, 1995.

Luo, M., Ding, L., and Perkel, D.J., An avian basal ganglia pathway essential for vocal learning forms a closed topographic loop, *J. Neurosci.*, 21, 6836–6845, 2001.

Margoliash, D., Acoustic parameters underlying the responses of song-specific neurons in the white-crowned sparrow, *J. Neurosci.*, 3, 1039–1057, 1983.

Margoliash, D., Preference for autogenous song by auditory neurons in a song system nucleus of the white-crowned sparrow, *J. Neurosci.*, 6, 1643–1661, 1986.

Margoliash, D., Functional organization of forebrain pathways for song production and perception, *J. Neurobiol.*, 33, 671–693, 1997.

Margoliash, D. and Fortune, E.S., Temporal and harmonic combination-sensitive neurons in the zebra finch's HVC, *J. Neurosci.*, 12, 4309–4326, 1992.

Margoliash, D. and Konishi, M., Auditory representation of autogenous song in the song-system of white-crowned sparrows, *Proc. Natl. Acad. Sci. U.S.A.*, 82, 597–600, 1985.

Marler, P., A comparative approach to vocal learning: song development in white-crowned sparrows, *J. Comp. Physiol. Psychol.*, 71, 1–25, 1970.

Marler, P., Categorical perception of a natural stimulus continuum: birdsong, *Science*, 244, 976–978, 1989.

Marler, P. and Peters, S., Subsong and plastic song: their role in the vocal learning process, in *Acoustic Communication in Birds*, Vol. 2, Kroodsma, D.E. and Miller, E.H., Eds., Academic Press, New York, 1982, pp. 25–50.

Marler, P. and Peters, S., The role of song phonology and syntax in vocal learning preferences in the song sparrow, *Melospiza melodia*, *Ethology*, 77, 125–149, 1988.

Matell, M.S. and Meck, W.H., Neuropsychological mechanisms of interval timing behavior, *Bioessays*, 22, 94–103, 2000.

McCasland, J.S., Interactions between auditory and motor activities in an avian song nucleus, *J. Neurosci.*, 7, 23–39, 1987.

McGeorge, A.J. and Faull, R.L.M., The organization of the projection of the cerebral cortex to the striatum in the rat, *Neuroscience*, 29, 503–537, 1989.

Meck, W.H. and Church, R.M., Abstraction of temporal attributes, *J. Exp. Psychol. Anim. Behav. Process.*, 8, 226–243, 1982.

Meck, W.H. and Church, R.M., Simultaneous temporal processing, *J. Exp. Psychol. Anim. Behav. Process.*, 10, 1–29, 1984.

Meck, W.H., Church, R.M., Wenk, G.L., and Olton, D.S., Nucleus basalis magnocellularis and medial septal area lesions differentially impair temporal memory, *J. Neurosci.*, 7, 3505–3511, 1987.

Medina, L. and Reiner, A., The efferent projections of the dorsal and ventral pallidal parts, of the pigeon basal ganglia, studied with biotynilated dextran amine, *Neuroscience*, 81, 773–802, 1997.

Mello, C.V. and Clayton, D.F., Song-induced ZENK gene expression in auditory pathways of songbird brain and its relation to the song control system, *J. Neurosci.*, 14, 6652–6666, 1996.

Miall, C., The storage of time intervals using oscillating neurons, *Neural Comput.*, 1, 359–371, 1989.

Middleton, F.A. and Strick, P.L., Basal ganglia and cerebellar loops: motor and cognitive circuits, *Brain Res. Rev.*, 31, 236–250, 2000.

Muller, C.M. and Leppelsack, H.J., Feature extraction and tonotopic organization in the avian forebrain, *Exp. Brain Res.*, 59, 587–599, 1985.

Nordeen, K.W. and Nordeen, E.J., Auditory feedback is necessary for the maintenance of stereotyped song in adult zebra finches, *Behav. Neural Biol.*, 57, 58–66, 1992.

Nottebohm, F., The origins of vocal learning, *Am. Naturalist*, 106, 116–140, 1970.

Nottebohm, F., Stokes, T.M., and Leonard, C.M., Central control of song in the canary, *Serinus Canaria, J. Comp. Neurol.*, 165, 457–486, 1976.

Okanoya, K. and Yamaguchi, A., Adult Bengalese finches (*Lonchura striata* var. *domestica*) require real-time auditory feedback to produce normal song syntax, *J. Neurobiol.*, 33, 343–356, 1997.

Oorschot, D.E., Total number of neurons in the neostriatal, pallidal, subthlamic and substantia nigra nuclei of the rat basal ganglia: a stereological study using the Cavalieri and optical disector methods, *J. Comp. Neurol.*, 366, 580–599, 1996.

Parent, A. and Cicchetti, F., The current model of the basal ganglia under scrutiny, *Mov. Disord.*, 13, 199–202, 1998.

Podos, J., Peters, S., Rudnicky, T., Marler, P., and Nowicki, S., The organization of song repertoires in song sparrow: themes and variations, *Ethology*, 90, 89–106, 1992.

Puelles, L., Kuwana, E., Puelles, A., Bulfone, A., Shimamura, K., Keleher, J., Smiga, S., and Rubenstein, J.L.R., Pallial and subpallial derivatives in the embryonic chick and mouse telencephalon, traced by the expression of the genes Dlx-2, Emx-1, Nkx-2.1, Pax-6, and Tbr-1, *J. Comp. Neurol.*, 424, 409–438, 2000.

Reiner, A., Functional circuitry of the avian basal ganglia: implications for basal ganglia organization in stem amniotes, *Brain Res. Bull.*, 57, 513–528, 2002.

Reiner, A., Medina, L., and Veenman, C.L., Structural and functional evolution of the basal ganglia in vertebrates, *Brain Res. Rev.*, 28, 235–285, 1998.

Rescorla, R.A., *Pavlovian Second Order Conditioning: Studies in Associative Learning*, Erlbaum, Hillsdale, NJ, 1980.

Ribeiro, S., Cecchi, G.A., Magnasco, M.O., and Mello, C.V., Towards a song code: evidence for a syllabic representation in the canary brain, *Neuron*, 21, 359–371, 1998.

Riebel, K. and Slater, J.B., Song type switching in the chaffinch, *Fringilla coelebs*: timing or counting? *Anim. Behav.*, 57, 655–661, 1999.

Rousseau, L. and Rousseau, R., Stop-reaction time and the internal clock, *Percept. Psychophys.*, 58, 434–448, 1996.

Sasaki, A., Wetsel, W.C., Rodriguiz, R.M., and Meck, W.H., Timing of the acoustic startle response in mice: habituation and dishabituation as a function of the interstimulus interval, *Int. J. Comp. Psychol.*, 14, 258–268, 2001.

Scharff, C. and Nottebohm, F., A comparative study of the behavioral deficits following lesions of various parts of the zebra finch song system: implications for vocal learning, *J. Neurosci.*, 11, 2896–2913, 1991.

Scharff, C., Nottebohm, F., and Cynx, J., Conspecific and heterospecific song discrimination in male zebra finches with lesions in the anterior forebrain pathway, *J. Neurobiol.*, 36, 81–90, 1998.

Schmidt, M.F. and Konishi, M., Gating of auditory responses in the vocal control system of awake songbirds, *Nat. Neurosci.*, 1, 513–518, 1998.

Schultz, W., Predictive reward signal of dopamine neurons, *J. Neurophysiol.*, 80, 1–27, 1998.

Schultz, W., Apicella, P., and Ljungberg, T., Responses in monkey dopamine neurons to reward and conditioned stimuli during successive steps of learning a delayed response task, *J. Neurosci.*, 13, 900–913, 1993.

Searcy, W.A., Nowicki, S., and Peters, S., Song types as fundamental units in vocal repertoires, *Anim. Behav.*, 58, 37–44, 1999.

Smith-Fernendez, A., Pieau, C., Reperant, J., Edoardo, B., and Wassef, M., Expression of Emx-1 and Dlx-1 homeobox genes define three molecularly distinct domains in the telencepahlon of the mouse, chick, turtle, and frog embryos: implications for the evolution of telencephalic subdivisions in amniotes, *Development*, 125, 2099–2111, 1998.

Soha, J.A. and Marler, P., Vocal syntax development in the white-crowned sparrow (Zonotrichia leucophrys), *J. Comp. Psychol.*, 115, 172–180, 2000.

Sossinka, R. and Bohner, J., Song types in the zebra finch *Poephilia guttata castanotis*, *Z. Tierpsychol.*, 53, 123–132, 1980.

Stevenson, J., Reinforcing effects of chaffinch song, *Anim. Behav.*, 15, 427–432, 1967.

Stevenson, J., Song as a reinforcer, in *Bird Vocalizations*, Hinde, R., Ed., Cambridge University Press, Cambridge, U.K., 1969, pp. 49–60.

Striedter, G.F., Homology in the nervous system: of characters, embryology, and levels of analysis, in *Homology, Novartis Symposium*, Bock, G.R. and Cardew, G., Eds., 222, 158–172, 1999.

Stripling, R., Kruse, A.A., and Clayton, D.F., Development of song responses in the zebra finch caudomedial neostriatum: role of genomic and electrophysiological activities, *J. Neurobiol.*, 48, 163–180, 2001.

Stripling, R., Volman, S.F., and Clayton, D.F., Response modulation in the zebra finch neostriatum: relationship to nuclear gene regulation, *J. Neurosci.*, 17, 3883–3893, 1997.

Swanson, L.W., Cerebral hemisphere regulation of motivated behavior, *Brain Res.*, 886, 113–164, 2000.

Tchernichovski, O., Mitra, P., Lints, T., and Nottebohm, F., Dynamics of the vocal imitation process: how the zebra finch learns its song, *Science*, 291, 2564–2569, 2001.

Todt, D. and Hultsch, H., How songbirds deal with large amounts of serial information: retrieval rules suggest a hierarchical song memory, *Biol. Cybern.*, 79, 487–500, 1998.

Vates, G.E., Broome, B.M., Mello, C.V., and Nottebohm, F., Auditory pathways of caudal telencephalon and their relation to the song system of adult zebra finches (*Taeniopygia guttata*), *J. Comp. Neurol.*, 366, 613–642, 1996.

Vates, G.E. and Nottebohm, F., Feedback circuitry within a song-learning pathway, *Proc. Natl. Acad. Sci. U.S.A.*, 92, 5139–5143, 1995.

Veenman, C.L., Wild, J.M., and Reiner A., Organization of the avian "corticostriatal" projection system: a retrograde and anterograde pathway tracing study in pigeons, *J. Comp. Neurol.*, 354, 87–126, 1995.

Vicario, D.S., Organization of the zebra finch song control system: 2. Functional-organization of outputs from nucleus robustus-archistriatalis, *J. Comp. Neurol.*, 309, 486–494, 1991.

Wang, X.-J., Synaptic basis of cortical persistent activity: the importance of NMDA receptors to working memory, *J. Neurosci.*, 19, 9587–9603, 1999.

Webster, K.E., Cortico-striate relations in the albino rat, *J. Anat.*, 95, 532–544, 1961.

Weisman, R., Brownlie, L., Olthof, M., Njegovan, C., Sturdy, C., and Mewhort, D., Timing and classifying brief acoustic stimuli by songbirds and humans, *J. Exp. Psychol. Anim. Behav. Process.*, 25, 139–152, 1999.

White, N., Mnemonic functions of the basal ganglia, *Curr. Opin. Neurobiol.*, 7, 164–169, 1998.

Wilson, C.J., The generation of natural firing patterns in neostriatal neurons, in *Progress in Brain Research: Chemical Signaling in the Basal Ganglia*, Arbuthnott, G.W. and Emson, P.C., Eds., Elesevier, Amsterdam, 1993, pp. 277–298.

Wilson, C.J., The contribution of cortical neurons to the firing patterns of spiny striatal neurons, in *Models of Information Processing in the Basal Ganglia*, Houk, J.C., Davis, J.L., and Beiser, D.G., Eds., MIT Press, Cambridge, MA, 1995, pp. 29–50.

Yu, A.C. and Margoliash, D., Temporal hierarchial control of singing in birds, *Science*, 273, 1871–1875, 1996.

17 Neuroimaging Approaches to the Study of Interval Timing

Sean C. Hinton

CONTENTS

17.1 Introduction 420
17.2 Structural Neuroimaging Techniques 420
    17.2.1 Autopsy 420
    17.2.2 X-ray 420
    17.2.3 Computerized Tomography without Contrast 422
    17.2.4 Computerized Tomography with Contrast 422
    17.2.5 Magnetic Resonance Imaging 422
17.3 Functional Neuroimaging Techniques 422
    17.3.1 Electroencephalography 422
    17.3.2 Magnetoencephalography 423
    17.3.3 Positron Emission Tomography 423
    17.3.4 Single-Photon Emission Computed Tomography 424
    17.3.5 Optical Recording 424
    17.3.6 Functional Magnetic Resonance Imaging 425
17.4 Neuroimaging of Human Interval Timing 426
    17.4.1 Experiment 1: FMRI of a Basic Peak-Interval Procedure Task 427
    17.4.2 Experiment 2: Whole-Brain Imaging and Group Averaging 431
    17.4.3 Experiment 3: ROI Analysis and a Motor Control Task 432
    17.4.4 Timing May Be Obligatory 433
17.5 Future Directions 435
Acknowledgments 436
References 436

17.1 INTRODUCTION

In 1890, William James framed a profound and enduring question: "To what cerebral process is the sense of time due?" Only in the last few decades have sophisticated imaging techniques been developed that allow us to begin to propose some tentative answers. This chapter provides an overview of neuroimaging methods and their application to research on interval timing. First, I will compare the various structural and functional neuroimaging methods, focusing on their strengths and limitations. Following is a brief review of some of the existing neuroimaging data on interval timing. Then, I will present three functional magnetic resonance imaging studies to illustrate a variety of different analysis approaches using this method. The chapter concludes with some thoughts on where functional imaging research on interval timing may go from here.

The primary neuroimaging techniques used to assess brain structure and function are presented in the top and bottom panels, respectively, of Table 17.1. These procedures are characterized in terms of their relative invasiveness and repeatability, how finely they can resolve neural activation in space and over time, the physical energy source they use, the physiological property that they measure, and their most significant limitations. The neuroimaging techniques and their properties will be discussed in greater detail below.

17.2 STRUCTURAL NEUROIMAGING TECHNIQUES

17.2.1 Autopsy

The most straightforward imaging technique, used for thousands of years, is the autopsy, derived from the Greek word *autopsia*, meaning "the act of seeing with one's own eyes." It is the ultimate in invasiveness and is nonrepeatable, but it is the only imaging technique in which one looks directly at the tissue of interest. The main disadvantage of the technique from a research perspective is that the participant must be dead, which precludes inferences about functional mechanisms.

17.2.2 X-ray

X-rays were discovered by Wilhelm C. Roentgen in 1895 and became the first relatively noninvasive neuroimaging technique. Unlike an autopsy, x-ray scans were repeatable and did no obvious damage to the tissue being imaged. However, researchers later determined that x-rays produce both acute local tissue damage and chronic genetic damage that increases the risk of cancer. These side effects restrict the use of x-ray exams to circumstances in which there is a strong clinical justification and argue against repeating exams unnecessarily. X-rays have a spatial resolution of about 3 mm. They are transmitted through soft tissue relatively unimpeded, although dense tissue like bone will to some degree both absorb and scatter them. The major limitation of x-rays as an imaging modality is that they act similarly to a light source transmitted through the entire thickness of the head. Consequently, subtle changes in tissue density may be masked by the shadow of intervening bone, so it is necessary to view a region of interest from multiple angles.

TABLE 17.1
Salient Characteristics of Structural and Functional Neuroimaging Methods

Neuroimaging Technique	Invasiveness	Repeatability	Spatial Resolution	Temporal Resolution	Energy Source	Physiological Measure	Other Limitations
Autopsy	*	+	N/A	N/A	N/A	N/A	Participant is dead
X-ray	++	++	3 mm	N/A	X-rays	Tissue density	Radiation, shadows
CT w/o contrast	++	++	3 mm	N/A	X-rays	Tissue density	Radiation
CT w/contrast	+++	++	3 mm	N/A	X-rays	Tissue density	Invasive, risky
MRI	+	+++	< 1 mm	N/A	Radio waves	[Water]	Metal fragments
EEG	+	+++	3 mm	~ 0.001 sec	Voltage	Electric potentials	Cortex, hippocampus
MEG	+	+++	3 mm	~ 0.001 sec	Current	Magnetic fields	Cortex, hippocampus
PET	+++	++	5 mm	~ 100 sec	Gamma rays	[Radionucleide]	Radioactivity
SPECT	+++	++	8 mm	~ 100 sec	Photons	[Radionucleide]	Radioactivity
Optical	+	+++	~ 1 mm	~ 1 sec	Photons	[Oxy/deoxy Hb]	Cortex
fMRI	+	+++	~ 1 mm	~ 1 sec	Radio waves	[Oxy/deoxy Hb]	Metal fragments

Note: Salient characteristics of structural and functional neuroimaging methods. Structural and functional methods are characterized in the upper and lower panels, respectively. Invasiveness and repeatability are rated on relative scales from low (+) to high (+++). * Autopsy, the gold standard or invasiveness is not rated. Spatial and temporal resolution refer to the smallest distance or duration that the technique is capable of sampling and resolving. Energy source is the physical phenomenon that the technique measures. Physiological measure is the property of the tissue or tracer that is measured (brackets indicate concentration; Hb = hemoglobin). Other limitations detail the primary drawbacks of the technique. N/A = not applicable.

17.2.3 COMPUTERIZED TOMOGRAPHY WITHOUT CONTRAST

Computerized tomography (CT) was developed in the early 1970s and is essentially a sophisticated x-ray exam. The technique makes use of mathematical algorithms to combine multiple x-ray images taken from different angles. The output is a tomographic image that represents a single slice through the brain. The procedure has characteristics similar to those of x-rays but is not as susceptible to shadows because the image is a reconstruction from multiple views.

17.2.4 COMPUTERIZED TOMOGRAPHY WITH CONTRAST

The development of contrast media that were opaque to x-rays greatly expanded the capabilities of CT. The contrast medium is injected into a blood vessel, and a CT scan is performed in a procedure called cerebral angiography. This technique allows visualization of aneurysms, tumors, or other abnormalities that have characteristic effects on blood flow. Because the contrast medium must be injected into the patient, this method is more invasive than a CT scan performed without contrast. There are also certain risks associated with using contrast agents, such as allergic reactions to the dyes used in cerebral angiography, as well as the usual risks associated with venipuncture.

17.2.5 MAGNETIC RESONANCE IMAGING

The most recently developed technique for viewing the structure of the brain is magnetic resonance imaging (MRI). It is less invasive and more repeatable than CT because it measures the response of hydrogen nuclei to radio frequency pulses in the presence of a large static magnetic field. There are no known risks of the technique itself, unlike the x-rays used in CT, so one may repeat studies with the same patient without any concerns about chronic exposure to harmful radiation. MRI's greatest advantages as an imaging modality are its submillimeter spatial resolution and its ability to resolve subtle differences in soft tissues, primarily based on their water content. The main safety concern with MRI is to avoid exposing patients to the scanner's strong magnetic field if they have any metal in their bodies, such as pins, surgical clips, or pacemakers. A typical MRI machine operates at 1.5 tesla, equivalent to 30,000 times the earth's magnetic field, and produces powerful attractive forces on metallic objects. In addition, the changing magnetic gradients that are used to generate images may induce electrical currents in metal objects that cause them to heat up.

17.3 FUNCTIONAL NEUROIMAGING TECHNIQUES

17.3.1 ELECTROENCEPHALOGRAPHY

Electroencephalography (EEG) has a long history in human psychophysiological research as a noninvasive and repeatable technique for measuring electrical fields generated by the brain (see Brannon and Roitman, this volume; Pouthas, this volume). It was only recently, however, that application of powerful computational methods made it possible to use EEG to localize the sources of these electrical potentials and create maps of their distribution on the cortex. EEG studies use electrodes placed on various areas of the scalp to record the electrical activity generated by the brain. The number of electrodes employed by researchers using EEG has increased over the years, and recent studies have used 128 or more electrodes. Using triangulation algorithms, one can locate the electrical field sources generating the EEG signal to within a few millimeters. Because the electrical signals generated by the brain are distorted somewhat as they pass through the skull and scalp, current techniques are probably close to the practical limit for the maximal spatial resolution that can be achieved with EEG. Another fundamental limitation of this technique is that only a highly organized, parallel arrangement of neurons is capable of generating a strong enough voltage difference to be recorded from the scalp. Consequently, using surface electrodes, EEG primarily measures electrical potentials from the cortex. However, what EEG lacks in spatial resolution it amply makes up for in its temporal resolution, which is on the order of 1 msec.

17.3.2 MAGNETOENCEPHALOGRAPHY

Magnetoencephalography (MEG) is analogous to EEG except that it uses special superconducting detectors to measure magnetic dipoles generated by the flow of electrical current within the brain. It has characteristics, advantages, and disadvantages similar to those of EEG, with the exception that the required equipment is much more expensive and is not widely available.

17.3.3 POSITRON EMISSION TOMOGRAPHY

Positron emission tomography (PET) creates images of brain activity using radioactive isotopes such as ^{15}O-labeled water. When such a substance is injected into the bloodstream, a PET scanner can measure changes in either blood flow or blood volume. Using radioactively labeled glucose analogs, metabolic activity itself may be imaged. The radioactive isotopes used in PET emit positrons, which are electron antimatter. A positron may travel up to a few millimeters from where it was emitted before it collides with an electron. The interaction between a positron and an electron creates a pair of high-energy gamma rays that travel in opposite directions. Collimated gamma detectors surround the participant's head and determine the line along which the gamma rays were emitted. Therefore, a single source is inferred to arise from the intersection of multiple gamma rays' axes of travel. A computer reconstructs the locations of all the source events to generate a map of activity within the brain. PET's ultimate spatial resolution of about 5 mm is determined by limitations on detector sensitivity and the physical process by which an event is generated. Because the technique requires injecting radioactive substances, it is a moderately invasive imaging modality, and the need to minimize participants' exposure to radioactivity limits the repeatability of PET scans. In addition, data must be collected over periods of about a minute or so because of constraints on the concentration of radioactive isotopes that may be used. Consequently, the temporal resolution of this method is quite poor, and each imaging run results in a single snapshot of activity. Finally, the radioisotopes used in PET must be created in a particle accelerator, and the practical issues associated with this additional equipment make PET scanners somewhat uncommon.

17.3.4 SINGLE-PHOTON EMISSION COMPUTED TOMOGRAPHY

Single-photon emission computed tomography (SPECT) is a technique analogous to PET except that it uses lower-energy radioisotopes. The photons emitted by the isotopes are directly detected, unlike in PET, where the proximal signal of emitted gamma rays is an energetic by-product of the collision of a positron with an electron. As in PET, SPECT is limited by the trade-offs between detector sensitivity, spatial resolution, and radioactivity exposure to the participant. These trade-offs, combined with variability due to photon scatter, limit spatial resolution in SPECT to about 8 mm. SPECT's dependence on radioactive isotopes means that it is somewhat invasive and limited in its repeatability. Another limitation of SPECT is that it is less sensitive than PET because SPECT uses lower-energy photons. Finally, fewer radioactive ligands have been developed for SPECT than for PET.

17.3.5 OPTICAL RECORDING

A relatively recent addition to the variety of functional neuroimaging tools is that of optical recording, which was developed in the 1980s and has been refined over the last few decades (Gratton and Fabiani, 2001). Optical recording arose as a noninvasive modification of an earlier technique that imaged neural changes over the course of milliseconds using voltage-sensitive dyes. It is similar to EEG in that an array of detectors is used to focus on the region of interest in the brain. Its temporal characteristics are similar to those of EEG and MEG, but it is based on a fundamentally different principle. In humans, a near-infrared laser projects a beam of light through the scalp and skull onto the cortex, where the beam is absorbed and scattered to varying degrees depending on the oxygenation state of hemoglobin in the blood. A photomultiplier then records the number of photons that are reflected back through the skull and scalp, and the transmission time gives information about the degree of scattering and absorption of the signal. Therefore, the optical signal actually measures the hemodynamic response to a neural event. The technique shows good temporal correspondence to event-related potentials recorded with EEG and good spatial correspondence with functional magnetic resonance imaging (Gratton et al., 1997). Further research has shown that the hemodynamic signal is proportional to the amplitude of the neuronal signal, providing additional support for the quantitative use of hemodynamic methods in the study of brain function (Gratton et al., 2000, 2001). In animal models, optical recording may be performed directly on the exposed cortex and may use voltage-sensitive dyes to visualize changes in neural activation dynamically. In this more invasive version of the method, optical recording is capable of measuring effects with a spatial resolution of 0.1 mm and a temporal resolution of milliseconds. Like EEG and MEG, this technique is limited to measuring functional activity in the cortex, but it provides another source of information on brain function using a different physical property of the tissue than the other imaging modalities.

17.3.6 FUNCTIONAL MAGNETIC RESONANCE IMAGING

Functional magnetic resonance imaging (FMRI) is a technique developed within the last decade or so that uses an MRI scanner to detect changes in blood oxygenation levels while participants perform some kind of sensory, motor, or cognitive task. Like standard MRI, it is noninvasive and highly repeatable because it measures the effects of radio frequency pulses on hydrogen nuclei in a static magnetic field. The signal strength of FMRI is a function of the magnitude of the static magnetic field, but it always involves a trade-off between spatial and temporal resolution. In current incarnations, FMRI can achieve spatial resolutions of less than a millimeter (Hyde et al., 2001; Menon and Goodyear, 1999) and a temporal resolution measured in milliseconds (Menon et al., 1998), although these extremes are obtained by sacrificing resolution in the other domain. With standard systems performing whole-brain imaging, spatial resolutions of a few millimeters and temporal resolutions of a few seconds are commonly obtained. The effect on which most FMRI depends is called blood oxygenation level-dependent (BOLD) contrast (Ogawa et al., 1990). BOLD contrast relies on choosing imaging parameters that are sensitive to the relative concentrations of oxy- and deoxyhemoglobin in the blood, which produce characteristic effects in FMRI because of their different magnetic susceptibilities. When neural activity increases, local blood flow gradually increases with a rise time of 4 to 6 sec. The signal returns to baseline after another 6 to 8 sec, and this extended time course of activation is termed the hemodynamic response. Although there remains some debate about the exact physiological processes that produce a hemodynamic response to neural stimulation, there is general agreement that the signal that FMRI measures corresponds in reliable ways to underlying neural processing (Logothetis et al., 2001). Indeed, most early FMRI experiments used a variety of simple sensory and motor tasks in which the outcome was predicted based on data from other techniques (Bandettini et al., 1992; Belliveau et al., 1991; Binder et al., 1993; Hammeke et al., 1994; Ogawa et al., 1990, 1992; Rao et al., 1993). The great strength of FMRI as an imaging modality is that it allows one to view physiological changes throughout the brain noninvasively and repeatedly with good spatial and moderate temporal resolution. These combined characteristics make FMRI ideally suited to determine the networks of brain regions that are involved in specific aspects of cognitive function.

The cascade of events that determines the FMRI signal and generates the signal for other functional neuroimaging techniques along the way is illustrated in Figure 17.1. Sensory, motor, or cognitive activity causes a localized increase in neural activity. The graded potentials and action potentials in the activated tissue result in increased metabolic demand for glucose and oxygen. The heightened metabolic demand is satisfied by a gradual increase in regional blood flow over the course of several seconds, termed the hemodynamic response. The increased blood flow delivers a relative excess of oxyhemoglobin, which decreases the regions' magnetic

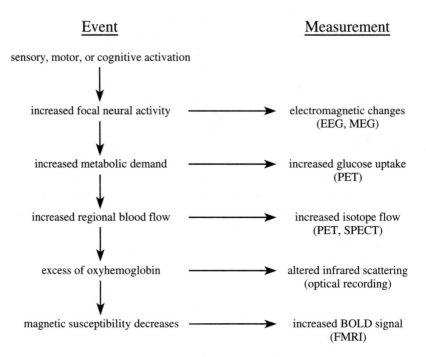

FIGURE 17.1 Diagram of the flow of physiological events and the associated signals measured by various functional neuroimaging techniques.

susceptibility and produces an increase in the measured BOLD signal. The BOLD signal is taken as indirect evidence of neural activity underlying the behavioral task or cognitive event of interest.

17.4 NEUROIMAGING OF HUMAN INTERVAL TIMING

A great variety of neuroimaging techniques, behavioral tasks, and data analyses have been applied to date in the study of timing, yet firm conclusions from this mass of data remain elusive (for reviews, see Harrington and Haaland, 1999; Macar et al., 2002). For example, some studies find that attention to time activates a left hemisphere fronto-parietal network (Coull and Nobre, 1998), while others argue that this function is served by a similar network in the right hemisphere (Rao et al., 2001). In spite of some conflicting findings in the human timing literature, many results are consonant with drug and lesion studies from the animal literature, which argue that the basal ganglia play a critical role in interval timing (Meck, 1996; Meck and Benson, 2002; see also Diedrichsen et al., this volume; Malapani and Rakitin, this volume; Matell et al., this volume).

A number of neuroimaging studies have been published examining motor timing (e.g., Gruber et al., 2000; Harrington et al., 1999), motor sequencing (Lepage et al., 1999), motor rhythms (Penhune et al., 1998), attention cued to a particular moment in time (Coull and Nobre, 1998), feedback about timing (Brunia et al., 2000), and discrimination of intervals in the millisecond range (Harrington et al., 1998; Jueptner et al., 1995; Maquet et al., 1996). However, only a handful of studies have examined timing of intervals longer than a few seconds. In addition, all of the studies mentioned above either used short durations or applied neuroimaging methods that lack the combined spatial and temporal resolution of FMRI.

Below, I present three experiments to illustrate some possible approaches to analysis of FMRI data that also provide converging evidence for the importance of the right putamen in interval timing. In all three experiments, participants were instructed not to count (for a discussion of this issue, see Hinton and Rao, 2002).

17.4.1 EXPERIMENT 1: FMRI OF A BASIC PEAK-INTERVAL PROCEDURE TASK

The first FMRI study of interval timing (Hinton et al., 1996) used the peak-interval procedure (Rakitin et al., 1998). This experiment tested reproduction of a single duration of 11 sec. Five fixed-time training trials were presented to participants immediately before each 4-min imaging run. The runs were composed of eight repetitions of a basic trial structure in which 10-sec intertrial intervals (ITIs) alternated with 20-sec signal periods that were timed, as illustrated in Figure 17.2. Four 7-mm axial echo-planar images were collected every second. The bottom surface of this volume was aligned with the horizontal plane encompassing the anterior and posterior commissures. The six male participants made two motor responses on each trial with their right hand: one when they felt that the 11-sec criterion time was near and another when they estimated that the 11-sec criterion time had elapsed. Trial types were distributed according to a Latin-square design, with one dimension being auditory and visual stimulus modalities and the other dimension foreground and background conditions. The foreground condition refers to the standard trial type described above, whereas the background condition is the reverse: the signal to be timed began when the stimulus went off, and the ITI was defined by the presence of the stimulus. The purpose of this manipulation was to attempt to minimize the contribution to the brain maps of sensory components, which should roughly cancel

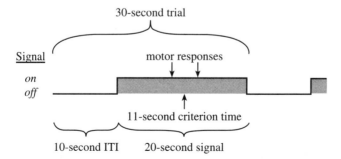

FIGURE 17.2 Illustration of the basic trial structure of Experiment 1 with 20-sec signal periods and 10-sec intertrial intervals. Participants were instructed to make two motor responses bracketing the criterion time of 11 sec.

out when the two trial types are averaged. The use of two different stimulus modalities to present the timing signal was intended to minimize modality-dependent processes, as the internal clock was assumed to be modality independent (however, see Penney, this volume).

The formal method to perform a group analysis of activated neural regions is to transform functional data from individual participants into a standard coordinate system, such as Talairach space (Talairach and Tournoux, 1988), and average the voxel-wise time series. Technical limitations prevented such an analysis for experiment 1. However, a simpler approach was used in an attempt to derive qualitatively similar data. Four participants completed the full study design, generating two runs of each of the four trial types. Activation maps were generated from these four participants by correlating their FMRI time course data with a square wave representing the period from 8 to 11 sec after signal onset. This particular period was chosen because the earliest responses typically occurred around 8 sec after signal onset; therefore, motor response activation should peak beyond 12 sec due to the hemodynamic delay. Individual correlation maps of activation during this period and structural brain images were aligned with each other. The correlation maps were spatially smoothed with a 3×3 center-weighted, spatial convolution and averaged across participants. The composite image was then thresholded at a correlation coefficient of 0.15, which is fairly conservative due to the smoothing and interparticipant averaging that preceded the thresholding step. One of the most reliably activated areas across all participants was the right putamen, as shown in Figure 17.3 (for additional details on the lateralization of interval timing, see Pouthas, this volume). This finding is consistent with drug and lesion data in rats and with this region's pivotal role in interval timing (e.g., Cohen et al., 2000; Gibbon et al., 1997; Matell et al., 2000; Meck, 1996).

As mentioned, averaging across participants' FMRI data is ordinarily performed by transforming each brain into some standardized space (e.g., Talairach and Tournoux, 1988). However, for purposes of visualizing activation in the basal ganglia and other centrally located structures, one may simply align participants' brains manually and then average across them, as was done here. While this is not the preferred method, it was necessary with the data from experiment 1 because software tools for proper averaging across participants were not widely available. The practical justification for such a procedure is that there is very little anatomical variability across brains in these central structures, so true interparticipant averaging would confer little advantage here.

Figure 17.4 illustrates this point with a montage of axial slices ranging from −70 to +75 mm in Talairach space with successive images separated by 5 mm. The montage is an average of the anatomical data of 14 male participants whose brains were aligned with respect to the anterior and posterior commissures and the longitudinal fissure. Note the fairly precise definition of the subcortical brain structures, such as the caudate, putamen, and thalamus. Many of the main white matter tracts also appear fairly distinct, primarily because of their size and more central location. However, there is substantial blurring apparent in the more distal cortical areas, where anatomical variability is greatest. Thus, aligning participants' brains with this

Neuroimaging Approaches to the Study of Interval Timing

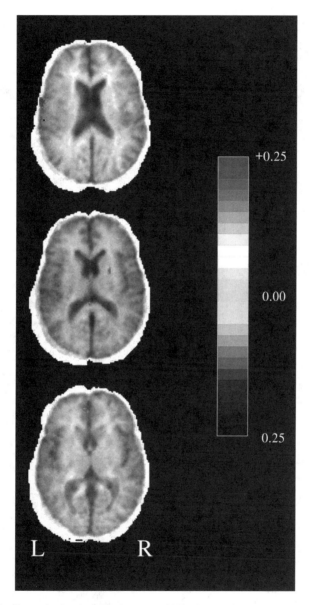

FIGURE 17.3 (See color insert following page 438.) Composite images from Experiment 1 showing 7-mm axial slices ranging from 0 to 21 mm above the AC-PC line with overlaid timing-related activation occurring during the 3 sec preceding the criterion time of 11 sec. Color scale indicates correlation coefficients between brain activation and a square wave covering the period described.

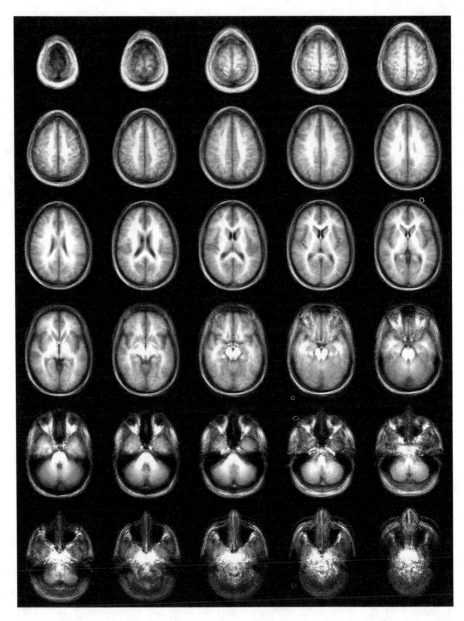

FIGURE 17.4 (See color insert.) Brains of 14 male participants were aligned along the AC-PC line and the longitudinal fissure and averaged to produce this axial montage of 5-mm slices ranging from $z = -70$ to $+75$ mm in Talairach coordinates.

method makes it possible to generate a reasonable average activation map, although it is more faithful toward the middle of the image. The likelihood of cortical activations aligning across participants is reduced because anatomical variability is greatest toward the edges of the brain.

Thus, this analysis approach produces a statistical threshold that varies spatially across the brain. It is approximately accurate for central structures that are well aligned, but it becomes increasingly more conservative than the nominal threshold for more distal structures, such as the cortex. Determining in a quantitative manner how the threshold varies with distance from medial structures is probably not possible, but use of this method means that cortical activations are less easily detected.

17.4.2 EXPERIMENT 2: WHOLE-BRAIN IMAGING AND GROUP AVERAGING

A later study similar in design to experiment 1 used whole-brain imaging and true group averaging across nine participants (unpublished data). The durations tested in this study were 7 and 17 sec. Unlike in experiment 1, participants made only a single motor response with their right index finger on each trial. Two fixed-time training trials were presented immediately before each block of ten testing trials. Eight blocks of each duration were presented in an alternating fashion, and the order of presentation was counterbalanced across participants. The signal duration of the testing trials for the 7-sec duration was 27 sec with a 3-sec ITI, and for the 17-sec duration it was 37 sec with a 3-sec ITI. Twelve 10-mm axial slices were collected every second.

All FMRI data were processed using AFNI (Cox, 1996). The collected volumes were motion corrected, transformed into standard stereotactic space (Talairach and Tournoux, 1988), and blurred with a Gaussian kernel of 4 mm full width, at half maximum to account for anatomical variability. A voxel-wise analysis of variance (ANOVA) was performed on the 7- and 17-sec data across participants for the period from 0 to 6 sec before the response time. The mean response times were 8.7 sec for the 7-sec task and 22.0 sec for the 17-sec task. The period before the response time was selected to minimize contamination by motor-related activation. The t-statistic map was thresholded at $P < .0001$. A cluster threshold of 100 µl and a nearest-neighbor threshold of 1 mm were applied to eliminate small, discontiguous activations. The result was rendered onto a three-dimensional brain with opacity set to 0.5. The data are displayed as a montage of 5-mm axial slices, as shown in Figure 17.5.

The montage shows several distinct activation foci within the caudate, putamen, and globus pallidus. Also apparent are midline activations of the supplementary motor area, anterior and posterior cingulate cortex, and precuneus, as well as bilateral activations of the insula. Right hemisphere foci include the middle frontal gyrus and inferior parietal lobe, which are areas that have been associated with attentional networks in general (Pardo et al., 1991; Posner and Peterson, 1990) and with attention to time in particular (Rao et al., 2001). Finally, there is some activation in left primary motor cortex, an unexpected finding.

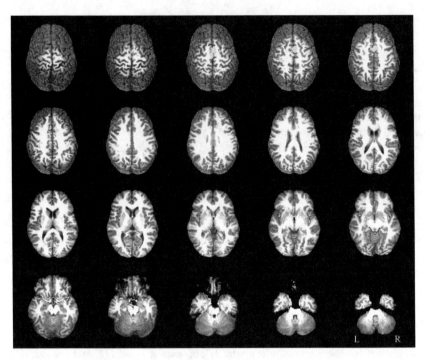

FIGURE 17.5 (See color insert.) Axial montage of activation from Experiment 2) during the period from 0 to 6 sec before the response time when 9 participants timed a 7- or 17-sec duration ($P < .0001$).

17.4.3 EXPERIMENT 3: ROI ANALYSIS AND A MOTOR CONTROL TASK

Another FMRI study similar to experiment 2 used a region-of-interest (ROI) analysis and two different durations to examine whether activation in the putamen could be reliably separated from that due to the motor response (unpublished data). The trial structure was the same as that in experiment 2. Each participant alternately performed a temporal reproduction task (similar to the one in experiments 1 and 2) and a motor task in which the timing of the response was cued to occur at an actual response time the participant had generated in the previous timing run. The timing and motor tasks alternated eight times in a session. In the motor control task, the attentional and sensory components of the task were similar, but participants were not required to time. They were instructed to attend to the central fixation cross and to respond when they saw it flash briefly. In both tasks, participants responded by pressing a button with their right index finger. An 11-sec duration was tested with six participants, and a 17-sec duration was tested with one participant. Twelve 10-mm axial slices were collected every second, and the imaging volume included the hand area of the sensorimotor cortex and the basal ganglia. The imaging data were motion corrected, and ROIs were traced for each participant on each slice for the left and right primary motor cortex (M1), primary somatosensory cortex (S1), and putamen. The motor response for each trial was time-locked to the criterion time, and like trial types were averaged across participants to extract time courses of activation for each ROI. Because of the relatively small number of participants and in order to better visualize differences by task and hemisphere, each time course was temporally smoothed using a 3-point moving average. The results of this study are displayed in Figure 17.6.

Activation in M1 for the 11-sec duration is shown in Figure 17.6a. The left M1, contralateral to the effector, was activated for both the timing and motor tasks to a similar extent and concurrently beginning just after the 11-sec criterion time. Figure 17.6b shows a similar pattern for the left primary sensory cortex. A smaller ipsilateral activation is apparent in both M1 and S1. In contrast, the pattern of activation in the putamen is distinctly different. Figure 17.6c shows that activation for the timing task begins in both the right and left putamen some seconds before the 11-sec criterion is reached, although the activation is much stronger in the right putamen, which is ipsilateral to the effector and therefore should not be engaged by motoric aspects of the task. For the motor task, the timing of activation in the putamen looks similar to the time courses in M1 and S1. However, whereas the response in M1 and S1 is highly lateralized to the contralateral left hemisphere, activation in the putamen is less lateralized and, if anything, is stronger in the ipsilateral right putamen. Another point to note is that the timing of the response in the putamen to the motor task looks as though it may be slightly delayed relative to those in S1 and M1, similar to their ipsilateral activations. This suggests the possibility of some kind of delayed feedback effect from the sensorimotor system to the putamen. The response in the putamen for the 17-sec criterion time is shown in Figure 17.6d. With this longer duration, it is clear that the putamen activation begins substantially before the motor response occurs for the timing task, and again the activation is stronger in the right than in the left putamen. For the motor task, the activation looks more like the hemodynamic response in the left M1 and S1.

Overall, what is striking about these data is how the time courses in M1 and S1 are more differentiated by hemisphere, whereas the putamen activation is more differentiated by the task and the timing of the activation. One hypothesis is that the putamen, and especially the right putamen, signals other areas of the brain in circumstances in which a precisely timed motor response is required to estimate when a particular short interval has elapsed. Whatever the exact role played by the putamen in interval timing, its time course of activation when the participant is timing a signal is clearly separable from and anticipates those of the motor and sensory cortices.

17.4.4 TIMING MAY BE OBLIGATORY

Indirect evidence for the involvement of the right putamen in interval timing comes from a study of motor sequencing (Harrington et al., 1999). In this experiment, participants were cued to make a series of specific finger movements within a 12-sec activation period. These motor periods were followed by a 12-sec period of rest. These two states alternated repeatedly in blocks. When the authors contrasted the activation periods with the rest periods, they observed a consistent pattern of BOLD signal in striate and sensorimotor cortices and the cerebellum. However, somewhat

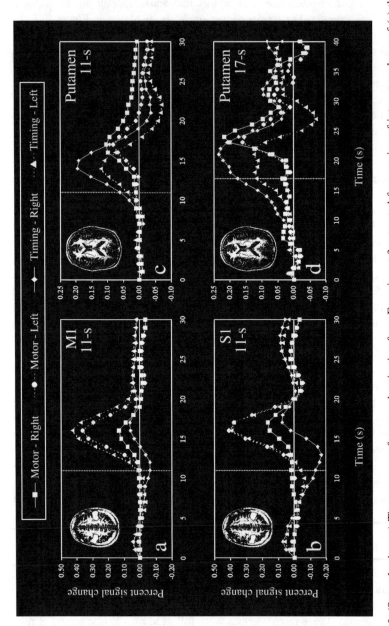

FIGURE 17.6 (See color insert.) Time courses of averaged activation from Experiment 3 extracted from region-of-interest analyses of (a) the primary motor cortex (M1), (b) the primary sensory cortex (S1), and (c) the putamen for an 11-sec signal. Panel (d) shows data from the putamen for a 17-sec signal. Dashed and solid lines respectively indicate activation measured from right and left hemisphere structures. Lighter and darker colors respectively indicate time courses of activation for the motor and timing tasks.

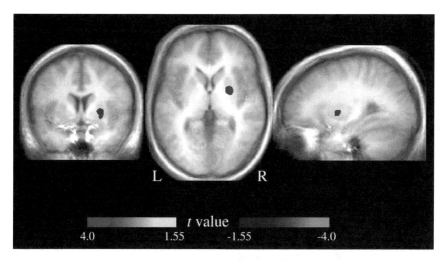

FIGURE 17.7 (See color insert.) Area in the right putamen of decreased MR signal intensity ($t < -1.55$) comparing a repetitive motor sequence task with rest (shown in coronal, axial, and sagittal views from left to right). This was the only area of decreased signal intensity reported in the experiment of Harrington et al. (1999). (Figure created with kind permission from D.L. Harrington and S.M. Rao.)

perplexingly, there was one brain region that was negatively activated, as shown in Figure 17.7: "Activation was greater during rest than repetitive sequencing in the right putamen, an unexpected result" (Harrington et al., 1999). The authors had no explanation for this observation, but when their experiment is viewed from the perspective of interval timing, their finding makes perfect sense. Participants were performing the complex motor movements during the activation period, as they had been instructed. During that period, their attention was fully engaged by the task. However, during the rest periods, there may not have been much else to occupy participants' minds other than predicting the onset of the next activation period. Thus, the individuals in this experiment may have inadvertently engaged in a timing task, albeit without any explicit instruction to do so. These data are consistent with a perceptual role of the putamen for precisely timing short intervals, as described for the timing task in experiment 3. The surprising finding of this motor sequencing study illustrates how pervasive interval timing behavior can be whenever there is temporal regularity in an organism's environment.

17.5 FUTURE DIRECTIONS

Ultimately, converging evidence from multiple behavioral tasks, analysis techniques, and imaging modalities will be required to present a coherent view of how the brain processes time.

In order to achieve that goal, the following questions, among others, will need to be addressed: What are the lower and upper boundaries of the interval timing mechanism, and how does it interact with and "hand off" functions to both milliseconds and circadian timing systems (Hinton and Meck, 1997b)? Might FMRI and EEG be combined to study the neural events that underlie timing in the millisecond range? Can magnets with higher field strength be used to test hypotheses about the neural substrates of timing on finer spatial and temporal scales? Will FMRI be able to identify different "internal clocks" for visual and auditory modalities (see Penney, this volume)? How do the multiple brain areas involved in interval timing interact to produce the subjective experience of the fluent passage of time? How reliable is the pattern of brain activation when different timing tasks are used? How does the scalar property arise from the operation of the neural mechanisms for interval timing (see Church, this volume; Gibbon et al., 1984)? Can timing of particular durations be associated with particular brain regions; for example, might there be a topographical representation of durations in different regions of the striatum (Meck et al., 1998)? What are the roles of the various neurotransmitters involved in timing (see Buhusi, this volume; Meck, 1996), and how do they interact to allow normal time perception as a function of developmental changes and normal aging (see Droit-Volet, this volume; Lustig, this volume)? To what extent is attention required to time short intervals (see Fortin, this volume; Pang and McAuley, this volume), and to what degree much may timing occur without conscious awareness? Is interval timing obligatory in healthy organisms when temporal regularities exist in the environment, and can striatal coincidence detection or cortical oscillations account for this phenomenon (see Matell et al., this volume; Meck, 2002)? These and other questions are likely to govern the direction of interval timing research for years to come in our pursuit of understanding how time flies (Hinton and Meck, 1997a).

ACKNOWLEDGMENTS

This chapter was supported by NIH Program Project grant P01 MH51358. The author gratefully acknowledges Stephen Rao, Frances Rauscher, and Warren Meck for helpful comments on earlier versions of this chapter.

REFERENCES

Bandettini, P.A., Wong, E.C., Hinks, R.S., Tikofsky, R.S., and Hyde, J.S., Time course EPI of human brain function during task activation, *Magn. Reson. Med.*, 25, 390–397, 1992.

Belliveau, J.W., Kennedy, D.N., McKinstry, R.C., Buchbinder, B.R., Weisskoff, R.M., Cohen, M.S. et al., Functional mapping of the human visual cortex by magnetic resonance imaging, *Science*, 254, 716–718, 1991.

Binder, J.R., Rao, S.M., Hammeke, T.A., Yetkin, F.Z., Jesmanowicz, A., Bandettini, P.A. et al., Functional magnetic resonance imaging of human auditory cortex, *Ann. Neurol.*, 35, 662–672, 1993.

Brunia, C.H.M., de Jong, B.M., Berg-Lenssen, M.M.C., and Paans, A.M.J., Visual feedback about time estimation is related to a right hemisphere activation measured by PET, *Exp. Brain Res.*, 130, 328–337, 2000.

Cohen, D., Matell, M.S., Meck, W.H., and Nicolelis, M.A.L., Role of the medial dorsal prefrontal cortex in a time perception task, *Soc. Neurosci. Abstr.*, 26, 365.16, 2000.

Coull, J.T. and Nobre, A.C., Where and when to pay attention: the neural systems for directing attention to spatial locations and to time intervals as revealed by both PET and fMRI, *J. Neurosci.*, 18, 7426–7435, 1998.

Cox, R.W., AFNI: software for analysis and visualization of functional magnetic resonance neuroimages, *Comput. Biomed. Res.*, 29, 162–173, 1996.

Gibbon, J., Church, R.M., and Meck, W.H., Scalar timing in memory, in *Annals of the New York Academy of Sciences: Timing and Time Perception*, Vol. 423, Gibbon, J. and Allan, L., Eds., New York Academy of Sciences, New York, 1984, pp. 52–77.

Gibbon, J., Malapani, C., Dale, C.L., and Gallistel, C.R., Toward a neurobiology of temporal cognition: advances and challenges, *Curr. Opin. Neurobiol.*, 7, 170–184, 1997.

Gratton, G. and Fabiani, M., The event-related optical signal: a new tool for studying brain function, *Int. J. Psychophysiol.*, 42, 109–121, 2001.

Gratton, G., Fabiani, M., Corballis, P.M., Hood, D.C., Goodman-Wood, M.R., Hirsch, J. et al., Fast and localized event-related optical signals (EROS) in the human occipital cortex: comparisons with the visual evoked potential and fMRI, *Neuroimage*, 6, 168–180, 1997.

Gratton, G., Goodman-Wood, M.R., and Fabiani, M., Comparison of neuronal and hemodynamic measures of the brain response to visual stimulation: an optical imaging study, *Hum. Brain Mapping*, 13, 13–25, 2001.

Gratton, G., Sarno, A., Maclin, E., Corballis, P.M., and Fabiani, M., Toward noninvasive three-dimensional imaging of the time course of cortical activity: investigation of the depth of the event-related optical signal, *Neuroimage*, 11, 491–504, 2000.

Gruber, O., Kleinschmidt, A., Binkofski, F., Steinmetz, H., and von Cramon, D.Y., Cerebral correlates of working memory for temporal information, *Neuroreport*, 11, 1689–1693, 2000.

Hammeke, T.A., Yetkin, F.Z., Mueller, W.M., Morris, G.L., Haughton, V.M., Rao, S.M. et al., Functional magnetic resonance imaging of somatosensory stimulation, *Neurosurgery*, 35, 677–681, 1994.

Harrington, D.L. and Haaland, K.Y., Neural underpinnings of temporal processing: a review of focal lesion, pharmacological, and functional imaging research, *Rev. Neurosci.*, 10, 91–116, 1999.

Harrington, D.L., Haaland, K.Y., and Knight, R.T., Cortical networks underlying mechanisms of time perception, *J. Neurosci.*, 18, 1085–1095, 1998.

Harrington, D.L., Rao, S.M., Haaland, K.Y., Bobholz, J.A., Mayer, A.R., Binder, J.R. et al., Specialized neural systems underlying representations of sequential movements, *J. Cognit. Neurosci.*, 12, 56–77, 1999.

Hinton, S.H. and Meck, W.H., How time flies: functional and neural mechanisms of interval timing, in *Time and Behaviour: Psychological and Neurobiological Analyses*, Bradshaw, C.M. and Szabadi, E., Eds., Elsevier, New York, 1997a, pp. 409–457.

Hinton, S.C. and Meck, W.H., The "internal clocks" of circadian and interval timing, *Endeavour*, 21, 3–8, 1997b.

Hinton, S.C., Meck, W.H., and MacFall, J.R., Peak-interval timing in humans activates frontal-striatal loops, *Neuroimage*, 3, S224, 1996.

Hinton, S.C. and Rao, S.M., "One-thousand, one ... one-thousand, two ... ": chronometric counting violates Weber's law in interval timing, *Psychonomic Bull. Rev.*, submitted.

Hyde, J.S., Biswal, B.B., and Jesmanowicz, A., High-resolution fMRI using multislice partial k-space GR-EPI with cubic voxels, *Magn. Reson. Med.*, 46, 114–125, 2001.

James, W., *Principles of Psychology*, Holt, New York, 1890.

Jueptner, M., Rijntjes, M., Weiller, C., Faiss, J.H., Timmann, D., Mueller, S.P. et al., Localization of a cerebellar timing process using PET, *Neurology*, 45, 1540–1545, 1995.

Lepage, R., Beaudoin, G., Boulet, C., O'Brien, I., Marcantoni, W., Bourgouin, P. et al., Frontal cortex and the programming of repetitive tapping movements in man: lesion effects and functional neuroimaging, *Cognit. Brain Res.*, 8, 17–25, 1999.

Logothetis, N.K., Pauls, J., Augath, M., Trinath, T., and Oeltermann, A., Neurophysiological investigation of the basis of the fMRI signal, *Nature*, 412, 150–157, 2001.

Macar, F., Lejeune, H., Bonnet, M., Ferrara, A., Pouthas, V., Vidal, F. et al., Activation of the supplementary motor area and of attentional networks during temporal processing, *Exp. Brain Res.*, 142, 475–485, 2002.

Maquet, P., Lejeune, H., Pouthas, V., Bonnet, M., Casini, L., Macar, F. et al., Brain activation induced by estimation of duration: a PET study, *Neuroimage*, 3, 119–126, 1996.

Matell, M.A., Chelius, C.M., Meck, W.H., and Sakata, S., Effect of unilateral or bilateral retrograde 6-OHDA lesions of the substantia nigra pars compacta on interval timing, *Soc. Neurosci. Abstr.*, 26, 650.7, 2000.

Meck, W.H., Neuropharmacology of timing and time perception, *Cognit. Brain Res.*, 3, 227–242, 1996.

Meck, W.H., Coincidence detection as the core process for interval timing in frontal-striatal circuits, *Neuroimage Program Suppl.*, 668, 2002.

Meck, W.H. and Benson, A.M., Dissecting the brain's internal clock: how frontal-striatal circuitry keeps time and shifts attention, *Brain Cognit.*, 48, 195–211, 2002.

Meck, W.H., Hinton, S.C., and Matell, M.S., Coincidence-detection models of interval timing: evidence from fMRI studies of cortico-striatal circuits, *Neuroimage*, 7, S281, 1998.

Menon, R.S. and Goodyear, B.G., Submillimeter functional localization in human striate cortex using BOLD contrast at 4 tesla: implications for the vascular point-spread function, *Magn. Reson. Med.*, 41, 230–235, 1999.

Menon, R.S., Luknowsky, D.C., and Gati, J.S., Mental chronometry using latency-resolved functional MRI, *Proc. Natl. Acad. Sci. U.S.A.*, 95, 10902–10907, 1998.

Ogawa, S., Lee, T.M., Kay, A.R., and Tank, D.W., Brain magnetic resonance imaging with contrast dependent on blood oxygenation, *Proc. Natl. Acad. Sci. U.S.A.*, 87, 9868–9872, 1990.

Ogawa, S., Tank, D.W., Menon, R., Ellermann, J.M., Kim, S.-G., Merkle, H. et al., Intrinsic signal changes accompanying sensory stimulation: functional brain mapping with magnetic resonance imaging, *Proc. Natl. Acad. Sci. U.S.A.*, 89, 5951–5955, 1992.

Pardo, J.V., Fox, P.T., and Raichle, M.E., Localization of a human system for sustained attention by positron emission tomography, *Nature*, 249, 61–64, 1991.

Penhune, V.B., Zatorre, R.J., and Evans, A.C., Cerebellar contributions to motor timing: a PET study of auditory and visual rhythm reproduction, *J. Cognit. Neurosci.*, 10, 752–765, 1998.

Posner, M.I. and Peterson, S.E., The attention system of the human brain, *Annu. Rev. Neurosci.*, 13, 25–42, 1990.

Rakitin, B.C., Gibbon, J., Penney, T.B., Malapani, C., Hinton, S.C., and Meck, W.H., Scalar expectancy theory and peak-interval timing in humans, *J. Exp. Psychol. Anim. Behav. Process.*, 24, 15–33, 1998.

Rao, S.M., Binder, J.R., Bandettini, P.A., Hammeke, T.A., Yetkin, F.Z., Jesmanowicz, A. et al., Functional magnetic resonance imaging of complex human movements, *Neurology*, 43, 2311–2318, 1993.

Rao, S.M., Mayer, A.R., and Harrington, D.L., The evolution of brain activation during temporal processing, *Nat. Neurosci.*, 4, 317–323, 2001.

Talairach, J. and Tournoux, P., *Co-planar Stereotaxic Atlas of the Human Brain*, Thieme, New York, 1988.

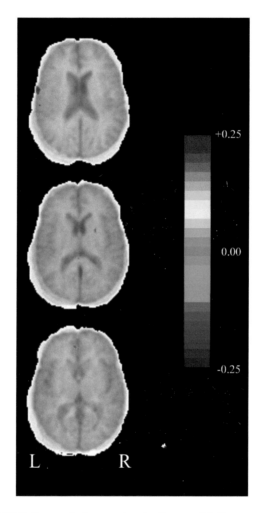

FIGURE 17.3 Composite images showing 7-mm axial slices ranging from 0 to 21 mm above the AC-PC line with overlaid timing-related activation occurring during the 3 sec preceding the criterion time of 11 sec. Color scale indicates correlation coefficients between brain activation and a square wave covering the period described.

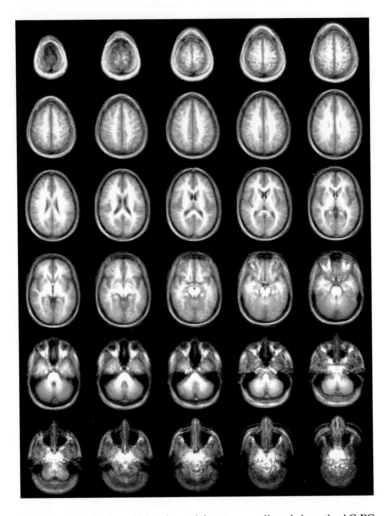

FIGURE 17.4 Brains of 14 male participants were aligned along the AC-PC line and the longitudinal fissure and averaged to produce this axial montage of 5-mm slices ranging from $z = -70$ to $+75$ mm in Talairach coordinates.

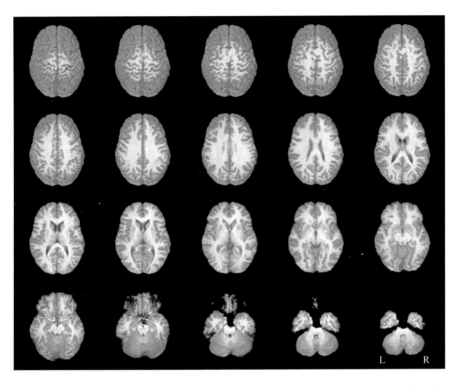

FIGURE 17.5 Axial montage of activation during the period from 0 to 6 sec before the response time when 9 participants timed a 7- or 17-sec duration ($P < .0001$).

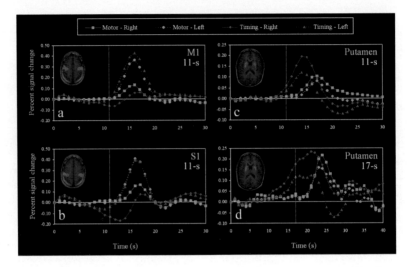

FIGURE 17.6 Time courses of averaged activation from region-of-interest analyses of (a) the primary motor cortex (M1), (b) the primary sensory cortex (S1), and (c) the putamen for an 11-sec signal. Panel (d) shows data from the putamen for a 17-sec signal. Dashed and solid lines respectively indicate activation measured from right and left hemisphere structures. Lighter and darker colors respectively indicate time courses of activation for the motor and timing tasks.

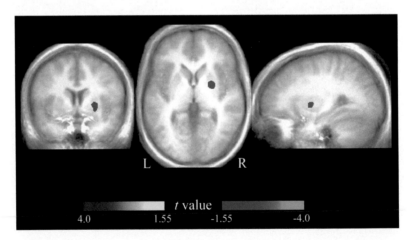

FIGURE 17.7 Area in the right putamen of decreased MR signal intensity ($t < -1.55$) comparing a repetitive motor sequence task with rest (shown in coronal, axial, and sagittal views from left to right). This was the only area of decreased signal intensity reported in the experiment of Harrington et al. (1999). (Figure created with kind permission from D.L. Harrington and S.M. Rao.)

18 Electrophysiological Evidence for Specific Processing of Temporal Information in Humans

Viviane Pouthas

CONTENTS

18.1 Introduction 439
18.2 Contingent Negative Variation and Time Estimation 440
    18.2.1 Contingent Negative Variation: Descriptive Data............................440
    18.2.2 CNV Amplitude and Temporal Processing 441
    18.2.3 CNV Resolution and Temporal Processing.....................................445
18.3 CNV Topography and Generators 448
    18.3.1 CNV Topography 448
    18.3.2 Some Insights on CNV Generators 448
    18.3.3 Electrophysiological Study of the CNV in Parkinson's Disease 448
    18.3.4 Event-Related Potentials and Positron Emission Tomography Analysis of CNV generators..449
18.4 Conclusion..453
References..453

18.1 INTRODUCTION

No sense organ is dedicated to time perception; therefore, the question of specific temporal processing mechanisms has been a recurrent matter of debate in the literature. Since Hoagland's (1933) original contribution suggesting that judgment of time was mediated by the reaction of some chemical pacemaker in the central nervous system, attention has been directed to the idea of a relationship between electroencephalographic (EEG) activity and the experience of time. A popular hypothesis was that the alpha rhythm of the electroencephalogram might provide the time base of an internal clock. The data, however, do not answer the question of whether there is a causal relation between time judgment and alpha frequency either in Werboff's (1962) or in Treisman's (1984) studies. The latter found much less variation in the alpha frequency than in the concurrent time productions. As pointed out by Surwillo (1966, p. 392) "factors other than the alpha rhythm are required to account for variations in the experienced time." More recently some studies have focused on the event-related desynchronization (ERD) of the EEG rhythm (Mohl and Pfurtscheller, 1991), which has the potential to become a promising line of research. Nevertheless, the most familiar electrophysiological correlate of time estimation is a slow negative wave, called the contingent negative variation (CNV). Therefore, the first section of this chapter will focus on data providing evidence that the CNV subtends processing of temporal information, and we will carefully consider how the amplitude and the time course of this slow wave could "reflect the cognitive activity leading up to the formation of a temporal judgment," as pointed out by Ruchkin et al. (1977, p. 454). The second part of this chapter will examine whether findings on CNV topography and sources corroborate data from lesion and neuropsychological studies aimed at determining the neural bases of a specific mechanism for temporal processing.

18.2 CONTINGENT NEGATIVE VARIATION AND TIME ESTIMATION

18.2.1 CONTINGENT NEGATIVE VARIATION: DESCRIPTIVE DATA

Walter et al. (1964) reported that when a warning stimulus indicates the presentation of an imperative stimulus after a constant foreperiod (1 or 2 sec), a negative potential shift gradually develops in the EEG over wide areas of the scalp during the interstimulus interval. This slow shift of negative polarity is fully developed whenever a subject is expecting the occurrence of a significant event in the next few seconds, as illustrated in Figure 18.1a. When the imperative stimulus is withdrawn, there is a progressive diminution of the CNV, as shown in Figure 18.1b to e, and when it is reinstated, the CNV is reestablished, as illustrated in Figure 18.1f.

The CNV changes to a biphasic waveform with two negative peaks when foreperiods exceed 3 to 4 sec (Loveless and Sanford, 1974). It is generally agreed that the early component, maximal over the frontal cortex, constitutes an index of the orienting response to the warning stimulus (e.g., Macar and Besson, 1985; Rohrbaugh et al., 1976). The second peak of the biphasic CNV, the terminal CNV, which has a more central distribution over motor areas of the cortex, has been linked to preparatory processes. It has been argued that short intervals in the typical range of 1 to 2 sec would not allow sufficient time for the CNV to develop its biphasic waveform with the initial and terminal CNV components, and that the monophasic waveform represents the consequence of overlapping components (Rohrbaugh and Gaillard, 1983).

The CNV has been associated with several cognitive processes, including expectancy and attention. In particular, it is thought to reflect the preparation or anticipation of a response (Birbaumer et al., 1990; Rockstroh et al., 1993). Several lines of evidence also support the hypothesis that temporal processing is reflected in the CNV (see Brannon and Roitman, this volume; Elbert et al., 1991; Macar and Besson, 1985; McAdam, 1966; Ruchkin et al., 1977). The one aspect of interval timing that

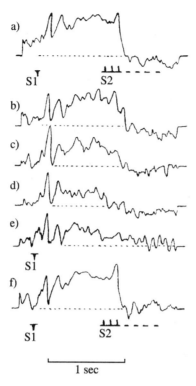

FIGURE 18.1 (a) CNV completely developed between the warning stimulus (S1) and the imperative stimulus (S2). (b to e) Progressive reduction of the CNV after withdrawal of the imperative stimulus. (f) CNV reinstatement after the restoration of the imperative stimulus. (Adapted from Walter, W.G., Cooper, R., Aldridge, V.J., McCallum, W.C., and Winter, A.L., *Nature*, 203, 380–384, 1964.)

has been investigated in CNV research is how this negative shift develops and how its amplitude and time course (i.e., negativity resolution) vary when participants have to estimate a given interval and press a key by themselves, in the absence of any imperative stimulus. For example, in the Macar et al. (1999) study, results showed that the longer the judgment or the production of a 2.5-sec target interval, the larger the CNV amplitude recorded was at a fronto-central site (electrode FCz). This illustrates that variation in temporal performance is reflected in slow brain potential changes. We will now examine results of some studies concerned with the relations between both CNV amplitude and resolution and the quality of time estimation performance in normal subjects and patients who suffer from time estimation deficits.

18.2.2 CNV Amplitude and Temporal Processing

The CNV has been shown to vary in amplitude depending on the level of training, accuracy, and precision of performance, as well as on mental state or disease. Not surprisingly, the picture of these variations given by the literature is not unequivocal, but relatively complex. McAdam (1966) reported a relationship between the progressive increase of the CNV amplitude over trials and the progressive learning of the temporal parameters of the task in the first stages of the experiment. Then the CNV decreased as temporal accuracy approached the maximum level of performance. Macar and Vitton (1980) proposed that the progressive development of the CNV might reflect the elaboration of an internal time reference, corresponding to the target duration. Once this reference is constituted, it may be used more automatically. Some parallels could be drawn with the results of a very recent experiment conducted by my research group (Pfeuty et al., 2003). Subjects had to compare the tempos of two isochronous tone sequences made up of either three or six intervals. CNV amplitude increased during the encoding phase over the frontal part of the scalp. Moreover, when the sequence was composed of six intervals, the results showed that the negativity stopped increasing after the third interval, as illustrated in Figure 18.2. We proposed that this increase in CNV amplitude might reflect the building of a memory trace of the reference interval and that beyond a critical number of intervals, the memory trace would be optimal and time intervals would no longer be estimated.

These data suggest that the level of brain activity may be related to the level of performance, with a relatively low activation corresponding to efficient processing. Some other experimental data corroborate this assumption. Ladanyi and Dubrovsky (1985) found smaller CNVs in accurate than in inaccurate subjects when they were performing a verbal estimation task. Similarly, Casini et al. (1999) showed that the level of CNV activity obtained with correct responses was lower than that obtained with incorrect responses in a task of identification of signal durations. High levels of CNV activity recorded when responses were wrong may reflect an inappropriate attention effort leading to defective encoding. This suggests that good temporal performance could be the result of an efficacious, but economic information-processing mechanism in the brain.

It could be assumed that during practice specific brain structures are progressively selected and interconnected in order to form a reduced network of neurons. Once this network is selected, correct responses may be produced almost automatically without demanding much attention, which would explain the observation of a low level of brain activity in well-practiced subjects. On the one hand, studies using dual-task paradigms have evidenced the crucial role of attention for temporal processing (see Brown, 1997; Fortin, this volume; Pang and McAuley, this volume). On the other hand, relationships between the level of attention and the CNV amplitude have been exemplified by electrophysiological studies outside the time domain. A correlation between the early component of the CNV and the number of checked letters in a letter cancellation test has been found, indicating significant differences in early CNV between low and high performance (Kropp et al., 2001). Similarly, in Filipovic et al. (2001) all segments of the CNV waveform were different in go and no-go conditions. The go CNV displayed a typical pattern of slow rising negativity reflecting the buildup of attentional resources necessary for adequate performance.

The effects of distraction on the CNV have also been investigated in elderly subjects (Michalewski et al., 1980). CNV amplitudes at central and parietal sites were comparable between young and old adults, while amplitude at the frontal sites

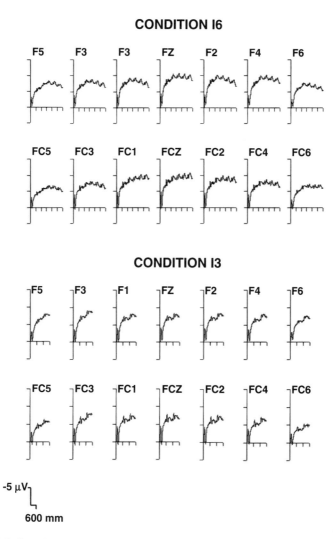

FIGURE 18.2 Grand mean ERPs over frontal and fronto-central electrodes recorded during tempo encoding in the six-interval condition (upper part of the figure) and in the three-interval condition (lower part of the figure). (Adapted from Pfeuty, M., Ragot, R., and Pouthas, V., *Psychophysiology*, 40, 1–8, 2003.)

was consistently reduced in the elderly compared to the young. These data are congruent with other reports (Dirnberger et al., 2000; Ferrandez and Pouthas, 2001). In our study, CNV amplitude in a duration discrimination task was lower in middle-aged adults (50 years) than in the young ones (20 years) when measured at frontal sites. However, it was higher when measured at more central sites, as shown in Figure 18.3. The finding of diminished frontal activity in older subjects suggests a process of selective cortical aging and possible cellular loss, which may be linked to performance deficits. We proposed that modifications of the topography of

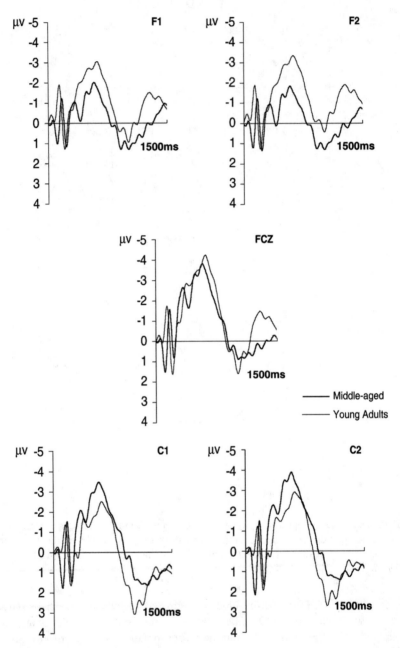

FIGURE 18.3 ERP waveforms at F1, F2, FCz, C1, and C2 sites for young adults (thin line) and middle-aged adults (thick line) in a duration discrimination task. (Adapted from Ferrandez, A.M. and Pouthas, V., *Neurobiol. Aging*, 22, 645–657, 2001.)

electrical activity in middle-aged and elderly adults could reflect a reorganization process, with old adults using more posterior areas to compensate for structural deficits in more frontal sites (Raz et al., 1997).

The amplitude of the CNV has also been shown to be significantly reduced in Parkinson's disease (PD) patients compared with age-matched controls (Ikeda et al., 1997; Praamstra et al., 1996; Pulvermuller et al., 1996; Wascher et al., 1997). Temporal performance of these PD patients is known to be impaired (Malapani et al., 1998; Pastor et al., 1992). Ikeda et al. (1997) found that the level of CNV reduction in PD patients, particularly over the frontal areas, was directly related to the severity of disease. Concurrently with temporal performance improvement, the amplitude of the CNV has been shown to increase following treatment with levodopa (Amabile et al., 1986; Oishi et al., 1995) or when patients receive bilateral subthalamic nucleus stimulation, as shown in a study by Gerschlager et al. (1999). These authors examined changes in the CNV, recorded from patients with Parkinson's disease when on and off bilateral subthalamic nucleus stimulation, and compared these data with the CNV of healthy control subjects. Without subthalamic nucleus stimulation, Parkinson's disease patients showed reduced CNV amplitudes over the frontal and fronto-central regions compared with control subjects, whereas with bilateral subthalamic nucleus stimulation, CNV amplitudes over the frontal and fronto-central regions were significantly increased. These results suggest that subthalamic nucleus stimulation in PD improves the cortical activity that underlies processing of temporal information (for additional details of the effects of subthalamic stimulation on interval timing in PD patients, see Malapani and Rakitin, this volume).

18.2.3 CNV Resolution and Temporal Processing

The amplitude of the CNV is not the only important parameter for assessing relationships between cortical activity and temporal judgments. Indeed, the resolution of this slow wave, i.e., its return to baseline, has also been taken into account. In experimental paradigms specifically designed to study time estimation, it has been shown that the return of the negativity to baseline started largely before the imperative stimulus (S2) and the motor response that are most often temporally close together (Ladanyi and Dubrovsky, 1985; Macar and Vitton, 1982; Ruchkin et al., 1977). Macar and Vitton (1982) asked whether the early resolution (ER) has to be referred to processes specifically involved in time estimation paradigms. Their experiment was aimed at checking whether the CNV presented an early resolution in a time discrimination paradigm, which dissociated cognitive timing processes from motor mechanisms more clearly than the S1-S2 paradigm more commonly used. In the latter, the authors observed a classical CNV during the delay that is interpreted as an index of expectancy. By contrast, in the task in which the subject had to collect temporal information, the CNV was followed by an early and prolonged resolution of negativity in the second half of the interval. Macar and Vitton (1982) concluded that this type of event-related potential (ERP) component, identified as a CNV-ER, appeared as a correlate of cognitive processes pertaining to the collection of temporal information.

An important issue is to determine the moment in time of the shift from negative to positive components of the CNV. The next three studies that we will describe address this question. Ladanyi and Dubrovsky (1985) investigated the CNV waves of subjects classified according to their performance in time estimation tasks in accurate and nonaccurate estimators. As mentioned above, accurate estimators have CNVs of lower amplitude and show a slower rise time to peak negativity than subjects with poor time estimation abilities. Most importantly for our purpose, the results revealed that subjects with a high degree of accuracy in time estimation tasks exhibited a faster resolution of the negativity. Moreover, the data also showed a positive correlation between delay in resolution of the CNVs and the degree of error in the subjects that overestimate time periods; i.e., the more the subjects overestimate time lapses in the behavioral tests, the more prolonged were their CNVs waves. The authors assumed that the development and resolution of the CNV may, in part, reflect time estimation processes in the nervous system. They concluded that CNV resolution is most likely related to the conscious decision to respond after the formation of the temporal judgment.

Another example can be found in a study by Ruchkin et al. (1977). Subjects had to reproduce a target interval of 900 msec after a click. The authors examined average waveforms synchronized to the click for three categories of interval reproduction times, i.e., 800 msec (752 to 828), 900 msec (892 to 908), and 1000 msec (972 to 1048). Average waveforms synchronized to the click for 800, 900, and 1000 msec evidenced a covariation between subjects' temporal judgments and the latency of the negative-to-positive amplitude shift of the concomitantly recorded ERPs. These results suggest that the differences between the latencies reflect differences in timing of the cognitive processes associated with the formation of a temporal judgment.

These data sets suggest that CNV resolution may relate, at least in part, to the conscious decision to respond after the formation of a temporal judgment. Therefore, the resolution of the CNV is not necessarily time-locked to an imperative stimulus. This is particularly well exemplified by results of one of our recent studies (Pouthas et al., 2000). We recorded ERPs in a matching-to-sample task in which subjects had to decide whether the duration of an light-emitting diode LED illumination was the same as or different from a standard duration (700 msec) previously memorized. As demonstrated in Figure 18.4, the CNV resolution occurred after the LED switched off. However, the moment of resolution occurrence depended on whether the duration of the standard had elapsed. For test durations shorter than the standard (i.e., 490 and 595 msec), the resolution intervened later after LED switch off, i.e., around 200 msec, compared to what happened with durations longer than the target (805 and 910 msec), i.e., around 100 msec, as shown in Figure 18.4. We proposed the following explanatory hypotheses to account for these results. On the one hand, slower resolution of the CNV would reflect the fact that subjects were waiting for the standard duration to elapse before making a decision. On the other hand, faster resolution would reflect the fact that after the standard duration was over, subjects had formed their judgment and anticipated the LED switch off.

This section has revealed that changes in the accuracy and precision of temporal performance are accompanied by reliable changes in brain activity as measured by

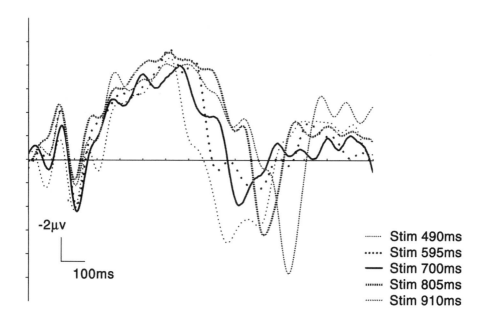
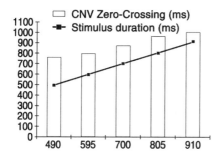

FIGURE 18.4 ERP waveforms recorded on the FCz electrode for the five stimulus durations (upper part of the figure). Diagram showing the delay between the moment in time of the LED switch off and the moment in time of the CNV return to baseline, i.e., zero-crossing (lower part of the figure). (Adapted from Pouthas, V., Garnero, L., Ferrandez, A.M., and Renault, B., *Hum. Brain Mapping*, 10, 49–60, 2000.)

event-related brain potentials, particularly by the CNV. As stressed by Macar et al. (1999), these changes should occur over the cerebral regions specifically concerned with temporal processing. ERPs reflect the rapidly changing electrical activity in the brain evoked by a stimulus or a cognitive event and permit us to discriminate between different stages in information processing. It is difficult, however, to determine the neural sources of ERP components. Therefore, in the following section, we report data from neuropsychological studies and from a combination of PET and ERP analyses that provide some evidence on the major contributing sources of the CNV.

18.3 CNV TOPOGRAPHY AND GENERATORS

18.3.1 CNV Topography

The primary source of evoked potentials is neuronal activity in the upper layers of the cortex. It is primarily the simultaneous occurrence of numerous excitatory postsynaptic potentials at apical dentrites of pyramidal neurons close to the recording electrode that causes changes in the evoked potentials at the surface of the head (Birbaumer et al., 1990). Most studies report that the CNV is mainly observed at fronto-central, central, and centro-parietal regions. However, it is sometimes difficult to disambiguate motor ERPs and cognitive ERPs in the CNV, whereas it has been suggested that cognitive preparation (reflected in portions of the CNV) is a psychological process that differs from motor preparation (Bereitschaftspotential (BP)). A study by Leynes et al. (1998) has provided separate records of brain activity during motor and cognitive preparations. The ERP topographies differed for the two types of tasks (cognitive vs. motor) used: a central CNV for the motor task and a frontal CNV for the cognitive task. These topographic effects suggest that the neural circuitries subserving the two preparatory processes differ. Hamano et al. (1997) pointed out that the CNV recorded from the scalp is the summation of several cortical potentials that have different origins and different functions. Besides, multiple cortical as well as subcortical regions have been suggested to participate in the generation of the CNV. Consequently, the exact generator or generators of the CNV in humans is still unclear.

18.3.2 Some Insights on CNV Generators

Using high-resolution spatiotemporal statistics and current source density, Cui et al. (2000) investigated the CNV topography. Their data suggest that the origin of the early CNV may rest in the frontal lobes. The authors argue that this is consistent with other previous results. For example, Lai et al. (1997) obtained the early CNV in the S1-S2 paradigm, which consisted of a frontal to fronto-polar midline negative potential, most likely associated with judgment and decision-making processes. Oishi and Mochizuki (1998) investigated the CNV paradigm with the rCBF method and found that there was a significant positive correlation between the amplitude of the early CNV and frontal blood flow. The generators of the CNV have been examined using subdural recordings in humans. They are said to include mesial frontal areas, the supplementary motor area, and the dorsolateral prefrontal cortex (Hamano et al., 1997; Ikeda et al., 1999; Lamarche et al., 1995). In sum, a growing body of evidence suggests that the frontal lobes appear to contain the main generators for the CNV. Nonetheless, surface recordings on the frontal part of the scalp may reflect the activity of a larger network. Results from studies recording the CNV of PD patients suggest that the cortical–basal ganglia–thalamo–cortical circuit plays a major role in the generation of the CNV. This possibility will be examined below.

18.3.3 Electrophysiological Study of the CNV in Parkinson's Disease

According to the model of basal ganglia–thalamo–cortical circuitry (Wichman and DeLong, 1996), a disease affecting basal ganglia function may lead to reduced outflow to the cortex. This view is supported by the analysis of the CNV recorded in patients with PD by Amabile et al. (1986). They found a correlation between CNV measures and pharmacological treatment after a washout period and 15 and 30 days after the start of treatment with L–DOPA. A small amount of CNV was observed in the nondrug condition, and an enhanced CNV was found during treatment with L–DOPA. This increase in the CNV following medication may reflect a functional recovery, perhaps partial, in the striato–thalamo–cortical connections that terminate in the upper frontal cortical layers, where they contact apical dentrites of pyramidal neurons (Amabile et al., 1986). Impaired activation of frontal cortical areas, including the supplementary motor area and prefrontal cortex, would result from impaired thalamo-cortical output of the basal ganglia.

Surgical interventions aimed at increasing basal ganglia–thalamic outflow to the cortex, such as electrical stimulation of the subthalamic nucleus with chronically implanted electrodes, have also been shown to improve cortical functioning, particularly within the frontal and premotor areas (Gerschlager et al., 1999). This is in accord with previous functional imaging studies showing increased activation of the supplementary motor area, cingulate cortex, and dorsolateral prefrontal cortex during effective stimulation of the subthalamic nucleus (Limousin et al., 1997).

Using EEG recordings, the results of Gerschlager et al. (1999) extend those of Limousin et al.'s (1997) positron emission tomography (PET) study. The latter study showed improved cortical activation only when the activity was averaged over a prolonged period during which the task was performed (i.e., a 90-sec acquisition period was used). The former study was able to show precisely when in time subthalamic nucleus stimulation improves the cortical activity, that is, during the preparatory period prior to the initiation of a response. Consequently, Gerschlager et al. (1999) assumed that subthalamic nucleus stimulation in PD improves the cortical activity that underlies cognitive processes associated with the preparation and organization of forthcoming responses. Combining EEG and PET data allows one to determine not only the areas involved in the processing of information, but also the time course of the activation of these areas.

18.3.4 EVENT-RELATED POTENTIALS AND POSITRON EMISSION TOMOGRAPHY ANALYSIS OF CNV GENERATORS

In order to question the specificity of the spatiotemporal organization of cerebral areas subserving the processing of stimulus duration, we combined positron emission tomography and event-related potential data recorded from subjects performing two visual discrimination tasks, one based on the duration of a visual stimulus and the other on the intensity of the same stimulus. Results of the PET study showed that the same network was activated in both tasks — right prefrontal cortex, right inferior parietal lobule, anterior cingulate cortex, left fusiform gyrus, and vermis (Maquet et al., 1996). We could not unambiguously conclude, however, that this pattern of activation was specific to the perception of a stimulus duration (see Hinton, this volume).

As stressed above, the PET method integrates radioactive tracer activity over a period of many seconds; therefore, the temporal resolution of this method is insufficient to determine when the different areas are active and then to determine in which processing stages these areas are involved. ERPs, on the other hand, reflect the rapidly changing electrical activity in the brain evoked by a stimulus or a cognitive event, providing the means to differentiate the electrical activity pertaining to the different stages of information processing (see Sakata and Onoda, this volume). It is difficult, however, to determine the neural generators of ERP components, the solution of the inverse problem not being unique. Therefore, we first carried a dipole modeling using a PET-seeded model (right prefrontal, right parietal, anterior cingulate, left and right fusiforms). Then, to obtain a better fit, two sources (cuneus and left prefrontal area) had to be added. This dipole modeling showed that the proportion of accounted variance was equivalent in both tasks (Pouthas et al., 2000). This indicates, consistent with the earlier PET findings, that duration and intensity dimensions of a visual stimulus are processed in the same cerebral areas.

However, ERPs revealed prominent differences between the time courses of the dipole activations for each task, particularly that of the probable generators of the late-latency ERP components. The magnitude waveforms of three dipoles — right frontal, anterior cingulate, and cuneus, observed when the intensity or the duration of a test stimulus equivalent to the memorized standard stimulus (i.e., 700 msec — 15 cd/m^2) has to be judged — are superimposed on Figure 18.5. The timing of activation of these dipoles largely differed between the two tasks. It must be stressed that the stimulus was physically the same and that only the instructions received by the subjects were different; i.e., evaluate either intensity or duration. We will focus on the duration task and the putative generator of the CNV. Importantly, in the duration task, the right prefrontal dipole was the most active during the CNV (from 400 to 1000 msec), whereas in the intensity task, it showed a low level of activity during this time range. Moreover, in the duration task, although the CNV began earlier (at 300 msec), the magnitude waveform of this dipole appeared to be very similar to that of the CNV, resolving its activity when the CNV resolved, as shown in Figure 18.6. This suggests that in such a matching-to-sample task, the right frontal area has an essential role in making a decision about the stimulus duration. Because the dipole located in this area was not very active in the intensity task, only following LED switch off, we assume that this role is specific to the temporal dimension of the stimulus.

This assumption is in accord with other neuroimaging data (e.g., Hinton and Meck, 1997; Hinton et al., 1996; Meck et al., 1998). In addition, neuropsychological data of Harrington et al. (1998) on patients with focal left or right lesions revealed that only patients with right lesions show time perception deficits. In a study designed to investigate how brain activity is lateralized during the encoding and recognition of a visual stimulus duration, results showed that the right frontal cortex was involved in both operations, suggesting that the involvement of right frontal structures is critical for time perception (Monfort et al., 2000). Another example of a right hemispheric bias for processing temporal information can be found in Damen and Brunia's (1987) experiment. In a typical warning-imperative stimulus paradigm, the CNV reflects both motor preparation and stimulus anticipation. In order to separate temporally motor and cognitive preparations, the authors used a

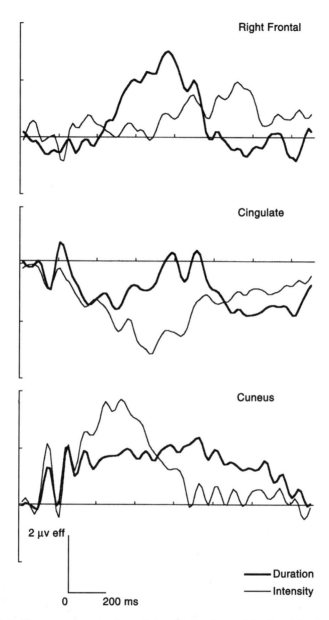

FIGURE 18.5 Time course activation of the dipoles located in the right frontal cortex, anterior cingulate cortex, and cuneus in the 100- to 1500-msec window after stimulus onset for the standard stimulus in both duration (thick line) and intensity (thin line) tasks. (Adapted from Pouthas, V., Garnero, L., Ferrandez, A.M., and Renault, B., *Hum. Brain Mapping*, 10, 49–60, 2000.)

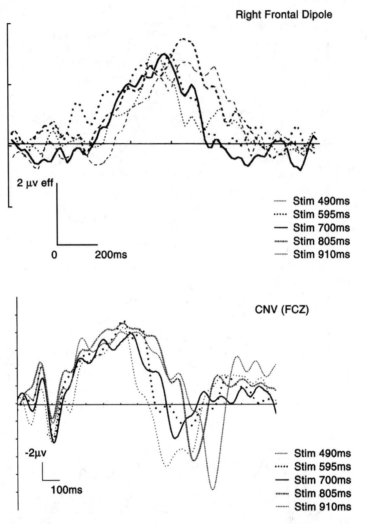

FIGURE 18.6 Time course activation of the right frontal dipole in the 100- to 1500-msec window following stimulus onset for the five stimulus durations. (Adapted from Pouthas, V., Garnero, L., Ferrandez, A.M., and Renault, B., *Hum. Brain Mapping*, 10, 49–60, 2000.)

feedback paradigm. In this paradigm participants press a key when they believe that a target interval has elapsed, and a few seconds later they receive feedback on the accuracy of their temporal estimation. The potential that occurs prior to the feedback stimulus, named stimulus preceding negativity (SPN), is very similar to the nonmotor CNV observed in S1-S2 paradigms that do not require a motor response. It would reflect a similar cognitive preparatory process. The results of Damen and Brunia (1987) showed that the SPN was larger over the right hemisphere, irrespective of the movement side. This pointed to a right hemisphere preponderance of the SPN source.

18.4 CONCLUSION

In sum, we have seen that variation in temporal performance is reflected in slow brain potential (CNV) changes. On the one hand, in normal subjects CNV amplitude progressively increases over progressive learning of temporal parameters and decreases as temporal accuracy approaches the maximum level of performance. On the other hand, CNV amplitude is consistently reduced in elderly adults compared to young ones and in Parkinson's disease patients compared with controls, particularly at the frontal sites. Moreover, latency of CNV resolution shows a positive correlation with accuracy in duration estimation, probably reflecting the conscious decision to respond after the formation of the temporal judgment. Studies in patients with Parkinson's disease suggest that the cortical–basal ganglia–thalamo–cortical circuit plays a major role in the generation of the CNV. The combination of EEG and PET data further indicates that cortical activity may underlie cognitive processes associated with the preparation and organization of forthcoming responses. Finally, there is some intriguing evidence of a right hemispheric bias for processing temporal information.

REFERENCES

Amabile, G., Fattapposta, F., Pozzessere, G., Albani, G., Sanarelli, L., Rizzo, P.A., and Morocutti, C., Parkinson disease: electrophysiological (CNV) analysis related to pharmacological treatment, *Electroencephalogr. Clin. Neurophysiol.*, 64, 521–524, 1986.

Birbaumer, N., Elbert, T., Canavan, A.G.M., and Rockstroh, B., Slow potentials of the cerebral cortex and behavior, *Physiol. Rev.*, 70, 1–41, 1990.

Brown, S.W., Attentional resources in timing: interference effects in concurrent temporal and nontemporal working memory tasks, *Percept. Psychophys.*, 59, 1118–1140, 1997.

Casini, L., Macar, M., and Giard, M.H., Relation between level of prefrontal activity and subject's performance, *J. Psychophysiol.*, 13, 118–125, 1999.

Cui, R.Q., Egkher, A., Huter, D., Lang, W., Lindinger, G., and Deecke, L., High resolution spatiotemporal analysis of the contingent negative variation in simple or complex motor tasks and a non-motor task, *Clin. Neurophysiol.*, 111, 1847–1859, 2000.

Damen E.J. and Brunia, C.H., Changes in heart rate and slow brain potentials related to motor preparation and stimulus anticipation in a time estimation task, *Psychophysiology*, 24, 700–713, 1987.

Dirnberger, G., Lalouschek, W., Lindinger, G., Egkher, A., Deecke, L., and Lang, W., Reduced activation of midline frontal areas in human elderly subjects: a contingent negative variation study, *Neurosci. Lett.*, 280, 61–64, 2000.

Elbert, T., Ulrich, R., Rockstroh, B., and Lutzenberger, W., The processing of temporal intervals reflected by CNV-like brain potentials, *Psychophysiology*, 28, 648–655, 1991.

Ferrandez, A.M. and Pouthas, V., Does cerebral activity change in middle-aged adults in a visual discrimination task? *Neurobiol. Aging*, 22, 645–657, 2001.

Filipovic, S.R., Jahanshahi, M., and Rothwell, J.C., Uncoupling of contingent negative variation and alpha band event-related desynchronization in a go/no-go task, *Clin. Neurophysiol.*, 112, 1307–1315, 2001.

Gerschlager, W., Alesch, F., Cunnington, R., Deecke, L., Dirnberger, G., Endl, W., Lindinger, G., and Lang, W., Bilateral subthalamic nucleus stimulation improves frontal cortex function in Parkinson's disease: an electrophysiological study of the contingent negative variation, *Brain*, 122, 2365–2373, 1999.

Hamano, T., Luders, H.O., Ikeda, A., Collura, T.F., Comair, Y.G., and Shibasaki, H., The cortical generators of the contingent negative variation in humans: a study with subdural electrodes, *Electroencephalogr. Clin. Neurophysiol.*, 104, 257–268, 1997.

Harrington, D.L., Haaland, K.Y., and Knight, R.T., Cortical networks underlying mechanisms of time perception, *J. Neurosci.*, 18, 1085–1095, 1998.

Hinton, S.H. and Meck, W.H., The "internal clocks" of circadian and interval timing, *Endeavour*, 21, 82–87, 1997.

Hinton, S.C., Meck, W.H., and MacFall, J.R., Peak-interval timing in humans activates frontal-striatal loops, *Neuroimage*, 3, S224, 1996.

Hoagland, H., The physiological control of judgments of duration: evidence for a chemical clock, *J. Genet. Psychol.*, 9, 267–287, 1933.

Ikeda, A., Shibasaki, H., Kaji, R., Terada, K., Nagamine, T., Honda, M., and Kimura, J., Dissociation between contingent negative variation (CNV) and Bereitschafspotential (BP) in patients with parkinsonism, *Electroencephalogr. Clin. Neurophysiol.*, 102, 142–151, 1997.

Ikeda, A., Yazawa, S., Kunieda, T., Ohara, S., Terada, K., Mikuni, N., Nagamine, T., Taki, W., Kimura, J., and Shibasaki, H., Cognitive motor control in human pre-supplementary motor area studied by subdural recording of discrimination/selection-related potentials, *Brain*, 122, 915–931, 1999.

Kropp, P., Linstedt, U., Niederberger, U., and Gerber, W.D., Contingent negative variation and attentional performance in humans, *Neurol. Res.*, 23, 647–650, 2001.

Ladanyi, M. and Dubrovsky, B., CNV and time estimation, *Int. J. Neurosci.*, 26, 253–257, 1985.

Lai, C., Ikeda, A., Terada, K., Nagamine, T., Honda, M., Xu, X., Yoshimura, N., Howng, S., Barrett, G., and Shibasaki, H., Event-related potentials associated with judgment: comparison of S1- and S2-choice conditions in a contingent negative variation (CNV) paradigm, *J. Clin. Neurophysiol.*, 14, 394–405, 1997.

Lamarche, M., Louvel, J., Buser, P., and Rektor, I., Intracerebral recordings of slow potentials in a contingent negative paradigm: an exploration in epileptic patients, *Electroencephalogr. Clin. Neurophysiol.*, 95, 268–276, 1995.

Leynes, P.A., Allen, J.D., and Marsh R.L., Topographic differences in CNV amplitude reflect different preparatory processes, *Int. J. Psychophysiol.*, 31, 33–44, 1998.

Limousin, P., Greene, J., Pollak, P., Rothwell, J., Benabid, A.L., and Frackowiak, R., Changes in cerebral activity pattern due to subthalamic nucleus or internal pallidum stimulation in Parkinson's disease, *Ann. Neurol.*, 42, 283–291, 1997.

Loveless, N.E. and Sanford, A.J., The impact of warning signal intensity on reaction time and components of the contingent negative variation, *Biol. Psychol.*, 2, 217–226, 1974.

Macar, F. and Besson, M., Contingent negative variation in processes of expectancy, motor preparation and time estimation, *Biol. Psychol.*, 21, 293–307, 1985.

Macar, F., Vidal, F., and Casini, L., The supplementary motor area in motor and sensory timing: evidence from slow brain potential changes, *Exp. Brain Res.*, 125, 271–280, 1999.

Macar, F. and Vitton, N., CNV and reaction time task in man: effects of interstimulus interval contingencies, *Neuropsychologia*, 18, 585–590, 1980.

Macar, F. and Vitton, N., An early resolution of contingent negative variation (CNV) in time discrimination, *Electroencephalogr. Clin. Neurophysiol.*, 54, 426–435, 1982.

Malapani, C., Rakitin, B., Levy, R., Meck, W.H., Deweer, B., Dubois, B., and Gibbon, J., Coupled temporal memories in Parkinson's disease: a dopamine-related dysfunction, *J. Cognit. Neurosci.*, 10, 316–331, 1998.

Maquet, P., Lejeune, H., Pouthas, V., Bonnet, M., Casini, L., Macar, F., Timsit-Berthier, M., Vidal, F., Ferrara, A., Degueldre, L., Delfiore, G., Luxen, R., Woods, R., Mazziota, J.C., and Comar, D., Brain activation induced by estimation of duration: a PET study, *Neuroimage*, 3, 227–242, 1996.

McAdam, D.W., Slow potential changes recorded from human brain during learning of a temporal interval, *Psychonomic Sci.*, 6, 435–436, 1966.

Meck, W.H., Hinton, S.C., and Matell, M.S., Coincidence-detection models of interval timing: evidence from fMRI studies of cortico-striatal circuits, *Neuroimage*, 7, S281, 1998.

Michalewski, H.J., Thompson, L.W., Smith, D.B., Patterson, J.V., Bowman, T.E., Litzelman, D., and Brent, G., Age differences in the contingent negative variation (CNV): reduced frontal activity in the elderly, *J. Gerontol.*, 35, 542–549, 1980.

Mohl, W. and Pfurtscheller, G., The role of the right parietal region in a movement time estimation task, *Neuroreport*, 2, 309–312, 1991.

Monfort, V., Pouthas, V., and Ragot, R., Role of frontal cortex in memory for duration: an event-related potential study in humans, *Neurosci. Lett.*, 286, 91–94, 2000.

Oishi, M. and Mochizuki, Y., Correlation between contingent negative variation and cerebral blood flow, *Clin. Electroencephalogr.*, 29, 124–127, 1998.

Oishi, M., Mochizuki, Y., Du, C., and Takasu, T., Contingent negative variation and movement-related cortical potentials in parkinsonism, *Electroencephalogr. Clin. Neurophysiol.*, 95, 346–349, 1995.

Pastor, M.A., Artieda, J., Jahanshahi, M., and Obeso, J.A., Time estimation and reproduction is abnormal in Parkinson's disease, *Brain*, 115, 211–225, 1992.

Pfeuty, M., Ragot, R., and Pouthas, V., Processes involved in tempo perception: a CNV analysis, *Psychophysiology*, 40, 1–8, 2003.

Pouthas, V., Garnero, L., Ferrandez, A.M., and Renault, B., ERPs and PET analysis of time perception: spatial and temporal brain mapping during visual discrimination tasks, *Hum. Brain Mapping*, 10, 49–60, 2000.

Praamstra, P., Meyer, A.S., Cools, A.R., Horstink, M.W., and Stegeman, D.F., Movement preparation in Parkinson's disease: time course and distribution of movement-related potentials in a movement precueing task, *Brain*, 119, 1689–1704, 1996.

Pulvermuller, F., Lutzenberger, W., Muller, V., Mohr, B., Dichgans, J., and Birbaumer, N., P3 and contingent negative variation in Parkinson's disease, *Electroencephalogr. Clin. Neurophysiol.*, 98, 456–467, 1996.

Raz, N., Gunning, F.M., Head, D., Dupuis, J.H., McQuain, J., Briggs, S.D., Loken, W.J., Thornton, A.E., and Acker, J.D., Selective aging of the human cerebral cortex observed *in vivo*: differential vulnerability of the prefrontal gray matter, *Cereb. Cortex*, 7, 268–282, 1997.

Rockstroh, B., Müller, M., Wagner, M., Cohen, R., and Elbert, T., 'Probing' the nature of CNV, *Electroencephalogr. Clin. Neurophysiol.*, 87, 235–241, 1993.

Rohrbaugh, J.W. and Gaillard, A.W.K., Sensory and motor aspects of contingent negative variation, in *Tutorials in ERP Research: Endogenous Components*, Gaillard, A.W.K and Ritter, W., Eds., Elsevier, Amsterdam, 1983, pp. 269–310.

Rohrbaugh, J.W., Syndulko, K., and Lindsley, D.B., Brain wave components of the contingent negative variation in humans, *Science*, 191, 1055–1056, 1976.

Ruchkin, D.S., McCalley, M.G., and Glaser, E.M., Event related potentials and time estimation, *Psychophysiology*, 14, 451–455, 1977.

Surwillo, W.W., Time perception and the 'internal clock': some observations on the role of the electroencephalogram, *Brain Res.*, 2, 390–392, 1966.

Treisman, M., Temporal rhythms and cerebral rhythms, in *Annals of the New York Academy of Sciences: Timing and Time Perception*, Vol. 423, Gibbon, J. and Allan, L., Eds., New York Academy of Sciences, New York, 1984, pp. 542–565.

Walter, W.G., Cooper, R., Aldridge, V.J., McCallum, W.C., and Winter, A.L., Contingent negative variation: an electric sign of sensorimotor association and expectancy in the human brain, *Nature*, 203, 380–384, 1964.

Wascher, E., Verleger, R., Vieregge, P., Jaskowski, P., Koch, S., and Kömpf, D., Responses to cued signals in Parkinson's disease: distinguishing between disorders of cognition and of activation, *Brain*, 120, 1355–1375, 1997.

Werboff, J., Time judgment as a function of electroencephalographic activity, *Exp. Neurol.*, 6, 152–160, 1962.

Wichmann, T. and DeLong, M.R., Functional and pathophysiological models of the basal ganglia, *Curr. Opin. Neurobiol.*, 6, 751–758, 1996.

19 Cerebellar and Basal Ganglia Contributions to Interval Timing

Jörn Diedrichsen, Richard B. Ivry, and Jeff Pressing

CONTENTS

19.1 Introduction 458
19.2 Review of Existing Studies 459
    19.2.1 The Production of Timed Sequences 459
    19.2.2 The Perception of Timed Events 462
19.3 Synchronization 464
    19.3.1 Components Involved in Synchronization 464
    19.3.2 Neural Structures Underlying Synchronization and Timing 466
19.4 Experimental Study 467
    19.4.1 Method 469
        19.4.1.1 Participants 469
        19.4.1.2 Procedure 469
    19.4.2 Data Analysis 470
    19.4.3 Results 470
19.5 Discussion 474
    19.5.1 Variability 475
    19.5.2 Error Correction Process 475
    19.5.3 Asynchrony 476
    19.5.4 Final Comments 477
Acknowledgments 478
References 478
Appendix 482

19.1 INTRODUCTION

Accurate timing is a ubiquitous aspect of mental processes. How does the central nervous system solve the demands involved in the temporal aspects of information processing? One solution would be that timing is handled by subsystems specialized for domain-specific processing. For example, the timing required for producing well-articulated speech would be solved by areas involved in speech production, whereas the timing demands for the coordination of manual actions would be controlled by brain areas that also control the force and spatial aspects of these movements. Alternatively, humans are capable of producing rather arbitrary behaviors that exhibit accurate timing. We can produce periodic movements over a considerable range of durations. These actions can be achieved with different parts of the body and, indeed, do not even require overt actions; we can covertly maintain an internal beat. We can also detect and judge rhythmicity in a wide variety of sensory signals. While our sense of rhythm may be most accurate for auditory events, we can readily detect temporal perturbations in a sequence of visual or tactile events. Thus, there likely exists some general system specialized to represent temporal information, a system that is recruited for tasks that require this form of computation.

There is ample evidence that temporal acuity correlates across different domains and behaviors (Ivry and Hazeltine, 1995). Furthermore, a direct relationship is observed between interval duration and temporal accuracy over a wide variety of intervals, organisms, and tasks (for a review, see Gibbon et al., 1997). These observations suggest at least one common underlying system for the representation of temporal information.

Three major challenges for research on temporal processing then become apparent. The first is primarily psychological, involving the distinction and characterization of different timing systems. What behaviors and perceptual skills share a common system and which functional domains engage domain-specific processes? For example, it has been proposed that a distinction can be made between repetitive movements for which timing is explicitly represented and repetitive movements in which temporal regularities are an emergent property (Ivry et al., in press; Robertson et al., 1999). A second distinction that has been considered is based on the idea that different systems may be engaged, depending on the temporal extent of the timed intervals (Gibbon et al., 1997; Ivry, 1996).

The second challenge is to generate a *process* model or models that make explicit the component operations involved in tasks that require temporal processing. For example, the scalar timing model specifies a series of component parts associated with the accumulation of clock pulses and the comparison of this sum to stored representations in long-term memory (Gibbon et al., 1997; Treisman, 1963). Similarly, the Wing–Kristofferson model postulates distinct processes that contribute to the variability observed during repetitive tapping tasks (Wing and Kristofferson, 1973). In this chapter, we consider models of this type, analyzing the psychological processes involved in synchronizing an internal timing mechanism to external, rhythmic events.

The last and ultimate challenge is to provide a mapping between psychological operations and neural circuits. This mapping need not be in a one-to-one correspondence.

While some operations may be localized to particular neural structures, it is possible that the operations we describe at a computational level of explanation are implemented in a distributed manner within the brain. Exploring this mapping not only provides a first step toward developing a mechanistic explanation at the neural level, but also can help shape our understanding of the psychological operations.

We will focus on the contribution of two subcortical structures that have been proposed to be the cornerstone of an internal timing system: the cerebellum and the basal ganglia (Ivry, 1997; Meck, 1996). Both structures form reciprocal loops with many cortical areas (Alexander et al., 1986; Middleton and Strick, 1997; Strick et al., 1995), which would enable them to provide the precise representation of temporal information across a range of different task domains. We first review neuropsychological studies that investigate the role of the cerebellum and basal ganglia in the production and perception of timed events. We then report a new experiment, examining the contribution of these structures to synchronization behavior.

19.2 REVIEW OF EXISTING STUDIES

While temporal regularities are manifest at many timescales, our review will be restricted to tasks involving intervals in the hundreds of milliseconds range. We opt for this limited range for two reasons. First, given the role of the cerebellum and basal ganglia in motor control, this range reflects the temporal extent of the component movements that form human actions such as walking, reaching, or speaking. Second, similar methodologies have been applied in studies looking at timing in this range, providing an empirical basis for comparison. By restricting our review to short intervals, we do not imply that the timing of intervals in the range of multiple seconds does not entail similar processes and neural structures as timing in the millisecond range. At present, we see this issue as one in need of further study (for discussion of this issue, see Gilden et al., 1995; Ivry, 1996; Malapani et al., 1998; Mangels and Ivry, 2001).

19.2.1 THE PRODUCTION OF TIMED SEQUENCES

Time production studies in this time span that have involved patients with either cerebellar lesions or Parkinson's disease (PD) are summarized in Table 19.1. The PD patients have been generally viewed as a model for studying basal ganglia dysfunction. Most of these studies have used a continuation task introduced by Wing and Kristofferson (1973). Trials begin with the presentation of a periodic signal, usually an auditory metronome. After an initial synchronization phase, the metronome is terminated and the participants are asked to continue tapping, attempting to separate each response by the interval specified by the metronome.

This task is appealing for its simplicity. The instructions are intuitive and comprehensible by people with a range of neurological or psychiatric disorders. The motor requirements are minimal. More important, this task has provided a process model for evaluating component sources of temporal variability in performance (Wing and Kristofferson, 1973). The model postulates two component sources of noise: a central clock that provides the timing signals for the series of successive

TABLE 19.1
Summary of Studies Involving Human Participants with Cerebellar Damage or PD on Repetitive Finger Tapping

Study (Pace)	Group	Condition	N	CD	Sig.	MD	Sig.	Symptom
Ivry et al. (1988) (550 msec)	Cerebellar, hemisphere	Impaired	4	37.0	*	17.8	—	
		Unimpaired		21.0		14.3		
	Cerebellar, vermal	Impaired	3	27.0	—	25.3	*	
		Unimpaired		23.0		13.0		
Franz et al. (1996) (400 msec)	Cerebellar	Impaired	4	22.0	*	12.0	—	
		Unimpaired		13.0		8.0		
Ivry and Keele (1989) (550 msec)	Control		21	24.3		11.0		
	Cerebellar		27	38.1	*	14.0	—	11 focal, 16 degenerative
	PD		29	27.7	—	9.3	—	On medication
	PD	Impaired	4	46.5	*	11.5	—	
		Unimpaired		25.6	—	9.4	—	
	PD	On meds	7	28.4	—	10.7	—	
		Off meds		27.7	—	11.8	—	
Pastor et al. (1992) (400 msec)	Control		20	9.0		6.1		
	PD		42	34.8	*	19.6	*	Off medication
Duchek et al. (1994) (550 msec)	Control		30	21.1		14.0		
	PD		20	24.2	—	10.4	*	HY = 1–2 On medication
O'Boyle et al. (1996) (550 msec)	Control		12	15.4		8.2		
	PD	Unimpaired	12	17.8	—	8.5	—	HY = 1.5
		Impaired		23.7	*	11.9	*	On medication
	PD	On	12	17.9	—	8.1	—	HY = 1.8
		Off		24.3	*	13.5	*	
Harrington et al. (1998a) (600 msec)	Control		24	27.0		15.0		
	PD		34	42.0	*	20.0	—	HY = 2.4 On medication

Note: N = number of participants; CD = standard deviation of clock component estimated with the Wing–Kristofferson model; MD = standard deviation of motor component estimated with the Wing–Kristofferson model; Sig. = significance (* indicates that the difference between the experimental and control group is statistically reliable); HY = Hohn and Yahr scale.

responses and a motor system that implements these responses. Based on a small set of assumptions, namely, that the component sources are independent of each other and that feedback mechanisms do not play a role in producing the next tap, estimates of the two components can be obtained from the autocovariance function of the intertap intervals. The assumptions of the model have received substantial empirical support (Ivry and Hazeltine, 1995; Wing, 1980).

Ivry and Keele (1989) provided the first large-scale study in which the performance of patients with different neurological disorders was assessed with the Wing–Kristofferson continuation task. Patients with lesions of the cerebellum exhibited significant increases in variability on this task. The increase was especially marked in the estimate of the clock component, and in a more detailed analysis of a subgroup of patients, the clock deficit was found to be associated with damage to the lateral regions of the neocerebellum (Ivry et al., 1988). In contrast, lesions centered in the medial cerebellum were associated with increases in the estimate of motor implementation noise. This double dissociation is in agreement with the anatomical projections of the output pathways of the cerebellum: the lateral regions primarily ascend to motor, premotor, and prefrontal regions of the cerebral cortex, while the medial regions innervate brain stem and spinal cord regions of the descending motor pathways. The finding of a clock deficit in patients with neocerebellar lesions has been replicated (Franz et al., 1996; Ivry et al., in press).

The results from experiments involving PD patients yield a more ambiguous picture. Two studies have reported that *medicated* PD patients perform similarly to age-matched controls in terms of overall variability during the continuation phase (Duchek et al., 1994; Ivry and Keele, 1989), whereas one study has shown a deficit in performance similar to that observed with cerebellar patients (Harrington et al., 1998a). An earlier study (Pastor et al., 1992) had also pointed to a deficit in non-medicated PD patients. However, each participant completed only a single trial per interval, rendering estimates of the variability components suspect. O'Boyle et al. (1996) found significant increases in estimates of clock and motor implementation for patients tested off L-dopa medication. On the other hand, Ivry and Keele (1989) reported no change in performance as a function of medication level.

Greater convergence is found in studies involving PD patients with unilateral symptoms either taking L-dopa medication (O'Boyle et al., 1996) or tested in a drug-naive state (Ivry and Keele, 1989; Wing et al., 1984). In these cases, the patients showed consistent increases in the estimate of clock variability when tapping with the impaired hand; the motor estimate was not significantly changed. For example, Keele and Ivry (1987) tracked one patient over a 2-week period during which he began L-dopa therapy. His performance showed a marked improvement over the test sessions with the decrease in variability solely associated with a reduction in the estimate of clock variability.

In sum, the neuropsychological studies suggest that disorders of the cerebellum and basal ganglia can result in increased variability during repetitive tapping. The increase may be isolated to either of the two component processes proposed in the Wing–Kristofferson model. The source of the increase following cerebellar pathology appears to be dependent on lesion location within the cerebellum. In PD, the deficit is less consistently observed, but when present, is almost always restricted to the clock component. Ivry and Hazeltine (1995) point out that the term *clock* is misleading in the Wing–Kristofferson model since this estimate refers to all sources of variability other than that associated with motor implementation. As such, they propose that the term *clock variability* should be replaced by *central variability*, encompassing all aspects of motor planning and preparation that occur prior to movement initiation. In this view, the cerebellum and basal ganglia might both add to central variability, but in distinct ways (Ivry and Keele, 1989).

Four studies have used functional imaging to investigate neural regions involved in the production of rhythmic movements. We defer our review of two of these to

Section 19.3.2. In the third study (Penhune et al., 1998), participants reproduced a sequence of isochronous intervals with the right hand, with the target pace indicated by either an auditory or visual metronome. Compared to just listening to the metronome, increased activation in the auditory condition was observed in the left globus pallidus and right anterior cerebellum (lobules V and VI). In the visual condition, activation was observed in the left lateral cerebellum (VIIa) in addition to the anterior cerebellar site. No basal ganglia activation was found in the visual condition. Another set of comparisons involved conditions in which participants produced either novel or well-practiced rhythmic patterns. Prominent cerebellar activation was found for the novel conditions, spanning left and right cerebellar hemispheres (VIIa and VIIb) and anterior and posterior vermal areas (III and IV, VIIIa and VIIIb). This pattern was similar for both auditory and visual stimulus conditions. The basal ganglia also showed a bilateral increase in activation during novel rhythms, but again, this increase was limited to the auditory condition. Kawashima et al. (2000) recently reported congruent results, in which they compared visually triggered and memory-timed finger taps. Again, this comparison was significant for the anterior cerebellum, but not for the basal ganglia. These results point to a consistent involvement of the cerebellum in the processing of temporal information. In contrast, the results are less consistent for the basal ganglia and are restricted to one sensory domain.

19.2.2 THE PERCEPTION OF TIMED EVENTS

Table 19.2 summarizes the performance of cerebellar and PD patients on time perception tasks. The table is restricted to studies in which participants judged the duration of a stimulus; we excluded papers in which the temporal judgment was based on gap detection or simultaneity (Artieda et al., 1992). Temporal acuity on duration discrimination tasks is conventionally assessed by varying the difference between a standard and a comparison stimulus. The participant's responses are fitted with a logistic distribution function to estimate the point of subjective equality (PSE) and variability, typically quantified as correct performance on 75% of the trials (corresponding to ±1 SD of the logistic distribution). The table also shows the variability data in a normalized form in which the SD has been divided by the PSE. Interestingly, the results indicate a considerable range in the Weber fractions across the studies, probably reflecting the different methodologies used to make the threshold estimates.

Deficits for cerebellar patients were reported by Ivry and colleagues in three studies (Casini and Ivry, 1999; Ivry and Keele, 1989; Mangels et al., 1998). Note that many of the same patients were tested in the 1998 and 1999 experiments and thus do not constitute independent samples. Nichelli et al. (1996) also reported elevated thresholds for intervals ranging from 100 to 600 msec. However, their group of patients with cerebellar degeneration performed comparable to controls for intervals ranging from 100 to 325 msec. As with the tapping data, the PD results are less consistent. Ivry and Keele (1989) observed no increase in the difference threshold in a group of medicated PD patients. In contrast, Harrington et al. (1998a) observed a significant impairment in a different group of medicated PD patients with intervals of 300 and 600 msec.

TABLE 19.2
Summary of Studies Involving Human Participants with Cerebellar Damage or PD on Perceptual Tasks Involving Duration Discrimination

Study	Standard Interval (msec)	Group	N	Threshold (1 SD)	K	Sig.	Symptom
Ivry and Keele (1989)	400	Control	21	19.2	0.05		
		Cerebellar	27	30.5	0.08	*	
		PD	28	21.0	0.05	—	
Mangels et al. (1998)	400	Control	14	31.5	0.08		
		Cerebellar	9	44.9	0.11	*	Unilateral
Casini and Ivry (1999)	400	Control		27.0	0.07		lesions
		Cerebellar		38.8	0.10	*	Unilateral
Nichelli et al. (1996)	100–600[a]	Control	13	29.6	0.11		lesions
		Cerebellar degenerative	12	48.9	0.17	*	CCA, 2 OPCA
Harrington et al. (1998a)	300 and 600	Control	24	53.0	0.09		
		PD	34	80.0	0.13	*	HY = 2.4

Note: Thresholds are 1 SD of a logistic distribution fit to the observer's response. K = Weber fraction of 1 SD/PSE; Sig. = significance (* indicates a statistically reliable difference between the experimental and control groups); HY = Hohn and Yahr scale; CCA = cortico cerebellar atrophy; OPCA = olivio pontine cerebellar atrophy.

[a] This task required the participants to classify single intervals as short or long; a standard interval was not presented on each trial. The PSEs were 274 and 282 for controls and cerebellar patients, respectively.

Jueptner et al. (1995, 1996) have conducted two positron emission tomography (PET) studies to investigate the neural correlates of duration discrimination, one with auditory stimuli (Jueptner et al., 1995) and a second with tactile stimuli (Jueptner et al., 1996). In both studies activation of the left inferior cerebellum (hemisphere of lobule VII) and vermis (VI and VII) was found during the duration discrimination task, compared to a control task in which the same stimuli were presented but the responses simply required alternating key presses. Increased blood flow was also seen bilaterally in the basal ganglia, but only in the experiment with auditory stimuli.

In summary, the patient and imaging studies of time production and perception yield a somewhat unsatisfying picture. The existing data indicate that cerebellar lesions are consistently associated with deficits on both production and perception tasks. While the initial studies involving PD patients reported no impairments, more recent reports have shown that PD patients also exhibit increased variability on time production and perception tasks. Thus, while Ivry and colleagues had argued that the dissociation between cerebellar and PD patients provided evidence of a specialized role for the cerebellum in the representation of temporal information, the current literature does not offer strong support for this dissociation.

Nonetheless, the fact that both groups can be impaired on similar tasks need not lead to the conclusion that temporal processing involves a distributed network that includes both the cerebellum and basal ganglia, as well as other structures such as the prefrontal cortex (Harrington et al., 1998b; Wittmann, 1999). It may well be that the analytic power of the tasks used in these studies is insufficient. As noted above, the estimate of clock variability in the Wing–Kristofferson model is really a composite of all nonmotor implementation sources of variability. Similarly, the perception tasks used in the patient studies have generally failed to provide a means for evaluating different sources of variability (Ivry and Hazeltine, 1995). Two exceptions are Mangels et al. (1998) and Casini and Ivry (1999), in which an attempt was made to separate the effects of timing from those associated with attention and working memory. In both studies, the results from the cerebellar group were consistent with an impairment in timing, whereas patients with prefrontal lesions were primarily influenced by the attentional demands of the tasks, temporal or nontemporal. It would be useful to apply a similar strategy in a direct comparison of cerebellar and PD patients.

19.3 SYNCHRONIZATION

Studying the synchronization of repetitive finger taps with a stream of regular external events has a long history in experimental psychology (Aschersleben and Prinz, 1995; Dunlap, 1910). Synchronization requires the ability to control motor output based on the prediction of external events (Hary and Moore, 1985, 1987). It is thought to entail both open- and closed-loop processes: the former in that the responses are generated in advance of the metronome signals, and the latter in that an error signal associated with the asynchrony between the responses and metronome signals is used to modify future responses. We first outline a general model of the hypothetical processes required for synchronized tapping and then turn to a review of previous attempts to link these processes to neural structures.

19.3.1 COMPONENTS INVOLVED IN SYNCHRONIZATION

A schematic of the component operations involved in synchronized tapping is presented in Figure 19.1. A clock-like system that represents the predicted interval emits a timing signal (t_k) for the next motor command. The timing of this command is set such that the resulting tap occurs approximately at the same time as the stimulus tone (s_k). The internal timing signal triggers a motor implementation process that adds a random motor delay component M_k conceptualized as an independent noise source (Wing and Kristofferson, 1973).

To ensure that the taps occur in simultaneity with the metronome, the system must contain a closed-loop component. Two types of mechanisms have been proposed (Mates, 1994a, 1994b). In period correction, the internal representation of the interval t is adjusted when a significant and consistent mismatch is detected between the metronome and the produced responses. This mechanism is thought to be relatively slow and dependent on a conscious perception of the mismatch of expected and perceived tempo (Repp, 2001).

The other process, phase correction, ensures that the error of the central clock does not accumulate over a number of taps. Such accumulation would lead to a loss of phase stability between the metronome signals and the responses. Phase correction

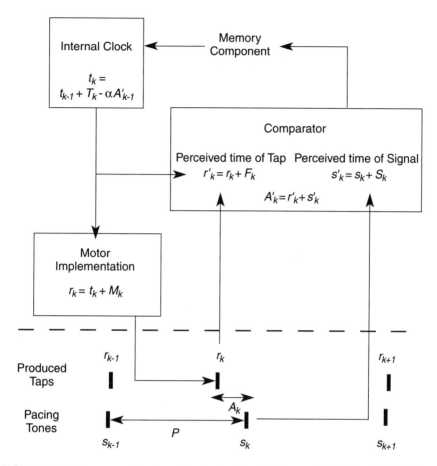

FIGURE 19.1 Process model of synchronization (see Vorberg and Schulze, 2002; Vorberg and Wing, 1996). Lowercase variables indicate time points of events, uppercase variables indicate the length of the intervals between events, and subscripted variables are conceptualized as random variables.* An external pacing signal occurs with period P at the time points s_k. An internal clock is emitting signals to the motor system at times t_k. In absence of error correction, the clock produces timing signals separated by the clock intervals T_k. The motor implementation process produces the kth taps at time r_k by adding a random motor delay M_k to the time of the internal timing signal. The real synchronization error A_k between the tap and the pacing signal is perceived (A'_k) by a comparator system. The perceived asynchrony is influenced by the perceptual delays, with which the comparator perceives the occurrences of the tap (F_k) and the pacing signal (S_k). The perceived asynchrony is then used to correct the next timing signal with a gain of α, the error correction parameter.

---

* For purposes of parameter estimation, we consider the perceptual delays F and S to be constants. Schulze and Vorberg (2002) showed that if F and S are regarded as random variables, the variance attached to these delays would inflate the variance estimations of motor delays. However, the main stationary characteristics of the model remain equivalent to the simplified model used here.

uses information about the asynchrony between the action and the external event to adjust the time of the next central command, thus compensating for this discrepancy. The simplest and normative method is a first-order linear correction, in which some fraction, α, of this synchronization error is used to adjust the interval before the next tap. Random errors are rapidly corrected when α is large, although the system would be overcompensating should α be greater than 1. Under some conditions, the adjustment process may take into account more than just the last asynchrony. In such cases, second-order error correction models are more appropriate. Futhermore, error correction may not always be linear (see Pressing, 1999).

Because the "real" asynchrony A_k is not readily available, the phase correction process has to estimate the time of occurrence of the metronome signals and the time of occurrence of the tap (Aschersleben and Prinz, 1995). For the time of the tap, proprioceptive reafferences from the action and the perceived consequences of the action, such as an audible click produced by the response key or the sound of the finger striking the table surface, could suffice. However, even without tactile, proprioceptive, visual, or auditory feedback, stable synchronization between actions and an auditory pacing signal can be accomplished. Billon et al. (1996) examined synchronization in a patient with a severe sensory neuropathy that rendered him functionally deafferented. This patient could maintain a stable phase relationship and correct for perturbations in a relatively normal manner. Thus, the estimate of the time at which an action occurs may also include the copy of the efferent signals (Haggard et al., 2002).

Alternative models have been proposed in which the clock is reset on every trial by a combination of the perceived tap and metronome signal, rather than through a modification process based on the perceived asynchrony (e.g., Hary and Moore, 1987). However, the former model has the advantage of capturing many characteristics of human performance in a parsimonious way (Mates, 1994b) and is more readily consistent with widespread evidence of tempo constancy in musical contexts. For example, humans tend to precede the pacing signal with the action by a constant amount of approximately 50 msec (Dunlap, 1910; Fraisse and Voillaume, 1971). From the framework of the model outlined in Figure 19.1, the phase lead of the taps over the metronome signals can be accounted for by longer delays in the perception of the action than in the perception of the pacing signal. This would lead to a perceived asynchrony that is close to zero, even though the actual asynchrony is negative (Aschersleben and Prinz, 1995; Fraisse, 1980).

19.3.2 NEURAL STRUCTURES UNDERLYING SYNCHRONIZATION AND TIMING

In the context of this chapter, two neuroimaging papers are of special interest (Jäncke et al., 2000; Rao et al., 1997) since they included a synchronization phase and a continuation phase during repetitive tapping. Both studies observed activation in the right anterior cerebellum (hemisphere of lobules I to VI, HI-VI) during the two phases, with the degree of activation approximately equal. A second cerebellar focus in the right inferior lateral hemisphere (H VIII to IX) was reported by Jânke et al. (2000) when the participants were paced by an auditory metronome. These authors failed to observe any activation in the basal ganglia. In contrast, Rao et al. (1997) reported significant activation in the putamen during the continuation phase only. Cortical areas were identified in both studies, especially during the synchronization phase, although these tended to be modality specific (e.g., primary or secondary auditory or visual regions).

Unfortunately, the design of these studies makes it difficult to draw strong conclusions. Rao et al. (1997) argue that neural correlates of an internal timing mechanism should be most activated during the continuation phase. Based on this, they argue that their results are consistent with a basal ganglia locus for timing and propose that the cerebellar activation reflects general contributions of this structure to sensorimotor control, consistent with the idea that the movements are similar during synchronization and continuation.

The assumption that internal timing is most pronounced during the continuation phase is suspect. All models of synchronization assume that the internal clock is engaged during paced and unpaced tapping. Indeed, it is difficult to see how participants would tap in advance of the metronome signals if they were not engaging in anticipatory timing. Mates et al. (1994) have shown that such anticipatory behavior holds for intervals up to around 2 sec (see also Fortin, this volume; Fortin and Couture, 2002). Beyond this duration, tapping becomes reactive, following rather than anticipating the tones.

If the demands on an internal timer are similar during the synchronization and continuation phases, then the cerebellar activation profile matches that expected of an internal timing process. However, this finding provides only weak support for a timing interpretation. As noted above, similar activation during both phases would also be expected of areas associated with the planning and execution of the finger movements, independent of whether these movements require the operation of an internal timing system. In sum, the cerebellar activation pattern during repetitive tapping is consistent with what would be expected if neural activity within this structure was involved in determining when each response should be produced, whereas activation within the basal ganglia is not consistent with a timing account. But the cerebellar activity could also be accounted for by hypotheses that do not involve internal timing.

19.4 EXPERIMENTAL STUDY

To investigate the role of the cerebellum and basal ganglia in sensorimotor synchronization, we conducted a study in which patients with cerebellar lesions or Parkinson's disease were asked to tap along with an auditory metronome. Unlike previous studies of tapping with these patient groups, we included only a paced phase, and the number of taps per trial (200) was considerably larger than in previous work. These long runs were essential for examining how the participants adjusted their behavior based on an error signal generated through the comparison of their own performance and the metronome signals. While previous studies have focused on component processes that are assumed to operate during unpaced tapping, the focus here was on the ability of these patients to use error correction during paced tapping.

We based our analysis on a linear error correction model (Vorberg and Wing, 1996). To separate the influence of the phase correction process from noise coming from the internal clock component and noise arising at stages of the motor implementation, we express the asynchrony on tap $k+1$ relative to the perceived asynchronies on the last two taps (see Figure 19.1):

$$A_{k+1} = A_k - \alpha A'_k - \beta A'_{k-1} + (T_k - P) + (M_{k-1} - M_k) \qquad (19.1)$$

While this formulation includes a second-order term based on the asynchrony preceding the adjusted tap by two taps, the first-order (AR1) forms of the models proved to be sufficient with most of the data sets. With a few exceptions, discussed in the results section, the estimates for β were close to zero. Thus, we will limit our focus to the first-order model.

Because the perceived asynchrony A' is the sum of the measured asynchrony A and the difference of the perceptual delays for the perception of the motor event (the tap), F_k, and auditory signal, S_k, Equation (19.1) becomes

$$A_{k+1} = (1 - \alpha)A_k - \alpha(F_k - S_k) + (T_k - P) + (M_{k+1} - M_k) \qquad (19.2)$$

We can solve for a stationary solution of the expected asynchrony by taking expected values, yielding

$$\overline{A} = \frac{\overline{T} - \overline{P}}{\alpha} - (\overline{F} - \overline{S}) \qquad (19.3)$$

where the bar denotes the average or expected value.

Thus the expected asynchrony is a direct function of the difference between mean perceptual delays (Aschersleben and Prinz, 1995), the deviation of the mean interval of the clock from the pacing interval, and the error correction parameter α. The stochastic properties of this and related models have been described extensively in several publications, along with different methods for estimating these parameters (Pressing, 1998; Pressing and Jolley-Rogers, 1997; Schulze and Vorberg, 2002; Vorberg and Schulze, 2002; Vorberg and Wing, 1996). To estimate the parameters, we established the equivalence between the first-order synchronization model and an ARMA(1, 1) process of the asynchronies (see Appendix). This approach, based on earlier work by Vorberg and Wing (1996), allows the estimation of the error correction parameter α, as well as of the motor and central variances, with standard estimation methods. In addition, second-order (AR2) models were also fit to the data, and the relative adequacy of the AR1 form of the model was tested against AR2 optional formulations.

Although the equations are linear, their application to real data requires considerable care for three main reasons. First, it is necessary to circumnavigate certain parameter ranges that exhibit near indeterminacy of a solution (Pressing, 1998; Schulze and Vorberg, 2002; Vorberg and Schulze, 2002). This effect was kept to a minimum by excluding trials in which the autocovariance function showed no significant deviation from zero for lags one through five since the model fit for these data would fall in the area of near indeterminacy. We also acquired converging results using the method of bins (Pressing, 1998). This method yields a more robust estimate of the error correction parameter, but requires that the size of the motor delay variance be specified *a priori*.

Second, the validity of the estimates is related to the number of intervals produced on each trail. Without error correction, stable estimates of clock and motor variability can be obtained with short runs of 20 to 40 taps (Wing and Kristofferson, 1973). However, such lengths are completely inadequate when error correction is involved. An order of magnitude longer is essential (at least 200 taps, depending on the consistency of the performer's control). Our run lengths were chosen with this in mind.

Third, the data analysis presumes that the same control process is used over the course of the run, a phenomenon that is described as stationarity. If a run is markedly nonstationary, then the estimation technique may exhibit significant biases. This is not a significant problem with younger control groups, but patient groups are selected precisely due to problems in their control processes, and nonstationarity may be more likely with them (Wilson et al., 2002). This issue can be handled in part by comparing parameter estimates in different sections of each run and discarding runs that are inconsistent. Runs with a notably poor model fit are also likely to be nonstationary.

19.4.1 METHOD

19.4.1.1 Participants

Three groups were tested. One group consisted of five patients with Parkinson's disease (age = 68.9 years, SD = 3.9; education = 16 years, SD = 1.5; time since diagnosis = 11.4 years, SD = 5.9). All patients were rated to have mild to severe (2 to 4) Parkinson's symptoms on the Hohn and Yahr scale. They were all on their regular dopaminergic replacement medication program at the time of testing.

The cerebellar group (N = 7; age = 65.9, SD = 10.1; education = 13.6, SD = 2.4) consisted of two patients with bilateral cerebellar degeneration and five patients with unilateral lesions due to either stroke (two) or tumor (three). Three of the unilateral patients had left-sided lesions, and two had right-sided lesions. These patients were tested in a chronic condition, at least 2 years after their neurological incident.

A control group of six elderly participants was also tested (age = 69.5, SD = 5.2; education = 14.0, SD = 1.8). These individuals reported no history of significant neurological disease or injury and were selected to match the patients in terms of age and education.

19.4.1.2 Procedure

Responses were made on a peripheral response device linked to a PC. The taps required flexion movements of the right or left index fingers on a piano-type key (2×10 cm), mounted parallel to the top surface of the response device. The tone generator in the PC was used to create the auditory metronome, with the pitch of the pacing tones fixed at 500 Hz.

The experimenter initiated the trial, triggering the onset of the auditory metronome. Participants were instructed to begin tapping when they had a good sense of the target pace, attempting to tap along with the metronome. Once the first response was detected, the metronome continued until it had completed another 200 cycles. At this point, the tones ceased. Feedback was then provided, indicating the target interval, the mean interval produced, and the variability of the interresponse intervals. The next trial began after a short rest period.

Two independent variables were manipulated. First, the target rate was either 500 or 900 msec. Second, participants used either the right hand alone, the left hand alone, or both hands, tapping bimanually. This chapter will only report data from the unimanual conditions.

The conditions were tested in separate blocks of four trials each, and the participants completed two blocks for each of the six conditions. Thus, the data set for the analyses consists of eight trials of approximately 200 intervals each. The order of the blocks was randomized across participants, although a complete counterbalancing was not possible given the small number of participants in each group. Rate was manipulated across sessions with half of the participants starting with testing at 500 msec and the other half starting with 900 msec. A short practice trial consisting of 20 paced intervals was included prior to the first block of each condition.

19.4.2 Data Analysis

The first ten taps were excluded to allow performance during synchronization to stabilize. The raw data were examined to screen for places in which a tap was missing or an extra response was recorded, and only intact segments of each trial were used for parameter estimation. Trials in which the asynchronies showed sudden drifts of the mean (for example, to an antiphase pattern) were excluded. These exclusions were far more frequent for the two patient groups (16%) than for the controls (3%). Parameter estimates from trials in which the autocorrelation function was degenerative, yielding indeterminacy of the model, were also excluded (9%). These trials were frequent for both the control and patient groups. Another effect to be considered is inconsistency between runs. Given the likelihood of stochastic variations in process control parameters, moderate effects of this kind are normative and can be addressed by averaging across runs. However, it is possible that performance changes over time due to learning. For example, with practice, Pressing (1999) observed an increase in the utilization of error correction and a decrease in the estimate of central variability. However, these effects occurred over a period of years. We assume that learning-related changes are minimal with the current design.

19.4.3 Results

Separate analyses of variance (ANOVAs) were performed to compare the performance of each patient group to that of the control participants for each of the dependent variables. The factors in these analyses were tempo (500 and 900 msec) and participant group. Based on preliminary analyses, we averaged the results over the two hands for the controls, PD patients, and two cerebellar degeneration patients. For the cerebellar patients with unilateral lesions, we compared performance with the impaired, ipsilesional hand to that of the control participants. We also ran a separate analysis in which we compared the performance of the ipsilesional and contralesional hands for the unilateral patients. Surprisingly, none of these within-subject comparisons were significant. These null results likely reflect two factors. First, the number of participants with unilateral cerebellar lesions was small (n = 5). Second, a couple of the patients showed little impairment on the task, consistent with their marked clinical recovery. Thus, our focus here is on the comparison of the patient groups to the controls.

As an overall measure of performance, we assessed the SD of the intervals (Figure 19.2a). This measure was significantly increased in the cerebellar group, $F(1, 11) = 7.11$, $P = .022$, but not in the PD group, $F(1, 9) = .44$, $P = .52$. The decomposition of the variance revealed that the SDs of the motor delays (Figure 19.2b) were not different between the control group and the two patient groups: $F(1, 11) = .001$, $P = .981$, for the cerebellar group; $F(1, 9) = .10$, $P = .753$, for the PD group. The effect of pace and the group by pace interaction were not significant in either comparison.

In contrast, the estimates for the SD of the clock intervals (Figure 19.2c) were significantly elevated for the cerebellar group, $F(1, 11) = 7.43$, $P = .019$. No such deficit was found for the PD patients, $F(1, 9) = .443$, $P = .52$. As expected from a linear increase of clock SD with interval length (Ivry and Hazeltine, 1995), the effect of pace on the clock SD was significant in both comparisons: $F(1, 11) = 36.6$, $F(1, 9) = 110.8$, $P < .001$. The group × pace interaction was not significant for either comparison: $F(1, 11) = .5$, $F(1, 9) = 2.6$. These results are congruent with the idea that the higher variability of the performance of cerebellar patients is due to deficits in a central timing mechanism.

One measure of the quality of error correction is the mean synchronization error (Figure 19.3). Similar to previous reports, the mean asynchrony for the control participants was negative, indicating that their taps anticipated the onset of the tones. The magnitude of this asynchrony was similar for the patients with cerebellar lesions and controls, $F(1, 11) = .09$, $P = .77$. The PD patients showed a greater negative asynchrony than the controls, $F(1, 9) = 7.99$, $P = .02$. That is, the responses of the PD patients preceded the tones to a greater extent than the responses of the controls. The effect of pace was significant in both the control/cerebellar ANOVA, $F(1, 11) = 9.23$, $P = .011$, and control/Parkinson's ANOVA, $F(1, 9) = 6.4$, $P = .031$. In both, the negative asynchrony was greater in the 900-msec condition than in the 500-msec condition. The group × pace interaction was not significant in either case.

The estimates for the error correction parameter α are shown in Figure 19.4. The analyses indicate that the estimates for the cerebellar patients were not different from those obtained for the controls, $F(1, 11) = .67$, $P = .80$. The error correction values were lower in the PD patients than in the controls, although this comparison did not reach significance, $F(1, 9) = 3.73$, $P = .085$. The error correction estimate increased in size from the faster to the slower pace, an effect that was significant

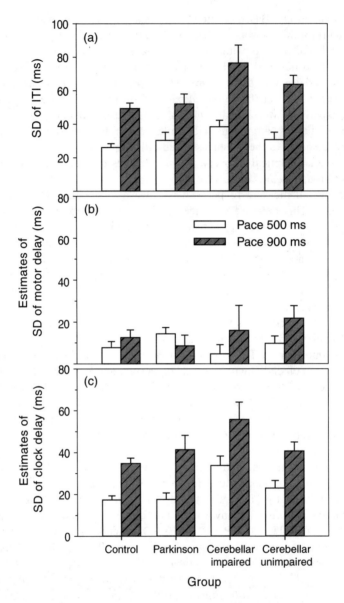

FIGURE 19.2 Variability of the intertap intervals for the control, PD, and cerebellar groups. For the unilateral cerebellar patients, separate values are shown when performing with the ipsilesional hand (impaired) or contralesional hand (unimpaired). A single value is included for the patients with bilateral cerebellar degeneration, based on the average of the two hands (and included in the impaired bar). (a) SD of the intertap intervals for the fast (500 msec) and slow (900 msec) pacing tones. The bottom two figures show the decomposition of the variability into estimates of the motor (b) and clock delay (c) components. Error bars indicate between-subject standard error.

Cerebellar and Basal Ganglia Contributions to Interval Timing

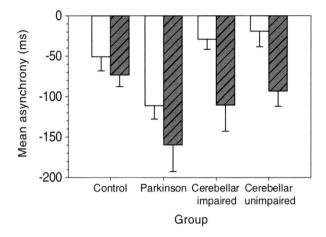

FIGURE 19.3 Mean asynchrony between the produced taps and the external pacing tone for the 500-msec (white bars) and 900-msec (gray bars) conditions. Negative values indicate that the taps occurred in advance of the tones. Conventions are as in Figure 19.2.

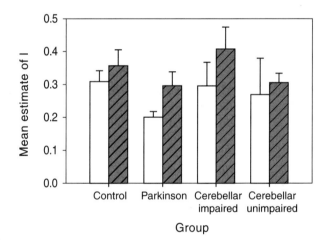

FIGURE 19.4 Mean estimates of the error correction parameter α for the 500-msec (white bars) and 900-msec (gray bars) conditions. Conventions are as in Figure 19.2.

when considering all three groups, $F(2, 15) = 6.83$, $P = .019$. The group factor did not interact with this effect (both F's < 1).

We also compared the error correction strategies in the groups by comparing the fraction of runs for which the linear, AR1 model provides a satisfactory fit in relation to the AR2 model. In the control group, 84.9% (SE = 2.8) of the runs were well fit by the AR1 model. This proportion did not significantly decrease in either the cerebellar group (79.4%, SE = 3.9) or the PD group (82.5%, SE = 5.5). Thus, based on the assessment of correction strategy as inferred by the validity of the AR1 model, it appears that the patient groups engaged in error correction in a way that was comparable to the control participants. The estimates of the α parameter and the higher mean asynchrony suggest that the PD patients might exhibit a quantitative deficit in error correction.

19.5 DISCUSSION

In this chapter, we have sought to evaluate the contribution of the basal ganglia and cerebellum to temporal processing, focusing on behaviors that require precise timing in the range of hundreds of milliseconds. As shown in our review of the existing literature, it has been difficult to dissociate the functions of these regions based on patient and neuroimaging studies. While cerebellar damage is consistently linked to deficits on both time production and time perception tasks, similar deficits are reported in some studies involving patients with Parkinson's disease.

At least three issues should be kept in mind when evaluating this state of affairs. First, the inferential nature of science is strengthened by considering results from various task domains and methodologies. Our evaluation should encompass a broad range of behavioral tasks and should also include computational, anatomical, pharmacological, and physiological evidence. Ivry (1997) has argued that the case for a cerebellar timing system provides a parsimonious account of the functions of this structure in a wide range of tasks, including many in which the demands on precise timing are more subtle than in tapping or time perception tasks. Computational models based on a detailed analysis of the architecture and physiology of the cerebellar cortex are also consistent with a specialized role of this structure in representing the temporal relationships between successive events (Fiala et al., 1996; Medina et al., 2000). The case for a basal ganglia role in internal timing has not been developed to the same extent. With the exception of the PD studies reviewed above, it has been primarily based on pharmacological and lesion studies in rats, and for the most part, this work has been conducted on tasks involving intervals that span many seconds (reviewed by Meck, 1996).

Second, there are limitations in inferring basal ganglia function from studies solely involving PD patients. While this degenerative disorder clearly produces a characteristic change in basal ganglia function, the loss of dopaminergic cells also has direct and indirect effects on other neural regions, including the frontal lobes. We have recently begun testing patients with unilateral basal ganglia lesions on time production tasks (Aparicio et al., 2002), and our preliminary results suggest that their performance is normal in repetitive tapping. It is possible, however, that this group, while seemingly better matched for comparison to patients with unilateral cerebellar lesions, will fail to provide insight into basal ganglia function due to recovery and reorganization following unilateral basal ganglia damage.

Third, many neuropsychological studies have been limited to either PD or cerebellar patients. There have been few efforts to directly compare the two groups of patients within the same experiment (but see Ivry and Keele, 1989). Such comparisons offer the best opportunity to test specific hypotheses concerning the differential contributions of neural structures, which collectively are recruited in the performance of specific tasks (Casini and Ivry, 1999; Mangels et al., 1998). In this way a mapping may be established between components of a psychological process model and the underlying neural substrates. In the current study, we compared two patient groups and investigated one aspect of performance involved in paced tapping, namely, the ability to use error correction processes in order to keep responses in synchrony with a pacing signal.

19.5.1 VARIABILITY

Consistent with earlier reports, the cerebellar group exhibited increased variability of the intertap intervals for both the 500- and 900-msec conditions. In contrast, we failed to observe a significant increase in the PD group at either rate. The PD patients were medicated, and previous results have suggested that the impairment in this group is especially marked when tested off medication or during the early stages of the disease process (Ivry and Keele, 1989; Wing et al., 1984). Nonetheless, despite their medication, the PD patients did exhibit clinical evidence of PD at the time of testing.

Estimates of clock and motor implementation variability were obtained through decomposition of the overall variability. All three groups exhibited similar estimates of motor variability, consistent with earlier work on these patient populations (Duchek et al., 1994; Harrington et al., 1998a; Ivry and Keele, 1989; Ivry et al., 1988). Notably, the increased variability in the cerebellar group was attributed to the clock component. Thus, these results again point to a central role for the cerebellum in the generation of the central signals related to the production of consistently timed responses (Franz et al., 1996; Ivry and Keele, 1989).

19.5.2 ERROR CORRECTION PROCESS

As a first step toward analyzing error correction in these patients, it is necessary to establish that a first-order linear error correction model provides an adequate account of the patients' performance. Given their neurological impairments and hypotheses concerning the role of the basal ganglia and cerebellum in online error correction (Flament and Ebner, 1996; Lawrence, 2000; Smith et al., 2000), we considered it possible that a qualitatively different strategy might characterize the performance. A second-order error correction model has been shown to provide a better fit under certain circumstances. For example, these higher-order models are more appropriate when expert musicians tap at a fast pace (Pressing and Jolley-Rogers, 1997). In our study, the proportion of runs that were adequately accounted for by a first-order error correction model was roughly equivalent across the groups. Thus, we infer that, qualitatively, both patient groups used a strategy similar to that of the control participants in how they used asynchrony information to adjust their performance. However, the results indicate a quantitative deficit in error correction for the PD patients. The estimates for the first-order parameter α tended to be lower for this group, although the result was only marginally significant. If this finding were replicated, it would suggest that a dopamine-related deficit in the striatum reduces the gain at which the error signal influences the next outgoing motor command (see also Malpani and Rakitin, this volume).

19.5.3 ASYNCHRONY

Another dissociation between the PD and cerebellar patients was observed in the asynchrony results. As is typically observed for synchronized tapping at intervals below 1 sec (Aschersleben and Prinz, 1995; Fraisse and Voillaume, 1971), the taps for all of the participants occurred prior to the tones. However, this asynchrony was greater for the PD patients (overall mean of 136 msec) than for both the controls and cerebellar patients (62 and 70 msec, respectively).

As outlined in the introduction, three different factors could contribute to the increase in the asynchrony (see Equation (19.3)). First, the increased asynchrony may be due to the fact that PD patients perceive their taps to occur later than do healthy individuals. The difference in perceptual delays for the perception of the tone and the tap influences the average asynchrony directly. For example, it has been shown that when participants tap with their foot, they precede the pacing tones by 50 msec more than when they tap with their finger (Aschersleben and Prinz, 1995). This effect may also be related to how PD patients integrate different sources of information to estimate the time at which the tap has occurred. A number of researchers have proposed a role for the basal ganglia in sensory integration, even though PD patients do not show obvious sensory impairments. To fully account for the observed difference between groups, one would have to posit that the PD patients perceived their taps to have occurred 66 msec later than the age-matched controls.

Second, negative asynchronies could result from an internal clock that is operating at a faster rate than the external pacing signal. A phase correction process would then prevent the error from accumulating across successive taps, but the faster rate of the internal clock would result in taps occurring prior to the tones. In contrast to the perceptual delay hypothesis, this explanation offers a parsimonious account of why the asynchronies were larger in the 900-msec condition for all of the groups. The mean error of the clock is likely to be proportional to the length of the timed interval, causing larger asynchronies for longer intervals.

Pharmacological studies in humans (Rammsayer, 1993) and rats (Meck, 1996) have suggested that the rate of an internal pacemaker may be altered by dopamine levels. However, in this work, the clock has been hypothesized to slow down when dopamine levels are low, not to speed up. Of course, we did not monitor dopamine levels, and we did not attempt a within-subject comparison in which the patients were tested both on and off their medication. It would be interesting to see if the mean asynchrony lead varied with medication level.

While the relationship of the asynchrony to dopamine is unclear, this behavioral change is reminiscent of the speeding up that is observed in PD patients when engaged in an extended action. For example, PD patients tend to speed up during unpaced, repetitive tapping (Ivry and Keele, 1989; O'Boyle et al., 1996). Similarly, although they have difficulty initiating locomotion, once started, their steps become smaller and marked by a faster cycle time. Such changes could be interpreted as reflecting a bias for an internal clock to operate faster when engaged repetitively.

Ivry and Richardson (2002) offer an alternative model that could account for the reduced cycle time. In their view, the basal ganglia operate as a threshold device, gating when centrally generated responses are initiated. They conceptualize the loss of dopamine as an increase in the threshold required to initiate a response. If we assume that this threshold drifts toward more normal levels with repetitive use, then the same input pattern will trigger a response at shorter latencies over successive cycles. Thus, the increased negative lag could reflect a change in a thresholding process, rather than a disturbance of sensory integration times, or a change in the operation of an internal timing process.

Third, as shown in Equation (19.3), the observed asynchronies caused by a difference between clock speed and the pacing rate will be modulated by α. Lower gains of the error correction process would allow the error to accumulate to a larger degree, resulting in larger asynchronies. Assuming that the internal timing mechanism runs too fast in all groups, the differences in α alone potentially could account for a substantial part of the observed differences between groups. In the current study, we can estimate the differences between internal clock speed and pacing signal for each group, given the values of the error correction parameter. It turns out that the differences in error correction can only partly explain the group differences in asynchronies. To fully account for this effect, we would have to posit an additional difference in clock speed, on average 20 msec faster for the PD patients.

At present, the results do not allow us to discriminate between accounts based on changes in perceptual delays or inaccuracies in the internal timing signal (caused by different clock speed or changes in threshold). Moreover, the relative contribution of reduced error correction to the increased asynchrony depends on the assumption that error correction behaves linearly. When the asynchronies deviate substantially from zero, as is the case for the PD patients, this assumption is likely to be violated (Pressing, 1999). Thus, converging evidence from independent methods is needed to distinguish between these factors.

19.5.4 FINAL COMMENTS

The current experiment points to differential contributions of the cerebellum and basal ganglia in the performance of synchronized tapping. Lesions in the cerebellum appear to perturb the internal timing mechanism, manifest as an increase in the noise of this system. In terms of overall variability, as well as the estimates of clock and motor noise, medicated PD patients performed comparably to the control participants.

Nonetheless, the PD patients differed from the controls on two measures, the error correction parameter α and mean asynchrony. Together, these findings suggest that these patients have difficulty in adjusting their movements based on sensory information. In contrast, no differences were observed between the cerebellar patients and controls on the measures of error correction. This null finding is rather surprising given the frequently suggested role of the cerebellum in the comparison of expected and actual sensory information for rapid error correction (Flament and Ebner, 1996; Kawato et al., 1987; but see Smith et al., 2000). Our results indicate that the cerebellar contribution is more of a feed-forward signal, indicating when the next response should be emitted. Online modulations of these timing signals may come from extracerebellar structures.

The role of the basal ganglia in error correction has been suggested in a very broad sense (Lawrence, 2000). One important distinction is between online adjustments that are used to ensure that the current movement is executed accurately and trial-by-trial information that is used to develop stable internal models for the production of future movements. Neither of these has been extensively tested. Smith et al. (2000) provide evidence of the role of the basal ganglia in online error correction of reaching movements, demonstrating that patients with Huntington's disease (HD) and asymptomatic HD gene carriers are impaired in correcting for external and self-generated perturbations of reaching movements. It remains to be seen how best to characterize error correction processes in synchronized tapping. While the adjustments appear to occur over time spans that are comparable to online error correction, the discrete nature of the taps and pacing signals may create conditions more akin to trial-by-trial error correction.

ACKNOWLEDGMENTS

This chapter is dedicated to Jeff Pressing, who died unexpectedly during the preparation of it. Jeff was attracted to the most difficult problems, and his creative intellect and passion always led to the most interesting insights. He was a great inspiration to many students and colleagues, and we are grateful for being able to count ourselves among them.

Preparation of this chapter was supported by National Institute of Health grants NS 30256 and NS 17778.

REFERENCES

Alexander, G.E., DeLong, M.R., and Strick, P.L., Parallel organization of functionally segregated circuits linking basal ganglia and cortex, *Annu. Rev. Neurosci.*, 9, 357–381, 1986.

Aparicio, P., Connor, B., Diedrichsen, J., and Ivry, R.B., The Effects of Focal Lesions of the Basal Ganglia on Repetitive Tapping Tasks, paper presented at the Cognitive Neuroscience Society meeting, San Fransisco, CA, 2002.

Artieda, J., Pastor, M.A., Lacruz, F., and Obeso, J.A., Temporal discrimination is abnormal in Parkinson's disease, *Brain*, 115, 199–210, 1992.

Aschersleben, G. and Prinz, W., Synchronizing actions with events: the role of sensory information, *Percept. Psychophys.*, 57, 305–317, 1995.

Billon, M., Semjen, A., Cole, J., and Gauthier, G., The role of sensory information in the production of periodic finger-tapping sequences, *Exp. Brain Res.*, 110, 117–130, 1996.

Casini, L. and Ivry, R.B., Effects of divided attention on temporal processing in patients with lesions of the cerebellum or frontal lobe, *Neuropsychology*, 13, 10–21, 1999.

Duchek, J.M., Balota, D.A., and Ferraro, F.R., Component analysis of a rhythmic finger tapping task in individuals with senile dementia of the Alzheimer type and in individuals with Parkinson's disease, *Neuropsychology*, 8, 218–226, 1994.

Dunlap, K., Reactions on rhythmic stimuli, with attempt to synchronize, *Psychol. Rev.*, 17, 399–416, 1910.

Fiala, J.C., Grossberg, S., and Bullock, D., Metabotropic glutamate receptor activation in cerebellar Purkinje cells as substrate for adaptive timing of the classically conditioned eye-blink response, *J. Neurosci.*, 16, 3760–3774, 1996.

Flament, D. and Ebner, T.J., The cerebellum as comparator: increases in cerebellar activity during motor learning may reflect its role as part of an error detection/correction mechanism, *Behav. Brain Sci.*, 19, 447–448, 503–527, 1996.

Fortin, C. and Couture, E., Short-term memory and time estimation: beyond the 2-second "critical" value, *Can. J. Exp. Psychol.*, 56, 120–127, 2002.

Fraisse, P., Les synchronisations sensori-motrices aux rythmes [The sensorimotor synchronization of rhythms], in *Anticipation et comportment*, Requin, J., Ed., Centre National, Paris, 1980.

Fraisse, P. and Voillaume, C., [The frame of reference of the subject in synchronization and pseudosynchronization], *Annee Psychol.*, 71, 359–369, 1971.

Franz, E.A., Ivry, R.B., and Helmuth, L.L., Reduced timing variability in patients with unilateral cerebellar lesions during bimanual movements, *J. Cognit. Neurosci.*, 8, 107–118, 1996.

Gibbon, J., Malapani, C., Dale, C.L., and Gallistel, C.R., Toward a neurobiology of temporal cognition: advances and challenges, *Curr. Opin. Neurobiol.*, 7, 170–184, 1997.

Gilden, D.L., Thornton, T., and Mallon, M.W., 1/f noise in human cognition, *Science*, 267, 1837–1839, 1995.

Haggard, P., Clark, S., and Kalogeras, J., Voluntary action and conscious awareness, *Nat. Neurosci.*, 5, 382–385, 2002.

Harrington, D.L., Haaland, K.Y., and Hermanowitz, N., Temporal processing in the basal ganglia, *Neuropsychology*, 12, 3–12, 1998a.

Harrington, D.L., Haaland, K.Y., and Knight, R.T., Cortical networks underlying mechanisms of time perception, *J. Neurosci.*, 18, 1085–1095, 1998b.

Hary, D. and Moore, G.P., Temporal tracking and synchronization strategies, *Hum. Neurobiol.*, 4, 73–79, 1985.

Hary, D. and Moore, G.P., Synchronizing human movement with an external clock source, *Biol. Cybern.*, 56, 305–311, 1987.

Ivry, R., Cerebellar timing systems, *Int. Rev. Neurobiol.*, 41, 555–573, 1997.

Ivry, R.B., The representation of temporal information in perception and motor control, *Curr. Opin. Neurobiol.*, 6, 851–857, 1996.

Ivry, R.B. and Hazeltine, R.E., Perception and production of temporal intervals across a range of durations: evidence for a common timing mechanism, *J. Exp. Psychol. Hum. Percept. Perform.*, 21, 3–18, 1995.

Ivry, R.B. and Keele, S.W., Timing functions of the cerebellum, *J. Cognit. Neurosci.*, 1, 136–152, 1989.

Ivry, R.B., Keele, S.W., and Diener, H.C., Dissociation of the lateral and medial cerebellum in movement timing and movement execution, *Exp. Brain Res.*, 73, 167–180, 1988.

Ivry, R.B. and Richardson, T., Temporal control and coordination: the multiple timer model, *Brain Cognit.*, 48, 117–132, 2002.

Ivry, R.B., Spencer, R.M., Zelaznik, H.N., and Diedrichsen, J., The cerebellum and event timing, in *Annals of the New York Academy of Sciences: New Directions in Cerebellar Research*, Highstein, S.M. and Thach, W., Eds., in press.

Jäncke, L., Loose, R., Lutz, K., Specht, K., and Shah, N.J., Cortical activations during paced finger-tapping applying visual and auditory pacing stimuli, *Brain Res. Cognit. Brain Res.*, 10, 51–66, 2000.

Jueptner, M., Flerich, L., Weiller, C., Mueller, S.P., and Diener, H.C., The human cerebellum and temporal information processing: results from a PET experiment, *Neuroreport*, 7, 2761–2765, 1996.

Jueptner, M., Rijntjes, M., Weiller, C., Faiss, J.H., Timmann, D., Mueller, S.P., and Diener, H.C., Localization of a cerebellar timing process using PET, *Neurology*, 45, 1540–1545, 1995.

Kawashima, R., Okuda, J., Umetsu, A., Sugiura, M., Inoue, K., Suzuki, K., Tabuchi, M., Tsukiura, T., Narayan, S.L., Nagasaka, T., Yanagawa, I., Fujii, T., Takahashi, S., Fukuda, H., and Yamadori, A., Human cerebellum plays an important role in memory-timed finger movement: an fMRI study, *J. Neurophysiol.*, 83, 1079–1087, 2000.

Kawato, M., Furukawa, K., and Suzuki, R., A hierarchical neural-network model for control and learning of voluntary movement, *Biol. Cybern.*, 57, 169–185, 1987.

Keele, S.W. and Ivry, R.I., Modular analysis of timing in motor skill, in *The Psychology of Learning and Motivation: Advances in Research and Theory*, Vol. 21, Bower, G.H. et al., Eds., Academic Press, San Diego, CA, 1987, pp. 183–228.

Lawrence, A.D., Error correction and the basal ganglia: similar computations for action, cognition and emotion? *Trends Cognit. Sci.*, 4, 365–367, 2000.

Malapani, C., Dubois, B., Rancurel, G., and Gibbon, J., Cerebellar dysfunctions of temporal processing in the seconds range in humans, *Neurorep. Int. J. Rapid Commn. Res. Neurosci.*, 9, 3907–3912, 1998.

Mangels, J.A. and Ivry, R.B., Time perception, in *The Handbook of Cognitive Neuropsychology: What Deficits Reveal about the Human Mind*, Rapp, B., Ed., Psychology Press/Taylor and Francis, Philadelphia, 2001, pp. 467–493.

Mangels, J.A., Ivry, R.B., and Shimizu, N., Dissociable contributions of the prefrontal and neocerebellar cortex to time perception, *Cognit. Brain Res.*, 7, 15–39, 1998.

Mates, J., A model of synchronization of motor acts to a stimulus sequence: I. Timing and error corrections, *Biol. Cybern.*, 70, 463–473, 1994a.

Mates, J., A model of synchronization of motor acts to a stimulus sequence: II. Stability analysis, error estimation and simulations, *Biol. Cybern.*, 70, 475–484, 1994b.

Mates, J., Mueller, U., Radil, T., and Poeppel, E., Temporal integration in sensorimotor synchronization, *J. Cognit. Neurosci.*, 6, 332–340, 1994.

Meck, W.H., Neuropharmacology of timing and time perception, *Brain Res. Cognit. Brain Res.*, 3, 227–242, 1996.

Medina, J.F., Garcia, K.S., Nores, W.L., Taylor, N.M., and Mauk, M.D., Timing mechanisms in the cerebellum: testing predictions of a large-scale computer simulation, *J. Neurosci.*, 20, 5516–5525, 2000.

Middleton, F.A. and Strick, P.L., Cerebellar output channels, in *The Cerebellum and Cognition*, Vol. 41, Schahmann, J.D., Ed., Academic Press, San Diego, CA, 1997, pp. 31–60.

Nichelli, P., Alway, D., and Grafman, J., Perceptual timing in cerebellar degeneration, *Neuropsychologia*, 34, 863–871, 1996.

O'Boyle, D.J., Freeman, J.S., and Cody, F.W., The accuracy and precision of timing of self-paced, repetitive movements in subjects with Parkinson's disease, *Brain*, 119, 51–70, 1996.

Pastor, M.A., Jahanshahi, M., Artieda, J., and Obeso, J.A., Performance of repetitive wrist movements in Parkinson's disease, *Brain*, 115, 875–891, 1992.

Penhune, V.B., Zatorre, R.J., and Evans, A.C., Cerebellar contributions to motor timing: a PET study of auditory and visual rhythm reproduction, *J. Cognit. Neurosci.*, 10, 752–766, 1998.

Pressing, J., Error correction processes in temporal pattern production, *J. Math. Psychol.*, 42, 63–101, 1998.

Pressing, J., The referential dynamics of cognition and action, *Psychol. Rev.*, 106, 714–747, 1999.

Pressing, J. and Jolley-Rogers, G., Spectral properties of human cognition and skill, *Biol. Cybern.*, 76, 339–347, 1997.

Rammsayer, T.H., On dopaminergic modulation of temporal information processing, *Biol. Psychol.*, 36, 209–222, 1993.

Rao, S.M., Harrington, D.L., Haaland, K.Y., Bobholz, J.A., Cox, R.W., and Binder, J.R., Distributed neural systems underlying the timing of movements, *J. Neurosci.*, 17, 5528–5535, 1997.

Repp, B.H., Processes underlying adaptation to tempo changes in sensorimotor synchronization, *Hum. Mov. Sci.*, 20, 277–312, 2001.

Robertson, S.D., Zelaznik, H.N., Lantero, D.A., Bojczyk, K.G., Spencer, R.M., Doffin, J.G., and Schneidt, T., Correlations for timing consistency among tapping and drawing tasks: evidence against a single timing process for motor control, *J. Exp. Psychol. Hum. Percept. Perform.*, 25, 1316–1330, 1999.

Schulze, H.H. and Vorberg, D., Linear phase correction models for synchronization: parameter identification and estimation of parameters, *Brain Cognit.*, 48, 80–97, 2002.

Shumway, R.H. and Stoffer, D.S., *Time Series Analysis and Its Applications*, Springer, New York, 2000.

Smith, M.A., Brandt, J., and Shadmehr, R., Motor disorder in Huntington's disease begins as a dysfunction in error feedback control, *Nature*, 403, 544–549, 2000.

Strick, P.L., Dum, R.P., and Picard, N., Macro-organization of the circuits connecting the basal ganglia with the cortical motor areas, in *Models of Information Processing in the Basal Ganglia*, Houk, J.C., Davis, J.L., and Beiser, D.G., Eds., MIT Press, Cambridge, MA, 1995, pp. 117–130.

Treisman, M., Temporal discrimination and the difference interval: implications for a model of the 'internal clock,' *Psychol. Monogr.*, 77, 1–13, 1963.

Vorberg, D. and Schulze, H.H., Linear phase-correction in synchronization: predictions, parameter estimation and simulations, *J. Math. Psychol.*, 46, 56–87, 2002.

Vorberg, D. and Wing, A., Modeling variability and dependence in timing, in *Handbook of Perception and Action*, Vol. 2, *Motor Skills*, Heuer, H. and Keele, S.W., Eds., Academic Press, San Diego, CA, 1996, pp. 181–262.

Wilson, S.J., Pressing, J., and Wales, R.J., Modeling rhythmic function in a musician poststroke, *Neuropsychologia*, in press.

Wing, A., The long and short of timing in response sequences, in *Tutorials in Motor Behavior*, Stelmach, G. and Requin, J., Eds., North-Holland, New York, 1980, pp. 469–484.

Wing, A.M., Keele, S., and Margolin, D.I., Motor disorder and the timing of repetitive movements, *Ann. N.Y. Acad. Sci.*, 423, 183–192, 1984.

Wing, A.M. and Kristofferson, A.B., Response delays and the timing of discrete motor responses, *Percept. Psychophys.*, 14, 5–12, 1973.

Wittmann, M., Time perception and temporal processing levels of the brain, *Chronobiol. Int.*, 16, 17–32, 1999.

APPENDIX

For the estimation of parameters, the linear first-order error correction model

$$A_k = A_{k-1} + (T_k - P + M_k - M_{k-1}) - \alpha A_{k-1} \qquad (19.4)$$

was reformulated as an ARMA(1, 1) process based on a Gaussian white noise series $w_1, \ldots, w_N$ with variance σ_w^2.

$$A_k = \phi A_{k-1} + w_k + \theta w_{k-1} \qquad (19.5)$$

The asymptotic autocovariance function of the process described in Equation (19.4) can be derived as (Vorberg and Schulze, 2002; Vorberg and Wing, 1996)

$$\gamma(k) = \begin{cases} \dfrac{\sigma_T^2 + 2\alpha\sigma_M^2}{1-(1-\alpha)^2}; & k=0 \\ \left[\dfrac{(1-\alpha)\sigma_T^2 + 2(1-\alpha)\alpha\sigma_M^2}{1-(1-\alpha)^2} - \sigma_M^2\right](1-\alpha)^{k-1}; & k>0 \end{cases} \qquad (19.6)$$

and the asymptotic autocovariance function for the ARMA(1, 1) model, described in Equation (19.5) (Shumway and Stoffer, 2000):

$$\gamma(k) = \begin{cases} \sigma_w^2 \dfrac{1+2\theta\phi+\theta^2}{1-\phi^2}; & k=0 \\ \sigma_w^2 \dfrac{(1+\theta\phi)(\phi+\theta)}{1-\phi^2}\phi^{k-1}; & k>0 \end{cases} \qquad (19.7)$$

From this we can extract the three equivalences

$$\theta = 1 - \alpha \qquad (19.8)$$

$$\theta = \sqrt{r(2+r)} - r - 1; \quad \text{with } r = \dfrac{\sigma_T^2}{2\sigma_M^2} \qquad (19.9)$$

$$\sigma_w^2 = -\dfrac{\sigma_M^2}{\theta} \qquad (19.10)$$

and conversely

$$\sigma_T^2 = \sigma_w^2 (1+\theta)^2 \qquad (19.11)$$

$$\sigma_M^2 = -\sigma_w^2 \theta \tag{19.12}$$

This reformulation has two important advantages. First, the estimation of parameters of the linear first-order error correction model can be accomplished using the standard methods for ARMA(1, 1) models. In practice, this was accomplished by using the ARMAX routine in MATLAB™ (System Identification Toolbox), which uses the iterative Newton–Raphson method to minimize the quadratic next-step prediction error. The second advantage is that the characteristics of the model only vary with ϕ and θ, but are homogenous across different levels of the parameter σ_w^2. For example, whereas optimal error correction α_{OPT} is a nonlinear function of σ_M^2 and σ_T^2 (Vorberg and Wing, 1996), it is a linear function of θ and independent of σ_w^2.

$$\alpha_{OPT} = 1 + \theta \tag{19.13}$$

The region of parameter space encompassing and near the region of optimal error correction is of importance in the estimation process since the model becomes unidentifiable here. The time series under this model becomes Gaussian white noise with autocovariance function

$$\gamma(k) = \begin{cases} \sigma_w^2; & k = 0 \\ 0; & k > 0 \end{cases} \tag{19.14}$$

Monte Carlo studies of this method have shown that the parameter values for α, σ_M^2, and can be validly estimated from simulated time series data produced following Equation (19.4). However, in the region surrounding the line of indeterminacy (Equation (19.13)), the estimates become unreliable (Schulze and Vorberg, 2002). In practice, we avoid this region by excluding trials in which $\hat{\gamma}(k)$ does not significantly deviate from zero for the lags $0 < k < 6$.

20 Interval Timing in the Dopamine-Depleted Basal Ganglia: From Empirical Data to Timing Theory

Chara Malapani and Brian C. Rakitin

CONTENTS

20.1 Introduction 486
20.2 Dopamine and Interval Timing...486
20.3 Temporal Memory and the Basal Ganglia 488
20.4 Modeling the DA-Dependent Timing Effects in PD....................................489
    20.4.1 SET and the PD Timing Deficits..489
    20.4.2 An Accumulator Model that Produces Migration to the Mean 491
    20.4.3 Simulations...493
    20.4.4 Modeling the Effects of PD...495
    20.4.5 Pros and Cons of the Accumulator Model of PD Timing Effects..496
20.5 Effects of Deep Brain Stimulation on PD Timing Errors............................497
    20.5.1 DBS in the STn of DA-Depleted PD Patients Corrects Migration..497
    20.5.2 Paradoxical Effects of GP DBS on Distorted Temporal Memory in PD 500
20.6 Functional Anatomy Account: A Revised Model of the Basal Ganglia.....501
    20.6.1 Effects of PD Stimulation on Distorted Temporal Memory 503
    20.6.2 Mechanisms of Action of DBS..503
    20.6.3 Origin of STn Overactivity in PD 504
References..506
Appendix A..513
Appendix B..514

20.1 INTRODUCTION

The research on timing and time perception conducted over the years in our laboratory has been driven by the assumption that a psychophysical model such as scalar expectancy theory (SET) would provide us with the advantages of a conceptual framework and the appropriate analytic tools to guide the search for the neurobiological mechanisms of the interval timing. We generated a body of data about the performance of patients with damage to different neural systems on the peak-interval (PI) timing production task, which yielded results amenable to analysis of timing accuracy, variability, and the scalar property using the SET framework (Malapani et al., 1998a, 1998b; Rakitin et al., 1998). A variety of patterns of temporal data generated by this procedure in animals have been found to be diagnostic of the level where a behavioral, pharmacological, or physiological manipulation acts in an information-processing model of interval timing (Gibbon and Church, 1984; Gibbon et al., 1984; Hinton and Meck, 1997; Meck, 1996). Our own findings argued strongly in favor of this approach for clinical research. We showed that simultaneous consideration of changes in both accuracy and variability of timing systems clearly distinguished deficits of temporal behavior resulting from damage to distinct brain regions (Malapani and Fairhurst, 2002).

However, there are major psychophysical findings of timing research that still remain poorly understood (Malapani and Fairhurst, 2002). More importantly, our level of understanding of both the neural basis of timing and the psychophysics of timing now exceeds our understanding of the connection between these two largely independent domains of investigation (Malapani et al., in press). As John Gibbon stated in his most recent paper, all these issues await for new theoretical developments in modeling timing and time perception (Gibbon and Malapani, 2002). Recent attempts made in that direction are quite promising (Matell and Meck, 2000, in press). Although the computations implemented by these studies appear more biologically plausible in translating psychophysical properties of timing systems, such as the scalar property, they do not reproduce the timing errors that have come to light in studies of Parkinson's disease (PD) patients.

In this chapter, we first describe the effects of dopamine (DA) depletion on timing performance previously reported in PD that current timing theories, including SET, cannot account for. We then discuss the way those findings force a rethinking of the theory that led us to propose a new computational model attempting to account for at least some of the timing errors seen in PD. This new model, although limited, might be an important contribution in the future search of valuable connections between neurobiology and psychology of timing behavior. The last part of this chapter contrasts the effects of DA to deep brain stimulation (DBS) in parkinsonian timing, which seem to challenge both functional anatomy and theoretical accounts of the basal ganglia involvement in timing and time representation. The questions we raise will be answered by future research.

20.2 DOPAMINE AND INTERVAL TIMING

Over the last two decades, advances have been made in specifying brain systems involved in temporal processing, and links have been established between cognitive processes underlying timing and particular pharmacological and physiological manipulations (Gibbon et al., 1997; Ivry, 1996; Rao et al., 2001). Extensive animal work has described the role of dopaminergic brain systems and the striato-frontal circuitry in timing behavior. Specifically, the output dopaminergic neurons originating in the substantia nigra pars compacta (SNc) and projecting to the striatum, as well as the striatum itself, are thought to implement the pacemaker–accumulator system in interval timing (Meck, 1986; Matell et al., this volume). The striatum is thought to serve as the accumulator by integrating the action potentials of the dopaminergic pacemaker cells. Data consistent to this hypothesis showed that striatal lesions eliminated timing in rats, and that timing did not recover after administration of L-Dopa to the lesioned striata. In contrast, timing in rats is restored when L-Dopa is applied to damage in the SNc, suggesting that those dopaminergic cells that survive the lesion act effectively again under L-Dopa supplementation (Hinton and Meck, 1997).

The method used to assess interval timing in the previously described animal studies was the peak-interval procedure (Catania, 1970). A human analog of the PI procedure was developed and used to test interval timing competence in both normal and brain-diseased human subjects (Malapani et al., 1998a, 1998b; Rakitin et al., 1998). This task first demonstrates a standard time interval a number of times, and then requires the subjects to reproduce that interval from memory throughout a block of trials. Reproduction of the interval consists of pressing a response key just before the expected end of the interval and releasing the key just afterward. Feedback is delivered after some of these trials either in the form of a histogram indicating whether the response was too short or too long (that is, within 15% of target time) or in the form of reminder trials (fixed-interval (FI) trials).

In animals tested with the PI procedure, the integrity of the striato-nigral dopaminergic system determines the pacemaker's speed in emitting pulses. In contrast, timing distortions in PD patients tested with the human analog of the PI task primarily involve temporal memory processes (Gibbon et al., 1997; Malapani et al., 1998b). Moreover, the memory-based distorted time production of PD patients is DA dependent in that it is alleviated with DA replacement therapy (Gibbon et al., 1997; Malapani et al., 2002a).

DA-dependent timing errors were evident in an experiment that asked PD patients to reproduce time intervals both on and off levodopa medication. When asked to produce two different intervals in separate blocks within a single experimental session, patients off medication reproduced the short value (8 sec) long and the long value (21 sec) short. We described this deficit as migration. It is as though there was a mutual attraction, or "coupling," in memory for these two target times. However, when the same patients, while unmedicated, reproduced only one interval (e.g., 21 sec) in a session, they reproduced that interval as too long, indicating a lengthening or "slowing" of temporal processing. In contrast, production on medication was accurate for both one and two intervals in a session. It was the dependence of the direction of DA-dependent timing errors on the number of intervals produced in a session, together with the fact that the observed shifts persisted despite abundant corrective feedback, that led us to attribute these PD timing inaccuracies to a memory failure. This hypothesis was also consistent with the pattern of gradual and rather persistent drifts in accuracy, which is suggested by extensive animal work to reflect deficits related to memory functions of the interval timing system (Meck, 1996).

20.3 TEMPORAL MEMORY AND THE BASAL GANGLIA

We followed up the memory failure hypothesis by investigating whether the deficits of PD patients result from dysfunction in either storing (writing to) or retrieving (reading from) temporal memory. PD patients were tested with a new design that withheld feedback on the second day of testing. Performance on the second day was therefore based solely on the retrieval of the information learned during the previous day. We used the standard PI timing procedure on the first day (the training session of the experiment) (Malapani et al., 1998a; 1998b; Rakitin et al., 1998). During the training session, patients learned and produced a short (6 sec) and a long (17 sec) interval, in separate blocks, with feedback provided in the same manner as in the previous study. On the following day (the testing session), subjects produced both intervals without further training and with no feedback.

We refer to this design as the encode–decode design (Malapani et al., 2002a; Rakitin et al., submitted). The key manipulation of the design in its initial form was the patients' drug state (ON or OFF medication) on the two successive days. PD patients were assigned to one of four experimental groups. The ON–ON group was provided with L-Dopa during both training and testing sessions. The OFF–OFF group was tested without L-Dopa for both sessions. The ON–OFF group was provided with L-Dopa during the training, but not the testing session. Finally, the OFF–ON group was tested without L-Dopa during the training, but with L-Dopa during the testing session. The idea was that by crossing PD patients' drug state with the availability of feedback, we could determine whether DA deficiency (associated with being in an OFF group) selectively affected memory storage (or encoding), retrieval (or decoding), or both, giving the design its name (Malapani et al., 2002a).

Two main effects are evident in these data. First, whenever subjects are OFF-drug, migration appears, replicating our earlier PD-related effect (Malapani et al., 1998b). That is, compared to the ON-drug state, PD patients tested OFF L-Dopa produce responses that are long compared to the 6-sec criterion and short compared to the 17-sec criterion. Migration was observed during both the training sessions, as seen in the OFF–OFF and OFF–ON groups' first-day performances, and the testing sessions, as seen in the OFF–OFF and ON–OFF groups' second-day performances. These results made it apparent that the migration effect was not dependent on the presence or absence of feedback, and hence was not likely attributable to a selective problem with memory encoding. This led us to attribute migration in PD patients to a DA-dependent dysfunction of retrieving temporal memories. A different type of timing deficit was observed in the OFF–ON groups' data from the testing session. In this case, patients substantially overproduced both target intervals. These data from the OFF–ON group contrast with the testing session data from the OFF–OFF group that clearly show strong migration. Apparently, restoring DA function during the testing session improves the retrieval deficit (manifested as migration) and allows a second, memory-encoding deficit associated with DA deficiency during the training session to be observed.

In addition to the direction of errors in accuracy, DA-dependent storage and retrieval distortions of temporal memories are also distinguishable by changes in timing variability. Psychophysical analysis of the data showed that the retrieval deficit, expressed with migration of productions, was associated with failure of the scalar property of timing variability. That is, the standard deviation of production response times — expressed as a proportion of the obtained, inaccurate means — is different for the two target intervals in all PD groups OFF L-Dopa, replicating our previous findings (Malapani et al., 1998b). The storage deficit, however, still reflects the scalar property of timing variability; that is, the standard deviation of response times rose in proportion to the magnitude of the timing errors. These results are illustrated in Figure 20.1 below.

To summarize, Figure 20.1 shows that DA deficiency leading to dysfunction in the basal ganglia results in two separable temporal memory deficits. Distortions that occur while storing multiple time intervals result in overestimation of all intervals during subsequent reproduction. In contrast, retrieving the trace of two or more different time intervals results in migration, a pattern of bidirectional errors such that reproductions of one of a pair of intervals are unusually close to the reproductions of the second. In addition to migration, DA-deficient retrieval is accompanied by a violation of the scalar property of timing variability, whereas with unidirectional shifts, the scalar property holds.

It is important to note here that estimation of continuous attributes other than time (e.g., line length) was spared in PD in an analogous encode–decode design (Malapani and Fairhurst, 2002; Malapani et al., 2002b). Moreover, patients with other kinds of brain damage, such as focal cerebellar lesions, did not show the temporal memory deficits seen in PD (Ivry, 1996; Malapani et al., 1998a, 2002a, b). Taken together, this evidence strongly suggests that the basal ganglia and their cortical targets are an important aspect of the neural basis of temporal memory in humans.

20.4 MODELING THE DA-DEPENDENT TIMING EFFECTS IN PD

20.4.1 SET AND THE PD TIMING DEFICITS

The encode–decode design combined with the ON–OFF method of determining whether performance is DA dependent has revealed two timing deficits in PD patients with distinct, well-documented psychophysical profiles. Those findings were novel and forced a rethinking of the theoretical account that has framed this research, i.e., the classical temporal information-processing model based on the SET framework, which is a largely linear stochastic system described by Gibbon et al. (1984).

Such a linear system can accommodate the unidirectional, proportional rightward shift seen in the OFF-ON groups ON testing (second day). One explanation for this effect is a perceptual deficit, corresponding to change in the SET clock-rate parameter. That is, the clock system (pacemaker and accumulator) may have been running at a different speed OFF L-Dopa compared to ON L-Dopa. More specifically, the clock would have had to run faster OFF L-Dopa than ON L-Dopa in order for the mismatch between stored and recalled accumulator representations of the target interval to yield

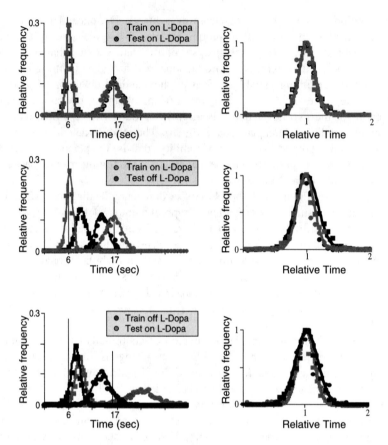

FIGURE 20.1 Migration on retrieval is associated with violation of the scalar property, while slowed encoding follows the scalar rule. Estimate distributions from training and testing sessions in real time (left panels) and time relative to the median time (right panels). The smooth curves are Gaussian fits to the data. The ON–ON group (top row) shows veridical estimates in both training and testing sessions ON DA supplementation. Migration occurs in the OFF state during both training (OFF–ON group, bottom row, black) and testing (ON–OFF group, middle row, black) sessions. The ON testing data from the OFF–ON group are right shifted (bottom row, gray). In the right panels all four estimation distributions of each group are shown in relative time. Functions in the ON–ON group (upper right panel) superpose (scalar property). This is not the case for the functions in the ON–OFF and OFF–ON groups, where in both cases increased variability (broader functions) is seen OFF L-Dopa (black distributions) as compared to the ON state (gray distributions). Increased variability in the OFF state was found to violate the scalar property (Malapani et al., 2002a).

response-times that are longer than the target interval. However, animal data suggest that dopamine depletion slows the clock (Meck, 1996), rather than speeding it up. Moreover, those results indicating a slowing of the clock with dopamine depletion are easier to reconcile with the findings of overall cognitive slowness known as "bradyphrenia" in dopamine-depleted PD patients (Malapani et al., 1994), than are suggestions of a speeded clock. Consequently, a clock–speed account of the lengthened estimates seems unlikely. A "slowed encoding" process was instead suggested to accommodate the double overestimates, such that veridical accumulations get translated into exaggerated (slowed) memory values (Gibbon and Malapani, 2002). The most likely SET implementation of this effect involves changes to the K* parameter, however, there is no corroborating data to indicate that DA modulates this parameter. While great difficulty remains in attributing specific neuromodulator action to specific SET parameters, SET itself has several information-processing components with parameters that can give rise to the "slowed encode" effect seen in PD.

The migration effect, however, cannot be reconciled with the scalar expectancy theory or other theories that posit a monotonic relationship between changes in the subjective experience of time and changes to model parameters that effect linear transformations of represented time. A great strength of SET is that cognitive sources of timing accuracy and variability are stated in time-independent terms. This formulation allows the scalar property of timing variability to emerge, but also makes modeling of duration-dependent effects, such as migration, difficult. As a result, new theories are necessary to further research a phenomenon that is emerging as central to understanding the internal clock.

Toward this end, we present a simple neural network model of a pacemaker–accumulator system developed by Miall (1996). This model can produce migration in mean time values by simultaneous adjustment of the model's two free parameters, as well as unidirectional effects of the sort exemplified by slowed encoding. Of special interest are the general properties of the model that can be abstracted from the overly simple network architecture and used to modify SET to account for the migration effect. In addition, this model has several other interesting features that suggest analogy to PD neuropathology and experimental tests of timing behavior in both healthy and pathological populations (Malapani et al., 2002a; in press).

20.4.2 AN ACCUMULATOR MODEL THAT PRODUCES MIGRATION TO THE MEAN

Miall (1989) developed two simple models capable of representing the passage of time within the activity of a simple neural network model. The beat frequency model supposes a bank of pulse generators with varying frequencies that serves as the internal time basis, and a register of neural nodes that responds to coincident activity, or "beats," among the pulse generators. This model has been employed by other researchers to model the interactions between cortical and basal ganglia structures involved in timing (Miall, 1996; Matell et al., this volume). Here we examine the properties of Miall's pacemaker–accumulator model. This model's architecture is similar to that of the structures of the same name found in the information-processing model of SET and other theories. This fact was the principle reason we chose the pacemaker–accumulator model over the beat frequency model for investigation.

Another reason was that unlike the beat frequency model, the pacemaker–accumulator model represents time continuously. That is, the beat frequency model is a time detector in that it signals the occurrence of a unique state of the time basis — namely, the co-occurrence of pulses from a fixed subset of a larger set of variable frequency oscillators. In contrast, the pacemaker–accumulator maps real time onto

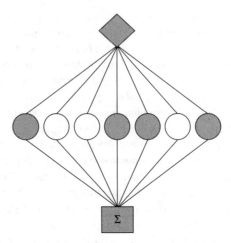

FIGURE 20.2 Architecture of Miall's pacemaker–accumulator artificial neural network. A Poisson pacemaker (indicated by the diamond) feeds simultaneously to an accumulator (indicated by circles) and is ultimately summed by a third process (marked as a square). The model discussed in the text included 1000 accumulator neurons. Accumulator nodes can transition from the off state (light circles) to the on state (dark circles) with each pulse of the pacemaker with a probability defined by the gain parameter, but they can also transition from the on state to the off state at any time with a probability defined by the decay parameter. (Adapted from Miall, R.C., *Time, Internal Clocks and Movement*, Vol. 115, Pastor, M.A. and Artieda, J., Eds., Elsevier/North-Holland, Amsterdam, 1996, pp. 69–94.)

subjective time upon each occurrence of a pacemaker pulse. The specific form of this mapping is of particular interest because it allows for both standard timing effects, such as the ability to subdivide intervals, and the more unique aspects of timing that are of interest here.

The conceptual design of the pacemaker–accumulator model is presented in Figure 20.2. The pacemaker consists of a pacemaker that emits discrete pulses. The pacemaker is a Poisson process in that the number of pulses emitted per unit time has both a central tendency and variability, and successive pulses are independent with respect to time. The accumulator consists of a bank of neural nodes that can be in either an active ON state or an inactive OFF state. Each pulse of the pacemaker has a certain probability of turning on a node that is currently off. This probability is referred to as the gain parameter because it determines the basic rate of accumulation. The corresponding probability of a node transitioning from the on state to the off state is referred to as the decay parameter. Transitions to the off state can occur between pacemaker pulses. These two parameters together determine the dynamical evolution of the total number of nodes that are in the on state. A summation process computes this total, which is the model's current estimate of real time.

This model was implemented in MATLAB code, listed in Appendix A. The function defined in that code returns the total number of active nodes per pacemaker pulse for an arbitrary number of simulations, *nruns*. The variable *nactive* stores and returns the results. The variable *steps* store the Poisson deviates (with mean *itsperstep*) that determine the number of main program loop iterations for each of the *nsteps* pacemaker pulses. The Poisson deviates are generated using a published algorithm implemented in the code listed in Appendix B. The main program consists of nested loops. The outer loop with counter variable *j* is iterated once for each simulation. Every iteration of the outer loop clears the storage variable *units*, used to represent the bank of nodes constituting the accumulator, and initiates a second, inner loop. The inner loop indexed with variable *i* iterates once for each pulse of the hypothetical pacemaker. Within this loop variable *ponmat* and *poffmat* are used to determine whether nodes change state using the gain (*pon*) and decay (*poff*) parameters, respectively. This is done by comparing a random array of uniform deviates on the interval 0 to 1 to one of the parameters, and then combining the results with the node bank's current state using Boolean operators. Finally, the number of active nodes in *units* is summed and then stored in the appropriate location of the output storage matrix.

20.4.3 SIMULATIONS

The results of a set of four simulations are presented in Figure 20.3. These simulations varied the values of the gain and decay parameters, but held constant the number of nodes in the accumulator. In the low-gain conditions, the gain parameter was set to 0.05, while in the high-gain conditions, the gain parameter was set to 0.1. In the low-decay conditions, the decay parameter was set to 0.0001, and in the high-decay conditions, the decay parameter was set to 0.0002. The number of accumulator nodes was fixed at 1000 for all four conditions. Each point on each of the four functions represents the means and standard errors of 1000 runs of the model, or iterations of the *j* index loop in the Appendix A code. These results are model psychophysical functions that relate the passage of real time (with arbitrary scale), represented by the occurrence of pacemaker pulses, to subjective time, represented as the total number of nodes active after each pacemaker pulse.

The basic performance characteristics of the model are evident in Figure 20.3. The most obvious is the curvilinear relationship between the passage of real time, as indicated by the pulsing of the pacemaker, and the value of subjective time indexed by the mean total number of active accumulator nodes. This is the result of the countervalent influence of the gain and decay parameters. Early on in the timing process, the gain parameter is the main determinant of accumulator activity because of its higher value and the availability of a large population of nodes currently in the off state. This can be seen in Figure 20.3, where the two functions with the higher gain value rise more steeply early in the interval than the two functions with the lower gain value. As more nodes become active over time, the cumulative probability of nodes' transitioning to the off state increases, thereby decelerating the per-pulse increase in active nodes. As the number of active nodes approaches the practical asymptote defined by the total number of accumulator nodes and the decay parameter, the process reaches dynamical equilibrium. This occurs because the few nodes that offset quickly turn back on due to the fact that the gain parameter is greater than the decay parameter. This feature can be seen in Figure 20.3, where the two functions with a lower decay parameter reach a higher asymptotic number of active nodes than the pair of functions with the higher decay parameter.

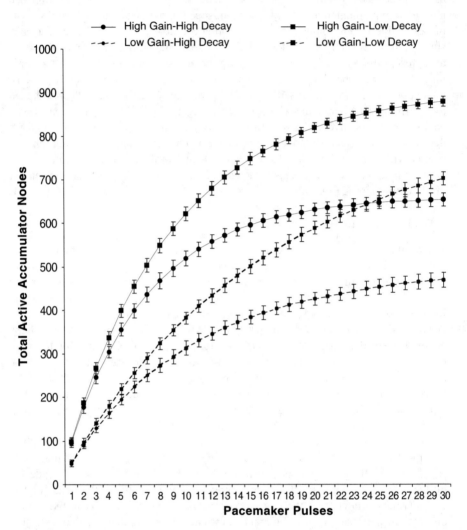

FIGURE 20.3 The results of four runs of 1000 simulations using Miall's accumulator model. Each function depicts the total number of active accumulator nodes as a function of the number of pacemaker pulses, for one of the four runs. Runs differed by the value of the gain and decay parameters. The two functions with high gains are indicated with solid lines. The gain parameter was set to 0.1 in the high-gain conditions. The two functions with low gains are indicated with dashed lines and employed a gain of 0.05. The two functions with high decays are indicated with circular markers. The decay parameter was set to 0.0002 in the high-decay conditions. The two functions with low decays are indicated with square markers and employed a gain of 0.0001.

It should be noted that no attempt to fix the crossover point at any particular point in the x-axis or at any particular point in the functions' curvatures has been attempted. Rather, the figure and data presented represent a direct replication of Miall's simulations along with a doubling of the gain and decay parameters. These were the first values chosen by us for investigation. Because the x-axis is essentially timescale invariant, that is, the rate of the pacemaker is arbitrary, and because the parameter space is effectively unbounded, it follows that the model can be made to produce any set of curvatures and crossover points at any point in the sequence of pacemaker pulses.

20.4.4 Modeling the Effects of PD

The curvilinear psychophysics of the pacemaker–accumulator model gives rise to the finding relevant to the duration-dependent timing effects stemming from PD. As can be seen in Figure 20.3, the high-gain–high-decay function crosses the low-gain–low-decay function. Consider a timing task where there are two target intervals, the durations of which fall on either side of the crossover point between the two functions. Now consider the case where the low-gain–low-decay function represents the psychophysical function associated with encoding time intervals in the OFF-medication state by PD patients, and the high-gain–high-decay function is associated with PD patients' decoding temporal memories in the OFF state. Such a mismatch between encoding and decoding will yield migration of mean production times because the memory of the shorter interval is now compared to a relatively slower clock, while the longer interval is compared to a relatively faster clock. The association of different psychophysical functions is of course motivated by the findings that encoding and decoding are associated with different behavioral effects in the experiments with PD patients employing the encode–decode design and the ON–OFF L-Dopa manipulation.

The fact that timing by PD patients in the ON conditions is accurate implies in the context of this model a similarity in the model's psychophysical functions associated with encoding and decoding temporal memories. However, the form of the slowed encode effect found in the OFF–ON group restricts the possible forms of the model's psychophysical functions. More specifically, if timing in the ON-drug condition is associated with a third set of model parameters (the first two being those that define the model's function for encoding and decoding temporal memories in the OFF-drug conditions), then we need only choose changes in parameters that give rise to a model decode–ON psychophysical function that accumulates more slowly than the putative encode–OFF function described above. This can be accomplished by assuming that encoding and decoding ON-drug are associated with the same very low gain–very low decay state of the pacemaker–accumulator. Such a system could give rise to accumulation that is slower for both target intervals, but is relatively faster over some range of longer time intervals.

To summarize, migration and slowed encode can be accommodated by Miall's pacemaker–accumulation model by assuming that the absence of DA increases the two parameters of the model, but does so more for encoding than for decoding. Although nothing in the model informs us of the potential origin of such a different effect of DA on encoding and decoding, we may surmise several potential factors contributing to this difference. Because encoding and decoding are dissociable cognitive functions, it is reasonable to assume that somewhat different brain circuitry underlies the two processes, allowing for somewhat different DA action. It could also be that the additional effort involved in the processing of feedback modulates the effect of DA on encoding relative to decoding.

20.4.5 Pros and Cons of the Accumulator Model of PD Timing Effects

As discussed earlier, SET and related models impose timescale independence as a central feature of the stochastic modeling of timing. This scale independence limits SET's capacity to explain duration-dependent timing errors. This simple pacemaker–accumulator model sheds light on the specific problematic features of SET. SET assumes that memory function in the accumulator as well as in short-term memory is linear in the sense that the function relating objective and subjective time is a straight line with an intercept of zero (barring the effects of attention on the latency to start and stop timing). The current model instead produces a curvilinear relationship between subjective and objective time, controlled by two parameters. If both parameters are under experimental control, such curvilinear functions can give rise to migration or other duration-dependent phenomena. If we divorce this mathematical fact from the implementation at hand, then we can modify other aspects of the SET information-processing model to serve as the source of migration. For example, the current working explanation of the cognitive mechanisms of migration assigns the effect to the short-term memory store, typically governed by the K* parameter in SET. The current findings suggest that SET can be modified to account for migration by bifurcating K* into two parameters, an exponent and a horizontal limit of a bounded-exponential function.

While the most important outcome from these simulations is the light it sheds on the manner in which limitations of SET can be overcome to account for migration, this model has certain features that interest us with respect to modeling timing in the basal ganglia. An important aspect of basal ganglia anatomy is the dichotomy between the direct and indirect pathways from the striatum to the globus pallidus. Both of these pathways originate in the striatum (Bejjani et al., 1997; Benabid et al., 1991, 1998; Limousin et al., 1995a, 1995b, 1998; Mitchell et al., 1986, 1995; Molinuevo et al., 2000; Pollak et al., 2002; Yelnik et al., 2000), rely on dopaminergic function (Young and Penney, 1984), and terminate at the globus pallidus (Albin et al., 1989; Crossman, 1989; DeLong, 1990; Filion et al., 1991; Miller and DeLong, 1987; Robertson et al., 1990, 1991; Tronnier et al., 1997). The net influence of activity in these pathways is opposite in that the direct pathway upregulates pallidal activity, while the indirect pathway downregulates pallidal activity. Thus, we suggest that this pacemaker–accumulator model that relies on competition between excitation and inhibition to produce the curvilinear psychophysics necessary for modeling the migration effect captures a crucial aspect of the brain anatomy thought to underlie timing in general. This analogy is consistent with Meck's assertion that the striatum implements the SET accumulator (Gibbon and Church, 1984; Gibbon et al., 1984) if one assumes that the striatum is the input node to an accumulator implemented by the entire striatal-pallidal system.

Although oversimplified in terms of the model's relation to brain anatomy, such implementations are still useful if they generate testable hypotheses. The most salient testable hypothesis derived from this model is the following: if the form of the psychophysical functions generated by the model is independent of the intervals being timed, then there will exist some set of two time values that do not migrate, but rather show monotonic errors. The direction of these errors will depend on their absolute values relative to the crossover point of the two psychophysical functions underlying encoding and decoding.

There are two major shortcomings of this model. First, the model does not currently predict the magnitude of the PD timing effects, only their directions. In the future we will use nonlinear searches of wider parameter spaces to determine the relative change necessary in the parameter to simultaneously account for both PD effects. Second, we have as yet made no effort to determine whether this new model and the proposed changes to SET negatively affect other critical psychophysical properties, especially the scalar property of timing variability. In general, alterations to the pacemaker–accumulator should not have a major impact on determining overall timing variability because the pacemaker–accumulator system contributes little variability itself and operates prior to the working memory and decision processing stages that contribute the majority of scalar variability (Gibbon and Church, 1984; Gibbon et al., 1984). However, we are offering a conceptual change in the relationship between subjective and objective time that entails an inequality in the ratio between two target times and the ratio between their corresponding internal representations. This could have profound effects on variability scaling, depending on the curvature of the functions relating subjective and objective time.

20.5 EFFECTS OF DEEP BRAIN STIMULATION ON PD TIMING ERRORS

20.5.1 DBS IN THE STN OF DA-DEPLETED PD PATIENTS CORRECTS MIGRATION

The finding of separable storage and retrieval mnemonic deficits restricted in the temporal domain in PD was followed up by a new set of experiments asking whether distinct neural pathways underlie those deficits. To this end, we applied the same encode–decode design in a series of studies testing PD patients OFF levodopa undergoing deep brain stimulation (DBS) either in the subthalamic nucleus (STn) or the pallidum. The effect of DBS is reversible, and the stimulation parameters (principally site and voltage) can be varied to achieve the optimal therapeutic result (Benabid et al., 1991, 1998). Moreover, the reversibility of DBS means that patients' cognitive performance can be assessed both with (ON) and without (OFF) DBS. This in turn presents the possibility of identifying specific patterns of impairment, including cognitive ones, associated with the stimulated basal ganglia substructures.

DBS in the STn, which alleviates parkinsonism (akinesia, bradykinesia) quite dramatically, also corrects the migration effect (Limousin et al., 1995a, 1995b; Pollak et al., 2002). The similar order of magnitude in correcting migration by STn DBS and L-Dopa replacement therapy contrasted to a clear difference between the dopaminergic and stimulation effect on encode (Molinuevo et al., 2000). The encode distortion (i.e., proportional overestimation of both time targets) that we described previously as slowed encode is not reversed by STn DBS. Moreover, overestimation occurs under STn DBS regardless of whether feedback is available. It is as though during encoding a lengthened estimate is laid down in memory, and it is not corrected by the presence of feedback. These results are illustrated in Figure 20.4.

In the same PD patients, stimulation has a clear effect on motor symptoms that mimics the dopaminergic effects, as can be seen in the right panel of Figure 20.4. This is an unexpected and quite striking finding, implying that STn stimulation (1) reverses slowness of motor, but not cognitive functions; and (2) fails to alleviate a cognitive (mnemonic) deficit, otherwise corrected by L-Dopa. The neural basis for slowed encode remains unclear and is not to be attributed to neural networks that would involve STn. Contrasts and similarities between the dopaminergic and DBS effects on both encode and decode sessions are illustrated in Figure 20.5.

Alleviating a deficit in temporal memory retrieval by altering neural activity in the STn is, on the other hand, a second unique and novel finding suggesting the role of a specific basal ganglia pathway through this particular nucleus, or the nucleus itself, in cognitive function. The STn stimulation is regarded as essentially correcting (attenuating) the abnormally high neural activity of this nucleus associated with the parkinsonian state (Limousin et al., 1998). The effects reported here were obtained with voltages below 5 V in all patients. Although the exact volume through which stimulation exerts its action is not known, it is generally accepted that at less than 5 V the effect is restricted to a few millimeters around the stimulation contact (Limousin et al., 1998). The anatomical localization of each of the stimulating contacts was carefully determined in each patient (for more details, see Bejjani et al., 1997). Only patients with verified implanted electrodes in the STn were included in the study. One might hence consider the observed effects to reflect modified (reduced) neural activity within the boundaries of the STn. This would suggest that normal activity levels in STn play a role in inhibiting a previously learned temporal memory when attempting to estimate a more recently learned memory.

Although overactivity of the STn is thought to be the key abnormality underlying parkinsonian symptoms in human and nonhuman primates, the evidence is almost exclusively restricted in motor deficits (Mitchell et al., 1986, 1995). STn overactivity is believed to result by overactivity of the striatal outputs to the lateral segment of the globus pallidus (GPe), the so-called "indirect striatal output pathway," due to the lack of dopaminergic inhibition (Crossman, 1989; Miller and DeLong, 1987; Robertson et al., 1990, 1991; Young and Penney, 1984). This implies that the retrieval process may involve the indirect output striatal pathway as a whole, rather than the STn itself.

The dopaminergic effect as conceived by the model implies that deprivation of dopaminergic nigro-striatal innervation reduces the positive feedback via the direct path and increases the negative feedback via the indirect path (described below in more detail). We postulated that the slowed encode may be produced by a delay for neurons in the direct pathway to either initiate decaying or, just after initiation,

Interval Timing in the Dopamine-Depleted Basal Ganglia

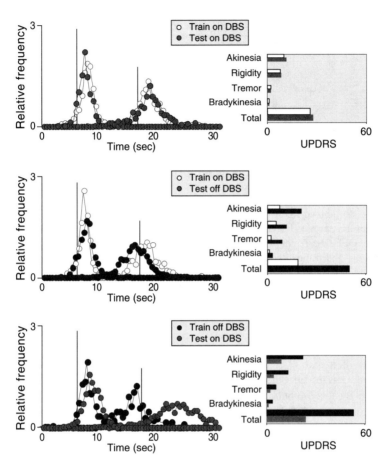

FIGURE 20.4 DBS in the STn of dopamine-depleted PD patients eliminates the retrieval, but not the encoding distortion. Estimate distributions in real time and time relative to the median time (left panels) and motor scores (right panels), for both training and testing days, in three groups of PD patients treated with DBS in the STn. The two groups trained ON DBS (ON–ON and ON–OFF) are shown in open circles (top and middle rows, respectively). Both groups show a slight, but reliable overestimation for both targets during training, while asymmetrical performance is seen during testing. Migration of estimates occurs in the testing session only OFF (ON–OFF group, middle panel, black points) but not ON DBS, indicating a beneficial effect of stimulation on retrieval. This is to be contrasted with the overestimation seen during testing stimulator ON (ON–ON group, upper panel, white points; OFF–ON group, lower panel, gray points). Training while stimulator OFF produces migration (OFF–ON group, lower panel, black points) that apparently masks the rightward shift, subsequently revealed in the testing session stimulator ON. The presence of a corrective mechanism, somehow effective under stimulation and in the presence of feedback, is implied by the different order of magnitude of the rightward shift as a function of whether training occurred with stimulator ON (ON–ON group, upper panel, gray) or OFF (OFF–ON group, lower panel, gray).

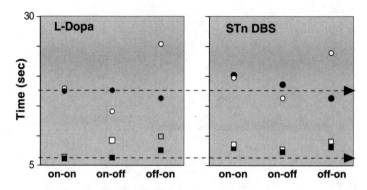

FIGURE 20.5 Contrasts and similarities between L-Dopa and DBS effects on temporal memory distortions in PD. Accuracy of subjective time estimates in PD groups trained and tested with the encode–decode design either ON and OFF DA supplementation (left panel) or ON and OFF DBS (right panel) in the STn. Accurate estimates are seen ON L-Dopa during both training and testing sessions, whereas estimates are rightward shifted with stimulator ON. Duration-dependent errors (migration), however, are similarly alleviated by drug replacement and functional surgery treatments. Slowed encode is revealed in both OFF–ON L-Dopa and DBS groups during testing, and the magnitude of the effect seems quite similar between these two groups, although reliably reduced in the presence of feedback during training, stimulator ON.

accelerate their gain over time. Either of these mechanisms would slow down the encode process of an otherwise normal (maybe cortical) accumulation of pulses generated by a hypothetical internal clock. Dysfunction of the indirect pathway, conceived to normally suppress conflicting memory patterns, would instead release commands (time is up) related to unwanted (not relevant to the context) memories producing the coupling — i.e., migration — effect seen on decoding. In other words, we considered the circuit of the basal ganglia as playing a role in both influencing the encode process via the direct pathway and suppressing potentially conflicting memories via the indirect pathway during retrieval. The results obtained with pallidal DBS failed to confirm this hypothesis.

20.5.2 Paradoxical Effects of GP DBS on Distorted Temporal Memory in PD

The STn DBS work was followed up with pallidal DBS (Benabid et al., 1991; Filion et al., 1991; Tronnier et al., 1997). Given the sophisticated DBS technique used by our collaborators, we were fortunate to identify distinct effects related to the altered neural activity in either the globus pallidus external capsule (GPe) or globus pallidus internal capsule (GPi) portions of the globus pallidus in PD, as compared to STn DBS. Details on the technique and clinical electrophysiological and anatomical data for each patient (N = 5) included in our study are published by Yelnik et al. (2000). The beneficial effects of the STn DBS were not reproduced when DBS targeted the GPi. Rather, inaccuracies in the form of migrated estimates were worse, whereas slowed encode was found to be highly variable among subjects. When in contrast, the plots targeting the external part of the pallidum (GPe) were the ones stimulated in the same patients, neither improvement nor aggravation was observed (data not shown). Those results are striking and yet difficult to reconcile with the proposed "classical" model of basal ganglia circuitry (Albin et al., 1989; DeLong, 1990). An initial explanation would be that the "classical" model is incomplete insofar as motor behavior is concerned and therefore may be inadequate to explain higher cognitive functions (Chesselet and Delfs, 1996; Levy et al., 1997; Marsden and Obeso, 1994; Obeso et al., 1997, 2000). It is to be considered, however, that the exact mechanism by which high-frequency stimulation (as DBS) acts on those nuclei is unclear (Yelnik et al., 2000).

A brief description of the "classical" model of basal ganglia function (to be challenged by the DBS results in both their contrast to the L-Dopa effects as well as in their intrinsic inconsistencies) is described in the functional anatomy section below. This is followed by a proposal for a revised model of cortico-striatal circuits.

20.6 FUNCTIONAL ANATOMY ACCOUNT: A REVISED MODEL OF THE BASAL GANGLIA

Overactivity of the STn is seen by the classical PD model as a key abnormality underlying parkinsonian symptoms in human and nonhuman primates (Albin et al., 1989; DeLong, 1990). However, the evidence not only remains controversial, but also is for the most part restricted in motor deficits. A cartoon illustration of this model is shown in Figure 20.6.

Like any model, the classical Albin–DeLong model of the basal ganglia is limited, because it is trying to simplify the complex reality (Chesselet and Delfs, 1996; Herrero et al., 1996b; Marsden and Obeso, 1994; Obeso et al., 2000). Anatomical studies have revealed that the D1 and D2 receptors are colocalized on striatal projection neurons (Aizman et al., 2000). Moreover, single-neuron labeling failed to identify direct-pathway striatal neurons that project only to the GPi (Bolam et al., 2000). The anatomy of the basal ganglia seems to be more complex (Betarbet et al., 1997; Damier et al., 1999) than the description of the model because of the back projections from the GPe to the striatum (Bevan et al., 1998; Kita et al., 1999) and the feed-forward projections from the cortex to the GABAergic interneurons in the striatum (Bolam et al., 2000). The output projections from the basal ganglia to the brain stem as well as direct projections from the STn to the SNc (Benazzouz et al., 2000; Hauber and Lutz, 1999a, 1999b; Pahapill and Lozano, 2000) might play a major role in the physiology of PD symptoms. The final prediction of the classical model of increased inhibitory pallidal output that would reduce the activity of the frontal cortex to cause akinesia in PD does not occur following methylphenyltetrahydropyridine (MPTP) administration (Goldberg et al., 2001). Finally, the physiological findings of human neurosurgery are different from the model. First, pallidotomy and STn lesions are also effective for the treatment of levodopa-induced dyskinesias (Bergman et al., 1990; Marsden and Obeso, 1994). Moreover, the firing rate in the GPi is not reduced in patients with hyperkinetic movement disorders, such as hemiballismus and dystonia (Lenz et al., 1998; Vitek, 2002; Vitek et al., 1997).

The adequacy of the model has also been questioned on its efficacy in taking into account higher cognitive functions (Pillon, 2002), known to involve the basal

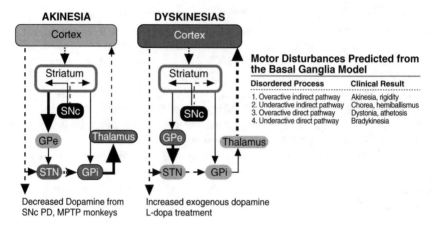

FIGURE 20.6 Classical model of the basal ganglia functional anatomy and its distortions in PD. The model postulates two separate systems within the motor circuit of the striato-pallidal complex. First, there is the direct putamino-medial pallidal (GPi) (GABA colocalized with substance P) and GPi-thalamic (GABA) system, while the indirect pathway would involve putamino-lateral pallidal (GPe) (GABA colocalized with enkephaline), GPe-suthalamic (GABA), subthalamo-GPi (glutamate), and GPi-thalamic (GABA) relays. Dopaminergic nigro-striatal input is believed to inhibit the indirect pathway, in contrast to its excitatory action on the direct system. The direct pathway would thus provide positive feedback to precentral motor fields, while the indirect pathway would contribute negative feedback. It is this dual modulatory role of the basal ganglia motor circuitry that would explain, according to the model, how the patient with PD treated with L-Dopa can be bradykinetic and exhibit dyskinesias at the same time.

ganglia circuitry and its frontal cortical targets (Brown and Marsden, 1991; Harrington et al., 1998; Jahanshahi et al., 1992; Malapani et al., 1994; Pillon et al., 1994). The fact that STn and pallidal stimulation aimed at specifically targeting the motor circuit affects cognitive function may imply that the location of the implanted electrodes is not sufficiently precise, although unlikely, at least in some cases (Bejjani et al., 1997, 1998; Yelnik et al., 2000). Indeed, pallidal dendritic domains are quite large, and the axonal plexi of afferents may span territories far beyond the proposed independent circuits (Parent and Cicchetti, 1998). An alternative hypothesis is that the cognitive and motor circuits are not as anatomically and functionally segregated as previously considered (Alexander and Crutcher, 1990). Studying cognitive deficits, along with motor disabilities, and the way they may be alleviated by the surgical treatments in PD, may thus be of major importance at our current stage of knowledge. Recent cognitive studies in stimulated PD patients call for careful development of animal models as well as integration of empirical animal and human data in remodeling the basal ganglia circuitry and its role in cognition (for a review, see Pillon, 2002).

Any attempt to suggest a new PD model should take into account (1) the effects of stimulation on both motor and cognitive functions, (2) the way stimulation acts to produce the observed effects, and (3) the pathways involved in the origin of abnormally high STn activation in parkinsonism.

20.6.1 EFFECTS OF PD STIMULATION ON DISTORTED TEMPORAL MEMORY

Our findings, in accordance with other recent cognitive research, support the view of the models' inadequacy to explain cognitive deficits seen in PD and the way they are affected by stimulation (Pillon, 2002), consistent with recent studies suggesting that functional surgery in PD could influence cognitive and affective processes (Malapani et al., submitted). The cognitive and behavioral outcome might depend on the exact location of the stimulation (Lombardi et al., 2000), suggesting that it may be crucial to evaluate changes between off and on stimulation on more specific behavioral and cognitive research tools (Ardouin et al., 1999; Jahanashi et al., 2000; Pillon et al., 2000; Saint-Cyr et al., 2000).

The model predicts that high-frequency stimulation in the STn would mimic the effects of DA replacement therapy overall (Bejjani et al., 1997; Limousin et al., 1995a, 1995b), which is not the case here (the encode deficit is alleviated by levodopa, but not by STn DBS). Moreover, if the STn overactivity is driven by a reduced neural activity in the GPe, as predicted by the model, then DBS in the GPe should produce effects similar to those of STn DBS. In contrast, DBS in the GPi would be expected to alleviate the coupling phenomenon on the decode and maybe the slowed encode, via its concomitant effect in reducing the excitation driven by the indirect pathway, allowing the inhibitory control of the direct one to take place. In sum, our results neither replicated the dopaminomimetic effect of STn DBS nor showed effects that would be easy to reconcile with the two functionally and anatomically segregated pathways, depicted by the classical model of basal ganglia circuitry (Albin et al., 1989; DeLong, 1990). Although these results may suggest that the cognitive and motor basal ganglia circuits are not as anatomically and functionally segregated as previously considered, some functional segregation does exist within the circuitry, since learning remains distorted with STn stimulation.

20.6.2 MECHANISMS OF ACTION OF DBS

A second important consideration relates to the mechanism of DBS in the parkinsonian brain that still remains unknown (Benabid et al., 2002). With the advancing use of DBS as the primary surgical tool not only in PD, but also in other basal ganglia diseases (i.e., dystonia, Huntington's disease) (Lenz et al., 1998), there is increasing interest in the mechanisms underlying its effects from both clinicians and researchers.

The STn stimulation is regarded as essentially correcting (attenuating) the abnormally high neural activity of this nucleus associated with the parkinsonian state (Dostrovsky and Lozano, 2002; Limousin et al., 1998). One could say that DBS induces a functional inhibition of a nucleus it targets, as its effects mimic what is functionally obtained when the nucleus is destroyed. Yet there are data suggesting that rather than inhibition, the stimulated site is activated (Benazzouz and Hallett, 2000; Vitek, 2002).

Although most recent evidence appears to weigh heavily in favor of DBS inhibiting neural activity and decreasing neuronal output in the case of the STn (Beurrier et al., 2001; Boraud et al., 1996; Wu et al., 2002), how stimulation acts on the pallidum — either the GPe or GPi — remains unclear. For instance, it might be the case that DBS has differential outcomes when applied in the STn or the GP. The anatomical (low vs. high neuronal density), biochemical (GABAergic vs. glutamatergic content), and electrophysiological (continuous vs. burst discharge) differences between the two nuclei may indeed challenge the generally accepted concept that high-frequency stimulation inactivates neurons in the region stimulated. An intriguing hypothesis is that DBS inhibits neural activity in the STn, but increases neurotransmission in the GP (Dostrovsky and Lozano, 2002; Yelnik et al., 2000).

20.6.3 Origin of STn Overactivity in PD

The STn plays a key role in the basal ganglia circuitry, and its hyperactivity may be a major factor in parkinsonian symptomatology (Marsden and Obeso, 1994). According to the classical PD model, it is the hypoactivity of the GPe that leads to increased STn activity, which in turn (in addition to the diminished influence of the direct striatal GABAergic input) induces the pathological overactivity of GPi (Figure 20.6). However, although stimulation of the STn induces a strong activation of GPe neurons, STn lesions result in a decreased activity of GPe neurons.

There is also physiological evidence suggesting that the STn is a driving force for not only the GPi, but also the GPe and even the SNc cells. In experiments in the normal rat, STn stimulation with the same parameters as in PD patients induced a transient but powerful inhibition (Benabid et al., 2002). As a consequence, a strong decrease in activity was recorded in the penducular pontine nucleus (PPN) (the equivalent of GPi in humans) and the SNc (Smith and Kieval, 2000). This increase in SNc activity, which should as a result produce more DA (Falkenburger et al., 2001), could be a possible mechanism of improvement of the PD symptoms. An increase of GP (equivalent to GPe in humans) is also recorded in rats after STn stimulation (Benazzouz et al., 1995). This could be due to a backfiring along the GPe-STn, and then the hyperexcited GPe, by its strong GABA projection, could in turn inhibit STn and GPi. However, experiments done recently in Benabid's laboratory (Benabid et al., 2002) have demonstrated that the destruction of GPe does not alter the effect of STn stimulation on GPi. Where then does this neural silencing arise from?

Apart from changing the tonic level of neural activity, other mechanisms such as resynchronization of the abnormal coupled neural circuits and cellular mechanisms (Bergman and Deuschl, 2002; Benabid et al., 2002; Bikson et al., 2001) may be involved in the stimulation effects. Coupled synchronized activity within otherwise (in normal conditions) segregated loops (Ni et al., 2001) may be induced under abnormal circumstances (DA depletion state in PD) as a result of either cellular local mechanisms in the STn-GPe axis (Vila et al., 1996) or dynamic circuitry alterations (cortico–striato–pallido–thalamic axis). This is indeed suggested by the double innervation pattern in the GPe (GABAergic inhibitory projections from the striatum together with dense glutamatergic excitatory input from STn). This in turn means that the resulting activity of GPe neurons is a counterbalancing of the activity of these two convergent pathways (Vila et al., 1996). A second interesting point is that axonal branches of a subthalamic fiber expand over groups of widely distributed pallidal neurons, whereas one striatal axon targets more restricted subsets of GPe cells (Herrero et al., 1996a). It is generally accepted that both projections converge on individual pallidal neurons. Thus, it seems apparent that the activity of the GPe neurons is balanced by the continuous interaction between two antagonistic inputs. This interaction is likely to be the cause of the decreased firing rate coupled with bursting activity recorded in pallidal neurons in monkeys with loss of dopaminergic neurons. Hence, parkinsonian signs may be a reflection of at least two different functional systems within the GP-STn axis (Plenz and Kital, 1999) that would, under abnormal circumstances (DA-depleted state), function as a unique coupled oscillating system.

Whether abnormal coupling occurs within the striato–pallidal axis or elsewhere, i.e., in cortical or thalamic sites (Calabresi et al., 1992, 1996), may depend upon the context and task (expectancy of reward, salience of the cue, goal-directed action) (Graybiel, 1998; Schultz, 1997). Of some importance is recent evidence suggesting that treatment with N-methyl-D-aspartate (NMDA) antagonists markedly affect both motor and cognitive functions in different animal species, including humans (Aultman and Moghaddam, 2001; Fredriksson et al., 2001). Those findings reveal the importance of DA-glutamate (GLU) interactions and synergetic effects underlying the generation of both motor and cognitive symptoms in PD (Blanchet et al., 1999). There may be unique ways that D-GLU interactions take place in the parkinsonian state (Carlsson and Svensson, 1990; Chase and Oh, 2000; Chase et al., 2000) in producing such abnormal coupling phenomena.

In the normal, non-DA-depleted striatum, for the most part glutamatergic transmission is driven via activation of L-alpha-amino-3-hydroxy-5-methylisoxazole-4-propionate (AMPA) receptors located on medium spiny neurons (Turski et al., 1991), as NMDA receptors are inactive at resting membrane potential. In Parkinson's disease, degeneration of the dopaminergic nigro-striatal pathway results in loss of dopamine D2 receptor-mediated modulation of striatal interneurons and decreased dopaminergic tone of medium spiny neurons (Greengard, 2001). Thus, stimulatory effects exerted by cortico-striatal and thalamo-striatal afferents are uninhibited, resulting in a shift in the membrane potential of striatal efferents to more depolarized potentials, which removes the magnesium block of NMDA receptors (Brotchie, 2000; Klockgether and Turski, 1993; Lange et al., 1997; Nash and Brotchie, 2002; Nash et al., 2000).

A cartoon illustration of a revised model that takes into account some of the aforementioned considerations is shown in Figure 20.7. We formulate two relatively simple hypotheses for the site(s)/pathway(s) involved in distorted storage and retrieval temporal memory processes seen in PD (Gillies and Arbuthnott, 2000). The first one posits that the distinct (direct–indirect) pathways do not really exist; dynamic functional reorganization of the system occurs under the parkinsonian state at a cellular rather than a circuitry level. Secondly, abnormal synchronized activity of the STn-GPe axis is suggested to be the major abnormality of DA depletion due to the persistent state of the negative cortico-striatal reinforcement signal. Segregation between loops is then abolished, and coupling phenomena occur, especially under conditions of nonreinforcement.

These ideas are to be taken as hypotheses to work on and are aimed mainly at highlighting the need for a careful reevaluation of current concepts of basal ganglia interactions. Studying cognitive deficits, along with motor disabilities in basal

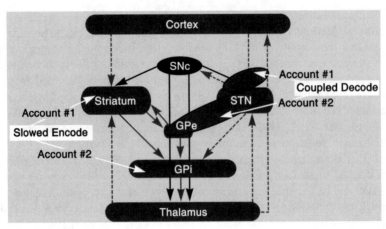

FIGURE 20.7 Timing in the basal ganglia: a revised model. This cartoon is a schematic illustration of a revised map where STn is shown as the driving force not only for both pallidal segments (GPe and GPi) but also for the SNc cells. Parkinsonian signs are accordingly conceived as a reflection of different functional systems that under abnormal circumstances (DA-depleted states) function as a unique coupled oscillating system. Coupled synchronized activity within otherwise (in normal conditions) segregated loops within the GP-STn axis may be induced as a result of a new balance between cellular local mechanisms and dynamic circuitry alterations (indicated as accounts 1 and 2, respectively). White arrows indicate what part of the circuitry would in each case mediate coupling on retrieval or slowed encode.

ganglia diseases, is of major importance at our current stage of knowledge and calls for careful development of animal models as well as integration of empirical animal and human data.

The DBS technique was useful in suggesting the neural pathway through the STn mediate temporal memory retrieval. But the way this occurs in finer detail remains elusive. Future research revolving around the attempt to identify the respective neural mechanisms underlying the time-specific distortions of learning and retrieval may bring new insights for the underlying neurobiological mechanisms associated with cognitive sequelae of mental disorders (i.e., PD, schizophrenia).

REFERENCES

Aizman, O., Brismar, H., Uhlen, P., Zettergren, E., Levey, A.I., Forssberg, H., Greengard, P., and Aperia, A., Anatomical and physiological evidence for D1 and D2 dopamine receptor colocalization in neo-striatal neurons, *Nat. Neurosci.*, 3, 226–230, 2000.

Albin, R.L., Young, A.B., and Penney, J.B., The functional anatomy of basal ganglia disorders, *Trends Neurosci.*, 12, 366–375, 1989.

Alexander, G.E. and Crutcher, M.D., Functional architecture of basal ganglia circuits: neural substrates of parallel processing, *Trends Neurosci.*, 13, 266–271, 1990.

Ardouin, C., Pillon, B., Peiffer, E., Bejjani, P., Limousin, P., Damier, P., Arnulf, I., Benabid, A.L., Agid, Y., and Pollak, P., Bilateral subthalamic or pallidal stimulation for Parkinson's disease affects neither memory nor executive functions: a consecutive series of 62 patients, *Ann. Neurol.*, 46, 217–223, 1999.

Aultman, J.M. and Moghaddam, B., Distinct contributions of glutamate and dopamine receptors to temporal aspects of rodent working memory using a clinically relevant task, *Psychopharmacologia*, 153, 353–364, 2001.

Bejjani, B., Damier, P., Arnulf, I., Bonnet, A.M., Vidailhet, M., Dormont, D., Pidoux, B., Cornu, P., Marsault, C., and Agid, Y., Pallidal stimulation for Parkinson's disease: two targets? *Neurology*, 49, 1564–1569, 1997.

Bejjani, B.P., Damier, P., Arnulf, I., Papadopoulos, S., Bonnet, A.M., Vidailhet, M., Agid, Y., Pidoux, B., Cornu, P., Dormont, D., and Marsault, C., Deep brain stimulation in Parkinson's disease: opposite effects of stimulation in the pallidum, *Mov. Disord.*, 13, 969–970, 1998.

Benabid, A., Benazzous, A., and Pollak, P., Mechanisms of deep brain stimulation, *Mov. Disord.*, 17 (Suppl. 3), 73–74, 2002.

Benabid, A.L, Benazzouz, A., Hoffmann, D., Limousin, P., Krack, P., and Pollak, P., Long-term electrical inhibition of deep brain targets in movement disorders, *Mov. Disord.*, 13 (Suppl. 3), 119–125, 1998.

Benabid, A.L., Pollak, P., Gervason, C., Hoffmann, D., Gao, D.M., Hommel, M., Perret, J.E., and de Rougemont, J., Long-term suppression of tremor by chronic stimulation of the ventral intermediate thalamic nucleus, *Lancet*, 337, 403–406, 1991.

Benazzouz, A., Gao, D.M., Ni, Z.G., Piallat, B., Bouali-Benazzouz, R., and Benabid, A.L., Effect of high-frequency stimulation of the subthalamic nucleus on the neuronal activities of the substantia nigra pars reticulata and ventrolateral nucleus of the thalamus in the rat, *Neuroscience*, 99, 289–295, 2000.

Benazzouz, A. and Hallett, M., Mechanism of action of deep brain stimulation, *Neurology*, 55, 13–16, 2000.

Benazzouz, A., Piallat, B., Pollak, P., and Benabid, A.L., Responses of substantia nigra reticulata and globus pallidus complex to high frequency stimulation of the subthalamic nucleus in rats: electro-physiological data, *Neurosci. Lett.*, 189, 77–80, 1995.

Bergman, H. and Deuschl, G., Pathophysiology of Parkinson's disease: from clinical neurology to basic neuroscience and back, *Mov. Disord.*, 17 (Suppl. 3), 28–40, 2002.

Bergman, H., Wichmann, T., and DeLong, M.R., Reversal of experimental parkinsonism by lesions of the subthalamic nucleus, *Science*, 249, 1436–1438, 1990.

Betarbet, R., Turner, R., Chockkan, V., DeLong, M.R., Allers, K.A., Walters, J., Levey, A.I., and Greenamyre, J.T., Dopaminergic neurons intrinsic to the primate striatum, *J. Neurosci.*, 17, 6761–6168, 1997.

Beurrier, C., Bioulac, B., Audin, J., and Hammond, C., High-frequency stimulation produces a transient blockade of voltage-gated currents in subthalamic neurons, *J. Neurophysiol.*, 85, 1351–1356, 2001.

Bevan, M.D., Booth, P.A., Eaton, S.A., and Bolam, J.P., Selective innervation of neostriatal interneurons by a subclass of neuron in the globus pallidus of the rat, *J. Neurosci.*, 18, 9438–9452, 1998.

Bikson, M., Lian, J., Hahn, P.J., Stacey, W.C., Sciortino, C., and Durand, D.M., Suppression of epileptiform activity by high frequency sinusoidal fields in hippocampal slices, *J. Physiol. Lond.*, 531, 181–191, 2001.

Blanchet, P.J., Konitsiotis, S., Whittemore, E.R., Zhou, Z.L., Woodward, R.M., and Chase, T.N., Differing effects of N-methyl-D-aspartate receptor subtype selective antagonists on dyskinesias in levodopa-treated 1-methyl-4-phenyl-tetrahydropyridine monkeys, *J. Pharmacol. Exp. Ther.*, 290, 1034–1040, 1999.

Bolam, J.P., Hanley, J.J., Booth, P.A., and Bevan, M.D., Synaptic organisation of the basal ganglia, *J. Anat.*, 96, 527–542, 2000.

Boraud, T., Bezard, E., Bioulac, B., and Gross, C., High frequency stimulation of the internal globus pallidus (GPi) simultaneously improves parkinsonian symptoms and reduces the firing frequency of GPi neurons in the MPTP-treated monkey, *Neurosci. Lett.*, 215, 17–20, 1996.

Brotchie, J.M., The neural mechanisms underlying levodopa-induced dyskinesia in Parkinson's disease, *Ann. Neurol.*, 47 (Suppl. 1), 105–114, 2000.

Brown, R.G. and Marsden, C.D., Dual task performance and processing resources in normal subjects and patients with Parkinson's disease, *Brain*, 114, 215–231, 1991.

Calabresi, P., Maj, R., Pisani, A., Mercuri, N.B., and Bernardi, G., Long-term synaptic depression in the striatum: physiological and pharmacological characterization, *J. Neurosci.*, 12, 4224–4233, 1992.

Calabresi, P., Pisani, A., Mercuri, N.B., and Bernardi, G., The corticostriatal projection: from synaptic plasticity to dysfunctions of the basal ganglia, *Trends Neurosci.*, 19, 19–24, 1996.

Carlsson, M. and Svensson, A., Interfering with glutamatergic neurotransmission by means of NMDA antagonist administration discloses the locomotor stimulatory potential of other transmitter systems, *Pharmacol. Biochem. Behav.*, 36, 45–50, 1990.

Catania, A.C., Reinforcement schedules and psychophysical judgments: a study of some temporal properties of behavior, in *The Theory of Reinforcement Schedules*, Schoenfeld, W.N., Ed., Appleton-Century-Croft, New York, 1970, pp. 1–42.

Chase, T.N. and Oh, J.D., Striatal dopamine- and glutamate-mediated dysregulation in experimental parkinsonism, *Trends Neurosci.*, 23 (Suppl. S), 86–91, 2000.

Chase, T.N., Oh, J.D., and Konitsiotis, S., Antiparkinsonian and antidyskinetic activity of drugs targeting central glutamatergic mechanisms, *J. Neurol.*, 247 (Suppl. 2), 36–41, 2000.

Chesselet, M.F. and Delfs, J.M., Basal ganglia and movement disorders: an update, *Trends Neurosci.*, 19, 417–422, 1996.

Crossman, A.R., Neural mechanisms in disorders of movement, *Comp. Biochem. Physiol.*, 93A, 141–149, 1989.

Damier, P., Hirsch, E.C., Agid, Y., and Graybiel, A.M., The substantia nigra of the human brain: I. Nigrosomes and the nigral matrix, a compartmental organization based on calbindin D(28K) immunohistochemistry, *Brain*, 122, 1421–1436, 1999.

DeLong, M.R., Primate models of movement disorders of basal ganglia origin, *Trends Neurosci.*, 13, 281–285, 1990.

Dostrovsky, J.O. and Lozano, A.M., Mechanisms of deep brain stimulation, *Mov. Disord.*, 17 (Suppl. 3), 63–68, 2002.

Falkenburger, B.H., Barstow, K.L., and Mintz, I.M., Dendrodentic inhibition through reversal of dopamine transport, *Science*, 293, 2465–2470, 2001.

Filion, M., Tremblay, L., and Bedard, P.J., Effects of dopamine agonists on the spontaneous activity of globus pallidus neurons in monkeys with MPTP-induced parkinsonism, *Brain Res.*, 547, 152–161, 1991.

Fredriksson, A., Danysz, W., Quack, G., and Archer, T., Co-administration of memantine and amantadine with sub/suprathreshold doses of L-Dopa restores motor behaviour of MPTP-treated mice, *J. Neural Transm.*, 108, 167–187, 2001.

Gibbon, J. and Church, R.M., Sources of variance in an information processing theory of timing, in *Animal Cognition*, Roitblat, H.L., Bever, T.G., and Terrace, H.S., Eds., Erlbaum, Hillsdale, NJ, 1984, pp. 465–488.

Gibbon, J., Church, R.M., and Meck, W.H., Scalar timing in memory, *Ann. N.Y. Acad. Sci.*, 423, 52–77, 1984.

Gibbon, J. and Malapani, C., Neural basis of timing and time perception, in *Encyclopedia of Cognitive Science*, Nature Publishing Group, London, in press.

Gibbon, J., Malapani, C., Dale, C., and Gallistel, C.R., Toward a neurobiology of temporal cognition: advances and challenges, *Curr. Opin. Neurobiol.*, 7, 170–184, 1997.

Gillies, A. and Arbuthnott, G., Computational models of the basal ganglia, *Mov. Disord.*, 15, 762–770, 2000.

Goldberg, J.A., Maraton, S., Boraud, T., Vaadia, E., and Bergman, H., The Discharge Synchrony of Neurons in the Primary Motor Cortex Increases in the Akinetic MPTP Treated Vervet Monkey, abstract presented at the VIIth meeting of the International Basal Ganglia Society (IBAGS), Waitangi, New Zealand, February 2001.

Graybiel, A.M., The basal ganglia and chunking of action repertoires, *Neurobiol. Learn. Mem.*, 70, 119–136, 1998.

Greengard, P., The neurobiology of slow synaptic transmission, *Science*, 294, 1024–1030, 2001.

Harrington, D.L., Haaland, K.Y., and Hermanowicz, N., Temporal processing in the basal ganglia, *Neuropsychology*, 12, 3–12, 1998.

Hauber, W. and Lutz, S., Dopamine D1 or D2 receptor blockade in the globus pallidus produces akinesia in the rat, *Behav. Brain Res.*, 106, 143–150, 1999a.

Hauber, W. and Lutz, S., Blockade of dopamine D2, but not of D1 receptors in the rat globus pallidus induced Fos-like immunoreactivity in the caudate-putamen, substantia nigra and entopeduncular nucleus, *Neurosci. Lett.*, 271, 73–76, 1999b.

Herrero, M.T., Levy, R., Ruberg, M., Javoy-Agid, F., Luquin, M.R., Agid, Y., Hirsch, E.C., and Obeso, J.A., Glutamic acid decarboxylase mRNA expression in medial and lateral pallidal neurons in the MPTP-treated monkey and patients with Parkinson's disease, *Adv. Neurol.*, 69, 209, 1996a.

Herrero, M.T., Levy, R., Ruberg, M., Luquin, M.R., Villares, J., Guillen, J., Faucheux, B., Javoy-Agid, F., Guridi, J., Agid, Y., Obeso, J.A., and Hirsch, E.C., Consequence of nigrostriatal denervation and L-dopa therapy on the expression of glutamic acid decarboxylase messenger RNA in the pallidum, *Neurology*, 47, 219–224, 1996b.

Hinton, S.C. and Meck, W.H., How time flies: functional and neural mechanisms of interval timing, in *Time and Behaviour: Psychological and Neuro-behavioural Analyses*, Vol. 120, Bradshaw, C.M. and Szabadi, E., Eds., Elsevier, New York, 1997, pp. 409–457.

Ivry, R.B., The representation of temporal information in perception and motor control, *Curr. Opin. Neurobiol.*, 6, 851–857, 1996.

Jahanshahi, M., Ardouin, C., Brown, R.G., Rothwell, J.C., Obeso, J., Albanese, A., Rodriguez-Oroz, M.C., Moro, E., Benabid, A.L., Pollak P., and Limousin-Dowsey, P., The impact of deep brain stimulation on executive function in Parkinson's disease, *Brain*, 123, 1142–1154, 2000.

Jahanshahi, M., Brown, R.G., and Marsden, C.D., The effect of withdrawal of dopaminergic medication on simple and choice reaction time and the use of advance information in Parkinson's disease, *J. Neurol. Neurosurg. Psychiatry*, 55, 1168–1176, 1992.

Kita, H., Tokuno, H., and Nambu, A., Monkey globus pallidus external segment neurons projecting to the neostriatum, *Neuroreport*, 10, 1467–1472, 1999.

Klockgether, T. and Turski, L., Toward an understanding of the role of glutamate in experimental parkinsonism: agonist-sensitive sites in the basal ganglia, *Ann. Neurol.*, 34, 585–593, 1993.

Lange, K.W., Kornhuber, J., and Riederer, P., Dopamine/glutamate interactions in Parkinson's disease, *Neurosci. Biobehav. Rev.*, 21, 393–400, 1997.

Lenz, F.A., Suarez, J.I., Metman, L.V., Reich, S.G., Karp, B.I., Hallet, T.M., Rowland, L.H., and Dougherty, P.M., Pallidal activity during dystonia: somatosensory reorganisation and changes with severity, *J. Neurol. Neurosurg. Psychiatry*, 65, 767–770, 1998.

Levy, R., Hazrati, L.N., Herrero, M.T., Vila, M, Hassani, O.K., Mouroux, M., Ruberg, M., Asensi, H., Agid, Y., Feger, J., Obeso, J.A., Parent, A., and Hirsch, E.C., Re-evaluation of the functional anatomy of the basal ganglia in normal and Parkinsonian states, *Neuroscience*, 76, 335–343, 1997.

Limousin, P., Krack, P., Pollak, P., Benazzouz, A., Ardouin, C., Hoffmann, D., and Benabid, A.L., Electrical stimulation of the subthalamic nucleus in advanced Parkinson's disease, *N. Engl. J. Med.*, 339, 1105–1111, 1998.

Limousin, P., Pollak, P., Benazzouz, A., Hoffmann, D., Broussolle, E., Perret, J.E., and Benabid, A.L., Bilateral subthalamic nucleus stimulation for severe Parkinson's disease, *Mov. Disord.*, 10, 672–674, 1995a.

Limousin, P., Pollak, P., Benazzouz, A., Hoffmann, D., Le Bas, J.F., Broussolle, E., Perret, J.E., and Benabid, A.L., Effect of parkinsonian signs and symptoms of bilateral subthalamic nucleus stimulation, *Lancet*, 345, 91–95, 1995b.

Lombardi, W.J., Gross, R.E., Trepanier, L.L., Lang, A.E., Lozano, A.M., and Saint-Cyr, J.A., Relationship of lesion location to cognitive outcome following microelectrode-guided pallidotomy for Parkinson's disease: support for the existence of cognitive circuits in the human pallidum, *Brain*, 123, 746–758, 2000.

Malapani, C., Deweer, B., and Gibbon, J., Separating storage from retrieval dysfunction of temporal memory in Parkinson's disease, *J. Cognit. Neurosci.*, 14, 311–322, 2002a.

Malapani, C., Dubois, B., Rancurel, G., and Gibbon, J., Cerebellar dysfunctions of temporal processing in the seconds range in humans, *Neuroreport*, 9, 3907–3912, 1998a.

Malapani, C. and Fairhurst, S., Scalar timing in animals and humans, *Learn. Motiv.*, 33, 156–176, 2002.

Malapani, C., Likhtik, D., Deweer., B.P., Benabid, F., Agid, Y., and Gibbon, J., Temporal memory retrieval is improved by deep brain stimulation in the subthalamic nucleus in Parkinson's disease, *Mov. Disord.*, submitted.

Malapani, C., Pillon, B., Dubois, B., and Agid, Y., Impaired simultaneous cognitive task performance in Parkinson's disease: a dopamine-related dysfunction, *Neurology*, 44, 319–326, 1994.

Malapani, C., Rakitin, B., Levy, R., Meck, W.H., Deweer, B., Dubois, B., and Gibbon, J., Coupled temporal memories in Parkinson's disease: a dopamine-related dysfunction, *J. Cognit. Neurosci.*, 10, 316–331, 1998b.

Malapani, C., Rakitin, B.C., Dube, K., Lobo, S., and Fairhurst, S., Disambiguating the role of cognitive strategy versus memory updating in an age-related time production deficit, *Cognit. Neurosci. Meet. Abstr.*, 30, 2002b.

Malapani, C., Rakitin, B.C., Fairhurst, S., and Gibbon, J., Neurobiology of temporal memory, *J. Cognit. Process.*, in press.

Marsden, C.D. and Obeso, J.A., The functions of the basal ganglia and the paradox of stereotaxic surgery in Parkinson's disease, *J. Neurol.*, 117, 877–897, 1994.

Matell, M.S. and Meck, W.H., Neuropsychological mechanisms of interval timing behaviour, *Bioessays*, 22, 94–103, 2000.

Matell, M.S. and Meck, W.H., Cortico-striatal circuits and interval timing: coincidence-detection of oscillatory processes, *Behav. Neurosci., Cog. Brain Res.*, in press.

Meck, W.H., Affinity for the dopamine D_2 receptor predicts neuroleptic potency in decreasing the speed of an internal clock, *Pharmacol. Biochem. Behav.*, 25, 1185–1189, 1986.

Meck, W.H., Neuropharmacology of timing and time perception, *Cognit. Brain Res.*, 3, 227–242, 1996.

Miall, C., The storage of time intervals using oscillating neurons, *Neural Comput.*, 1, 354–371, 1989.

Miall, R.C., Models of neural timing, in *Time, Internal Clocks and Movement*, Vol. 115, Pastor, M.A. and Artieda, J., Eds., Elsevier/North-Holland, Amsterdam, 1996, pp. 69–94.

Miller, W.C. and DeLong, M.R., Altered tonic activity of neurons in the globus pallidus and subthalamic nucleus in the primate MPTP model of parkinsonism, in *The Basal Ganglia II*, Carpenter, M.B. and Jayaraman, A., Eds., Plenum, New York, 1987, pp. 415–427.

Mitchell, I.J., Cross, A.J., Sambrook, M.A., and Crossman, A.R., Neural mechanisms mediating 1-methyl-4-phenyl-1,2,3,6-tetrahydropyridine-induced parkinsonism in the monkey: relative contribution of the striatopallidal and striatonigral pathways are suggested by 2-deoxyglucose uptake, *Neurosci. Lett.*, 63, 61–65, 1986.

Mitchell, I.J., Hughes, N., Carroll, C.B., and Brotchie, J.M., Reversal of parkinsonian symptoms by intrastriatal and systemic manipulations of excitatory amino acid and dopamine transmission in the bilateral 6-OHDA lesioned marmoset, *Behav. Pharmacol.*, 6, 492–507, 1995.

Molinuevo, J.L., Valldeoriola, F., Tolosa, E., Rumia, J., Valls-Sole, J., Roldan, H., and Ferrer, E., Levodopa withdrawal after bilateral subthalamic nucleus stimulation in advanced Parkinson disease, *Arch. Neurol.*, 57, 983–988, 2000.

Nash, J.E. and Brotchie, J.M., Characterisation of striatal NMDA receptors involved in the generation of parkinsonian symptoms: intrastriatal microinjection studies in the 6-OHDA-lesioned rat, *Mov. Disord.*, in press.

Nash, J.E., Fox, S.H., Henry, B., Hill, M.P., Peggs, D., McGuire, S., Maneuf, Y., Hille, C., Brotchie, J.M., and Crossman, A.R., Antiparkinsonian actions of ifenprodil in the MPTP-lesioned marmoset model of Parkinson's disease, *Exp. Neurol.*, 165, 136–142, 2000.

Ni, Z.G., Bouali-Benazzouz, R., Gao, D.M., Benabid, A.L., and Benazzouz, A., Intrasubthalamic injection of 6-hydroxydopamine induces changes in the firing rate and pattern of subthalamic nucleus neurons in the rat, *Synapse*, 40, 145–153, 2001.

Obeso, J.A., Rodriguez, M.C., and DeLong, M.R., Basal ganglia pathophysiology: a critical review, *Adv. Neurol.*, 74, 3–18, 1997.

Obeso, J.A., Rodriguez-Oroz, M.C., Rodriguez, M., DeLong, M.R., and Olanow, C.W., Pathophysiology of levodopa-induced dyskinesias in Parkinson's disease: problems with the current model, *Ann. Neurol.*, 47 (Suppl. 1), 22–34, 2000.

Pahapill, P.A. and Lozano, A.M., The pedunculopontine nucleus and Parkinson's disease, *Brain*, 123, 1767–1783, 2000.

Parent, A. and Cicchetti, F., The current model of basal ganglia organization under scrutiny, *Mov. Disord.*, 13, 199–202, 1998.

Pillon, B., Neuropsychological assessment for management of patients with deep brain stimulation, *Mov. Disord.*, 17 (Suppl. 3), 116–122, 2002.

Pillon, B., Ardouin, C., Damier, P.H., Krack, P., Houeto, J.L., Klinger, H., Bonnet, A.M., Pollak, P., Benabid, A.L., and Agid, Y., Neuropsychological changes between "off" and "on" STN or GPi stimulation in Parkinson's disease, *Neurology*, 55, 411–418, 2000.

Pillon, B., Michon, A., Malapani, C., Agid, Y., and Dubois, B., Are explicit memory disorders of progressive supranuclear palsy related to damage of striato-frontal circuits? Comparison with Alzheimer's, Parkinson's, and Huntington's diseases, *Neurology*, 44, 1264–1270, 1994.

Plenz, D. and Kital, S., A basal ganglia pacemaker formed by the subthalamic nucleus and external globus pallidus, *Nature*, 400, 677–682, 1999.

Pollak, P., Fraix, V., Krack, P., Moro, E., Mendes, A., Chabardes, S., Koudsie, A., and Benabid, A.L., Treatment results: Parkinson's disease, *Mov. Disord.*, 17 (Suppl. 3), 75–83, 2002.

Rakitin, B.C., Gibbon, J., Penney, T.B., Malapani, C., Hinton, S.C., and Meck, W.H., Scalar expectancy theory and peak-interval timing in humans, *J. Exp. Psychol. Anim. Behav. Process.*, 24, 15–33, 1998.

Rakitin, B.C., Stern, Y., and Malapani, C., Timing deficits in aging are duration-dependent, submitted.

Rao, S.M., Mayer, A.R., and Harrington, D.L., The evolution of brain activation during temporal processing, *Nat. Neurosci.*, 4, 317–323, 2001.

Robertson, R.G., Clarke, C.A., Boyce, S., Sambrook, M.A., and Crossman, A.R., The role of striatopallidal neurones utilizing gamma-aminobutyric acid in the pathophysiology of MPTP-induced parkinsonism in the primate: evidence from [^{3}H]flunitrazepam autoradiography, *Brain Res.*, 531, 95–104, 1990.

Robertson, R.G., Graham, W.C., Sambrook, M.A., and Crossman, A.R., Further investigations into the pathophysiology of MPTP-induced parkinsonism in the primate: an intracerebral microdialysis study of gamma-aminobutyric acid in the lateral segment of the globus pallidus, *Brain Res.*, 563, 278–280, 1991.

Saint-Cyr, J.A., Trepanier, L.L., Kumar, R., Lozano, A.M., and Lang A.E., Neuropsychological consequences of chronic bilateral stimulation of the subthalamic nucleus in Parkinson's disease, *Brain*, 123, 2091–2108, 2000.

Schultz, W., Dopamine neurons and their role in reward mechanisms, *Curr. Opin. Neurobiol.*, 7, 191–197, 1997.

Smith, Y. and Kieval, J.Z., Anatomy of the dopamine system in the basal ganglia, *Trends Neurosci.*, 23 (Suppl. S), 28–33, 2000.

Tronnier, V.M., Fogel, W., Kronenbuerger, M., and Steinvorth, S., Pallidal stimulation: an alternative to pallidotomy? *J. Neurosurg.*, 87, 700–705, 1997.

Turski, L., Bressler, K., Rettig, K.J., Löschmann, P.A., and Wachtel, H., Protection of substantia nigra from MPP+ neurotoxicity by N-methyl-D-aspartate antagonists, *Nature*, 349, 414–418, 1991.

Vila, M., Levy, R., Herrero, M.T., Faucheux, B., Obeso, J.A., Agid, Y., and Hirsch, E.C., Metabolic activity of the basal ganglia in parkinsonian syndromes in human and nonhuman primates: a cytochrome oxidase histochemistry study, *Neuroscience*, 71, 903–912, 1996.

Vitek, J.L., Mechanisms of deep brain stimulation: excitation or inhibition, *Mov. Disord.*, 17 (Suppl. 3), 69–72, 2002.

Vitek, J.L., Bakay, R.A., and DeLong, M.R., Microelectrode-guided pallidotomy for medically intractable Parkinson's disease, *Adv. Neurol.*, 74, 183–198, 1997.

Wu, Y.R., Levy, R., Ashby, P., Tasker, R.R., and Dostrovsky, J.O., Does stimulation of the GPi control dyskinesia by activating inhibitory axons? *Mov. Disord.*, 16, 208–216, 2002.

Yelnik, J., Damier, P., Bejjani, B.P., Francois, C., Geervais, D., Dormont, D., Arnulf, I., Bonnet, A.M., Cornu, P., Pidoux, B., and Agid, Y., Functional mapping of the human globus pallidus: contrasting effect of stimulation in the internal and external pallidum in Parkinson's disease, *Neuroscience*, 101, 77–87, 2000.

Young, A.B. and Penney, J.B., Neurochemical anatomy of movement disorders, *Neurol. Clin.*, 2, 417–433, 1984.

APPENDIX A

```
function [nactive]=miallt1(nunits,pon,poff,nsteps,itsperstep,
nruns);
%function [nactive]=miallt1(nunits,pon,poff,nsteps,itsperstep,
nruns)
%
%Implements Miall's population accumulator model of timing,
%with a
%poisson distribution of iterations per time step.
%
%Input parameters are:
% nunits -> the number of neurons in the network
% pon -> probability of an off unit turning on per time step
% poff -> prob. of an on unit turning off per iteration of model
% nsteps -> number of time steps processed per run
% itsperstep -> mean number of model iteration per time step
% nruns -> number of repetitions of whole model
%
%Output parameters are:
% nactive -> matrix containing nruns columns, and nsteps rows
%    each row is the number of units active at one time step.

%Modification history:
% BCR 12/01/98 - created first version
% BCR 4/1/02 - reformatted for publication

nactive=zeros(nsteps,nruns);
steps=poissondev(itsperstep,nsteps*nruns);
steps=reshape(steps,nsteps,nruns);
for j=1:nruns
  units=zeros(nunits,1);
  for i=1:nsteps
    ponmat=rand(size(ponmat))<=pon;
    units=units|ponmat;
    poffmat=rand(length(find(units==1)),steps(i,j))<=poff;
    units(find(units==1))=units(find(units==1))-
(sum(poffmat')>0)';
    nactive(i,j)=sum(units);
end
```

APPENDIX B

```
function pdevs=poissondev(xm,n);
%function pdevs=poissondev(xm,n);
%
%Generates n random deviates drawn from a poisson distribution.
%Algorithm modified from "Numerical Recipes in C," pp.293-295.
%
%Modification History:
%  BCR 11/25/98 - created first version
%  BCR 04/02/02 - reformatted for publication

pdevs=zeros(1,n);
%if xm < 120
   g=exp(-xm);
   for i=1:n,
      j=1;
      t=cumprod(rand(1,2*xm));
      em=find(t<g);
      while isempty(em)
           j=j+1;
           t=t(length(t)).*cumprod(rand(2*xm));
           em=find(t<g);
      end
      pdevs(i)=em(1)+((j-1)*(2*xm))-1;
   end
```

21 Overview: An Image of Human Neural Timing

Penelope A. Lewis and R. Chris Miall

CONTENTS

21.1 Introduction 515
21.2 Are There Multiple Systems for Interval Timing? 516
    21.2.1 Interval Duration as a Defining Factor 516
        21.2.1.1 Attention and Duration 516
        21.2.1.2 The Motor System and Duration 517
    21.2.2 Hypothesis: Two Systems for Time Measurement 518
        21.2.2.1 The Automatic Timing System 518
        21.2.2.2 The Cognitively Controlled Timing System 518
    21.2.3 Supporting Evidence from the Neuroimaging Literature 519
        21.2.3.1 Significance of the Meta-Analysis 520
        21.2.3.2 Possibility of Confounds 522
    21.2.4 Summary 523
21.3 Neuroimaging and the Time Measurement System 524
    21.3.1 The Functional Anatomy of Time Measurement 524
    21.3.2 Imaging the Timer Components 525
21.4 Concluding Remarks 527
References 528

21.1 INTRODUCTION

The chapters in this book have described current ideas about the functional and neural mechanisms involved in timing behaviors and the temporal judgments of intervals. We optimistically conclude that current and future imaging techniques will soon allow a detailed understanding of the neural circuits involved in interval timing. We can, however, envisage two pitfalls that might slow progress if not treated with caution. The first is the probability that multiple mechanisms are involved in time measurement, and that these are functionally and anatomically discrete. If unrecognized, such duplicity of mechanisms could lead to extreme confusion regarding the locus and function of neural timing systems. The second pitfall is associated with the inherent limitations of neuroimaging techniques and the implications of these for investigations of time measurement. We believe that due to the sluggish and indirect nature of some techniques, such as functional magnetic resonance imaging (fMRI) and positron emission tomography (PET), and the spatial imprecision of others, such as electroencephalography (EEG) and magnetoencephalography (MEG), neuroimaging results must be interpreted with great caution when used to investigate a delicate system such as that used for time measurement.

We will address these pitfalls by raising two general questions: First, does all human interval timing depend on the same basic neural system, or are fundamentally different processes used in different timing tasks? Second, to what extent can we expect functional imaging techniques to be useful in describing the detailed function of the mechanisms involved?

21.2 ARE THERE MULTIPLE SYSTEMS FOR INTERVAL TIMING?

Let us start by considering the evidence for multiple timers. A number of authors have suggested that different mechanisms may be used for different types of time measurement, including a distinction between implicit and explicit timing mechanisms (see Hopson, this volume). The length of a measured duration, whether it is timed or defined via movement, and the degree of awareness associated with the temporal judgment have all been suggested as factors that determine which system is used. Further, different authors have independently suggested that the degree of awareness involved in a timing judgment depends on the length of the duration measured, and that the extent to which the motor system is used is determined by the same factor (e.g., Ivry, 1996, 1997; Rammsayer, 1999). To our knowledge, however, no framework has yet combined all three factors into a unified model. We will here propose precisely such a model, adding one further important task characteristic: the *continuousness* of timing.

21.2.1 INTERVAL DURATION AS A DEFINING FACTOR

A variety of different observations have suggested that measurement of intervals in the milliseconds range draws upon a different timer than the measurement of intervals in the multiseconds range. These include differential psychophysical characteristics for temporal measurements at the two duration ranges (Gibbon et al., 1997), differential responses to pharmacological agents (Mitriani et al., 1977; Rammsayer, 1993, 1999; Rammsayer and Vogel, 1992), differential impairment by dual-task scenarios (Rammsayer and Lima, 1991), and differential impairment by specific brain lesions (Clarke et al., 1996). Most recently, some of our own neuroimaging data have shown that different brain regions are active during timing of 0.6- and 3-sec intervals using the same task (Lewis and Miall, submitted).

21.2.1.1 Attention and Duration

A number of studies have suggested that the measurement of intervals longer than 1 sec requires cognitively controlled and attended processing, while measurement of intervals in the milliseconds range does not require direct attention. These include works showing that active processing in working memory is only required during the timing of longer intervals (Fortin, this volume; Fortin and Breton, 1995; Fortin et al., 1993; Rammsayer and Lima, 1991); temporal processing in the milliseconds range is unaffected by level of arousal (Rammsayer, 1989; Rammsayer and Vogel, 1992), but does depend on sensory processes (Rammsayer and Lima, 1991); and pharmacological agents, such as LSD and mescaline, know to interfere with cognitive processing, disrupting the timing of multiple seconds but not of milliseconds (Mitriani et al., 1977). On the basis of these findings, at least two authors (Mitriani et al., 1977; Rammsayer, 1999) have separately suggested that intervals in the milliseconds range are measured more or less automatically, while intervals in the multiseconds range require active processing under direct cognitive control.

21.2.1.2 The Motor System and Duration

Because the durations used in movement, for instance, in muscle phasing and coordination, fall within the subsecond range, it has been suggested that the timers used to measure these intervals may be located within the motor system. One candidate structure for such involvement is the cerebellum. Observations that the cerebellum is frequently active in tasks involving measurement of subsecond intervals (Belin et al., 2002; Coull et al., 2000; Coull and Nobre, 1998; Jancke et al., 2000b; Jueptner et al., 1995, 1996; Kawashima et al., 2000; Lutz et al., 2000; Maquet et al., 1996; Parsons, 2001; Penhune and Doyon, 2002; Penhune et al., 1998; Rao et al., 1997; Roland et al., 1981; Schubert et al., 1998; Schubotz et al., 2000; Schubotz and von Cramon, 2001) and that cerebellar lesions lead to deficits in this type of movement-related timing (Ivry et al., 1988) have led to the idea that this structure may contain subsecond specific timers (see Diedrichsen et al., this volume; Hazeltine et al., 1997; Ivry, 1996). Further, network models of the cerebellum have shown that the structure could feasibly measure subsecond intervals in a number of different ways (De Zeeuw et al., 1998; Guigon et al., 1994; Medina et al., 2000; Perrett et al., 1993). However, the idea that the cerebellum is *exclusively* involved in movement-related timing, or for that matter, in the measurement of subsecond intervals, has been rejected due to evidence showing cerebellar involvement both in perceptual (i.e., nonmotor) timing (Casini and Ivry, 1999; Ivry and Keele, 1989; Nichelli et al., 1996) and in timing of intervals as long as 21 sec (Malapani et al., 1998; Nichelli et al., 1996).

Other regions of the motor system, for instance, the premotor cortex, could also be involved in time measurement. One possible mechanism (Lewis and Miall, 2002) for such involvement is the predictable activity of buildup cells, which has been shown by others (Matsuzaka et al., 1992) to increase or decrease during movement preparation. Central pattern generators (CPGs) offer another possibility. They are known to produce rhythmic activity with periods ranging from under 60 msec to several seconds (Arshavsky et al., 1997) for all manner of rhythmic motor activity, especially locomotor, respiratory, and chewing actions. Brain stem and spinal cord CPGs are modulated by top-down control (Armstrong, 1988) and have projections to cerebral regions (Arshavsky et al., 1978). They therefore have the potential to elicit fMRI-measurable activity in the cortex and cerebellum; cortical pattern generators are also a possibility.

21.2.2 Hypothesis: Two Systems for Time Measurement

21.2.2.1 The Automatic Timing System

We propose that if an interval is measured again and again without change or interruption (as in self-paced finger tapping or perception of an isochronous rhythm), the temporal measurement can be performed by an automatic circuit, which does not require overt attention. This idea is in keeping with a loose interpretation of the motor program concept (Schmidt, 1982), which suggests that all of the information needed for an overlearned movement can be stored in such a way that, once selected and initiated, the movement is essentially performed automatically. Hence, it might be necessary to attend the first cycle or two of temporal production or perception in order to select the appropriate timing mechanism and set it running, but after that, attention should be required only when there is a mismatch between interval and expectation.

Studies of overlearned movement support this model because they have shown that explicit attention is not required for performance of these "automatic" movement tasks (Passingham, 1996). If attention is not required for the movement, then it cannot be required for the related temporal measurements. We therefore propose that a timing system exists for the measurement of brief intervals that are produced continuously and via movement, as in paced finger tapping or execution of other overlearned motor programs. This system likely recruits timing circuits within the motor system that can act without attentional modulation; we will therefore refer to it as the *automatic* timing system. CPGs would provide an ideal mechanism for the automatic system because they are characterized by continuous rhythmic output. The proposed timing mechanisms of the cerebellum would be similarly appropriate to measurement of intervals in automatic movement, as the cerebellum seems to have an important role in automated actions (Nixon and Passingham, 2000).

21.2.2.2 The Cognitively Controlled Timing System

Although the automatic system may be very handy for the nonattended measurement of time under certain very predictable conditions, it is unlikely to serve in all circumstances. For a start, automatic timing may only be possible when the interval in question is repeated over and over without stopping because unpredictable breaks in the sequence may mean that attention is required to restart or reset the timer for each new epoch. Furthermore, there may be limitations on the maximum duration length that the timers used by this system can conveniently measure (De Zeeuw et al., 1998; Guigon et al., 1994; Medina et al., 2000; Perrett et al., 1993). Finally, if the timers of the automatic system lie within the motor cortex or cerebellum, then they may be preferentially used for measurement of intervals that are part of a movement. We suggest, therefore, that intervals longer than a second or so, measured as discrete events rather than as part of a predictable sequence, and not defined by movement, are not appropriate for the automatic system and must draw instead upon a directly attended framework, which we will refer to as the *cognitively controlled* timing system.

Analogous to the overlapping use of the motor system for motor control and timing, we imagine that the cognitively controlled system may use neural circuits that are typically invoked for other cognitive operations, but can be recruited, when appropriate, for storing and processing information for temporal processes. Hence, we envisage that the cognitively controlled timing system draws on flexible, multipurpose cognitive modules within the prefrontal and parietal cortex, and thus shows overlap in functional imaging experiments with many other cognitive tasks. Following from the conclusions of Rammsayer (1999) and Mitriani et al. (1977) that cognitively controlled timing draws on active working memory and attention, we might therefore predict the involvement of the premotor cortex (PMC) or dorsolateral prefrontal cortex (DLPFC), both of which are known for working memory processing (Petrides, 1994; Smith and Jonides, 1999), and of some portion of the attentional system, currently thought to comprise the parietal, anterior cingulate, and frontal cortex (for a review, see Coull, 1998).

21.2.3 SUPPORTING EVIDENCE FROM THE NEUROIMAGING LITERATURE

If our hypothesis is correct and activity in the automatic and cognitively controlled systems can be measured using neuroimaging techniques, then an analysis of the existing neuroimaging literature should show dissociation in the brain areas activated by time measurement tasks with different characteristics. We have recently undertaken such an analysis, including all neuroimaging studies of primate time measurement known to us (Belin et al., 2002; Brunia and de Jong, 2000; Coull et al., 2000; Coull and Nobre, 1998; Gruber et al., 2000; Jancke et al., 2000a; Jueptner et al., 1995, 1996; Kawashima et al., 1999, 2000; Larasson et al., 1996; Lejeune et al., 1997; Lewis and Miall, 2002, submitted; Lutz et al., 2000; Macar et al., 2002; Maquet et al., 1996; Onoe et al., 2001; Penhune et al., 1998; Rao et al., 1997, 2001; Roland et al., 1981; Rubia et al., 1998, 2000; Sakai et al., 1999; Schubotz et al., 2000; Schubotz and von Cramon, 2001; Tracy et al., 2000). Two relevant studies were excluded because the complete results were not reported (Parsons, 2001; Schubert et al., 1998), two because they examined learning specific activities (Ramnani and Passingham, 2001; Penhune and Doyon, 2002), and one because it dealt with noncontrol subjects (Volz et al., 1999).

To test our hypothesis, it is necessary to examine how the pattern of activity observed in each study relates to the characteristics of the task performed. Accordingly, we have categorized the studies in three ways: (1) according to whether a duration greater than 1 sec was measured, (2) according to whether the measured duration was defined by movement, and (3) according to whether timing was continuous or occurred in discrete episodes. We listed all brain areas that were activated by these studies and recorded which studies showed activity in each. To be inclusive, we used the most lenient subtraction presented (for instance, test vs. rest) rather than a more rigorous control condition, as in Coull and Nobre (1998). In papers presenting multiple data sets, each independent set was included as a distinct study (Coull and

Nobre, 1998; Jancke et al., 2000b; Lewis and Miall, in preparation; Rao et al., 1997; Rubia et al., 1998, 2000; Sakai et al., 1999). Finally, we performed a meta-analysis, using all of this information to determine the percentage of studies with certain task characteristics that showed activity in any given area.

The results of the meta-analysis are shown in Table 21.1. Brain areas are listed across the top row, with the laterality of each area listed just below. To reduce the complexity of this table, only those areas that were active in at least 40% of the studies in one of our categories are shown; thus many areas reported to be active in a minority of studies are not included. Different combinations of studies are dealt with in rows 1 to 9, with the relevant category of task characteristics indicated to the left of each row. Thus, row 1 deals with all studies in the review, while row 2 deals only with studies in which any two out of three task characteristics are associated with the cognitively controlled timing system. Rows 3 to 5 deal with pairings of task characteristics associated with the cognitively controlled system. Rows 6 to 9 follow a similar model, but deal with studies in which task characteristics are associated with the automatic system. The remainder of the table shows the percentage of the studies in each category (row) that report activity in each brain area, with more commonly activated regions shaded more darkly.

21.2.3.1 Significance of the Meta-Analysis

The first row of Table 21.1 shows no strong consensus regarding the areas involved in time measurement. Only the bilateral supplementary motor area (SMA) and cerebellum are active in more than 50% of studies, and no area is active in more than 61% of studies. The remainder of the table, however, shows clearly that a different set of areas is active during tasks associated with the cognitively controlled timing system from that active during tasks associated with the automatic timing system.

Tasks associated with the automatic timing system most commonly elicit activity in the bilateral SMA and sensorimotor cortex. The right hemispheric cerebellum, PMC, superior temporal gyrus, and to some extent, the left hemispheric basal ganglia and thalamus are also frequently activated in these tasks, though they do not appear so commonly if intervals longer than 1 sec are measured (see row 9). Activity associated with these tasks is also observed in the occipital cortex under some conditions and to a lesser extent, in the right interior parietal cortex. Interestingly, the DLPFC and the remainder of the parietal cortex rarely activate in tasks associated with automatic timing. In tasks associated with cognitively controlled timing, however, the right hemispheric DLPFC activates more commonly than any other area. The left hemispheric cerebellum and PMC, and right hemispheric intraparietal sulcus (IPS) are also very frequently active during all combinations of cognitively controlled tasks. A number of other areas activate commonly in more specific conditions: the bilateral SMA, left IPS, and right PMC so long as the interval measured is longer than one second (see row 4), the right insula, ventrolateral prefrontal cortex (VLPFC), inferior parietal, and left DLPFC so long as timing occurs in discrete epochs (see row 2), and the right basal ganglia so long as timing does not rely upon movement (see row 3).

TABLE 21.1
Summary of the Results from Our Meta-Analysis of the Neuroimaging Literature on Time Measurement[a]

Summary of the Results from our Meta-Analysis of Neuroimaging Literature on Time Measurement

	DLPFC R	IPS R	Insula R	CB Lat. L	PMC L	IPS L	DLPFC L	Basal G. R	VLPFC R	F. Pole R	Inf Par. R	Cing. R	CB Med R	PMC R	Basal G. L	CB Lat. R	Occip. L	Thalamus L	SMA L	Occip. R	SMA R	S. Temp. R	M1 L	S1 L	row
all studies	34	26	29	50	46	26	29	24	40	26	49	31	29	46	32	54	32	24	58	24	61	29	45	33	1
Cognitively Controlled Timing System																									
any 2 cognitive char.	53	47	42	58	63	42	37	32	47	26	53	26	21	53	32	47	26	21	53	16	53	21	21	5	2
lg + n_mt	67	44	33	56	78	44	22	44	22	33	33	22	22	56	44	56	11	11	44	0	44	22	11	11	3
lg + disc	64	55	64	45	55	45	45	27	55	27	55	36	18	45	36	36	36	18	55	27	55	27	27	0	4
disc + n_mt	71	43	57	71	43	14	57	43	71	71	71	43	57	29	29	29	29	14	29	0	29	14	0	0	5
Automatic Timing System																									
any 2 automatic char.	13	0	13	19	25	6	19	13	31	25	44	38	25	38	31	38	38	25	56	31	63	38	69	63	6
sh + mt	0	0	0	14	14	0	0	0	14	14	29	14	29	57	43	57	43	43	86	43	86	71	100	100	7
sh + seq	11	0	0	22	22	0	11	11	22	22	33	22	33	56	44	56	33	44	67	33	67	67	89	89	8
mt + seq	7	0	14	14	21	7	14	7	29	21	43	36	21	36	29	36	43	21	64	36	71	36	71	64	9

Note: lg = interval longer than 1 sec; sh = interval shorter than 1 sec; n_mt = interval not defined by movement; mt = interval defined by movement; disc = interval measured noncontinuously (discretely); seq = interval measured continuously (sequentially); CB lat. = lateral cerebellum; F. pole = frontal pole; basal G. = basal ganglia; operc. = frontal operculum; inf. par. = inferior parietal lobe; S. par. = superior parietal lobe; occip. = occipital lobe; S. temp. = superior temporal lobe; M1 = primary motor cortex; S1 = primary sensory cortex.

[a] The table is explained in detail in the text.

Perhaps the most important observation to make regarding these results is that the patterns seen when studies are divided based on combinations of task characteristics produce a more coherent picture than when all studies are averaged together. If these studies truly all draw on the same time measurement mechanism, then we might expect a stronger consensus regarding the areas involved than what is shown in row 1. Because different networks appear to be activated by tasks with different combinations of characteristics, this meta-analysis strongly supports the possibility of duplicitous mechanisms for time measurement.

Looking more closely at the specific areas activated, we see that several prefrontal regions believed to contain flexible cognitive modules (Duncan, 2001) are associated with the cognitively controlled tasks, but remain inactive during automatic tasks. These include the DLPFC, VLPFC, IPS, and to a large extent, inferior parietal. Also interesting is the observation that many regions of the motor system (the SMA, sensorimotor cortex, left basal ganglia, right PMC, and right cerebellum) commonly activate during automatic tasks. This pattern supports the hypothesis that what we have termed automatic timing may rely upon mechanisms located within the motor system itself. That some of these areas (bilateral SMA and right PMC) are also commonly activated in association with the cognitively controlled tasks suggests that use of the cognitively controlled system does not preclude involvement of modules from the automatic system. Before reading too much into these patterns, however, it is important to consider whether the observed activity is all truly associated with timing mechanisms, or whether some of it might be due to non-timing-related confounders.

21.2.3.2 Possibility of Confounds

Because we have reported the most inclusive contrast from each study in our analysis, much of the activity we describe may be due to movement or other task-related nontiming behaviors. Observations that the auditory, visual, and primary sensorimotor cortices are frequently activated in association with automatic timing tasks, for instance, should not necessarily be interpreted as support for the direct involvement of these areas in time measurement, because auditory or visual stimuli and movement in the tasks may have elicited this activity. Based upon the analysis presented thus far, it is impossible to determine whether activities are due to temporal processing or confounding factors. By looking more closely at some of the studies reviewed, however, we can begin to address this question.

If regions of the motor system are active even in those studies of timing where very little movement or movement preparation (or in some cases, no movement or movement preparation at all) occurred during scanning, then we can safely conjecture that their involvement is not merely motor associated, although we cannot rule out the possibility that motor imagery may be involved. This is the case for activity in the right cerebellar hemisphere (Belin et al., 2002; Jueptner et al., 1996; Larasson et al., 1996; Roland et al., 1981; Sakai et al., 1999; Schubotz and von Cramon, 2001), SMA (Gruber et al., 2000; Larasson et al., 1996; Schubotz and von Cramon, 2001), and left basal ganglia (Larasson et al., 1996; Parsons, 2001; Schubotz and von Cramon, 2001) during tasks requiring covert decisions, memory encoding, memory rehearsal of rhythms, or detection of oddballs but not movement. Because this activity is not due to movement, it may be genuinely linked to timing.

Likewise, several studies have described activity in the temporal cortex during time measurement tasks involving no auditory cues (Coull et al., 2000; Larasson et al., 1996; Rao et al., 2001). Others have shown auditory activity during task phases that come after the cessation of auditory cues, such as continuation of tapping after auditory synchronization (Rao et al., 1997) or memory encoding after presentation (Sakai et al., 1999). It has been suggested (Rao et al., 1997) that this activity may be associated with auditory imagery used for the task, so the observation that the right hemispheric superior temporal cortex is one of the most commonly activated areas during tasks that would be expected to draw on the automatic timing system may well mean that the timing of these intervals frequently draws on auditory imagery. By contrast, the lack of studies in which the occipital cortex is activated in response to tasks that do not involve visual stimuli makes it unlikely that the activity observed here is associated with temporal processing.

Because the tasks associated with the cognitively controlled system are quite different from those associated with the automatic system, it could be argued that activity unique to these tasks is due to some form of confounder. Looking carefully at the literature, however, we see that these regions activate even when a more complete cognitive subtraction is used (Lewis and Miall, 2002, in preparation; Rao et al., 2001); hence, their involvement very likely relates directly to temporal processing. Because these areas include regions known for involvement in both working memory (DLPFC) and attention (IPS and inferior parietal lobe), this observation conforms to predictions regarding the cognitively controlled system (for further details concerning activations specific to interval timing, see Hinton, this volume; Hinton and Meck, 1997; Morell, 1996; Pouthas, this volume).

21.2.4 SUMMARY

This section has explained why we believe that different mechanisms are recruited for the measurement of time in different tasks. Both an automatic timing system, which is used to measure subsecond intervals when these are measured continuously via movement, and a cognitively controlled system, which is recruited for temporal measurements that cannot easily be performed by the automatic system (i.e., those of suprasecond durations, measured discontinuously, and not via movement), have been described. Evidence from lesion studies as well as from studies of motor circuitry suggests that the motor system could perform the task of the automatic system, while the flexible cognitive modules of the prefrontal and parietal cortices are more suited to the task of the cognitively controlled system. Hence, we have hypothesized that there may be a dissociation in functional locus for these two systems. A meta-analysis of existing neuroimaging studies of time measurement has shown that when the literature is taken as a whole, there is no strong consensus regarding the areas most commonly involved. If the studies are divided based on the characteristics of the task performed, however, a clear dissociation is seen between areas activated by automatic-associated and cognitive control–associated tasks. The former frequently activate parts of the motor system (SMA, sensorimotor cortex, cerebellar hemisphere, PMC, and basal ganglia) as well as the superior temporal gyrus, but only rarely activate the prefrontal or parietal cortices. The latter frequently activate the prefrontal and parietal cortices (DLPFC, VLPFC, inferior parietal, and IPS), with additional activity in the cerebellum and SMA and PMC, among other areas. This analysis supports the possibility that functionally and anatomically distinct systems for time measurement exist within the human brain and illustrates how failure to recognize this multiplicity can lead to confusion in the literature. Future attempts to investigate the neural locus of time measurement should therefore take the possibility of multiple systems into account, both when choosing a task to study and when interpreting their findings or the findings of others.

21.3 NEUROIMAGING AND THE TIME MEASUREMENT SYSTEM

We, like many others, are using functional magnetic resonance imaging as a tool to study human timing. Our survey of the literature shows that more than 25 imaging papers of interval timing have been produced so far, and like any other topic to which neuroimaging has been applied, we expect many more to follow in the next few years. Therefore, in this section we aim to discuss some of the conceptual limits to the imaging of human timing and explore ideas about what these studies may be expected to achieve. In the limit, any single functional imaging technique on its own (whether fMRI, PET, MEG, or EEG) is unlikely to be sufficient. As the bulk of this book has demonstrated, these imaging techniques must be complemented by patient studies, lesion experiments, drug interventions, and electrophysiological recording studies, spanning the range from system to cellular analyses. Nevertheless, neuroimaging by itself will prove an important tool (see Hinton, this volume; Pouthas, this volume).

21.3.1 THE FUNCTIONAL ANATOMY OF TIME MEASUREMENT

The first level of imaging analysis is to simply identify the areas involved in timing tasks (see Hinton, this volume; Meck, this volume; Meck and Benson, 2002; Sakata and Onoda, this volume). In many of the studies we have reviewed, this is achieved by using block analysis of timing tasks contrasted with nontiming control conditions. In these studies, activity is measured in blocks of 30 or more seconds at a time, so there is no real temporal resolution to the data. More specific localization of timing components can be achieved with event-related imaging techniques, but there are also clear limits on this technique, as we will describe below. The second level of attack is the use of imaging techniques to explore interactions between the timing subsystems or to approach the neural mechanism of their functions. At this level, we must consider whether the operation of each component in a time measurement system depends on neural mechanisms that we can actually detect. The most basic measures afforded by functional imaging studies are the changes in the activity of neural populations from one moment to another. In PET and fMRI, these are detected using the resulting changes in local blood flow or oxidation; thus, if a component of the time measurement system does not cause a significant change in metabolic cost, we may not detect its presence.

The most obvious example of this problem is the time-dependent process or "clock" central to the timing system, perhaps a "ticking" oscillator or similar circuit: if the clock is always ticking, but other components (e.g., the accumulator) only intermittently use its output, it may be very difficult to detect this process using neuroimaging. One solution may be to selectively speed or slow the clock, independent of all other neural processes (Meck, 1996), and detect the changing activity that correlates with these alterations. However, it is possible that a neural clock circuit could be accelerated or slowed without leading to gross change in metabolic load: if the duty cycle (active to inactive states) is kept constant, then the main metabolic costs (e.g., dendritic processing and some contribution to ionic pumping across the membrane after spike activity) could be nearly identical in a cell or a circuit oscillating slowly or rapidly. Because we do not yet know what form of clock ticking, if any, is used in the timing process, we cannot predict whether the changes in neural activity associated with changes in clock speed would be imageable.

MEG and EEG techniques complement PET and fMRI with regard to temporal precision, as they can detect neural activity in the millisecond range (for a discussion of how EEG and PET techniques can be used to inform each other, see Pouthas, this volume). Hence, for example, these techniques would be invaluable for detecting a rhythmically active clock, as they could differentiate between the signals of different clock rates. However, these techniques also have their limitations, as both depend on the synchronous activity of a group of aligned neurons (or rather, their dendritic processes) and are insensitive to currents in tissue that are oriented perpendicular or tangential to the scalp respectively. They are also insensitive to deep brain sources. One could certainly imagine time measurement processes that would be invisible to MEG or EEG.

21.3.2 IMAGING THE TIMER COMPONENTS

Bearing knowledge of the limiting characteristics of neuroimaging techniques in mind, let us think about the basic components of the scalar expectancy theory (or scalar timing) model (see Church, this volume; Gibbon et al., 1984) and ask how we can identify the mechanisms and the neural loci of each. The various components are the time-dependent process (the pacemaker), the local memory stores (the accumulator and the reference memory), and the comparator, as well as sensory input and modulatory output systems. Temporal information processing would also include the attentional system and the cognitive output structures or the motor systems using information from the timer.

For much of the imaging literature, the sensory input systems are treated as items of secondary interest. PET and fMRI depend on contrasting different behavioral states, and thus any process in common to the two states is not visualized. Hence, it is typical to attempt to balance the contribution of systems of secondary interest between the timing task and the control (baseline) task. Sensory inputs or motor outputs, if carefully balanced, do not confound the final results of the imaging study. However, this strategy has the implicit danger that it may obscure data suggesting that the timing functions actually depend on the specific sensory structure. Thus, if the time-dependent process is active from the start of the sensory stream, then it will be nearly impossible to distinguish between these two using functional neuroimaging. However, because interval timing is easily achieved across the gaps between delimiting stimuli, or by using stimuli in different modalities, this should not pose a real problem for the investigation of non-sensory-specific timing systems. The possibility that parts of the motor system may be obligatory components of some timing operations is less easy to dismiss.

In this vein, we have discussed above and other authors in this book have highlighted the fact that motor areas of the brain (cerebellum, basal ganglia, and premotor cortical areas) are strong candidates for involvement in interval timing tasks (see in this volume, Diedrichsen et al.; Hinton; MacDonald and Meck; Malapani and Rakitin; Matell et al.; Pang and McAuley). If these circuits are recruited only for some timing operations, such as those in which repetitive motor outputs are needed (e.g., rhythmic tapping), then separation of motor timing and motor execution becomes very difficult. The wealth of evidence suggesting that imagined movement or mental rehearsal does activate the motor system compounds this difficulty because implicit use of motor systems to measure time, even without active movement, could cause neural activity. In the limit, we should perhaps ask if the attempt to separate timing from movement is sensible, if indeed the movement, or its internal rehearsal or planning, is what is actually used as the timing signal.

To approach this problem, it would be useful to know whether different motor timing circuits were selectively recruited for specific timing tasks. It seems likely that the neural operations involved in selecting or recruiting pattern-generating circuits during the first epoch of repeated, subsecond interval measurements would be detected by current functional imaging techniques. We have evidence (Lewis et al., in preparation) that this is the case, as areas known to be involved in movement selection show activity at the onset of different rhythm epochs, but appear inactive during the immediately following rhythm production. We believe that timing circuits are therefore actively recruited, or adjusted to the target intervals, but then continue to cycle with little additional cost.

Let us think about what this observation might mean at a finer level. In an earlier model of the neural mechanism of timing (Miall, 1989), it was suggested that different neural oscillators could be selected and combined to provide an interval timing system. Only the weighted output of the multiple oscillators could be said to encode a specific interval: many oscillators were active in each interval, and the selectivity of the system was generated by synaptic weighting of a subset of these to some output unit. Hence, the activity of this output unit, excited at the critical moment by the synchronous activity of its oscillating inputs, would easily be detectable, but the ongoing activity of the population of oscillators would not (for an extension of this model, see Matell et al., this volume; Meck, this volume). Miall (1989) proposed that additional neural machinery might be used to synchronize the oscillators at the start of each timed interval, but beyond that, the system could free-run with no additional metabolic cost. Again, this suggests that the neural activation required to start, select, or synchronize the oscillator system might be imageable, but its ongoing activity would be hard to detect. If the oscillator population itself became active at the start of each interval, from an inactive state, this should also be detectable, but this scenario seems unlikely.

The accumulator as described by Gibbon et al. (1984) is probably the most easily detectable timing element, as by definition its activation changes throughout each interval and must be reset. A naïve viewpoint might therefore be that the bulk of the imaging data produced so far reflects the activation of this accumulator circuit. However, using carefully designed baseline conditions, it should be possible to dissociate the accumulation process from related events such as the comparison or decision processes. The reference memory store in which the previous intervals are recorded might also seem easy to image, as it would accumulate traces of the previous intervals, changing with experience of the target interval. It is striking to us that cortical prefrontal areas are prominently active in cognitively controlled timing tasks: these may be the systems in which a trace of activity is set up and changes throughout the timed interval. Overlap of the observed regions with areas known to be involved in working memory is also important.

In contrast, it is likely that EEG and MEG techniques would be poor for studying the accumulator activity. A basic model of the accumulator (Miall, 1993) and a recent, more elaborate model (Koulakov et al., 2002) suggest that it may be made up of a population of independently active cells, and thus would not have the synchronous behavior necessary to cause a large signal. Some MEG analysis techniques have made use of the switch from synchronous activity in the idle state to desynchronized activity in an active state (Singh et al., 2002), and this could prove useful.

Lastly, the comparator function would appear to be difficult to detect. As a singular event at the end of each interval, comparison would contribute rather little to the overall signal within a typical block design imaging study. Event-related imaging designs would allow temporal separation of different events within the timing task if their occurrence could be varied with respect to each other. In such studies, the blood oxygenation signals are correlated with specific event times, for example, with the onset and offset of each interval, as long as these events are themselves uncorrelated (Buckner et al., 1996). Unfortunately, the comparator process will almost always be time-locked to other events, such as the initiation of whatever action is required at the end of the trial. For example, if the subject was asked to respond at the end of the target interval, the comparison operation, the transfer of that interval to reference memory, the resetting of the accumulator, possibly the stopping of the pacemaker clock, and the initiation of the response would all be very close in time. Better temporal differentiation using MEG or EEG, where events that cannot be decoupled could be temporally ordered, might provide a solution to this problem if it were clear what the order of their occurrence must be (see Sakata and Onoda, this volume).

21.4 CONCLUDING REMARKS

In summary, we can envisage some specific problems in functional imaging of human timing systems at present. Many of these are rooted in the intrinsic limits of PET and fMRI, with their dependency on sluggish and indirect measures of relative change in blood flow and oxygenation levels, rather than direct neural measures. EEG and MEG techniques in turn have limits in spatial resolution and are relatively insensitive to central brain structures. However, several features of the review presented in Table 21.1 encourage us. First, it is now clear that functional imaging can detect multiple areas of the human brain associated with time measurement tasks, and while the variation among different experiments is high, some areas activate consistently in association with timing tasks having specific characteristics and are thus strong candidates for further exploration. Second, we have argued that much of the variation between the experiments may be due to the diversity of timing tasks used — these tasks may even be drawing on quite separate systems. We can draw an analogy here with functional localization within the visual cortex. When an obvious, but inappropriate stimulus (such as a natural scene) is used, one gets the impression that the visual areas of the brain are horribly difficult to distinguish functionally. When a more appropriate stimulus is used (moving bars of light or drifting fields of dots), the organization becomes apparent. Thus, when appropriate timing tasks are used, the relationships between neural locus, neural mechanism, and timing behavior may become clear. Third, the enormous expansion of techniques now available to tackle the problem of time measurement, including imaging, multielectrode unit recording, drug studies, gene knockouts, transcranial magnetic stimulation, and the rest, means that a combined approach is both feasible and fruitful.

We believe that the most immediate future goal in the study of time measurement is to determine which timing systems are used under which specific circumstances. Once this has been established, the imposing arsenal of techniques at our command will facilitate further examination of the detailed functioning of each time measurement system. We are therefore optimistically confident that with the aid of these methods, we will soon succeed not only in finding the timers for which we search, but also in understanding how they work.

REFERENCES

Armstrong, D.M., The supraspinal control of mammalian locomotion, *J. Physiol.*, 405, 1–37, 1988.

Arshavsky, Y.I., Deliagina, T.G., and Orlovsky, G.N., Pattern generation, *Curr. Opin. Neurobiol.*, 7, 781–789, 1997.

Arshavsky, Y.I., Gelfand, I.M., Orlovsky, G.N., and Pavlova, G.A., Messages conveyed by spinocerebellar pathways during scratching in the cat: I. Activity of neurons of the lateral reticular nucleus, *Brain Res.*, 151, 479–491, 1978.

Belin, P., McAdams, S., Thivard, L., Smith, B., Savel, S., Zilbovicius, M., Samson, S., and Samson, Y., The neuroanatomical substrate of sound duration discrimination, *Neuropsychologia*, 40, 1956–1964, 2002.

Brunia, C.H.M. and de Jong, D.M., Visual feedback about time estimation is related to a right hemisphere activation measured by PET, *Exp. Brain Res.*, 130, 328–337, 2000.

Buckner, R.L., Bandettini, P.A., O'Craven, K.M., Savoy, R.L., Petersen, S.E., Raichle, M.E., and Rosen, B.R., Detection of cortical activation during averaged single trials of a cognitive task using functional magnetic resonance imaging, *Proc. Natl. Acad. Sci. U.S.A.*, 93, 14878–14883, 1996.

Casini, L. and Ivry, R.B., Effects of divided attention on temporal processing in patients with lesions of the cerebellum or frontal lobe, *Neuropsychology*, 13, 10–21, 1999.

Clarke, S., Ivry, R., Grinband, J., Roberts, S., and Shimizu, N., Exploring the domain of the cerebellar timing system, in *Time, Internal Clocks, and Movement*, Vroon, G.E.S.P.A., Ed., Elsevier, New York, 1996, p. 257.

Coull, J.T., Neural correlates of attention and arousal: insights from electrophysiology, functional neuroimaging and psychopharmacology, *Prog. Neurobiol.*, 55, 343–361, 1988.

Coull, J.T., Frith, C.D., Buchel, C., and Nobre, A.C., Orienting attention in time: behavioural and neuroanatomical distinction between exogenous and endogenous shifts, *Neuropsychologia*, 38, 808–819, 2000.

Coull, J.T. and Nobre, A.C., Where and when to pay attention: the neural systems for directing attention to spatial locations and to time intervals as revealed by both PET and fMRI, *J. Neurosci.*, 18, 7426–7435, 1998.

De Zeeuw, C.I., Simpson, J.I., Hoogenraad, C.C., Galjart, N., Koekkoek, S.K., and Ruigrok, T.J., Microcircuitry and function of the inferior olive, *Trends Neurosci.*, 21, 391–400, 1998.

Ducan, J., An adaptive coding model of neural function in prefrontal cortex, *Nat. Rev. Neurosci.*, 2, 820–829, 2001.

Fortin, C. and Breton, R., Temporal interval production and processing in working memory, *Percept. Psychophys.*, 57, 203–215, 1995.

Fortin, C., Rousseau, R., Bourque, P., and Kirouac, E., Time estimation and concurrent nontemporal processing: specific interference from short-term-memory demands, *Percept. Psychophys.*, 53, 536–548, 1993.

Gibbon, J., Church, R.M., and Meck, W.H., Scalar timing in memory, in *Annals of the New York Academy of Sciences: Timing and Time Perception*, Vol. 423, Gibbon, J. and Allan, L., Eds., New York Academy of Sciences, New York, 1984, pp. 52–77.

Gibbon, J., Malapani, C., Dale, C.L., and Gallistel, C.R., Towards a neurobiology of temporal cognition: advances and challenges, *Curr. Opin. Neurobiol.*, 7, 170–184, 1997.

Gruber, O., Kleinschmidt, A., Binkofski, F., Steinmetz, H., and von Cramon, C.Y., Cerebral correlates of working memory for temporal information, *Neuroreport*, 11, 1689–1693, 2000.

Guigon, E., Grandguillaume, P., Otto, I., Boutkhil, L., and Burnod, Y., Neural network models of cortical functions based on the computational properties of the cerebral cortex, *J. Physiol. Paris*, 88, 291–308, 1994.

Hazeltine, E., Helmuth, L.L., and Ivry, R.B., Neural mechanisms of timing, *Trends Cognit. Sci.*, 1, 163–169, 1997.

Hinton, S.H. and Meck, W.H., The "internal clocks" of circadian and interval timing, *Endeavour*, 21, 82–87, 1997.

Ivry, R., Cerebellar timing systems, *Int. Rev. Neurobiol.*, 41, 555–573, 1997.

Ivry, R.B., The representation of temporal information in perception and motor control, *Curr. Opin. Neurobiol.*, 6, 851–857, 1996.

Ivry, R.B. and Keele, S.W., Timing functions of the cerebellum, *J. Cognit. Neurosci.*, 1, 134–150, 1989.

Ivry, R.B., Keele, S.W., and Diener, H.C., Dissociation of the lateral and medial cerebellum in movement timing and movement execution, *Exp. Brain Res.*, 73, 167–180, 1988.

Jancke, L., Loose, R., Lutz, K., Specht, K., and Shah, N.J., Cortical activations during paced finger-tapping applying visual and auditory pacing stimuli, *Cognit. Brain Res.*, 10, 51–66, 2000a.

Jancke, L., Shah, N.J., and Peters, M., Cortical activations in primary and secondary motor areas for complex bimanual movements in professional pianists, *Cognit. Brain Res.*, 10, 177–183, 2000b.

Jueptner, M., Flerich, L., Weiller, C., Mueller, S.P., and Hans-Christoph, D., The human cerebellum and temporal information processing: results from a PET experiment, *Neuroreport*, 7, 2761–2765, 1996.

Jueptner, M., Rijntjes, M., Weiller, C., Faiss, J.H., Timmann, D., Mueller, S.P., and Diener, H.C., Localization of a cerebellar timing process using PET, *Neurology*, 45, 1540–1545, 1995 (see comments).

Kawashima, R., Inoue, K., Sugiura, M., Okada, K., Ogawa, A., and Fukuda, H., A positron emission tomography study of self-paced finger movements at different frequencies, *Neuroscience*, 92, 107–112, 1999.

Kawashima, R., Okuda, J., Umetsu, A., Sugiura, M., Inoue, K., Suzuki, K., Tabuchi, M., Tsukiura, T., Narayan, S.L., Nagasaka, T., Yanagawa, I., Fujii, T., Takahashi, S., Fukuda, H., and Yamadori, A., Human cerebellum plays an important role in memory-timed finger movement: an fMRI study, *J. Neurophysiol.*, 83, 1079–1087, 2000.

Koulakov, A.A., Raghavachari, S., Kepecs, A., and Lisman, J.E., Model for a robust neural integrator, *Nat. Neurosci.*, 5, 775–782, 2002.

Larasson, J., Gulayas, B., and Roland, P.E., Cortical representation of self-paced finger movement, *Neuroreport*, 7, 463–468, 1996.

Lejeune, H., Maquet, P., Bonnet, M., Casini, L., Ferrara, A., Macar, F., Pouthas, V., Timsit Berthier, M., and Vidal, F., The basic pattern of activation in motor and sensory temporal tasks: positron emission tomography data, *Neurosci. Lett.*, 235, 21–24, 1997.

Lewis, P.A. and Miall, R.C., Brain activity during non-automatic motor production of discrete multi-second intervals, *Neuroreport*, 13, 1–5, 2002.

Lewis, P.A. and Miall, R.-C., Differential brain activity during the measurement of .6 and 3 seconds, submitted.

Lewis, P.A., Wing, A.M., Pope, P., Praamstra, P., and Miall, R.C., Brain activity correlates differentially with increasing temporal complexity of rhythms during selection, synchronisation, and continuation phases of paced finger tapping, in preparation.

Lutz, K., Specht, K., Shah, N.J., and Jancke, L., Tapping movements according to regular and irregular visual timing signals investigated with fMRI, *Neuroreport*, 11, 1301–1306, 2000.

Macar, F., Lejeune, H., Bonnet, M., Ferrara, A., Pouthas, V., Vidal, F., and Maquet, P., Activation of the supplementary motor area and of attentional networks during temporal processing, *Exp. Brain Res.*, 142, 475–485, 2002.

Malapani, C., Dubois, B., Rancurel, G., and Gibbon, J., Cerebellar dysfunctions of temporal processing in the seconds range in humans, *Neuroreport*, 9, 3907–3912, 1998.

Maquet, P., Lejeune, H., Pouthas, V., Bonnet, M., Casini, L., Macar, F., Timsit Berthier, M., Vidal, F., Ferrara, A., Degueldre, C., Quaglia, L., Delfiore, G., Luxen, A., Woods, R., Mazziotta, J.C., and Comar, D., Brain activation induced by estimation of duration: a PET study, *Neuroimage*, 3, 119–126, 1996.

Matsuzaka, Y., Aizawa, H., and Tanji, J., A motor area rostral to the supplementary motor area (presupplementary motor area) in the monkey: neuronal activity during a learned motor task, *J. Neurophysiol.*, 68, 653–662, 1992.

Meck, W.H., Neuropharmacology of timing and time perception, *Cognit. Brain Res.*, 3, 227–242, 1996.

Meck, W.H. and Benson, A.M., Dissecting the brain's internal clock: how frontal-striatal circuitry keeps time and shifts attention, *Brain Cognit.*, 48, 195–211, 2002.

Medina, J.F., Garcia, K.S., Nores, W.L., Taylor, N.M., and Mauk, M.D., Timing mechanisms in the cerebellum: testing predictions of a large-scale computer simulation, *J. Neurosci.*, 20, 5516, 2000.

Miall, R.C., The storage of time intervals using oscillating neurones, *Neural Comput.*, 1, 359–371, 1989.

Miall, R.C., Neural networks and the representation of time, *Psychol. Belg.*, 33, 255–269, 1993.

Mitriani, L., Shekerdijiiski, S., Gourevitch, A., and Yanev, S., Identification of short time intervals under LSD25 and mescaline, *Act. Nerv. Sup.*, 19, 103–104, 1977.

Morell, V., Setting a biological stopwatch, *Science*, 271, 905–906, 1996.

Nichelli, P., Alway, D., and Grafman, J., Perceptual timing in cerebellar degeneration, *Neuropsychologia*, 34, 863–871, 1996.

Nixon, P.D. and Passingham, R.E., The cerebellum and cognition: cerebellar lesions impair sequence learning but not conditional visuomotor learning in monkeys, *Neuropsychologia*, 38, 1054–1072, 2000.

Onoe, H., Komori, M., Onoe, K., Takechi, T., Tuskada, H., and Watanabe, Y., Cortical networks recruited for time perception: a monkey positron emission tomography (PET) study, *Neuroimage*, 13, 37–45, 2001.

Parsons, L.M., Exploring the functional neuroanatomy of music performance, perception, and comprehension, in *The Biological Foundations of Music*, Vol. 930, Zatorre, R.J. and Peretz, I., Eds., New York Academy of Sciences, New York, 2001, pp. 211–230.

Passingham, R.E., Attention to action, *Philos. Trans. R. Soc. Lond. B Biol. Sci.*, 351, 1473–1479, 1996.

Penhune, B.B. and Doyon, J., Dynamic cortical and subcortical networks in learning and delayed recall of timed motor sequences, *J. Neurosci.*, 22, 1397–1406, 2002.

Penhune, V.B., Zatorre, R.J., and Evans, A.C., Cerebellar contributions to motor timing: a PET study of auditory and visual rhythm reproduction, *J. Cognit. Neurosci.*, 10, 752–766, 1998.

Perrett, S.P., Ruiz, B.P., and Mauk, M.D., Cerebellar cortex lesions disrupt learning-dependent timing of conditioned eyelid responses, *J. Neurosci.*, 13, 1708–1718, 1993.

Petrides, M., Frontal lobes and working memory: evidence from investigations of the effects of cortical excisions in nonhuman primates, in *Handbook of Neuropsychology*, Vol. 9, Boller, F. and Grafman, J., Eds., Elsevier Science, Amsterdam, 1994, pp. 59–82.

Rammsayer, T., Dopaminergic and serotoninergic influence on duration discrimination and vigilance, *Pharmacopsychiatry*, 22 (Suppl. 1), 39–43, 1989.

Rammsayer, T.H., On dopaminergic modulation of temporal information processing, *Biol. Psychol.*, 36, 209–222, 1993.

Rammsayer, T.H., Neuropharmacological evidence for different timing mechanisms in humans, *Q. J. Exp. Psychol.*, 52B, 273–286, 1999.

Rammsayer, T.H. and Lima, S.D., Duration discrimination of filled and empty auditory intervals: cognitive and perceptual factors, *Percept. Psychophys.*, 50, 565–574, 1991.

Rammsayer, T.H. and Vogel, W.H., Pharmacologic properties of the internal clock underlying time perception in humans, *Neuropsychobiology*, 26, 71–80, 1992.

Ramnani, N. and Passingham, R.E., Changes in the human brain during rhythm learning, *J. Cognit. Neurosci.*, in press.

Rao, S.M., Harrington, D.L., Haaland, K.Y., Bobholz, J.A., Cox, R.W., and Binder, J.R., Distributed neural systems underlying the timing of movements, *J. Neurosci.*, 17, 5528–5535, 1997.

Rao, S.M., Mayer, A.R., and Harrington, D.L., The evolution of brain activation during temporal processing, *Nat. Neurosci.*, 4, 317–323, 2001.

Roland, P.E., Skinhoj, E., and Lassen, N.A., Focal activations of human cerebral cortex during auditory discrimination, *J. Neurophysiol.*, 45, 1139–1151, 1981.

Rubia, K., Overmeyer, S., Taylor, E., Brammer, M., Williams, S., Simmons, A., Andrew, C., and Bullmore, E., Prefrontal involvement in "temporal bridging" and timing movement, *Neuropsychologia*, 36, 1283–1293, 1998.

Rubia, K., Overmeyer, S., Taylor, E., Brammer, M., Williams, S.C., Simmons, A., Andrew, C., and Bullmore, E.T., Functional frontalisation with age: mapping neurodevelopmental trajectories with fMRI, *Neurosci. Biobehav. Rev.*, 24, 13–19, 2000.

Sakai, K., Hikosaka, O., Miyauchi, S., Takino, R., Tamada, T., Iwata, N.K., and Nielsen, M., Neural representation of a rhythm depends on its interval ratio, *J. Neurosci.*, 19, 10074–10081, 1999.

Schmidt, R., More on motor programs, in *Human Motor Behavior: An Introduction*, Kelso, J., Ed., Lawrence Erlbaum Associates, Hillsdale, NJ, 1982, pp. 189–235.

Schubert, T., von Cramon, D.Y., Niendorf, T., Pollmann, S., and Bublak, P., Cortical areas and the control of self-determined finger movements: an fMRI study, *Neuroreport*, 9, 3171–3176, 1998.

Schubotz, R.I., Friederici, A.D., and von Cramon, D.Y., Time perception and motor timing: a common cortical and subcortical basis revealed by fMRI, *Neuroimage*, 11, 1–12, 2000.

Schubotz, R.I. and von Cramon, D.Y., Interval and ordinal properties of sequences are associated with distinct premotor areas, *Cereb. Cortex*, 11, 210–222, 2001.

Singh, K.D., Barnes, G.R., Hillebrand, A., Forde, E.M.E., and Williams, A.L., Task-related changes in cortical synchronisation are spatially coincident with the haemodynamic response, *Neuroimage*, 16, 103–114, 2002.

Smith, E. and Jonides, J., Storage and executive processes in the frontal lobes, *Science*, 283, 1657–1661, 1999.

Tracy, J.I., Faro, S.H., Mohamed, F.B., Pinsk, M., and Pinus, A., Functional localization of a "time keeper" function separate from attentional resources and task strategy, *Neuroimage*, 11, 228–242, 2000.

Volz, H.P., Nenadic, I., Gaser, C., Rammsayer, T., Hager, F., Sauer, H., and Volz, H., Processing of temporal information in the human brain assessed with FMRI, *Neuroreport*, 12, 313–316, 2001.

Afterword
Timing in the New Millennium: Where Are We Now?

The vivid and diverse nature of current research on interval timing and the speed with which it has progressed over the last 20 years are superbly illustrated in the present book. By combining animal and human data derived from such complementary approaches as experimental psychology, ecology, neurobiology, and formal modeling, *Functional and Neural Mechanisms of Interval Timing* portrays an impressive state of the art and suggests a bridging among research fields that have developed independently. It therefore leads to a set of questions: What are the consensual issues in research on interval timing at the start of the 21st century? What are the major advances and the ever-pending questions? What are the stumbling blocks that indicate the paths we should take in our future research?

CONSENSUAL ISSUES AND MAJOR ADVANCES

First of all, the book masterfully demonstrates how the scalar timing model, or scalar expectancy theory (SET) (Gibbon et al., 1984), fuels many distinct fields of interval timing research. Initially grounded on animal studies using conditioning schedules for timing (see Church, this volume) and counting (see Brannon and Roitman, this volume), SET has inspired ecological foraging theories and the analysis of other natural behaviors (see Bateson, this volume; Hills, this volume; MacDonald and Meck, this volume), has invaded the ontogenetic and aging fields (see Droit-Volet, this volume; Lustig, this volume), and provides a foundation to the attentional models developed in the frame of human timing (see Fortin, this volume; Pang and McAuley, this volume). As any vivid and comprehensive model, it also elicits alternative and contradictory views (see Crystal, this volume; Hopson, this volume; Malapani and Rakitin, this volume; Matell et al., this volume; Meck, this volume), reminding us of the fragility of consensus, which could someday dissolve. Will appealing new models freed from the pacemaker–accumulator postulate become prominent in the next few years? Future research will decide. Major advances in the field of neural bases will certainly be crucial in selecting the most plausible formal model(s).

Another issue that has received increasing consensus since the 1980s concerns the role of attention in timing performance. It is extensively documented in the present volume (see Buhusi, this volume; Fortin, this volume; Lewis and Miall, this volume; Lustig, this volume; Pang and McAuley, this volume; Penney, this volume) and inspires interpretations even in unexpected experimental contexts (e.g., genetic manipulations — see Cevik, this volume). An impressive set of data has been obtained in dual-task paradigms, showing that compared to full attention conditions, divided attention produces a systematic bias to shorten time estimation, as if the quantity of pulses stored by the accumulator (within the SET framework) were reduced (for neuronal correlates of divided attention in the frontal motor cortex, see Pang and MacAuley, this volume). This bias appears in both humans (for a review, see Brown, 1997) and animals (Lejeune et al., 1999), and cross-species similarities are also evident when gaps or breaks occur during the target interval (see Buhusi, this volume; Fortin, this volume). Dual-task and break studies bear clear analogy: during the target interval, an attention shift toward a concurrent task seems to produce an interruption in the pulse accumulation process (Macar et al., 1994; Rousseau et al., 1984). Similar effects are observed whether an interfering event takes place or is simply expected (Casini and Macar, 1997; see Fortin, this volume). Such data have important implications for modeling the functioning of the accumulator. They indicate that the concept of a flickering switch (Lejeune, 1998; see Penney, this volume; cf. switch modes — see Buhusi, this volume) placed under attentional control during the entire target interval has much greater plausibility than the idea of a switch that merely delimits this interval (Zakay and Block, 1995). Note that interpreting the attentional bias within the framework of accumulator models does not imply that accumulator-free concepts may not account for it. However, in most accumulator-free models, attentional bias has not been considered explicitly, as it should be in view of its high consistency and theoretical importance.

Remarkable advances have also been made concerning the neural bases of timing, due to research on brain lesions and pharmacological manipulations and, obviously, to the development of brain imaging methods (see Hinton, this volume). Two cerebral networks are now identified as having plausible timing functions: the striato-thalamo-frontal pathways involved in dopaminergic systems (see Matell et al., this volume; Malapani and Rakitin, this volume; Meck, 1996, this volume), and the cerebello-thalamo-frontal pathways (see Diedrichsen et al., this volume; Meck, this volume). Strong debates exist between tenants of each system. In this line, the current literature contains incompatible assumptions, such as the possibility that the two networks subserve different components of the timer (Gibbon et al., 1997), or that all components involve tight interactions within single striatal neurons (see Matell et al., this volume).

A FEW PROBLEMS

Admittedly, technical limitations slow progress on the neural bases of interval timing. Current brain imaging methods, however exciting they may be, are unlikely to detect a clock that is supposed to be ticking continuously (see Lewis and Miall, this volume). Or, taken differently, the clock may have been detected long ago, if it consists of aperiodic neural events such as firing bursts in neuronal populations or of oscillatory phenomena that are ubiquitous in organisms (for a review, see Jacklet, 1989) and possibly modulate information processing (Burle and Bonnet, 1999). The point is evidently how these events are selected and stored during the target interval, and how they are repeated on each target occurrence. As a consequence, such processes may be more accessible to experimentation (e.g., encode–decode mechanisms) (see Malapani and Rakitin, this volume). In the same line of reasoning, it is notable that effects on clock speed cannot be detected after genetic manipulations because individuals readapt to their own timing speed (see Cevik, this volume).

Other limitations come from theory rather than methods. Current speculation as expressed in the timing literature often deals with structure and function as if they were inescapably linked. Although such links might be decisive (see Matell et al., this volume), a common timing mechanism may also be subserved by various structures, and various timing mechanisms may coexist in one or several structures. Thus, on many occasions, explicitly disentangling structure and function would help clarify the debate. Taking into account the structural characteristics of the systems discussed may also prove useful. Part of the literature concerning modality effects on timing illustrates these points (see Penney, this volume). Admittedly, postulating that distinct switch–accumulator modules may time signals in each sensory modality implies different neural implementation of core timer components, but does not imply different timing mechanisms. Now, consider, for example, that stimulus coding involves a stage of physico-chemical transduction in visual, but not in auditory receptors; it is, therefore, impossible to decide whether the visual module might be less efficient than the auditory one in encoding time, or whether visual, compared to auditory input, might less efficiently trigger one common timing mechanism implemented in a central structure.

Unfounded distinctions may also be misleading. For example, separating motor tasks from cognitively controlled ones (see Lewis and Miall, this volume) is risky, because a number of motor tasks, even rhythmic ones, require controlled attention or may, at least, be performed with greater accuracy when attention is involved. Cognition is pervasive; it may take place at so-called elementary behavioral levels. Similarly, making a distinction between processes occurring in millisecond vs. multisecond ranges may be questionable, because attentional effects in dual tasks transcend this classification (Macar et al., 1994). A significant outcome of brain imaging reviews on timing is that certain structures are activated irrespective of such distinctions, in particular the supplementary motor area (see Lewis and Miall, this volume; Macar et al., 2002), which, via the thalamus, receives major output from the striatum.

NEW CHALLENGES

Showing that controlled attention may be implicated in temporal parameters and have determinant effects on timing performance was an exciting challenge 20 years ago. Now that this challenge has been won, another, complementary one is forthcoming, namely, the outcome of prolonged learning, which leads to automatic performance. Data obtained in timing (Casini et al., 1999; see Pouthas, this volume) as well as in other tasks suggest that efficient processing may be related to low rather than high levels of cortical activity. Whether this reflects transfer toward subcortical pathways or concentration on narrower, very specific neural foci is an open question. Training in most experimental tasks is generally rather brief. How does cortical activation evolve after extended training, and above all, how do temporal representations evolve? Data on motor control show that bimanual transfer is obtained in early learning phases, but not after long-term practice limited to one hand (Hikosaka et al., 1999). Can we expect analogous effects on timing, suggesting that a novel temporal representation is first available to different effectors and sensory modalities, but becomes specific to the behavioral context and to the motor program it created once performance is automatized?

The role of context itself is another urgent question, possibly underestimated so far. Context seems to have a marked influence on timing (e.g., the effects of gaps and modality — see Buhusi, this volume; Penney, this volume), as on many other types of processing. A meta-analysis considering contextual parameters may enlighten interpretation of various data sets, in the same manner as attentional findings have stimulated new insights into various theoretical frameworks, and may continue doing so if they are systematically taken into account in forthcoming research (for new questions raised by the dopaminergic involvement in attention sharing, see Buhusi, this volume). An interesting approach would also be to consider how context might have determined the level of attention paid to relevant vs. irrelevant parameters in a number of tasks.

With respect to brain structures, a current issue that is likely to be clarified in the near future is whether the striato-frontal and the cerebello-frontal circuits have complementary functions in timing tasks. A related question lies in the sequential activation of the anatomical pathways involved in timing, not only between the distinct functional modules of timing models, but even within each module itself. Attempts to describe activation sequences during information processing are still rather timid, but might soon be successful, thanks to a combination of brain imaging techniques and their continuous improvement (e.g., the increasingly fine temporal resolution of functional magnetic resonance imaging (fMRI) and its co-registration with electroencephalography (EEG)) (see Hinton, this volume; Sakata and Onoda, this volume). Finally, separating the components of the timer, and relating each one to dedicated brain structures, is currently a hot topic, though by no means easy to solve (see Lewis and Miall, this volume; Rao et al., 2001) or even mandatory (see Matell et al., this volume). This raises particularly delicate problems when considering that certain components may unfold simultaneously (as for comparison and encoding during a test interval) or may have a very brief time course (as for decision).

PERSISTING QUESTIONS

Especially tricky is the question of whether timing involves a single dedicated mechanism or different mechanisms that can be selected on the basis of duration ranges, current context, or other parameters. Timing in the microsecond range is known to involve delay lines, at least in barn owls, bats, and electric fish (for a review, see Carr, 1993). Coincidence detection mechanisms are considered more suitable to longer-duration ranges (see Mattel et al., this volume; Meck, this volume; Miall, 1992). In fact, various possibilities that account adequately for temporal encoding seem plausible in view of the great diversity of functional phenomena displayed by the central nervous system, whether these phenomena consist of oscillations, multiple convergent inputs, spatio-temporal potentiation, or sustained changes in membrane potentials, among others. The argument of biological plausibility is primordial in selecting models of timing mechanisms. Surprisingly, it is most often evoked in favor of accumulator-free models (e.g., coincidence detector mechanisms fitting with striatal architecture — Matell and Meck, 2000; see Meck, this volume; Meck and Benson, 2002; or delay lines fitting with cerebellar architecture — Moore, 1992), although accumulator-like mechanisms can be traced at various organic levels. For example, increased activity as a function of time is typically observed in neuronal or electroencephalographic recordings during motor preparation (for a review, see Requin et al., 1991). Moreover, in temporal production and discrimination tasks, the level of activity in certain cortical areas is higher when subjects overestimate rather than underestimate the target (Macar and Vidal, 2002; Macar et al., 1999). These data are entirely compatible with the hypothesis that in dedicated timing networks, the level reached in some type of accumulator mechanism determines subjective duration.

Now, how do we reconcile the idea of an accumulator under attentional control and that of implicit timing, pervasive whenever temporal regularities exist in the environment (see Hinton, this volume)? An interesting possibility is that we have a unique timing system that is activated in all cases, but the output of which is labile, and rapidly fades away from memory whenever we process nontemporal parameters (Zakay, 1989), as in many everyday life situations. In these cases, timing judgments would then be based on any information less sensitive to fast decay, such as the amount of work performed during a past interval. Behavioral data showing that in dual-task paradigms the directional error bias typical of prospective judgments (implying preliminary awareness of the timing task) is reversed for retrospective ones (obtained without preliminary warning) strongly suggest that it is not the timer output that is used in the latter case. The idea that the timer is nevertheless activated systematically, even when its output cannot be used, is strengthened by considering the ubiquity of anticipatory behavior. The rule is that a response to a significant event is primed by the earliest index that announces this event (cf. classical conditioning, which reflects basic computational mechanisms displayed by complex as well as very simple organisms). Anticipation supposes the coding of temporal order and leads to accurate timing after the repetition of relatively few trials, even when there is no need for accuracy in the current context (cf. Gallistel and Gibbon, 2000, 2001). Why, then, would timing not be subtended by a nonspecialized population of neurons? Subsets of these neurons might subsequently be selected depending on the prevailing context (see Hopson, this volume).

Certainly, this hypothesis does not reflect the dominant trend in the literature; rather, the timer is thought to be located within key structures of the central nervous system. However, even the dopaminergic effects on timing, which clearly highlight the role of striatal structures, might be viewed as reflecting the pervasive influence of dopaminergic systems on all basic functions, rather than as an argument for specific localization of a central timer in the striatum. Following this line of reasoning, choosing a general mechanism, such as learning, as the principle that guides the development of temporal dynamics within the neural network as a whole is certainly appealing (see Hopson, this volume), though error gradient backpropagation mechanisms seem farther from biological reality than simpler learning algorithms, such as reinforcement (Watkins, 1989) or self-organizing maps (Kohonen, 1987). An even less demanding hypothesis might be to transpose the results achieved in robotics with the automatic synthesis of behaviors to the temporal dimension. Here, no learning algorithm is required, because learning is a border effect, resulting from the mere repetition of behavior (Touzet, 1999).

In sum, although the outcome of research on the neural bases of interval timing, as well as the new propositions it has inspired, is to be counted among the most significant advances of the last decade or so, the debate on whether a central timer can or cannot be identified is likely to persist for the foreseeable future. Perhaps this apparent paradox comes from the fact that the neural structures that are now acknowledged as crucial in temporal processing are involved in multiple functions and subtend elementary to highly complex behaviors. Progress on the functional organization of the brain has likely rendered archaic the idea of localizing the internal clock in a single structure devoted solely to interval timing.

ACKNOWLEDGMENTS

The author is grateful to Claude Touzet for his comments on formal models and to Jennifer Coull for her editorial assistance.

REFERENCES

Brown, S.W., Attentional resources in timing: interference effects in concurrent temporal and nontemporal working memory tasks, *Percept. Psychophys.*, 59, 1118–1140, 1997.

Burle, B. and Bonnet, M., What's an internal clock for? From temporal information processing to temporal processing of information, *Behav. Process.*, 45, 59–72, 1999.

Carr, C.E., Processing of temporal information in the brain, *Annu. Rev. Neurosci.*, 16, 223–243, 1993.

Casini, L. and Macar, F., Effects of attention manipulation on perceived duration and intensity in the visual modality, *Mem. Cognit.*, 25, 812–818, 1997.

Casini, L., Macar, F., and Giard, M.-H., Relation between the level of prefrontal activity and subject's performance: a comparison between a timing and a semantic task, *J. Psychophysiol.*, 13, 117–125, 1999.

Gallistel, C.R. and Gibbon, J., Time, rate, and conditioning, *Psychol. Rev.*, 107, 289–344, 2000.

Gallistel, C.R. and Gibbon, J., Computational versus associative models of simple conditioning, *Curr. Directions Psychol.*, 10, 146–150, 2001.

Gibbon, J., Church, R.M., and Meck, W.H., Scalar timing in memory, in *Annals of the New York Academy of Sciences: Timing and Time Perception*, Vol. 423, Gibbon, J. and Allan, L., Eds., New York Academy of Sciences, New York, 1984, pp. 52–77.

Gibbon, J., Malapani, C., Dale, C.L., and Gallistel, C.R., Toward a neurobiology of temporal cognition: advances and challenges, *Curr. Opin. Neurobiol.*, 7, 170–184, 1997.

Hikosaka, O., Nakahara, H., Rand, M.K., Sakai, K., Lu, X., Nakamura, K., Miyachi, S., and Doya, K., Parallel neural networks for learning sequential procedures, *Trends Neurosci.*, 22, 464–471, 1999.

Jacklet, J.W., Ed., *Neuronal and Cellular Oscillators*, Basel & Dekker, New York, 1989.

Kohonen, T., *Self-Organization and Associative Memory*, 2nd ed., Springer Series in Information Sciences, Vol. 8, Springer Verlag, Berlin, 1987.

Lejeune, H., Switching or gating? The attentional challenge in cognitive models of psychological time, *Behav. Process.*, 44, 127–145, 1998.

Lejeune, H., Macar, F., and Zakay, D., Attention and timing: dual-task performance in pigeons, *Behav. Process.*, 45, 141–157, 1999.

Macar, F., Grondin, S., and Casini, L., Controlled attention sharing influences time estimation, *Mem. Cognit.*, 22, 673–686, 1994.

Macar, F., Lejeune, H., F., Bonnet, M., Ferrara, A., Pouthas, V., Vidal, F., and Maquet, P., Activation of the supplementary motor area and of attentional networks during temporal processing, *Exp. Brain Res.*, 142, 539–550, 2002.

Macar, F. and Vidal, F., Time processing reflected by EEG surface Laplacians, *Exp. Brain Res.*, 145, 403–406, 2002.

Macar, F., Vidal, F., and Casini, L., The supplementary motor area in motor and sensory timing: evidence from slow brain potential changes, *Exp. Brain Res.*, 125, 271–280, 1999.

Matell, M.S. and Meck, W.H., Neuropsychological mechanisms of interval timing behaviour, *Bioessays*, 22, 94–103, 2000.

Meck, W.H., Neuropharmacology of timing and time perception, *Cognit. Brain Res.*, 3, 227–242, 1996.

Meck, W.H. and Benson, A.M., Dissecting the brain's internal clock: how frontal-striatal circuitry keeps time and shifts attention, *Brain Cognit.*, 48, 195–211, 2002.

Miall, R.C., Oscillators, predictions and time, in *Time, Action and Cognition: Towards Bridging the Gap*, Macar, F., Pouthas, V., and Friedman, W.J., Eds., Kluwer Academic Publishers, Dordrecht, Netherlands, 1992, pp. 215–227.

Moore, J.W., A mechanism for timing conditioned responses, in *Time, Action and Cognition: Towards Bridging the Gap*, Macar, F., Pouthas, V., and Friedman, W.J., Eds., Kluwer Academic Publishers, Dordrecht, Netherlands, 1992, pp. 229–238.

Rao, S.M., Mayer, A.R., and Harrington, D.L., The evolution of brain activation during temporal processing, *Nat. Neurosci.*, 4, 317–323, 2001.

Requin, J., Breener, J., and Ring, C., Preparation for action, in *Handbook of Cognitive Psychophysiology: Central and Autonomic Nervous System Approaches*, Jennings, J.R. and Coles, M.G.H., Eds., Wiley & Sons, New York, 1991.

Rousseau, R., Picard, D., and Pitre, E., An adaptative counter model for time estimation, in *Annals of the New York Academy of Sciences: Timing and Time Perception*, Vol. 423, Gibbon, J. and Allan, L., Eds., New York Academy of Sciences, New York, 1984, pp. 639–642.

Touzet, C., Programming Robots with Associative Memories, paper presented at Proceedings of International Joint Conference on Neural Networks, Washington, D.C., July 10–16, 1999.

Watkins, J.C.H., Learning from Delayed Rewards, Ph.D. thesis, King's College, Cambridge, England, 1989.

Zakay, D., Subjective time and attentional resource allocation: an integrated model of time estimation, in *Time and Human Cognition: A Life Span Perspective*, Levin, I. and Zakay, D., Eds., North Holland, Amsterdam, 1989, pp. 365–397.

Zakay, D. and Block, R., An attentional-gate model of prospective time estimation, in *Time and the Dynamic Control of Behavior*, Richelle, M., De Keyser, V., d'Ydewalle, G., and Vandierendonck, A., Eds., Presses de l'Université, Liège, Belgium, 1995, pp. 167–178.

Françoise Macar
Centre National de la Recherche Scientifique
Laboratoire de Neurobiologie de la Cognition
Marseille, France
October 7, 2002

Index

A

Accumulator, 496. *See also* Pacemaker-accumulator systems
  models, 527
  modules, 223
  in scalar timing theory, 8–9
Accuracy, 446
Acetylcholine, xxvi
Adaptive behavior, xx, 115
African fish eagles, 99
African mouthbreeder, 84
Age, sensitivity to signal duration and, 193
Albin-DeLong model, 501
Alces alces, 101
Alligators, 99
L-alpha-amino-3-hydroxy-5-methylisoxazole-4-propionate, 505
Alzheimer's disease, xxi, xxxi
American osprey, 101
Animal timing, vii, 78
  assay, 79
  behaviors associated with, 85
    ecological studies, 89
  fundamental mechanisms, 85
Anterior cingulate cortex, 449, 519
Anterior forebrain pathway, 398, 399, 402
  circuitous root, 409
  interaction between auditory template and, 410
Apis mellifera, 84
Aquila chrysaetos, 101
Arabidopsis, 85
Arbitrary numeron hypothesis, 161
  principles, 161–162
    cardinality, 162
    one-to-one, 161
    order irrelevance, 162
    stable-order, 161
Area-restricted search, 94
Arithmetic reasoning. *See* Number(s); Numerosity(ies)
Artificial intelligence, 43
Artificial neural network, 54
Assignment-of-credit problem, 43, 44
Associative learning, xx, 299
Asynchrony, 476

Attention, viii, xxi, 280, 351
  age differences, 264
  automatic aspects, 265
  circadian influence, 88
  CNV and, 440
  controlled, 265, 282
  deficit. *See also* Attention deficit hyperactivity disorder
    signal duration and, xix
  definitions, 352
  divided, 352, 358
    neurons, 359
  dopamine and, 321
  in older adults, 266, 276
  models, 203
  neuropsychological evaluation, 282
  selective, 352
  switch hypothesis, 357, 359
  time-sharing, 254, 325
  working, 283
Attention deficit hyperactivity disorder, 309
  model, 309, 312
Attentional switch hypothesis. *See* Attention, switch hypothesis
Auditory numerical discrimination, 149
Auditory stimuli, vii, 213
  ERP, vii, 213
  response time, 220
  visual *vs.,* 196, 214, 276–277
Autopsy, 420
  invasiveness, 421
  limitations, 421
  repeatability, 421
Avian communication, 96

B

Backpropagation algorithm, 36, 57
Basal ganglia, xxxi, 459, 467, 526
  cognitive function, 501
  damage, xxi, xxxi
  dysfunction model, 459
  information processing, xxix
  model, 501, 502
  neuroimaging, 520, 521
  timing in, 506

541

Basal ganglia-thalamic outflow, 449
Bayesian foraging behavior, 100
Beat frequency model, 491
Behavior, foraging. *See* Foraging
Behavior, timing and, vii
Behavioral ecology, 114
Behavioral theory of timing, 5, 25, 30
   connectionist models, viii, 26, 30–31
   diffusion model, 26
   models, viii, 26, 30–31
   revised connectionist variant, 26
   scaling in, 53
   striatal timing process *vs.*, 385–386
   time steps proportional to reinforcement interval, 53
Bird(s)
   foraging, 116
   Hermit hummingbird, 118
   hummingbird(s), 116, 118
   songs (*See* Birdsongs)
   vocal learning, 395
      auditory feedback and, 403
      comparison signal and, 404, 405, 408
      crystallization of song type, 395
      sensorimotor integration phase, 395
      sensory phase, 395
      telencephalon areas associated with, 398
Bird's own song, 402, 410
Birdsongs, 394
   bouts, 394
   defined, 395
   motifs, 394
   notes, 394
   repertoire, 394
   syllable, 394
      as time marker, 396
   types, 394
Blood oxygenation level-dependent (BOLD) contrast, 425
Bradyphrenia, 490
Brain damage, xxvii
Brain imaging studies, 358
Break-and-run response, 79, 84
Break location effect, 237, 534
   attentional shifts and, 242
   in cue and non-cue conditions, 250, 251
   defined, 238
   expectancy and, 252, 253
   experimental confirmation, 247
   in high-frequency conditions, 243, 245, 247
      pulse accumulation, 247
   in low-frequency conditions, 243, 246, 247
      pulse accumulation, 247
   prebreak duration and, 255
   rate of pulse accumulation and, 242
   time discrimination, 245, 247
   time-sharing interpretation, 240
Broadcast theory of timing, 72

C

Caenorhabditis elegans, xx, 90, 94
Canary, 396
Canis lupus, 101
Carassius auratus, 83, 84
Catechol-o-methyltransferase, xxvi
Caudal nucleus of the neostriatum, 403
Central pattern generators, 517, 518
Cerebellum, xxx, xxxi, 339, 459, 467, 526
   blood flow, xxvi
   lesions, xxx–xxxi
   neuroimaging, 520, 521
   right anterior, 466
Cerebral cortex, xxx, 339
   anterior cingulate, 449, 519
   attentional system, 519
   blood flow, xxvi
   cortical-basal ganglia-thalamo-cortical circuit, 448
   cortico-striato-thalamo-cortical circuit, 385
   damage, xxx, 357, 359
   dorsolateral prefrontal, xxx, 409, 448, 519
      neuroimaging, 521
   electrophysiological recordings, 360
   frontal, 519
   parietal, 519
   premotor, 519, 521, 526
      neuroimaging, 521
   reference memory and, 357
   theta rhythm activity, xxviii, 344–345
   ventrolateral prefrontal, 520, 521
Chemotaxis, 97
Children, xxiii
   generalization gradients, 192
   internal clock in, 200
   scalar timing theory model adapted to, 187
   temporal bisection, 183. *See also* Temporal bisection
   temporal bisection procedure, 186, 188
   temporal tasks, 183
Chimpanzee, 148, 149, 150
Choice, 4
Cicadas, 99
Circadian timing, xv, 65–69, 225
   age-related differences, 279
   analysis, 69
   attention and, 88
   light and, 86
   melantonin and, 88

Index 543

memory and, 88
molecular circuitry, xv
oscillator, 66, 69
pacemaker, 90
period gene, 298
temperature and, 87
ultradian behaviors and, 87
Classical conditioning, 4
Clock(s), 25–29, 263, 525
goal, 55
learning model and, 24
location, 537
pacemaker-accumulator systems, 28
single protein, 93
speed
dopamine and, 320, 321, 333, 487
temperature and, 83
state-based, 25
trace-based, 27
Clock gene, 88, 298
Clock variability, 461
Cognition, vii
aging and, 285–288
development of, xxiii
Combination sensitivity, 402
Computed tomography
with contrast, 422
energy source, 421
invasiveness, 421
limitations, 421
physiological measure, 421
repeatability, 421
risks associated with, 422
spatial resolution, 421
without contrast, 422
energy source, 421
invasiveness, 421
limitations, 421
physiological measure, 421
repeatability, 421
spatial resolution, 421
Computer simulations, 11–12, 20–22
Conditioned stimulus, xx
Conditioned stimulus-unconditioned, 102
Conditioning, xx
real-time, 5
standard *vs.* real-time, 5
theories, 5
Connectionist models, viii, 26
behavioral theory of timing, 30–31
neural mechanisms involved, 30
oscillator rate, 54
strengths, 30
Contingent negative variation (CNV)
amplitude

in discriminative tasks, 443
in Parkinson's disease, 445
increase during progressive learning, 442
variations in, 441
attention and, 440
biphasic, 440
defined, 440
effect distractions on, 442
ERP analysis, 447–448, 449–452
expectancy and, 440
generators, 448
ERP and PET analyses, 449–452
nonmotor, SPN *vs.*, 452
in Parkinson's disease, 445, 448–449
PET analysis, 447–448, 449–452
resolution, 445
degree of error and, 446
formation of temporal judgment and, 446
temporal processing in, 440, 445
terminal, 440
Continuous reinforcement schedule, 44
Corophium spp., 118
Corophium volutator, 118
Cortical-basal ganglia-thalamo-cortical circuit, 448
Cortico-striatal modules, 398, 400
parallel, 405, 406
as pattern detector, 406
Cortico-striato-thalamo-cortical circuit, xxx, 385
Counting model, xxiv
mode control, xxiv
Cross talk, cortico-striato-thalamo-cortical circuit, 385
Cross-validation, 16
Crustaceans, 118
Cyclic AMP, element binding transcription factor, 302

D

Death-feigning behavior, 98
Decay constant, 39
Decay hypothesis, 321
Decay timers, time fields *vs.*, 94
Decision making, viii
Deep brain stimulation
L-dopa *vs.*, 500
mechanism of action, 503–504
Delayed match-to-sample processes, 386
Discriminative tasks, xxi, 116–117
break location effect in, 245, 247
duration, CNV amplitude in, 443
EEG during, 340
ERP and, 342

numerical, 149, 153, 154, 157
stimuli, interfood interval as, 116–117
Dopamine system, 226, 284, 356
asynchrony and, 476
attention and, 321
clock speed and, 320, 321, 333, 487
deep brain stimulation vs., 486
evidence for role in attention-sharing, 329
feedback and activity of, xxix
genetic modification, xxv, 308
in PD, 489
Dopamine transporter, 308
Dopaminergic neurons, midbrain, 397, 398, 404, 405
Dorsolateral anterior thalamic nucleus, 399
Dorsolateral prefrontal cortex, xxx, 409, 448, 519
neuroimaging, 521
Doves, parent, 98
Down's syndrome, xxxi
Drosophila melanogaster, 83, 86, 94, 298, 304
pattern sensitive neurons, 96
Drug abuse, xxv
Dual-task conditions, 266, 322, 352, 534

E

Electroencephalography, 340, 422–423, 525
alpha frequency, 439, 440
of CNV, 449
energy source, 421
fMRI and, 536
invasiveness, 421
learning and, 343
limitations, 421, 423
MEG vs., 423
PET and, 449
physiological measure, 421
repeatability, 421
spatial resolution, 421
temporal resolution, 421
Encode-decode mechanisms, 489, 535
Energy intake, foraging and long-term net rate of, 117
Episodic memory, 91
ERP. *See* Event-related potential (ERP)
Errors(s), 43, 446
correction model for synchronization studies of PD, 475
dopamine-dependent, 487
general timing model, 35–36, 43
Escherichia coli, 94, 97
Ethology, 114
European starling, 116, 120, 128, 133–134
Event-related desynchronization, 440

Event-related potential (ERP), xxi, 170, 171
from auditory stimuli, 342
in CNV, 447–448
cognitive *vs.* motor, 448
defined, 340
in duration discriminative tasks, 444
motor *vs.* cognitive, 448
primary source, 448
recording of, 341
Expectancy, 252, 253
attentional shifts and, 255
CNV and, 440
theory, 29, 123
uncertainty and, 251
Eye blink classical conditioning, xxxi

F

Falco tinnunculus, 89
Feedback, 275, 403, 487
anticipation of, xxvii
dopaminergic and, xxix
intertrial interval, xviii
Filled-gap simulation, 51
Fixed-interval procedure, xv, xvi, 12–13, 79, 354
defined, 39
history, 39
initialization, 12
as rapid behavioral screen, 300
simulation, 41, 42
Flycatchers, spotted, 119
Food -anticipatory behavior, 67
Foraging, 115
basis of analysis, 116
birds, 116
energy intake and, 117
insect, 101
marginal value theorem in, 130
optimal theory of, 99, 117–123
models, 117, 137
Yoccoz model and, 122
patch residence time, 120, 128
risk sensitive, 136
temporal factors in, 101, 119
temporal predictability of food source and, 118
time-based moving-on rule, 119
universality, 116
frq expression, 87
Fruitfly, 83
Functional magnetic resonance imaging (fMRI), 171, 425–426, 524, 525
advantages, 425
BOLD signal in, 425, 426, 433

Index 545

EEG and, 536
energy source, 421
invasiveness, 421
limitations, 421, 428
physiological measure, 421
region-of-interest analysis in, 432
repeatability, 421
spatial resolution, 421
temporal resolution, 421
Fusiform gyrus, left, 449

G

Gap procedure, 322, 324
  attention sharing in, 325, 326
  behavioral manipulations nontemporal parameters, 328
  comparative studies, 327
  intensity, 329
Gap timing, 45, 55
  fixed duration simulations, 50
  fixed offset simulations, 48, 49
  fixed onset simulations, 47, 48
  lack of learning in, 49
  parametric study of, 47
  shift in peak response time in, 45
  stop *vs.* reset pattern, 45
Gene expression, 87
Gene knockouts, 305
  behavioral phenotyping, 307–308
  dopamine-transporter, 308
  inducible and tissue-specific, 306–307
Gene targeting, 305
  quantitative trait loci, 313
Generalization gradients, 188, 189, 190, 191
  animal *vs.* human adult, 192
  children *vs.* adult, 192
Genetic algorithms, 56
Germline mutagens, 298
Glaucous gulls, 99
Globus pallidus, 397, 500
Golden eagles, 101
Goldfish, 83, 84
Goodness-of-fit, 14, 15
Guillemot fledglings, 99

H

Habituation, 299, 300
  neural basis, 300
Haliatus vocifer, 99
Haloperidol, xxix, 221, 332
Hermit hummingbird, 118

Hippocampus, 91, 339
  as working memory buffer, 344
  electroencephalography, 92
  long--term potentiation in, 102
  neurophysiology, 91–92
  reference memory and, 357
  role, 103
  synapse, 102
  theta wave, xxviii, 344–345
Honeybees, 84, 88
  waggle dance, 97
Hummingbird(s)
  hermit, 118
  rufous, 116
  wild, 118
Huntington's disease, xxi, xxxi, 305

I

Indirect striatal output pathway, 498
Inferior parietal lobule, right, 449
Information-processing, xix
  basal ganglia and, xxix
  elements, xxix
  model, 263, 340
    internal clock, 319, 320
    major components, brain areas corresponding to, 339
Insect communication, 96
Intertrial interval, xviii, 41, 116
  feedback enhancement, xviii
Internal clock, 24, 194. *See also* Clock(s)
  animal *vs.* human, 318
  cognitive models, 318, 319
  evolution of, 114–115, 116
  historical perspectives, 318
  human *vs.* animal, 318
  information-processing model, 319, 320
  in children, 200
  in older adults, 267
  in schizophrenia, 226
  models, 195–196, 236, 318
  speed, 196, 267
  time base of, 439
Internal marker hypothesis, 212
Interval timing
  adaptation and, xx
  attention and, 264. *See also* Attention
  in cognitive development, xxiii
  defined, 24
  dopamine and, 486–488
  drug abuse and, xxv
  EEG correlation, 340
  example, xv

experiments with older adults, 277
  problems with, 286
fMRI studies, 427
inaccuracy and imprecision in, 114
learning and, xx
long, 64–65
models, 209–210, 372–374
  basic functions, 124
  hippocampal role in, 345–346
  memory-based, 211
  process, 459
  reference memory component, 283
multiple systems, 516
mutants, 301
neural basis, xxviii
neuroimaging, 426
neuropharmacology and, xxv
performance on neuropsychological tests and, 285
procedures, 4
quantitative and information processing account, xix
scalar features, 54, 61
short, 61, 63, 352
  oscillator interpretation, 63
striatal role in, 387
temperature and, 83
ubiquitousness, 57–58
use, xv

J

Japanese macaque, 103

K

Kestrels, 89

L

Larus hyperboreus, 99
Lateral magnocellular nucleus of the anterior neostriatum, 399, 402, 409
Latin square design, 10
Learning, xx, 411
  algorithm, 35, 36, 132, 133
    backpropagation algorithm *vs.,* 36
    modification, 54
  associative, xx, 299
  avian vocal, 395
    auditory feedback and, 403
    comparison signal and, 404, 405, 408
    crystallization of song type, 395

sensorimotor integration phase, 395
sensory phase, 395
telencephalon areas associated with, 398
CNV amplitude during progressive, 442
constant
  in nonreinforcement, 40
  in reinforcement, 40
during reinforcement, 40
EEG and, 343
hippocampal role in, 91
lack of, gap timing in, 49
model, 2, 5, 315
  clocks and, 24
  criteria, 24
nonassociative, 299
Lever pressing, 383
Linear-timing hypothesis, 62
  target intervals to test, 63
Local maxima, 62
Long interval timing, 64–65
  assessment of, 65
Long-term energy intake, 117
Long-term potentiation-mediated plasticity, 406

M

Macaca fuscata, 103
Macaque, Japanese, 103
Magicicada spp., 99
Magnetic resonance imaging, xxi, 422
  advantages, 422
  energy source, 421
  functional. *See* Functional magnetic resonance imaging
  invasiveness, 421
  limitations, 421
  physiological measure, 421
  repeatability, 421
  safety concerns, 422
  spatial resolution, 421
Magnetoencephalography, 423, 525
  EEG *vs.,* 423
  energy source, 421
  invasiveness, 421
  limitations, 421
  physiological measure, 421
  repeatability, 421
  spatial resolution, 421
  temporal resolution, 421
Marginal value theorem, 120, 130
Marijuana, xxvi
MATLAB, 11–12, 20–22
  code, 20–22

Index

MATLAB code, 492
Melantonin, 88
Memory, viii, xxi, 80, 187
   age-related differences in, 284, 285
   basal ganglia and, 488–489
   circadian influence, 88
   electrical dissection of formation, 89–90
   embedded, context-specific, 3
   encoding process, 199, 522
   episodic, 91
   hippocampal role in, 91
   in older adults, 273
   mutant, 301
   reference, 138, 282, 283
      age-related differences in, 284
      anatomy of, 357
      defined, 133
      representation for options, 134, 135
      representation of different temporal stimuli in, 125
   retrieval, 9
   semantic, 91
   spatial, 91
   storage, 9, 13
      constant, xxvii
      speed, xxvii
   working, 344
Methamphetamine, xxix, 196, 221, 321, 332
   abuse, xxvi
N-methyl-D-aspartate, 102, 409, 505
Methylphenyltetrahydropyridine, 501
Midbrain dopaminergic neurons, 397, 398, 404
Migration effect, 491, 497
Mismatched negativity, 170
Mode-control model, xxiv, 163–164
   event mode, 163
   neural network model *vs.*, 167
   stop mode, 163
Monkeys, 151
Moose, 101
Motor system, 517, 526
   tasks, cognitively controlled *vs.*, 535
mRNA, 85
Multiple-oscillator theory of timing, 5, 72, 526
Multiple switch-accumulator framework, 222
Multiple-timescale model, 5, 27
Mus musculus, 304
Muscicapa striata, 119
Mutagenesis, 298
   inducible, 307
   saturation, 299
   tissue-specific, 307
Mutant(s), 300
   isolating interval timing, 301
   memory, 301

N

N-methyl-D-aspartate, 102, 409, 505
Neural networks, 33
   artificial, 54
   model, 165–166
      mode-control model *vs.*, 167
   number representation and, 165–166
   recurrent, 33, 34
Neurodegenerative diseases, xxxi
Neuroimaging studies, 519. *See also* Positron emission tomography*, etc.; specific techniques, e.g.,* Functional magnetic resonance imaging
   limitations, 534–535
   meta-analysis, 520, 521, 523
   significant outcome, 535
Neuropharmacology, xxv
Neurospora, 85
Neurospora crassa, 86
Neurotoxicants, xv
Nonassociative learning, 299
Number(s), 144. *See also* Numerosity(ies)
   abstract concepts, 149
   addition, 155, 159, 160
   auditory discrimination and, 149
   discrimination, 157
      auditory, 149
      spontaneous, 154
      Weber's law and, 153
   large, 157
   object-file representation, 163
   production paradigm, 147
   representation, 147
      arbitrary numeron hypothesis, 161–162
      connectionist timing model, 164–165
      mode-control model, 163
      neural basis of, 170
      neural network model, 165–166
      nonverbal, 161–166
      object-file, 163
      running speed and, 148
      subtizing hypothesis, 163
   subtraction, 155, 159, 160
Numeron, hypothesis, 161. *See also* Arbitrary numeron hypothesis
Numerosity(ies)
   absolute, 145
   development and, 168
   in chimpanzees, 148, 149
   in human infants, 155
   in monkeys, 151–153
   in pigeons, 146
   in raccoons, 146
   in rats, 145, 149

making relative judgments, 150
ordinal relationships between, 159
single, 145
symbols and, 148

O

Object-file model, 163
  subtizing hypothesis vs., 166
One-to-one principle, 161
Operant conditioning, 4
Operant psychology, 114
Optical recording, 424–425
  energy source, 421
  invasiveness, 421
  limitations, 421
  physiological measure, 421
  repeatability, 421
  spatial resolution, 421
  temporal resolution, 421, 424
Optimal foraging theory, 99
Optimality model, 137
Order selectivity, 402
Oscillator(s)
  circadian timing, 66, 69
  interpretation of short interval timing, 63
  multiple, 5, 72, 526
  rate, connectionist models, 54
Oscillator models, viii
  modified, 30
Osprey, 95, 101

P

Pacemaker, 262
  amodal central, 222
  circadian timing, 90
  counter models, vii, viii
  in scalar timing theory, 8
  rate, effect of drugs on, 356
Pacemaker-accumulator systems, 8, 28, 186, 210
  model, 491, 492
    pros and cons, 496
  neural and neurochemical substrates, 221
  striatal implementation, 487
Pallidum, 397, 398, 501
Pallium, 397
Pandion haliaetus, 101
Parasitoid wasp, 83
Parent doves, 98
Parietal cortex, 519
Parkinson's disease, xxi, 284, 445, 448–449
  deep brain stimulation in, 497

dopaminergic system in, 489
model, 502
  as model for basal ganglia dysfunction, 459
subthalamic nucleus overactivity in, 498, 501, 504–506
time perception tasks and, 462, 463
Parrot, 148
Parsimonious timing, 43–44, 45
Parus caeruleus, 116
Pavlovian temporal conditioning, 4
Peak-interval procedure, xviii, xix, 41, 220, 278, 319, 341, 487. *See also* Gap timing
  description, 354
  intertrial interval, 41
  simulations, 42, 43
Peak time, xvi
  after cerebral cortex damage, 357
  following hippocampal damage, 357
  in gap trials, 322
  reward and, 81
  shift in, 45, 82
per expression, 87
Periodic behavior, 63
Phaethonis superciliosus, 118
Physostigmine, xxvi
Pigeons, 95, 130–131, 136
  numerosity in, 147, 150
Pineal gland, 88
Point of subjective equality, 169
Poisson pacemaker, 198
Positron emission tomography, 171, 423–424, 524, 525
  of CNV, 447–448, 449–450
  EEG and PET, 449
  energy source, 421
  invasiveness, 421
  limitations, 421, 423, 449–450
  physiological measure, 421
  repeatability, 421, 423
  spatial resolution, 421, 423
  SPECT vs., 424
  temporal resolution, 421
Posterior motor pathway, 399
Precision, 446
Predator avoidance, 98–99
Predator-satiating mechanism, 99
Predatory behavior, 95
Prefrontal cortex, right, 449
Probabilistic models, viii
Probe trials, 81
Proportional training, 6
Psychoactive drugs, xv

Index 549

Q

Quantitative trait loci, 313

R

Raccoons, 146
Rats, 89, 145
Real-time conditioning, 5
Redshanks, 118
Region-interest analysis, 432
Reinforcement
  fixed-interval schedules, 7, 55, 116 (*See also* Fixed-interval procedures)
  learning constant during, 40
  rate of, 84
Repetitive tapping, 466. *See also* Synchronization
Response rate, as function of relative time, 15
Reward(s)
  peak time and, 81
  ratio schedules and, 137
Risk-sensitive foraging, 100
Risk sensitivity, 133–134
Robustus archistriatalis, 399, 401
  muscle map, 401
Rufous hummingbird, 116

S

Scalar expectancy theory, xix, 29, 123, 353
  accumulator, 496
  alternative views, 533
  central tenet, 264
  cognitive model, 25
  components, 354
  contradictory views, 533
  information processing elements, xxix
  limitations, 496, 497
  mathematical formulations, 282
  migration effect, 491
Scalar property, 6
  failure, 53
  simulation, 52
Scalar timing theory, vii, xix, xxvii, 3–22. *See also* Scalar expectancy theory
  basis, 194
  defined, 123
  features, 70, 72
  learning algorithm, 132, 133
  modality-specific switch-accumulator modules, 223
  model, 10, 127
    adapted to children, 187
  advantages of, 126
  application to foraging, 126
  application to optimal foraging problems, 126
  assumptions of, 138
  flexibility, 138
  initialization of, 12
  predictions and preferences from, foraging, 136
  three-stage information-processing, 211
  modification, 134, 187, 223
    choice scenario, 134, 135, 136
  pacemaker-accumulator systems in, 8, 28, 186, 210
  parameters, 16
  predictions, 14
  relationship between procedure and model, 10
  scale transform idea, 4
Schizophrenia, 225, 287
  attention dysfunction, 226
  temporal processing, 226
Selasphorus rufus, 116
Semantic memory, 91
Sensory modality, 196
Sequencing, 387
Short interval timing, 61
  oscillator interpretation, 63, 72
  principles, in analysis of circadian timing, 69
Side bias, 100
Signal duration, xxii
  age and sensitivity to, 193
  attention deficit and, xix
Signal-to-noise ratio, 97
Simulations. *See also* Computer simulations
  fixed-interval, 41
  fixed offset, 48, 49
  fixed onset, 47, 48
Single-bucket model, 27
Single-photon emission tomography, 424
  energy source, 421
  invasiveness, 421
  limitations, 421, 424
  PET *vs.,* 424
  physiological measure, 421
  repeatability, 421
  spatial resolution, 421, 424
  temporal resolution, 421
Skinner box, 114
Sodium-dependent high-affinity choline uptake, xxviii
Songbirds, 96, 393
Spatial memory, 91
  sequential, 91

Spectral timing model, 28
   strengths, 29
   weaknesses, 29
Spectral timing theory, 5
Spermophilus lateralis, 90
SPN. *See* Stimulus preceding negativity (SPN)
Spotted flycatchers, 119
Stable-order principle, 161, 167
Starlings, 89
   European, 116, 120, 127
   foraging behavior, 116, 120, 128
   risk sensitivity, 133–134
State-based clocks, 25
   connectionist model, 26
   diffusion model, 26
Stem cells, 305
Stimulus onset asynchrony, 222
Stimulus preceding negativity (SPN), 452
Stochastic counting cascades, 72
Streptopelia risioria, 98
Striatal beat frequency model, 343, 375, 406–407
   decision stage output function hypothesized by, 378
   predictions, 381
Striato-frontal circuitry, xxiii, 487
Striatum, xxx, 383
   avian *vs.* mammalian, 398
   GABAergic inhibitory projections from, 504
   in interval timing, 387
   in sequencing, 387
   of telencephalon, 397
   mammalian *vs.* avian, 398
   neuronal activity related to stress, 384
   timing in, 384, 385–386
   winner-takes-all process in, 386
Stroop task, 359
Sturnus vulgaris, 89, 116, 120
Subitizing, 162, 163
Substantia nigra pars compacta, xxix, 487
Subthalamic nucleus
   glutamatergic excitatory input, 504
   overactivity, 498, 501, 504–506
   stimulation of, 497, 498
Subtizing hypothesis
   object-file model *vs.,* 166
Suprachiasmatic nucleus, 90
Switch, in scalar timing theory, 8–9
Switch hypothesis, attentional. *See* Attention, switch hypothesis
Synchronization, 459
   error correction model for PD studies, 475
   process model, 464–466

T

Taenopyggia guttata, 400
Talairach space, 428
Tapping task, xxxi
Target intervals, 63
tau mutant, 88, 90
Telencephalon
   anterior forebrain pathway, 398, 399
   cortico-striatal modules, 398, 400
   high vocal center, 399
   midbrain dopaminergic neurons, 397, 398, 404
   parts of, 397
   posterior motor pathway, 399
Temperature, 83
   circadian timing and, 87
Temporal bisection procedure, xxii, 183, 221, 274
   auditory *vs.* visual stimuli, 197
   children *vs.* adults, 186, 188
   description, 341
   experimental phases, 184
   psychophysical functions, 184
      variations in, 187, 188
   sensory modality, 196–197
   visual *vs.* auditory stimuli, 197
Temporal discrimination, xxi, 341. *See also* Discriminative tasks
   measure of, 6
Temporal memory, xxvi
   brain damage and, xxvii
   index, xxviii
   theta rhythm activity and, xxviii
Temporal processing, 226, 440, 445
Tetrahydrocannabinol, xxvi
Thalamus, 398, 404
   neuroimaging, 520, 521
Theory(ies). *See also specific theories*
   evaluation, 16
Theta rhythm activity. *See* Hippocampus, theta wave
Theta wave. *See* Hippocampus, theta wave
Tiger beetles, 101
Tilapia macrocephala, 84
Time estimation, electrophysiological correlate of, 440. *See also* Contingent negative variation
Time fields, decay timers *vs.,* 94
Time horizons, 101
Time judgment tasks, 272, 281
Time markers, 396, 407
Time measurement, 524–527
Time perception, vii, 78
   age differences in, 261
   attentional models, 203

Time processing, viii
Timescale invariance, 6, 51, 303
Timing
  animal event, 78
  behavioral theory, 25, 30
  continuousness, 516
  information-processing models, 263
  model (*See* Timing model(s))
Timing errors, absolute size of, 43
Timing intervals
  mediation, xv
  molecular circuitry, xv
Timing model(s). *See also specific models, e.g., Behavioral theory of timing, models*
  dopamine-dependent, deficient, 489
  error, 35–36, 43
  generalized, 372–374
    clock stage, 372
    decision stage, 372
    features, 372
    information-processing for, 373
    striatal beat frequency *vs.*, 375
  layers, 34, 35
  neural, 374–375
  parameters, 37
  scaling in, 52
  striatal beat frequency, 375
    generalized *vs.*, 375
Timing systems, 518–519
  automatic, 518
  cognitively controlled, 518–519
  complementary roles of striato-frontal and cerebello-frontal circuits in, 536
  without auditory stimuli, 523
Trace-based clocks, 27
  multiple-timescale model, 27
  single-bucket model, 27
  spectral timing model, 28
Transgenic animals, 305
Treisman's repetitive stimulation technique, 196
Trichogramma dendrolimi, 83
Tringa totanus, 118

U

Uncertainty
  expectancy and, 251

in psychological research, 244
Unconditioned stimulus, xx
Uria lomvia, 99

V

Ventrolateral prefrontal cortex
  neuroimaging, 520, 521
Verbal estimation task, 221
Vermis, 449
Visual stimuli, vii, 213
  auditory *vs.*, 196, 214, 276–277
  response time, 220

W

Wasp, parasitoid, 83
Weber's law, xxviii, 6, 80, 103, 122
  ecological consequences, 82
  number discrimination and, 153
Wheel-running activity, 68
Wing-Kristofferson continuation task, 460, 461
Wolves, 101

X

X-ray, 420
  energy source, 421
  invasiveness, 421
  limitations, 421
  physiological measure, 421
  repeatability, 421
  spatial resolution, 421

Y

Yoccoz model, 12

Z

Zebra finch, 400, 403, 410
Zeitgedachtnis, 88